2022 IEEE International Conference on Integrated Circuits, Technologies and Applications (ICTA 2022)

Xi'an, China
28 – 30 October 2022

IEEE Catalog Number: CFP22VVL-POD
ISBN: 978-1-6654-9270-6

**Copyright © 2022 by the Institute of Electrical and Electronics Engineers, Inc.
All Rights Reserved**

Copyright and Reprint Permissions: Abstracting is permitted with credit to the source. Libraries are permitted to photocopy beyond the limit of U.S. copyright law for private use of patrons those articles in this volume that carry a code at the bottom of the first page, provided the per-copy fee indicated in the code is paid through Copyright Clearance Center, 222 Rosewood Drive, Danvers, MA 01923.

For other copying, reprint or republication permission, write to IEEE Copyrights Manager, IEEE Service Center, 445 Hoes Lane, Piscataway, NJ 08854. All rights reserved.

*** This is a print representation of what appears in the IEEE Digital Library. Some format issues inherent in the e-media version may also appear in this print version.

IEEE Catalog Number:	CFP22VVL-POD
ISBN (Print-On-Demand):	978-1-6654-9270-6
ISBN (Online):	978-1-6654-9269-0
ISSN:	2831-395X

Additional Copies of This Publication Are Available From:

Curran Associates, Inc
57 Morehouse Lane
Red Hook, NY 12571 USA
Phone: (845) 758-0400
Fax: (845) 758-2633
E-mail: curran@proceedings.com
Web: www.proceedings.com

Table of Contents

A passive balancing algorithm for multi-cells.. 1

Kaikai Wu, Hongyi Wang, Shucai Wang

Prediction of Key Metrics of Stacked Nanosheet nFETs using Genetic Algorithm-based Neural Networks 3

Haoqing Xu, Weizhuo Gan, Lei Cao, Huaxiang Yin, Zhenhua Wu

Optical Receiver Front-End for 50G PON in 40nm CMOS 5

Nianquan Ran, Jia Li, Shuaizhe Ma, Yuye Yang, Wanqing Zhao, Hao Li, Dan Li

A Fully-Connected and Area-Effcient Ising Model Annealing Accelerator for Combinatorial Optimization Problems 7

Yukang Huang, Dong Jiang, Yongkui Yang, Enyi Yao

A 77-101GHz 6-Bit Vector-Modulated Phase Shifter with Low RMS Error in 65nm SOI CMOS 9

Qingzhe Zhang, Keping Wang

A Fast Sampling Open-circuit Voltage Algorithm for Piezoelectric Energy Harvesting........ 11

Ying Yu, Xufeng Liao, Lianxi Liu

A Wideband High-linearity Input Buffer Based on Cascade Complementary Source Follower 13

Tian Feng, Dengquan Li, Jiale Ding, Shubin Liu, Yi Shen, Zhangming Zhu

Low-Latency FPGA Design and Implementation of Hermitian Matrix Inversion Based on Partitioned Systolic Array for Massive MIMO 15

Ke Han, Daokun Li

Accurate 3DIC thermal simulation for BEOL influence study 17

Hao Yang, Bin Yan, Jianjun Sun, Jian Pang, Guangyao Li, Ouyang Keqing, Shuqiang Zhang

Formation Mechanism of high Ni content $(Cu,Ni)_6Sn_5$ in Cu/Sn/Ni microbump for solid state aging.. 20

Haiyang Yu, C.R.Kao

LIPFD-NPU: Low-overhead Instruction-driven Permanent Fault Detection for Neural Processing Unit.. 22

Pengfei Wu, Zheng Wang, Zhiming Pan, Weilun Wang

Large Suppression to Lateral Charge Migration (LCM) Related Error Bits in Charge-Trap TLC 3D NAND Flash.. 24

Kenie Xie, Peng Guo, Fei Chen, Binglu Chen, Xiaotong Fang, Jixuan Wu, Xuepeng Zhan, Jiezhi Chen

A Q-Band Low-Noise Amplifier in 40- nm CMOS for Q/V-band satellite communications.......
.... 26

Qin Tian, Dixian Zhao

A V-band Power Amplifier for Satellite Communications in 40-nm CMOS....................... 28

Hengzhi Wan, Dixian Zhao

First Demonstration of High PAE Performance Using InGaN Channel HEMT for 5G RF Applications.. 30

Hao Lu, Likun Zhou, Longge Deng, Ling Yang, Bin Hou, Xiaohua Ma, Yue Hao

A 41-GHz 19.4-dBm PSAT CMOS Doherty Power Amplifier for 5G NR Applications.......... 32

Zheng Li, Zixin Chen, Qiaoyu Wang, Jian Pang, Atsushi Shirane, Kenichi Okada

A 120nW, 121kHz, -20~ 100℃ CMOS Relaxation Oscillator with Digital Current Comparator and On-Chip Voltage and Current Reference 34

Renwei Chen, Yifei Zhang, Chenchang Zhan

A 4.2-to-5.6 GHz Transformer-Based PMOS-only Stacked-gm VCO in 28-nm CMOS......... 36

Mingkang Zhang, Zihao Zhu, Yueduo Liu, Zehao Zhang, Rongxin Bao, Jiahui Lin, Haoyu Zhuang, Jiaxin Liu, Xiong Zhou, Shiheng Yang, Qiang Li

A Multiplying Delay-Locked Loop design with low jitter and high linearity...................... 38

Jiahao Hu, Zhongxian Huang, Baoxing Duan, Qing Li, Ziqi Song, Dian He

Cooperative surface-activation strategy for low-temperature Cu/SiO$_2$ hybrid bonding.... 40

Qiushi Kang, Ge Li, Fanfan Niu, Chenxi Wang

A DLL-Based Offset Calibration Loop Technology for Wake-Up Receivers 42

Yuhang Xie, Xufeng Liao, Xincai Liu, Lianxi Liu

An Efficient FPGA Design for Fixed-point Exponential Calculation.................................... 44

Weiyi Zhang, Chun Zhang, Liting Niu, Fasih Ud Din Farrukh, Hanjun Jiang

A 6-18GHz Low-Noise Amplifier Using Noise Canceling Technique in 130-nm CMOS PD-SOI....
.... 46

Jialong Xue, Tenghao Zou, Hao Xu, Tingting Han, Mi Tian, Weiqiang Zhu, Zhijian Li, Na Yan

A CGP-based Efficient Approximate Multiplier with Error Compensation........................ 48

Qiao Shen, Renyuan Zhang, Hao Zhang, Hao Cai, Bo Liu, Jian Xiao

Photon-Memristive Device for Neuromorphic Computing... 50

Yuqing Fang, Qingxuan Li, Tianyu Wang, Jialin Meng, QingQing Sun, David Wei Zhang, Lin Chen

An Input Buffer with 85dB SFDR for High-Speed Pipeline ADC... 52

Cece Huang, Yuanfu Zhao, Yafei Ji, Xin Yang, Tieliang Zhang, Weixin Gai

TSV Defects Classification with Machine Learning Approaches 54

Haitao He, Changhao Luo, Junchen Dong, Yudi Zhao, Min Miao, Kai Zhao

SAUST: A Scheme for Acceleration of Unstructured Sparse Transformer.......................... 56

Yifan Song, Shunpeng Zhao, Song Chen, Yi Kang

Improve the Robustness of Diffusive Memristor based True Random Number Generator via Voltage-to-Time Transformation... 58

Haoyang Li, Yuyang Fu, Tianqing Wan, Yifan Lu, Ling Yang, Yi Li

A statistics-based background capacitor mismatch calibration algorithm for SAR ADC.... 60

Zhiqiang Luo, Peng Wang, Fule Li, Chun Zhang, Zhihua Wang

TEDOP: A Tiny Event-Driven Neural Network Hardware Core Enabling On- Chip Spike-Driven Synaptic Plasticity.. 62

Cong Shi, Sihao Chen, Haibing Wang, Zhengqing Zhong, Ping Li, Junxian He, Tengxiao Wang, Jianyi Yu, Min Tian

A 92.7% Peak Efficiency 48/1V DSD Power Converter with 102mV Droop and 1.6 μ s Settling Time for a 1A/10ns Load Transient 64

Yongchao Zhang, Zhuoqi Guo, Zhongming Xue, Zhuoneng Li, Xihao Liu, Shangzhou Zhao, Dexuan Lv, Mengqi Duan, Li Geng

A 14.39ppm/kPa Stress Sensor with Low Temperature-drift and High Linearity for turbulence Stress.. 66

Lanxiang Xiao, Lei Chen, Fengwei An

A 30W and 95% Efficiency Class-E Wireless Power Transfer Transmitter with Vector Algorithm Control.. 68

Shangzhou Zhao, Zhongming Xue, Yuhao Xiong, Zhuoneng Li, Xihao Liu, Yongchao Zhang, Zhuoqi Guo, Li Geng

A Novel Segmented Temperature Monitor for Adaptive MRAM....................................... 70

Yu' ang Wu, Mingyang Zhou, Hao Cai

A High Reliability Sensing Amplifier for Hybrid MTJ/CMOS Circuits................................. 72

Jiawei Fu, Pengcheng Wu, Hao Cai

A 0.7-2.5GHz NB-IoT/GNSS/BLE Hybrid PLL with PA Pulling Mitigation and Out-of-Band Phase Noise Reduction... 74

Jiahao Zhao, Xuansheng Ji, Su Han, Ziwei Wang, Woogeun Rhee, Zhihua Wang

A High-Density Large-Ratio Fuse Based Oxide Devices for One-time- programmable Memory Applications.. 76

Xuecheng Cui, Dong Liu, Jifang Cao, Xiao Yu, Bing Chen

A 210nA Quiescent Current Bandgap Reference with 5mA Load Capability Using Shared Error Amplifier 78

Binwei Yang, Renwei Chen,Chenchang Zhan

Implementation of Polynomial Fitted Poly-Harmonic Distortion Model with Frequency Defined Device 80

Xiaoqiang Tang, Jialin Cai

Dynamic Sensor Arrays Based on Solution-Processed Metal Oxide Semiconductor Thin-Film Transistors.. 82

Bowen Zhu

A 22.8 GHz to 32.8 GHz Compact Power Amplifier with a 15 dBm Output P_{1dB} and 36.5% Peak PAE in 65-nm CMOS 84

Huabing Liao, Haikun Jia, Xiangrong Huang, Bao Shi, Wei Deng, Baoyong Chi, Zhihua Wang

A Fully Synthesizable Injection Locked PLL with Dual-DCO Frequency Tracking in 55nm CMOS 86

Xuanchi Yu, Yan Chen, Gaofeng Jin, Fei Feng, Xun Luo, Xiang Gao

DFT Architecture for Click-Based Bundled-Data Asynchronous Circuits 88

Ruimin Zhu, Zeyang Xu, Yuhao Huang, Shanlin Xiao, Zhiyi Yu

Towards Near LLC Speed STT-MRAM Sensing Using Reconfigurable Clock Trimming....... 90

Xiaoyun Tian, Zhongjian Bian, Hao Cai

Characterization and Modeling of Trapping Effects in GaAs Enhanced HEMT under High Input Dynamic Range 92

Lei Huang, Huanpeng Wang, Qingzhi Wu, Shuman Mao, Yuehang Xu

A Broadband 20W GaN High Power Amplifier for Ku-band satellite communication........ 94

Yujie Liu, Zhilong Xiao, Shiquan Zhu, Huanpeng Wang, Shuman Mao, Qingzhi Wu, Ruimin Xu, BO Yan, Yuehang Xu

Hardware Based RISC-V Instruction Set Randomization... 96

Sheng Zuo, Junjie Zhuang, Yao Liu, Mingyu Wang, Zhiyi Yu

A 28nm, 4.69TOPS/W Training, 2.34 μ J/Image Inference, on-chip Training Accelerator with Inference-compatible Back Propagation.. 98

Haitao Ge, Weiwei Shan, Yicheng Lu, Jun Yang

A 20W Ka-Band Dual-Port Power Amplifier for Communication Satellites...................... 100

Chi Chen, Kuan Hu, Weilin Luo, Kang Yin, RuiYuan Kang, Ying Zhao, Fei Yang

A Voltage Error Quantizer For Digital Low Dropout Regulators With Fast Transient Response and Low Steady- State Error... 102

Kaize Zhou, Dejian Li, Chongfei Shen, Yuxuan Du, Zhuo Chen, Weiwei Shan

A 200-Gb/s PAM-4 Feedforward Linear Equalizer with Multiple- Peaking and Fixed Maximum Peaking Frequencies in 130nm SiGe BiCMOS.. 104

Zhengzhe Jia, Taiyang Fan, Dongfan Xu, Dongshen Zhan, Linxuan Hu, Zhengyang Zhang, Yanchao Wang, Chun-Zhang Chen, Xuhui Liu, Hanming Wu, Quan Pan

An Analog-Assisted Digital LDO with 0.37mV Output Ripple and 5500x Load Current Range in 180nm CMOS... 106

Luhua Lin, Bowen Wang, Woogeun Rhee, Zhihua Wang

A 2.06μW/MHz, 5.05-MHz, -40-125 ℃ , 22ppm/ ℃   Relaxation Oscillatior with Single Comparator Control... 108

Yifan Yao, Chuqi Chen, Chenchang Zhan

A 4 to 5GHz Digitally Controlled Ring Oscillator with 100kHz Resolution using Noise Cancellation Technology in 40nm CMOS... 110

Shuyue Fang, Jinrui Hu, Haigang Feng, Xinpeng Xing, Han Wang, Lei Yang

An IMPLY-based Memristive Multiplier for Computing-in-Memory Systems with Weight-Stationary CNN Acceleration... 112

Wenhui Liang, Jiarui Xu, Yuansheng Zhao, Zixuan Shen, Guoyi Yu, Yuhui He, Chao Wang

A Novel Concept of using Double Threshold Voltage Coupling to Improve the linearity of AlGaN/GaN HEMTs for millimeter-wave applications.. 114

Pengfei Wang, Minhan Mi, Sirui An, Xiang Du, Xiaohua Ma, Yue Hao

Photoresponses and Memory Effects in Optoelectronic Synaptic Devices Based on CdSe Quantum Dots and Poly(3-hexylthiophene).. 116

Zhicheng Li, Zhulu Song, Zhaojin Wang, Jiayun Sun, Kai Wang

A Current Domain Computing-in- Memory SRAM Macro with Hybrid IAF-SAR ADC for Signal Margin Enhancement... 119

Tianqi Xu, Shumeng Li, Fukun Su, Xian Tang

A Low Frequency Drift LC-DCO with Wide Temperature Range 121

Jinrui Hu, Shuyue Fang, Haigang Feng, Xinpeng Xing, Han Wang, Lei Yang

A High PSR and Fast Transient Response Output-Capacitorless LDO using Gm-Boosting and Capacitive Bulk-Driven Feed-Forward Technique in 22nm CMOS................................... 123

Heng Liu, Dongxu Li, Xian Tang

ROPY-SLAM: a Heterogeneous CPU- FPGA System for Simultaneous Localization and Mapping
.... 125

Weiyi Zhang, Liting Niu, Chaoyang Ding, Yiyang Wang, Fasih Ud Din Farrukh, Chun Zhang

A Single-input Dual-output Three- level Buck Converter for SoC Applications................ 127

Zhuoneng Li, Zhongming Xue, Chenglong Liang, Yongchao Zhang, Mengqi Duan, Shangzhou Zhao, Xihao Liu, Zhuoqi Guo, Li Geng

Dynamically Reconfigurable Memory Address Mapping for General- Purpose Graphics Processing Unit 129

Weiliang Chen, Zhaoshi Li, Leibo Liu, Shaojun Wei

Strain-regulated flexible molecular sensors enabled by 2D PtTe$_2$.................................. 131

Zhehan Wang, Dingxuan Kang, Jiayi Chen, Xiao Xu, Xu Jing, Li Tao

$BaFe_{12}O_{19}$ based Ferroelectric Memristor for Applications of True Random Number Generator 133

Ziyang Chen, Miaocheng Zhang, Zixuan Ding, Aoze Han, Xingyu Chen, Xinpeng Wang, Lei Wang, Hao Zhang, Yi Tong

Memristor-based Digital Circuits for Realizing the Pavlov's Associative Neural Network.. 135

Yu Wang, Yi Liu, Jiayu Bao, Yu Yan, Ertao Hu, Xiang Wan, Rongqing Xu, Hao Zhang, Yi Tong

A Split-Ring Resonator with Interdigitated Electrodes Aimed at the Dielectric Characterization of Liquid Mixtures (Invited Paper).. 137

Giovanni Gugliandolo, Xiue Bao, Haoyun Yuan, Jinkai Li, Juncheng Bao, Giovanni Crupi, Nicola Donato

Two-bit multi-level spin orbit torque MRAM with the fully one-step write operation... 142

Chenyi Wang, Min Wang, Zhaohao Wang, Weisheng Zhao

3D-NWA: A Nested-winograd Accelerator for 3D CNNs.................................... 144

Huafeng Ye, Huipeng Deng, Jian Wang, Mingyu Wang, Zhiyi Yu

Long Short-Term Memory Networks for Behavioral Modeling of A GaN Sequential Power Amplifier... 146

Peng Chen, Yucheng Yu, Chao Yu

A Cap-Less High PSR and Low Output Noise Low-Dropout Regulator for Cryogenic Applications.. 148

Lingyun Liu, Chenglong Liang, Zhuoqi Guo, Zhongming Xue, Li Geng

All-Digital Full-Precision In-SRAM Computing with Reduction Tree for Energy-Efficient MAC Operations.. 150

Dengfeng Wang, Zhi Li, Chengjun Chang, Weifeng He, Yanan Sun

A Compact 60 GHz LNA with 22.7-dB Gain and 4.4-dB NF in 40nm CMOS 152

Jiacong Ke, Guangyin Feng, Yanjie Wang

A Fully-integrated 110-GHz Wideband Direct-detection Imaging Unit MMIC Integrating Balanced Power Detector and Log-periodic Antenna.. 154

Jinyu Xie, Xiaojun Bi

A 250Mbps 100kV/μs CMTI On-Chip Double-Isolated Transformer-Based Digital Isolator........ 156

Zhiyong Xiong, Dongfang Pan, Guolong Li, Lin Cheng

Implementation of CNN Hetero-geneous Scheme Based on Domestic FPGA with RISC-V Soft Core CPU... 158

Hailong Wu, Jindong Li, Xiang Chen

A TWN Inspired Speaker Verification Processor with Hardware-friendly Weight Quantization... 160

Xuanhao Zhang, Haige Wu, Renyuan Zhang, Zihang Xu, Hao Zhang, Bo Liu, Hao Cai

A 64Gb/s PAM-4 Digital Equalizer With Tap-Configurable FFE and Partially Unrolled DFE in 28nm CMOS... 162

Xinjie Feng, Yongzhen Chen, Youzhi Gu, Jiangfeng Wu

A Tunable Monopole Antenna for 5G Communication Applications............................... 164

Liangfan Chen, Lu Zhao, Zihao Chen

PPBAM: A Preprocessing-based Power-Effcient Approximate Multiplier Design for CNN.......... 166

Yi Hu, Tao Huang, Run Run, Li Yin,Guolin Li, Xiang Xie

A 0.58-pJ/bit 56-Gb/s PAM-4 Optical Receiver Frontend with an Envelope Tracker for Co-Packaged Optics in 40-nm CMOS... 168

Yue Yu, Da Ming, Min Tan

A Novel Fold-Back Current Limiting Protection used in Sub-threshold LDO for Wireless Sensor Applications.. 170

Ziyue Chen, Yihui Shi, Ao Hu, Jiarui Xu, Guoyi Yu, Chao Wang

A Wideband and High Output Swing Analog Frontend Circuit for FMCW LiDAR............. 172

Yiyun Xie, Youze Xin, Bing Zhang, Ruipeng Yang, Yaoxin Li, Li Geng

A 0.3 V-4 V Input Voltage Range, 0.7 V Cold Start Boost Converter with 1 V Internal Voltage Supply Generator by Using 0.18 μm CMOS Process for Energy Harvesting Application 174

Zheng Lu, Shiquan Fan, Weiqing Ma, Ying Xie, Li Geng

An AOT Buck Converter with Adaptive TON Extender Achieving 2.5 μ s Settling Time in 4A Load Transient 176

Zhuang Zhang, Hanyu Shi, Danzhu Lu, Peng Cao, Zhiliang Hong

A 1000 fps Spiking Neural Network Tracking Algorithm for On-Chip Processing of Dynamic Vision Sensor Data 178

Chi Zhang, Lei Kang, Xu Yang, Guanghao Guo, Peng Feng, Shuangming Yu, Liyuan Liu

A Low Supply Sensitivity CMOS Temperature Sensor Using Dynamic- Distributing-Bias Circuit 180

Shichong Zhai, Wenchang Li, Jian Liu, Tianyi Zhang

A Reconfigurable SRAM Computing- in-Memory Macro Supporting Ping- Pong Operation and CIM pipeline for Multi-mode MAC operations................................. 182

Kanglin Xiao, Xiaoxin Cui, Xin Qiao, Xin'an Wang, Yuan Wang

Floorplanning and Power/Ground Network Design for A Programmable Vision Chip..... 184

Haozhe Xu, Siyuan Wei, Nan Qi, Peng Wu, Jian Liu, Nanjian Wu, Liyuan Liu, Shuangming Yu

An Analytical Model for doping effect in charge-plasma-based TFET............................ 186

Wenbo Li, Qian Xie, Zheng Wang

A Charge Pump Based 1.5A NMOS LDO with 1.0~6.5V Input Range and 110mV Dropout Voltage.. 188

Yifa Wang, Tong Wu, Jianping Guo

IPOCIM: Artificial Intelligent Proces- sor with Adaptive Ping-pong Compu- ting-in-Memory Architecture.. 190

Liang Chang, Chenglong Li, Xin Zhao, Shuisheng Lin, Jun Zhou

CVD Monolayer tungsten-based PMOS Transistor with high performance at Vds = -1 V   192

Xin Wang, Yanqing Wu

Efficient AVS3 Intra Prediction Hardware Design for Real-time Applications.................   194

Yucheng Jiang, Haifeng Guo, Junhao Zheng, Jingsheng Wang, Songping Mai

Thermal Fatigue Analysis of Microbumps in a 3D TSV Integration Device......................   196

Yuqing Lu, Jun Wang

A 224-Gb/s Inverter-Based TIA with Interleaved Active-Feedback and Distributed Peaking in 28-nm CMOS for Silicon Photonic Receivers   198

Sikai Chen, Jintao Xue, Leliang Li, Guike Li, Zhao Zhang, Jian Liu, Liyuan Liu, Binhao Wang, Yingtao Li, Nan Qi

A 0.3-μW,2.1-μVrms Neural Recording Chopper Amplifer with Low Noise DC-Servo-Loop........
....   200

Yuchen Bao, Weijian Chen, Zhixian Li,Yongsen Chen, Yanhan Zeng

A 56Gb/s De-serializer with PAM-4 CDR for Chiplet Optical-I/O.....................................   202

Yunqi Yang, Ming Zhong, Qianli Ma, Ziyi Lin, Leliang Li, Guike Li, Liyuan Liu, Jian Liu, Nanjian Wu, Haikun Jia, Xinghui Liu, Nan Qi

Author Index...   204

Proceedings of
2022 IEEE International Conference on Integrated Circuits, Technologies and Applications
(ICTA)

28-30 October 2022

Sponsored By

IEEE MTT-S Nanjing Chapter
Beijing Institute of Electronics

Supported By

Xi'an Jiaotong University
Beijing Association for Science and Technology

Patrons

IEEE SSCS Guangzhou Chapter
EDS Beijing Chapter
IEEE Circuits and Systems Society (CAS) Xi'an Chapter
Beijing Academy of Science and Technology

ICTA2022 CONFERENCE COMMITTEE

General Chairs:
Nanning Zheng, XJTU, China

TPC Co-Chairs:
Zhiyi Yu, Sun Yat-sen University, China
Xiaoyan Gui, XJTU, China

Steering Committee:
Dixian Zhao (Chair), Southeast University, China
Zhihua Wang, Tsinghua, China
Fujiang Lin, USTC, China
Nanjian Wu, IS-CAS, China
Liyuan Liu, IS-CAS, China

International Advisors:
Jan Van der Spiegel, University Penn, USA
Christian Enz, EPFL, Switzerland
Yann Deval, University of Bordeaux, France
Hoi-Jun Yoo, KAIST, Korea
Yong Lian, IEEE CASS, USA
Ke Wu, IEEE MTT-S, Canada
Albert Wang, IEEE EDS, USA
Makoto Ikeda, University of Tokyo, Japan
James Hwang, Lehigh University, USA

Local Arrangement Chair:
Li Geng, XJTU, China

Invited Program Chair:
Haiding Sun, USTC, China

Tutorial Chair:
Bo Zhao, Zhejiang University, China

Keynote Speakers Chair:
Lin Cheng, USTC, China

Industry Papers Chair:
Maliang Liu, Xidian University, China

Student Papers Chair:
Hongbin Sun, XJTU, China
Xian Tang, Tsinghua University, China

Program/Sessions Chair:
Weiwei Shan, Southeast University, China

Awards Committee Chair:
Zheng Wang, UESTC, China

Publication/Booklet Chair:
Ying Zhang, BJAST, China

Website Chair:
Qing Wu, BIE, China

Finance Chair:
Jiayue Zhang, BIE, China

Sponsorship/Exhibition Chair:
Dan Li, XJTU, China

Conference Secretariat:
Fanfang Zeng, BIE, China
Xiaohan Zhang, BIE, China
Tianye Jin, BIE, China

Technical Program Committee:

SC1: RFIC and MMIC
Co-Chair:
Bo Zhao, Zhejiang University, China
Zheng Wang, UESTC, China

Members:
Xiaojun Bi, HUST, China
Haigang Feng, Tsinghua Shenzhen Institute, China
Wei Deng, Tsinghua University, China
Na Yan, Fudan University, China
Keping Wang, Tianjing University, China
Haikun Jia, Tsinghua University, China
Xiang Gao, Zhejiang University, China
Shengxi Diao, East China Normal University, China
Shiheng Yang, UESTC, China
Jian Pang, Tokyo Institute of Technology, Japan
Yun Wang, Fudan University, China

SC2: Analog and Mixed-Signal ICs
Co-Chair:
Sai-Weng Sin, University of Macau, China
Lin Cheng, USTC, China

Members:
Zhijie Chen, Beijing University of Technology, China
Jianping Guo, Sun Yat-sen University, China
Jiaxin Liu, Tsinghua University, China
Yuxuan Luo, Zhejiang University, China
Wanyuan Qu, Zhejiang University, China
Chenchang Zhan, SUSTc, China
Xian Tang, Tsinghua University Shenzhen, China
Xiong Zhou, UESTC, China
Liang Qi, Shanghai Jiaotong University, China
Mo Huang, University of Macau, China
Haoyu Zhuang, UESTC, China
Dongfang Pan, USTC, China
Bing Yuan, Xidian University, China

SC3: Digital ICs and Memory
Co-Chair:
Jun Zhou, UESTC, China
Zhiyi Yu, Sun Yat-Sen University, China

Members:
Shouyi Yin, Tsinghua University, China
Weiwei Shan, Southeast University, China
Chao Wang, HUST, China
Jun Lin, Nan Jing University, China
Shanlin Xiao, Sun Yat-Sen University, China
Cong Shi, Chongqing University, China
Chang Liang, UESTC, China
Yanan Sun, Shanghai Jiaotong University, China
Lin Li, Hisilicon, China
Longyang Lin, SUST, China
Hao Cai, Southeast University, China
Yanhan Zeng, Guangzhou University, China

SC4: Wireline ICs
Co-Chair:
Nan Qi, IS-CAS, China
Feng Zhang, IME-CAS, China

Members:
Quan Pan, SUSTech, China

Zhao Zhang, IS-CAS, China
Binhao Wang, OPMCAS, China
Xiaoyan Gui, XJTU, China
Dan Li, XJTU, China
Min Tan, HUST, China
Xuqiang Zheng, IME-CAS, China
Ziqiang Wang, Tsinghua University, China
Yong Chen, Macau University, China

SC5: Modeling, EDA and Testing
Co-Chair:
Yuehang Xu, UESTC, China
Jun Liu, HDU, China

Members:
Liang Zhou, Shanghai Jiaotong University, China
Giovanni Crupi, University of Messina, Italy
Kai Lv, Shanghai Simchip Technology Group Co. Ltd, China
Yunqiu Wu, UESTC, China
Peng Chen, Southeat University, China
Yonghao Jia, NWPU, China
Letian Huang, UESTC, China
Yang Lu, Xidian University, China

SC6: Device and Process Technologies
Co-Chair:
Bin Gao, Tsinghua University, China

Members:
Xingsheng Wang, HUST, China
Jiezhi Chen, Shandong University, China
Hong Wang, Xidian University, China
Yijiao Wang, Beihang University, China
Binjie Cheng, Synopsys, UK
Xiao Gong, National University of Singapore, Singapore
Can Li, University of Hong Kong, China
Zongwei Wang, Peking University, China
Bing Chen, Zhejiang University, China
Jie Liang, Shanghai University, China

SC7: Packaging and Hybrid Integration
Co-Chair:
Chenxi Wang, Harbin Institute of Technology, China
Jun Wang, Fudan University, China

Members:
Jintang Shang, Southeast University, China
Tao Hang, Shanghai Jiaotong University, China
Qidong Wang, IM-CAS, China
Yingxia Liu, City University of Hong Kong, China
Xingchang Wei, Zhejiang University, China
Jian Cai, Tsinghua University, China
Min Miao, BISTU, China
Sheng Sun, UESTC, China
Liguo Sun, USTC, China
Rongxiang Wu, UESTC,China
Chengqiang Cui, GDUT, China
Daquan Yu, Xiamen University, China
Wenhua Yang, Anhui University, China
Ran He, Huawei Technologies Co., China

SC8: Sensors and Applications
Co-Chair:
Fengwei An, SUSTech, China

Members:
Guoyi Yu, HUST, China
Quan Chen, SUSTech, China
Gengzhen Qi, Sun Yat-Sen University, China
Xu Liu, Beijing University of Technology, China
Xiwei Huang, Hangzhou Dianzi University, China
Yongfu Li, Shanghai Jiaotong University, China
John Deepu, University College Dublin, Ireland
Xingyuan Tong, XUPT, China
Jian Zhao, Shanghai Jiaotong University, China
Kameng Lei, University of Macau, China
Peng Feng, IS-CAS, China

SC9: IC Based Applications
Co-Chair:
Zheng Wang, SIAT, China
Enyi Yao, SCUT, China

Members:
Zihao Chen, Harbin Institute of Technology (Shenzhen), China
Yongkui Yang, SIAT, China
Aijiao Cui, Harbin Institute of Technology, China
Xin Lou, Shanghai Technology University, China
Tao Tang, Zhejiang lab, China
Chen Lang, Sydnicon RF, Australia

Hao Zhang, Gusu Laboratory of Materials, China

SC10: Emerging Technologies and Applications
Co-Chair:
Haiding Sun, USTC, China

Members:
Li Tao, Southeast University, China
Mengmeng Li, IME-CAS, China
Zhaoliang Liao, USTC, China
Kai Wang, SUSTech, China
Qiaoling Tong, HUST, China
Wei Cao, University of California, Santa Barbara, USA
Hong Zhou, Xidian University, China
Zhengguo Xiao, USTC, China

SC11: Intelligent Robots
Co-Chair:
Chun Zhang, Tsinghua University, China
Qi Peng, Xidian University, China

Members:
Songping Mai, Tsinghua Shenzhen, China
Zhongyi Chu, Beihang University, China
Qingguo Zhou, Lanzhou University, China
Gang Chen, Sun Yat-Sen University, China
Yixin Zhao, Dongguan University of Technology, China

WELCOME MESSAGE-GENERAL CHAIRS

Nanning Zheng
Xian Jiaotong University, China

As the conference chair, I sincerely welcome you to the 2022 IEEE International Conference on Integrated Circuits, Technologies and Applications(ICTA 2022) being held here in the beautiful and historic city of Xi'an, Shaanxi, China. The semiconductor industry shows a continue growing in the past years. The Semiconductor Industry Association (SIA) announced that the global sale of semiconductor products in 2021 reached 555.9 billion U.S. dollars, the highest-ever annual total and an increase of 26.2% compared to 440.4 billion U. S. dollars in the year of 2020. And China remained the largest individual market for semiconductors, with sales totaling $192.5 billion in 2021, an increase of 27.1%. With the application of 5G, IoT, AI and Automotive, semiconductors have a major impact on the development of industry, as well as daily life. Various new IC technologies emerge and are equipped to new products.

China is now becoming a semiconductor hub for academia, industry and market. However, young Chinese researchers, in particular students, lack the opportunity to attend international conferences, especially since the COVID-19. International counterparts also wish to experience firsthand the fast-growing semiconductor sector in China. To this end, ICTA was founded by a group of Chinese scholars and is held annually in China. It has been a broad yet advanced forum for IC designs, technologies and applications from worldwide. The ICTA held in the last 5 years achieve a success and attendees shows great passion of participations. To celebrate the 5 year anniversary of ICTA, top 10 ICTA paper contributors will be selected based on the number of the accepted papers in the past 5 years.

I would like to express my gratitude to the members of the Steering Committee. They gave valuable advice which made this year ICTA a successful event. My thanks go to the members of the Local Organization Committee for their excellent work in organizing the conference. My gratitude is particularly given to the subcommittee chairs, members and their affiliations of the Technical Program

Committee. It is through their efforts that the 2022 ICTA has received many submissions, and it is also with their contributions that the conference has succeeded in a selection of high-quality papers.

Finally, let us express our best wishes for the success of ICTA-2022, as well as for successes in future ICTA conferences.

Conference Chair of ICTA-2022, Nanning Zheng

WELCOME MESSAGE- TECHNICAL PROGRAMME CO-CHAIRS

Zhiyi Yu
Sun Yat-sen University, China

Xiaoyan Gui
Xi'an Jiaotong University

It's our great honor to welcome all of you to 2022 International Conference on Integrated Circuits Technologies and Applications (ICTA). Five years ago, some experts proposed and organized ICTA. The goal is to make ICTA a conference based in China with a high quality and international standard like ISSCC and ASSCC. We believe the ICTA is so special, because it provides an opportunity for exchanging information and ideas in the advanced IC field for Chinese researchers and engineers, and it also provides a good platform for foreign experts to communicate and know about the IC academic and industry in China.

This year we received 198 paper submissions, among which 101 high quality papers will be presented in lecture or poster forms. Technical Program Committee (TPC) consists of a balanced mix of both academia and industry. TPC members worked hard to review all the papers and organized a TPC meeting to select the papers. We believe that the selected papers are with high quality and will draw interests from both industry and academia.

Along with the exciting conventional conference lectures and posters, we have prepared the great quality plenary talks, keynotes, and tutorials.

Plenary talks are given by Dr. Wei Tsao from HiSilicon Technogies, China, and Professor Yukinori Ono from Shizuoka University, Japan on the first day. Dr. Wei Tsao will present on "Chiplet Technique in Future: Package, Interconnection and Power Supply", covering challenges and solutions in this new technology roadmap. Professor Yukinori Ono from Shizuoka University, Japan, will give a talk on "Control of Electronic Charges and Currents in Nano-Scald Silicon", discussing promising properties of Si-MOSFET for future quantum nano-electronics.

On Saturday morning, Prof. Peng ZHOU from Fudan University covers "The Road for 2D

Semiconductor in Silicon Age", and Prof. Kai KANG from University of Electronic Science and Technology of China presents "Wideband CMOS mm-wave circuits design for 5G communications". On Saturday afternoon, Prof. Qidong WANG from Institute of Microelectronics of the Chinese Academy of Sciences focuses on "Advanced Packaging for Chiplet", and Prof. Sai-Weng SIN from University of Macau talks on "The Historical Development of Data Converters; and What We Can Do". On Sunday morning, Prof. Hong ZHOU from Xidian University presents "Near Ideal GaN Schottky Diode for High Efficiency and High Power Wireless Power Transfer Application", and Prof. Chunmeng DOU from Institute of Microelectronics of Chinese Academy of Sciences talks on "Design Highly Efficient RRAM Computing-In-Memory AI Chips Leveraging the Interplay between Device, Circuit, and System".

We organize six tutorials for educating students, engineers and researchers in the region. They are "Joint Radar-communication CMOS Transceiver: From System Architecture to Circuit Design" by Prof. Wei DENG, "Design of Low-Power PLL and CDR Integrated Circuits for High Speed Wireless/Wireline Communication" by Prof. Zhao ZHANG, "Overcoming the Transimpedance Limit: A Tutorial on Design of Low-Noise TIA" by Prof. Dan LI, "Design of integrated circuits for high temperature applications" by Prof. Yimeng ZHANG, "Break the Memory Wall: Cross-Layer Co-Design for Energy Efficient AI Processors" by Prof. Chixiao CHEN and "Memristive devices for analog computing" by Prof. Peng LIN.

The excellent conference preparation and management are conducted by local organizing committee. The competitive technical program is a result of hard work of the TPC members. I'd like to deeply appreciate of their outstanding efforts here.

Please enjoy the conference by participating in the interactive discussions in and outside the sessions and share your opinions on the promising "Integrated Circuits and Systems for Intelligent & 6G Society".

Technical Program Committee Co-Chair, Zhiyi Yu and Xiaoyan Gui

A passive balancing algorithm for multi-cells

Kaikai Wu, Hongyi Wang, Shucai Wang

School of Microelectronics, Xi'an Jiaotong University

Abstract—**For reducing the impact of large voltage difference between batteries in battery packs, a battery cell balancing algorithm is proposed in this paper. Based on the 0.18 μm process, the algorithm has been integrated into a 7-cell lithium battery charge and discharge protection chip. The test results show that the algorithm can reliably reduce the voltage difference between batteries and help to prolong the service life of battery packs.**

Keywords—*voltage difference, battery packs, passive balancing algorithm, lithium battery protection chip*

I. INTRODUCTION

With the development of society, rechargeable lithium-ion batteries are more and more widely used in the automotive field [1]. For realizing high-voltage applications, lithium batteries are often realized in series [2,3]. However, due to the characteristics of self-discharge rate, capacity and impedance, there are always slight differences between lithium batteries [4]. After numerous charge-discharge cycles, these differences eventually lead to voltage differences across the cells within the battery pack. Therefore, it is often necessary to balance these batteries in series [5-7].

In this paper, a passive balancing algorithm is proposed for a seven-cell lithium battery pack in series. The battery voltage is balanced by diverting some of the charging current external to the selected battery through a bypass resistor. The other batteries are receiving the full charging current and the selected battery has a limited charging current.

II. DESIGN AND IMPLEMENTATION OF THE PASSIVE BALANCING ALGORITHM

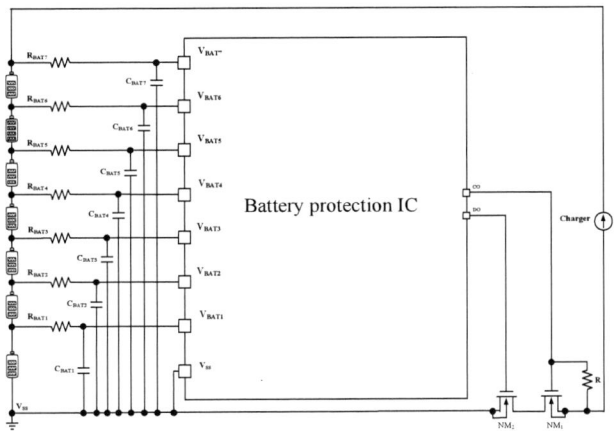

Fig.1. Seven-cell lithium battery protection system

The application of the designed seven-cell battery protection chip is shown in Figure 1, which contain seven batteries, RC filters and FETs. As shown in Figure 1, the voltage of the sixth battery is very high, while the voltage of other batteries is normal. The battery protection chip often stops charging because of the sixth battery, resulting in the loss of other battery capacity. Therefore, the battery balancing circuit is designed to ensure that the voltage of all batteries is almost the same,

The research was sponsored jointly by Foundation of Key Laboratory of Shanghai Jiao Tong University-Xidian University, Ministry of Education (Project No.LHJJ/2020-05).

ensuring that the battery pack capacity is normal.

A. Implementation methods of passive balancing

The balancing circuits designed in this paper are shown in Figure 2, which mainly include battery voltage transfer circuit, comparators, balancing switches, balancing driver, balancing logic circuit, and some basic circuits, such as bandgap, LDO, charger detection, charging and discharging driver, oscillator circuit, etc.

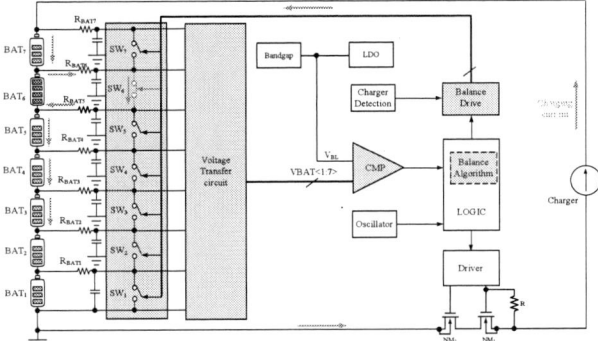

Fig.2. The balance circuit of the chip

The balancing algorithm designed in this paper is realized by analog loop control. In the charging state, when the battery voltage of a battery (For example, BAT_6) is higher than V_{BL} after conversion, the corresponding balancing switch SW_6 will be turned on, thereby bypassing the charging current of this battery, while other batteries receive all the charging current, and finally the voltages of other batteries catch up with this battery voltage slowly.

B. External balancing circuit

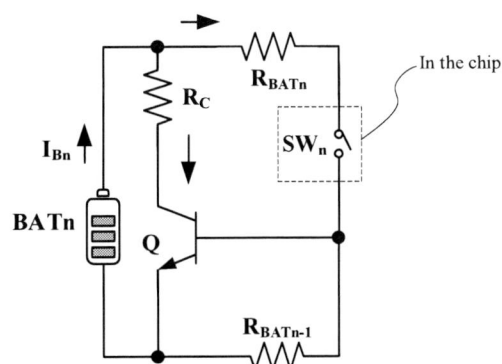

Fig. 3. External balancing circuit

When the chip internal switch of a battery in Figure 2 is turned on, the internal balancing current is calculated as shown below:

$$I_{Bn}=V_{BATn}/(R_{BATn}+R_{BATn-1}+R_{SWn}) \qquad (1)$$

Assume that the battery voltage is 4V, R_{BATn} and R_{BATn-1} are the filter resistance of the battery respectively, the resistance value is usually 1 KΩ, the internal switching impedance of the chip is 100 Ω, then the balancing current is 1.9mA. The balancing current is too small, resulting in low balancing efficiency. Therefore, the corresponding external balancing circuit is designed, as shown in Figure 3. The transistors Q and R_C are added, and their balancing current is as follows:

$$I_{Bn}=V_{BATn}/R_C+V_{BATn}/(R_{BATn}+R_{BATn-1}+R_{SWn}) \qquad (2)$$

Under the same battery voltage, if the R_C resistance is 47Ω, the balancing current is 87mA. Compared with internal balance, the balancing current is much higher.

C. Passive balancing algorithm timing

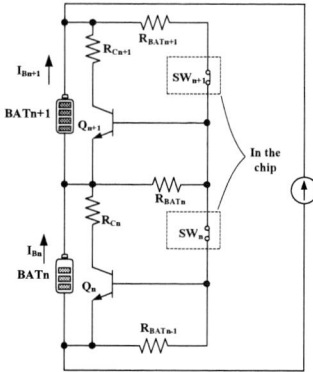

Fig. 4. External balancing circuit

When two adjacent batteries are balanced at the same time in Figure 4, the balancing current of BATn+1 and BATn are shown below:

$$I_{Bn+1} = (V_{BATn} + V_{BATn+1}) / (R_{BATn+1} + R_{BATn-1} + R_{SWn} + R_{SWn+1}) \quad (3)$$

$$I_{Bn} = I_{Bn+1} + V_{BATn}/R_{Cn} \quad (4)$$

We can find that the balancing current is not the same, the balancing current of BATn+1 is small, and its voltage rises faster than that of BATn. Thus, the strategy of balancing odd and even batteries separately is adopted, that is, the first, third, fifth and seventh batteries are balanced at the same time, and the second, fourth and sixth batteries are balanced at the same time. The control sequence of the balancing switch is shown in Figure 5. Check the battery voltage before balancing, and then decide whether to balance the battery.

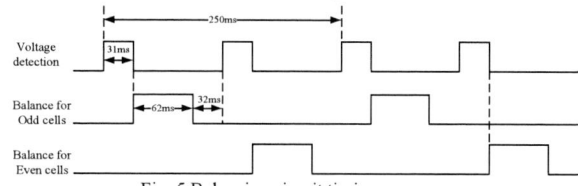

Fig. 5 Balancing circuit timing sequence

III. EXPERIMENTAL RESULTS

The passive balancing algorithm proposed in this paper is successfully integrated into a multi-cell lithium battery protection chip. The micrograph of the chip is shown in Figure 6.

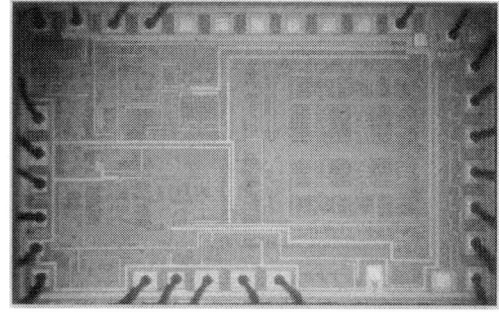

Fig.6. Micrograph of the chip

Figure 7 shows the balancing process of 7 batteries. Table I shows the change of battery voltage before and after balancing. It can be found that the voltage difference between BAT1 and BAT2 before balancing was 687mV, and 14mV after balancing. Meanwhile, the maximum voltage difference of all batteries is only 18mV. Therefore, the balancing algorithm designed in this paper can effectively control the battery voltage and prevent the voltage difference between the batteries in the battery pack from being too large. Table II provides a comparison of this work with similar ICs from industry and academia.

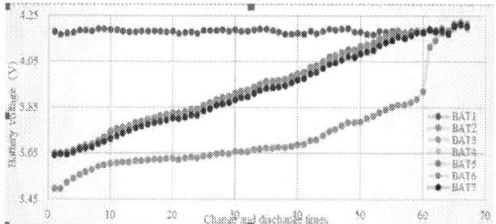

Fig.7. The balancing process of the chip

TABLE I. INITIAL VOLTAGES AND BALANCED VOLTAGES OF 7 CELLS

BATX	BAT1	BAT2	BAT3	BAT4	BAT5	BAT6	BAT7
Initial (V)	4.18	3.493	3.648	3.644	3.65	3.65	3.642
Balanced(V)	4.203	4.217	4.213	4.199	4.208	4.2	4.204

TABLE II. COMPARISON WITH OTHER DESIGNS

Property	This work	[2]	[5]	[6]
Balancing switch	Internal	Internal	External	External
Number of series Cells	7	7	4	5
Balanced Max. ΔV_{BAT}	18mV	around 10mV	<25mV	32mV

IV. CONCLUSION

This paper proposes and implements a balancing algorithm, which is integrated into 7 battery protection chips and verified by 0.18 μm process. The test result shows the battery voltage before and after the balanced of 7 batteries, which shows that the proposed algorithm is helpful to prolong the service life of the battery pack.

REFERENCES

[1] F. A. Machado, P. J. Kollmeyer, D. G. Barroso and A. Emadi, "Multi-Speed Gearboxes for Battery Electric Vehicles: Current Status and Future Trends," in IEEE Open Journal of Vehicular Technology, vol. 2, pp. 419-435, 2021, doi: 10.1109/OJVT.2021.3124411.

[2] V. B. Vulligaddala et al., "A 7-Cell, Stackable, Li-Ion Monitoring and Active/Passive Balancing IC With In-Built Cell Balancing Switches for Electric and Hybrid Vehicles," in IEEE Transactions on Industrial Informatics, vol. 16, no. 5, pp. 3335-3344, May 2020, doi: 10.1109/TII.2019.2953939.

[3] Kadirvel K., Carpenter J., Huynh P., et al. A Stackable, 6-Cell, Li-Ion, Battery Management IC for Electric Vehicles With 13, 12-bit Σ Δ ADCs, Cell Balancing, and Direct-Connect Current-Mode Communications [J]. IEEE Journal of Solid-State Circuits, 2014, 49(4):928-934.

[4] Wu K K, Wang H Y, Chen C, et al. Battery voltage transfer method for multi-cells Li-ion battery pack protection chips[J]. Analog Integrated Circuits and Signal Processing, 2021:1-12.

[5] I. C. Guran, L. A. Perişoară, A. Florescu and D. I. Săcăleanu, "4-Cell Passive Battery Management System for Automotive Applications," 2021 IEEE 27th International Symposium for Design and Technology in Electronic Packaging (SIITME), 2021, pp. 338-341, doi: 10.1109/SIITME53254.2021.9663604.

[6] A. Lasić, Ž. Ban, B. Puškarić and V. Šunde, "Supercapacitor Stack Active Voltage Balancing Circuit Based on Dual Active Full Bridge Converter with Selective Low Voltage Side," 2020 IEEE 11th International Symposium on Power Electronics for Distributed Generation Systems (PEDG), 2020, pp. 627-636, doi: 10.1109/PEDG48541.2020.9244352.

2022 IEEE International Conference on
Integrated Circuits, Technologies and Applications

Prediction of Key Metrics of Stacked Nanosheet nFETs using Genetic Algorithm-based Neural Networks

Haoqing Xu[1,2], Weizhuo Gan[1,2], Lei Cao[1,2], Huaxiang Yin[1,2], Zhenhua Wu[1,2,*]

[1] Institute of Microelectronics, Chinese Academy of Sciences
[2] University of Chinese Academy of Sciences
[*] email: wuzhenhua@ime.ac.cn

Abstract—**In this paper, we demonstrate the prediction of important figures of merit (FoMs) including threshold voltage (Vth), subthreshold swing (SS), on-state (Ion) and off-state (Ioff) current, of vertically stacked lateral nanosheet field-effect-transistors (NSFET) using 1) an artificial neural network generated by genetic algorithm (GA) and 2) a conventional multi-layer neural network (NN). Our work shows that the trained GA-based NN has a great capability of predicting FoMs with an average of coefficients of determination at 0.992, which is better than that of the trained multi-layer neural network at 0.987. Additionally, GA-based NN has a significant reduction of calculation time by 80% compared with that of multi-layer NN under the same computing power, which indicates the possibility to reduce the computational cost by using the auto-machine learning approach for TCAD simulation.**

Keywords—*Nanosheet, TCAD Simulation, auto-machine learning, deep learning.*

I. INTRODUCTION

The devices scale down in keep with Moore's Law in the past few decades [1]. In recent years, stacked nanosheet FETs are introduced with good gate control capability [2], [3]. However, some specifications for 3-nm node NSFET were not determined yet [4]-[6]. To analyze the geometry variation effect on the figures of merit and optimize the geometry parameters, time-consuming TCAD for Design of Experiment (DOE) is required. Previous work has demonstrated that machine learning is a good approach for reducing computation costs with high accuracy in predicting electrostatic potential [7], current-voltage (I-V) and capacitance-voltage (C-V) curves [8], threshold voltage [9], and other figures of merit [10], [11]. In this work, we use an auto-machine learning approach to generate networks with the GA, predicting FoMs of the state-of-art 3nm technology node stacked nanosheet, and showing the comparative prediction accuracy with multi-layer neural networks and significant reduction in the calculation time.

II. TCAD SIMULATION AND DATA GENERATION

3D n-type three-layer-stacked NSFET is built using GTS framework TCAD software [12]. The transfer characteristics of various NSFETs are calculated using carrier mobility model MINIMOS-6 with phonon scattering, ionized impurity scattering, and surface roughness scattering [13]-[15].

Fig. 1 shows the calibration of the 3-nm node nanosheet FET measurements in Ref.[6]. The gate length Lg, spacer length Lsp, sheet width W, sheet height H, and nanosheet rounding corner radius R are shown in Table I Calibration column. The drain voltage is 0.65V in both calibration and data generation and the gate voltage is swept from 0.1V to 0.7V. The Id-Vg curve in the subthreshold area is calibrated by adjusting the gate metal work function, where Vth and SS could be matched with the reference [6]. Then the carrier saturation velocity is adjusted for a calibrated Ion.

To generate the FoMs, the DOE is performed with the variation of geometric parameters as listed in Table I Data Generation column. Five geometric parameters including Lg, W, H, R, and Lsp, are split into various values with steps. All the 2400 simulation tasks are performed and FoMs are extracted from the generated Id-Vg curves.

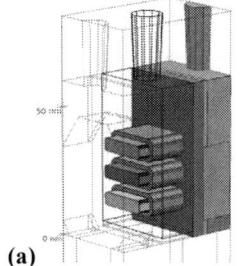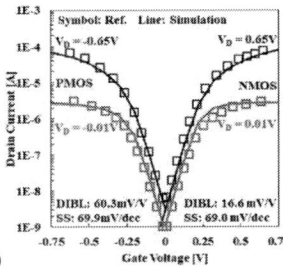

(a)   (b)

Fig. 1. (a) NSFET structure used in this article and (b) calibrated drain current-gate voltage (Id-Vg) curve with [2].

A sensitivity analysis of the features and labels is carried out (see Fig. 2), where Pearson Correlation Coefficients are calculated between each pair of features and labels [8]. Fig. 2 shows the Lg has a strong impact on all FoMs due to the short channel effect, whereas the Lsp and R show little impact on FoMs. Since Ion is proportional to effective channel width, the W and Ion show a positive correlation. Vth has a negative correlation with H because of poor control for thicker nanosheets.

TABLE I. GEOMETRIC PARAMETERS FOR DATA GENERATION

	Calibration	Data Generation (min, max, step)
Lg [nm]	15	(9, 30, 3)
W [nm]	20	(10, 25, 5)
H [nm]	5	(4, 8, 1)
R [nm]	1	(0, 2, 1)
Lsp [nm]	5	(5, 15, 2.5)

	Ioff	Ion	SS	Vth
SheetWidth	0.2	0.9	0.02	-0.3
SheetHeight	0.3	-0.02	0.4	-0.6
SheetRadius	-0.02	-0.01	-0.008	0.01
SpacerLength	-0.009	-0.1	-0.05	0.03
GateLength	-0.3	-0.2	-0.5	0.5

Fig. 2 Sensitivity analysis of TCAD samples using Pearson Correlation Coefficients.

III. MACHINE LEARNING AND FOMS PREDICTION

A. Genetic algorithm-based NN

GA was first introduced by John Holland as an optimization method based on natural selection theory [16]. The GA is evolved by calculating the fitness function of each chromosome and the best-fitted individuals are selected for the next generation. The three evolution operations, that is, selection, crossover, and mutation, are applied to all the individuals to generate the best solution. In this work, GA is used to synthesize the best model to predict FoMs, whose workflow is shown in Fig. 3. The GA starts with generating the initial population randomly from the gene pool. All the chromosomes are evaluated, and the fittest ones are kept and selected for crossover and mutation. Thus, the new generation is generated, and the fitness function is checked. The selection, crossover, and mutation will repeat till the fitness function meets the criteria and the fittest ones are employed for prediction tasks.

The whole perdition workflow is composed of three parts, that is, model design, hyper-parameter optimization, and training and test. In the model design part, the maximum number of five hidden nodes is used. The size of the population per generation is set to be 16 and the maximum number of generations is 10. The mutation probability is kept at 10% while the crossover with the possibility of 90% is applied. The model performance is evaluated based on the coefficient of determination R2 criteria. After the GA, the hyper-parameter is fine-tuned. The dataset is randomly divided into three parts: 80% for training, 10% for cross-validation, and 10% for testing. The neural networks are trained and tested, and the performance of the selected model is evaluated by R2. Table II shows the prediction performance of the top-5 GA-based neural network models, all of which show good prediction capability with the coefficient of determination R2 higher than 0.98.

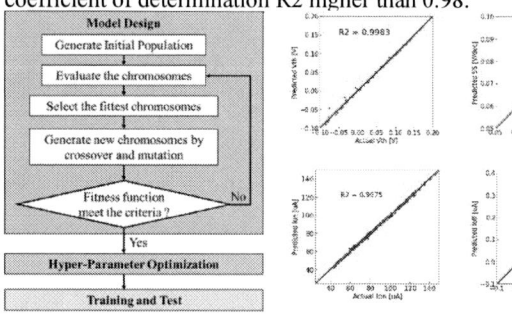

Fig. 3 Overall workflow of GA is shown in model design part.

Fig. 4 Scatter plots showing good matches between the predicted and actual values of (a)V_{th}, (b)SS, (c)I_{on}, and (d)I_{off} by the GA-based NN with average R^2 at 0.9921.

TABLE II. PREDICTION PERFORMANCE OF DIFFERENT GA-BASED MODELS

Models	Hidden nodes (No.)	R^2 (train set)	R^2 (test set)
Model 1	3	0.9961	0.9921
Model 2	4	0.9962	0.9917
Model 3	2	0.9965	0.9904
Model 4	4	0.9956	0.9896
Model 5	3	0.9960	0.9895

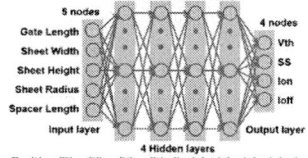

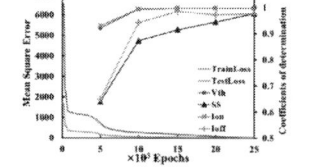

Fig. 5 NN used in this work. The output vector Y is a function of input X operated by weight W, bias b and σ.

Fig. 6 The loss function on the train (red) and test (blue) set by the red line and bule line respectively. Prediction accuracy R^2 are shown with symbols.

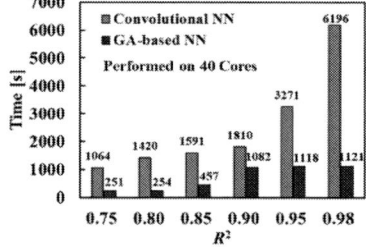

Fig. 7 Benchmarking of computational time cost for the two networks under the same hardware, showing the GA-based NN could reduce computation cost significantly.

The best-fitted model (Model 1 in Table II) is composed of a Lasso model with Least Angle Regression (LARS) and a standard scaler, which are integrated and a feature union model is employed. The average R2 is 0.9921 and it could be seen that all the FoMs are well-predicted by the GA-based neural network in Fig. 4, where the predicted values and actual values are highly matched.

B. Benchmark with Conventional Multi-layer NNs

Fig. 5 shows the basic structure of a multi-layer neural network built and trained for predicting NSFET FoMs. The NN is composed of a single input layer (a row vector with the size of 5×1), a single output layer (a row vector with the size of 4×1), and four hidden layers with 100 nodes for each of them, which are determined as a result of comparing performance of

networks with various width and depth. The activation function σ is the sigmoid function. The NN is implemented in Python with the machine learning framework PyTorch [17] and trained with the backpropagation (BP) algorithm.

The training process starts with data preprocessing. The cleaned data is normalized by a standard scaler and all the 2074 samples are split to 80%, 20% for training and testing sets respectively. The mean square error (MSE) is used as the loss function to evaluate the accuracy of the trained NN in the training process, which is optimized by the Adam optimizer with the learning rate at $5.0×10^{-6}$. The batch size is 16 and the number of epochs is 25000. Fig. 6 shows the MSE for training and testing sets as a function of epochs. The NN is determined when the MSE of both sets converges. R2 of the predicted values and labels increases in the training process and reach above 0.95 for all the FoM at 25000 epochs, which is shown in Fig. 6 as well. The R2 is 0.9872 at 2500 epochs.

C. Discussion

Both the trained multi-layer NN and the GA-based NN perform well in NSFET FoMs prediction. The average R2 of 0.9872 (by GA) and 0.9921 (by ANN) indicate a quite close match between the predicted and the simulated values of FoMs of 3-nm node NSFET for both neural networks. For the multi-layer ANN, the R2 of each FoM reach higher than 0.97 as the number of epochs increases to 25000, where Ioff and SS have lower R2 due to their low sensitivity to the selected five device parameters. Whereas the Vth and Ioff, who show great dependence on H and W, have better-predicted values under the same number of epochs.

Additionally, a benchmark of time cost is performed for the two neural networks as is shown in Fig. 7. Both networks are trained and tested with 40 Intel Xeon Gold 6230 CPUs on our in-house high-performance computation platform and the time cost for multi-layer NN and GA-based NN are 6196 and 1121 seconds respectively, which shows an 80% computational cost reduction for GA-based NN. Considering the higher average R2 of FoMs predicted by GA-based neural networks, the GA-based neural network shows great potential in machine learning augmented-TCAD.

IV. CONCLUSION

In this work, NSFET FoMs (Vth, Ion, Ioff, and SS) are predicted by a GA-based NN and a conventional multi-layer NN. The dataset for the training and testing network is carried out from TCAD where 2400 three-layer-stacked nanosheet nFETs with various geometric parameters are built up and simulated. Both the multi-layer NN and GA-based NN show the good capability of FoMs prediction with average R2 at 0.9872 and 0.9921, respectively. The computational cost is reduced by 80% for GA-based NN compared with the multi-layer NN, indicating GA-based auto-machine learning approach has great potential in machine learning-augmented TCAD.

ACKNOWLEDGMENT

This work was supported in part by NSFC under Grant No. 91964202. The authors would like to thank Intelligence Qubic for providing the auto-machine learning platform DarwinML®.

REFERENCES

[1]Auth, Chris, et al. 2012 VLSI. IEEE, 2012. [2]Loubet, N., et al. 2017 VLSI. IEEE, 2017. [3]Lee, Y. M., et al. 2017 IEDM. IEEE, 2017. [4]Yakimets, D., et al. 2017 IEDM. IEEE, 2017. [5]Bardon, M. Garcia, et al. 2018 VLSI. IEEE, 2018. [6]Kim, Hyunsuk, et al. *IEEE TED* 67.4 (2020): 1537-1541. [7]Han, Seung-Cheol et al. 2019 SISPAD. IEEE, 2019. [8] Xu, Haoqing, et al. *IEEE TED* 69.7 (2022):3568-3574. [9]Carrillo-Nuñez, Hamilton, et al. *IEEE EDL* 40.9 (2019): 1366-1369. [10]Ko, Kyul, et al. *IEEE TED* 66.10 (2019): 4474-4477. [11]Ko, Kyul, et al. *IEEE TED* 67.4 (2020): 1575-1580. [12]*Global TCAD Solutions.* [Online]. [13]Yao, Jiaxin, et al. *IEEE JEDS* 6 (2018): 841-848 [14]Huo, Qiang, et al. *IEEE TED* 67.3 (2020): 907-914. [15]Huo, Qiang, et al. *IEEE JEDS* 8 (2020): 295-301. [16]Holland, John Henry. *MIT press*, 1992. [17]Paszke, Adam, et al. *arXiv preprint* arXiv:1912.01703 (2019).

978-1-6654-9270-6/22 $31.00 © 2022 IEEE

Optical Receiver Front-End for 50G PON in 40nm CMOS

Nianquan Ran, Jia Li, Shuaizhe Ma, Yuye Yang, Wanqing Zhao, Hao Li, Dan Li

Faculty of Electronic and Information Engineering, Xi'an Jiaotong University, Xi'an, China

Abstract—With the determination of the 50Gb/s PON communication network standard, there is a large demand for 50Gb/s PON chips. In this paper, we design a 50Gb/s transimpedance amplifier (TIA) chip with very low power consumption, which greatly reduces the manufacturing cost by adopting the 40nm standard CMOS process. In the high gain mode, the transimpedance gain is 66.0dBΩ and the bandwidth is 30.4GHz. In the low gain mode, the transimpedance is 52.4dBΩ and the bandwidth is 34.1GHz. The input signal range can reach 2mA at most and the maximum differential output swing is 440mVpp. The receiver front-end circuit consumes 23.4mW, and the energy efficiency is 0.47pJ/bit.

Keywords—40nm CMOS, transimpedance amplifier, 50Gb/s PON, low power

I. INTRODUCTION

In last year, the International Telecommunication Union adopted the 50Gb/s passive optical network (PON) international standard and the optical network unit (ONU) will adopt the NRZ data modulation mode [1]. It is expected to be commercially available in 2025. Because the standard CMOS process has the characteristics of low cost and high integration, it is more conducive for commercial application. And low power consumption is conducive to reducing the cost of use [2].

This paper describes a 50Gb/s NRZ receiver front-end for 50Gb/s PON ONU scenarios. Fig. 1 is the conceptual diagram of this scenario. At the TIA stage, the value of the feedback resistor Rf is increased by means of a three-stage core amplifier, so as to reduce the equivalent input noise of the later stage. At the same time, use a lower supply voltage can reduce energy consumption and improve the efficiency of data transmission. Through shunt inductance peaking and negative capacitance techniques, the bandwidth is further extended.

II. CIRCUIT DESIGN

A. TIA architecture

The photo diode (PD) converts optical signal into the electrical signal, which is amplified by a transimpedance amplifier. The TIA has two gain modes for large input dynamic range. A level shifter (LS) is followed to increase the input nodes DC bias voltage of the single to differential amplifier (S2D). Then, the S2D converts the single ended signal to differential. The main amplifier (MA) of current mode logical (CML) pulls the signal to a large swing, and finally we realize the impedance matching by the 50Ω resistors of the output buffer.

In this work, the influence of input and output parasitic capacitance and inductance is also fully considered. The specific circuit block diagram is shown in Fig. 2.

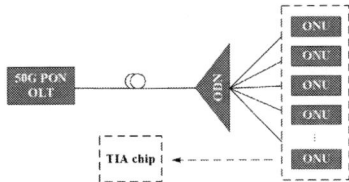

Fig. 1.   Conceptual diagram of 50G PON.

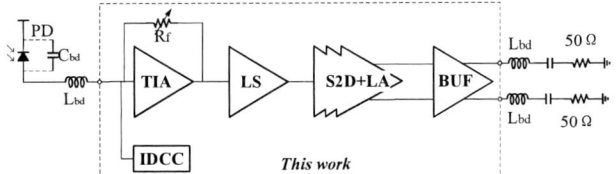

Fig. 2.   Optical Receiver block diagram.

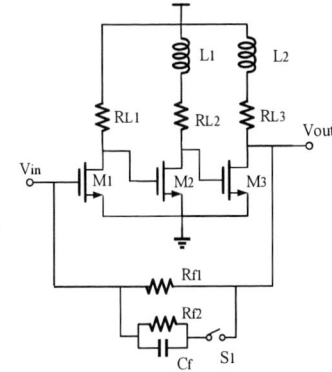

Fig. 3.   Schematic of three-stage TIA.

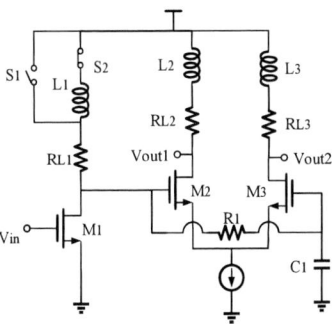

Fig. 4.   Schematic of level shifter and single-to-differential amplifier.

B. TIA Front-end and LS

According to the analysis of the second-order feedback TIA, the trade-off between gain and bandwidth is

$$f_{-3dB} = \frac{\sqrt{2}A_0}{2\pi R_F C_D} \qquad (1)$$

After making the trade-off between stability and gain, we finally choose the three-stage core amplifier TIA structure, as show in Fig. 3. In order to reduce power consumption, we reduce the current flowing through the input transistors, which makes the noise increase. But this trade-off is considered acceptable.

When the input is a large signal, the noise performance is not important. So we can reduce the gain of TIA to ensure the undistorted transmission. When a large input signal comes, the switch S1 is turned on to reduce the value of feedback resistor, and the frequency response becomes smooth again by reducing the inductance value of the latter stage.

In order to increase the driving capability of the circuit, TIA is followed by a level shifter. And the single ended signal is converted into differential through S2D, as show in Fig. 4.

C. MA and BUF

Fig. 5 shows the circuit diagram of the two-stage main amplifier. In order to amplify the weak signal, it is necessary to further improve the gain without reducing the bandwidth. The gain expansion is realized with appropriate transistor sizes and load resistances, and the bandwidth is expanded by using shunt inductance peaking. Cascading two CML differential amplifiers provide a total gain of 4.3dB. In order to obtain a smooth transfer function in low gain mode, the inductance of the shunt inductors in LS and the second stage MA is set to variable mode. When the mode changes, it changes along with the feedback resistance of TIA.

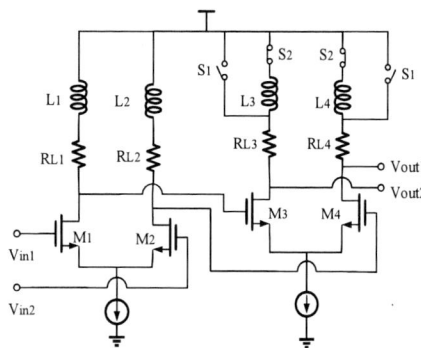

Fig. 5. Schematic of the main amplifier.

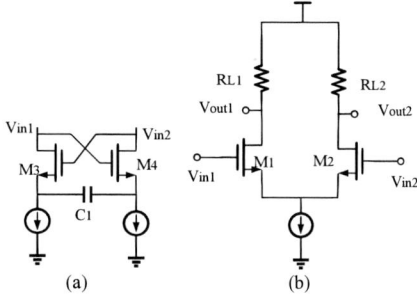

(a)   (b)

Fig. 6. Schematic of output buffer stage: (a)negative capacitance, (b)output buffer.

Fig. 6 shows the circuit of the output buffer stage, which aims to achieve 50Ω impedance matching. In order to achieve a large output swing, the large tail current is required, which makes the size of the input transistors relatively large and introduces a large parasitic capacitor. In order not to affect the matching degree and avoid bandwidth reducing caused by large parasitic capacitance, the negative capacitance technology is added at the input node to offset the influence of some parasitic capacitance.

III. SIMULATION RESULTS

The receiver front-end circuit is designed in 40nm CMOS process. The whole circuit consumes 26mA current under 0.9V voltage supply. Energy efficiency is 0.47pJ/bit. The PD capacitance is 50fF, bonding wire inductance is 200pH and the capacitance of PAD is 40fF. As shown in Fig. 7, the receiver front-end total gain is 66.0dBΩ with the bandwidth of 30.4GHz in the high gain mode and 52.4dBΩ with the bandwidth of 34.1GHz in the low gain mode. The equivalent input noise current is 4.48μArms, and input noise density is 25.7pA/√Hz.

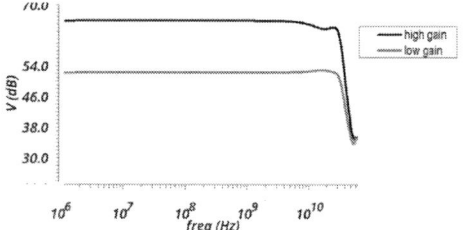

Fig. 7. Simulated receiver transfer functions.

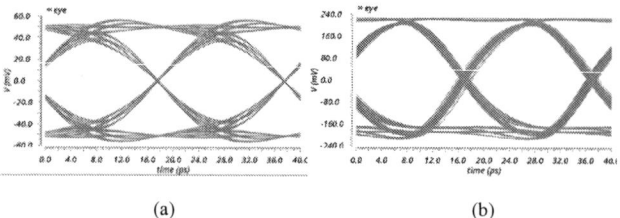

(a)   (b)

Fig. 8. Simulated 50Gb/s NRZ output eye diagram with input current at: (a)50uApp and (b)2mApp.

TABLE I.   PERFORMANCE COMPARISON OF HIGH-SPEED OPTICAL RECEIVERS

Parameters	[3]	[4]	[5]	This work*
Technology	250nm BiCMOS	180nm BiCMOS	40nm Bulk CMOS	40nm CMOS
Data rate(Gb/s)	56(NRZ)	50(NRZ)	50(PAM4)	50(NRZ)
BW(GHz)	43	35	23.4	30.4/34.1
Gain(dBΩ)	66	70	66/46	66.0/52.4
Noise(μArms)	2.08	5.7	3.2	4.48
Supply(V)	3.3	3.3	3.3	0.9
Power(mW)	205	319	76	23.4
Noise (pA/sqrt(Hz))	11.3	30.4	20.9	25.7
Input overload (mApp)	N/A	N/A	2	2
Max. Output swing(mV)	N/A	N/A	400	440

*Simulation.

The 50Gb/s NRZ output eye diagrams are shown in Fig. 8. When the input is a 50Gb/s, 50uA amplitude signal, its eye diagram is shown in Figure 8 (a) and the jitter is 500fs. When the input is a large signal of 50Gb/s and 2mA amplitude, the eye diagram is shown in Figure 8 (b) and the jitter is 2 ps.

In Table 1, we compare this work with optical receivers front-end circuits published in recent years. From this table, we can find that in this work, even if noise indicator is sacrificed to some extent, the power consumption is far lower than that of other works for comparison. And this work has the potential to be applied to a 50Gb/s PON network.

IV. CONCLUSION

A 50Gb/s optical receiver analog front-end circuit is designed for 50Gb/s PON ONU scenario in 40nm CMOS technology, with very low power supply voltage and low power consumption. Two gain modes are adopted to realize the undistorted signal transmission in a large input dynamic range.

REFERENCES

[1] ITU-T Recommendation G.9804.3, "50-Gigabit-capable passive optical networks (50G-PON): Physical media dependent (PMD) layer specification," Sep. 2021; https://www.itu.int/rec/T-REC-G.9804.3-202109-I

[2] D. Li et al., "A Low-Noise Design Technique for High-Speed CMOS Optical Receivers," in IEEE Journal of Solid-State Circuits, June 2014, vol. 49, no. 6, pp. 1437-1447.

[3] .G. Dziallas, A. Fatemi, A. Peczek, L. Zimmermann, A. Malignaggi and G. Kahmen, "A 56-Gb/s Optical Receiver With 2.08-μA Noise Monolithically Integrated into a 250-nm SiGe BiCMOS Technology," in IEEE Transactions on Microwave Theory and Techniques, Jan. 2022., vol. 70, no. 1, pp. 392-401.

[4] T. Takemoto et al., "A 50-Gb/s High-Sensitivity (−9.2 dBm) Low-Power (7.9 pJ/bit) Optical Receiver Based on 0.18-μm SiGe BiCMOS Technology," in IEEE Journal of Solid-State Circuits, May 2018, vol. 53, no. 5, pp. 1518-1538 .

[5] Y. Liu et al., "A 50Gb/s PAM-4 Optical Receiver with Si-Photonic PD and Linear TIA in 40nm CMOS," 2020 IEEE International Symposium on Circuit and Systems (ISCAS), 2020, pp. 1-4.

A Fully-Connected and Area-Efficient Ising Model Annealing Accelerator for Combinatorial Optimization Problems

Yukang Huang[1], Dong Jiang[1], Yongkui Yang[2], Enyi Yao[1*]
[1]School of Microelectronics, South China University of Technology, Guangzhou 511442, China
[2]Shenzhen Institute of Advanced Technology, Chinese Academy of Science, Shenzhen, China

Abstract—The combinatorial optimization problem is ubiquitously in our daily life and typically inefficient for modern Von Neumann architecture-based computer. Targeting for various combinatorial optimization problems, this paper presents a 10K-bit area-efficient architecture of the domain specific accelerator based on fully-connected Ising model using an FPGA platform. The proposed system is based on simulated annealing algorithm with a spin pre-selection scheme to prevent the system to be trapped in the local minimum and increase the convergence efficiency, which is more easily and efficiently to be hardware implemented. Using max-cut problem as the experiment benchmark, the proposed hardware architecture achieves an acceleration of $50,000\times$ compared with the software simulation result.

Keywords—Combinatorial optimization, Ising model, simulated annealing, hardware accelerator, FPGA

I. INTRODUCTION

Ising model based simulated annealing algorithms have been widely used for solving different types of combinatorial optimization problems such as max-cut, max-flow and traveling salesman problem (TSP), which are usually belong to NP-hard or NP-complete problems and notoriously difficult to tackle for modern computer with general non-heuristic method. Inspired by quantum theory and technology, D-Wave System Inc. developed the world's first quantum annealing computer with potential applications in field of logistics, portfolio optimization, drug discovery, materials sciences, scheduling, fault detection, traffic congestion, supply chain management, etc. [1]. However, the requirement for ultra-low temperature operation environment, decoherence effect and limited multi-body interference for quantum computing finally raise the difficulty for D-Wave to be widely applied in our daily life. At the same time, with the rapid development of semiconductor manufacturing technology, various kinds of hardware accelerators have also been proposed for these emerging smart computer architectures, as well as the hardware solvers for combinatorial optimization problems.

CMOS based Ising model annealing machine is gaining more attention in academia and industrial because of its normal working environment requirement and mature manufacturing technology. Over the past decade, many groups have worked on different implementation of the Ising machines using lattice, King's graph or Hexagon based structures with spins connected in a local manner with its adjacent neighbors [2]–[4]. These architectures are compact and scalable, but application limited due to their sparse spin-to-spin connectivity. In [5], Fujitsu developed an Ising model based digital annealer on the FPGA platform with 1024 fully-connected spins. With a complete spin-to-spin connectivity, at each step, only one spin is allowed to be updated resulting in a low computation efficiency. Furthermore, with at most one spin update strategy also decreases

the opportunity for the system to escape local minimum for the case where the energy wall is beyond 2 succeeded steps.

In this article, a novel fully-connected Ising model based digital accelerator specific for combinatorial optimization problems is proposed. Max-cut problem is utilized as the benchmark to evaluate the performance of this work. It is it quite commonly used in the layout automatic place and route process for VLSI design. Based on simulated annealing algorithm, a specific hardware implementation for acceleration is described with less computation resource and high iteration efficiency.

II. ARCHITECTURE

The architecture of the proposed fully-connected Ising model-based hardware accelerator for combinatorial optimization problems is shown in Fig. 1. When a Monte Carlo iteration is started, the Global Random Number Generation (RNG) block will randomly select a particular spin through the Local MUX to get the corresponding local energy change ΔE_i with its current state x_i and local field value h_i by the Local Energy Units (LEUs). The initial values of the local energies are pre-calculated and loaded from the memory when the system is initialized and the un-solved problem is already mapped to the hardware. At the same time, a random noise r will be generated by the local random number generation module within the Flip Possibility Calculation unit (FCU). The random noise r is utilized by the Linear Approximation Unit (LAU) and multiplied with the temperature value set by the Temperature Update Unit (TUU). Then the Acceptance Desision Unit (ADU) will take the multiplied product to compare with the selected local energy change to determine weather this spin should be updated in the following Spin Update Unit (SUU). If the spin is flipped by the SUU, the new state value x_j will be fed-back through the Globle MUX to the corresponding LEUs to update the local field value h_i before next iteration cycle.

The RAM is used for the storage of the weight matrix with a size of $N \times N$ where N is the total number of the spins in the design. For simplicity, binary weight values are adopted in this paper. Hence, each bit of the RAM represents the connection relationship of two different spins. RAM outputs every spin connection corresponding to the input address.

Consider the case when the spin x_j is flipped, energy of the system is only affected by the connection between the spin x_j and any other one, which indicates calculating the local field values for all the other unrelated spin is unnecessary. The local field h_i can be updated via the corresponding W_{ij} instead of the whole matrix. For the next cycle, system energy change by flipping any spin state can also be calculated in the LEU block with the new updated local field value h. Then a random spin selection after that is added to identify which spin with corresponding local energy suitable to be updated during this cycle. In this article, Ising model with 10K fully connected spins is implemented with 10K LEUs and only one single ADU and FCU respectively. The proposed system architecture puts the spin selection after the LEUs while for the implementation of conventional simulated annealing algorithm, we have to put it after flip possibility calculation with totally 10K ADUs and FCUs accordingly.

III. PERFORMANCE

During the experiment, the proposed simulated annealing algorithm based method is implemented and compared with

*Email:yaoenyi@scut.edu.cn

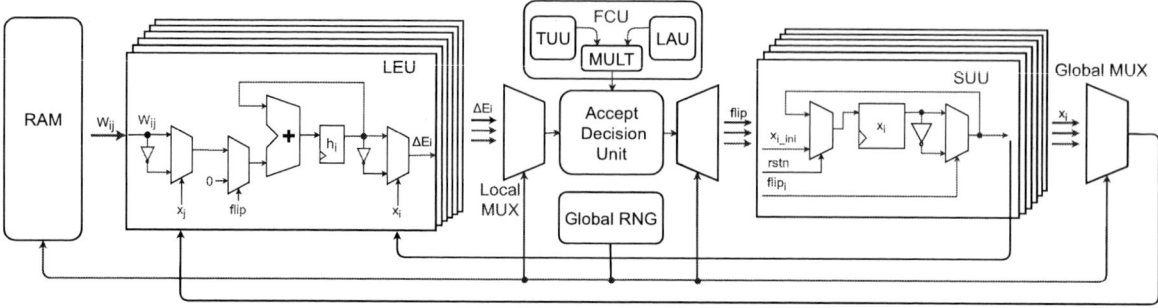

Fig. 1: System architecture of the proposed Ising model based combinatorial problem solver.

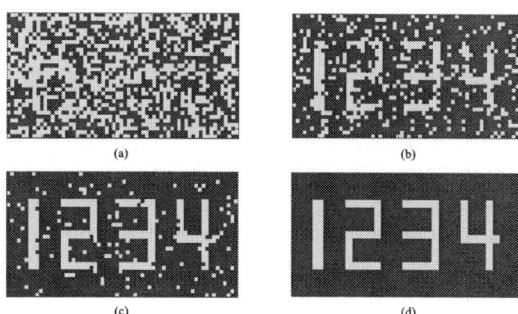

Fig. 2: Solving a max-cut problem with proposed system to reveal the numbers.

conventional method for the max-cut problem. The max-cut problem is defined as partitioning a undirected graph into two subsets to maximize the weighted edges between the vertices of the two subsets. One example of the max-cut problem is to reveal the hided pattern during the annealing process, which is shown in Fig. 2. To compare with conventional simulated annealing algorithm, G-set benchmark (G51) is selected with 1000 vertices and 5909 edges [6]. The best cut value using proposed meithod is 3824, which is 99.4% of the ever reported best value (3846) [7].

To further accelerate the optimization speed, a hardware fully connected combinatorial optimization problem solver is designed with 10K spins using a FPGA platform (Xilinx xc7v2000tfhg 1761) since FPGA could provide a flexible and fast hardware design and evaluation environment. Hardware implementation result shows that about 231K LUTs and 185K registers are utilized for the 10K spins. 1138 BRAMs are used to store the spin values and interaction coefficients that is unchanged during optimization. We compare the speed performance of the hardware accelerator with a general purpose CPU. The algorithms are simulated using Intel i7-10700 processor with 8 cores, 16 threads and the clock frequency of 2.9 GHz. During annealing process, the system energy change with time. The proposed hardware implementation achieves about $50,000\times$ speedup than the software optimization.

Table I details the comparisons with some prior works. Compared to quantum annealing using superconductor, the proposed work could be operated at the room temperature. Furthermore, with an implementation of 10K-spin Ising model accelerator, the proposed design achieves an all-to-all connection architecture with an higher accuracy, faster speed and

efficient hardware resource utilization.

TABLE I. Comparison with related works

Design	Technology	Spin Number	Connection topology	Power
[1]	Quantum	128	Chimera	25 kW
[2]	CMOS 40 nm	9×16K	King's graph	-
[3]	CMOS 65 nm	480	King's graph	86 μW
[4]	CMOS 65 nm	560	Hexagon	23 mW
[5]	FPGA	1024	All-to-all	-
[8]	Optical	100	All-to-all	-
This work	FPGA	10K	All-to-all	-

IV. CONCLUSION

In this paper, we present a large scale of Ising model annealing accelerator design with 10K all-to-all spins connected for combinatorial problem. Compared to conventional implementation approach, the proposed method could achieve a higher accuracy, faster optimization speed and less hardware resource consumption.

REFERENCES

[1] Z. Bian, F. Chudak, W. G. Macready, and G. Rose, "The ising model: Teaching an old problem new tricks," *D-Wave Syst.*, vol. 2, 2010.

[2] T. Takemoto, K. Yamamoto, C. Yoshimura, M. Hayashi, M. Tada, H. Saito, M. Mashimo, and M. Yamaoka, "4.6 a 144kb annealing system composed of 9×16kb annealing processor chips with scalable chip-to-chip connections for large-scale combinatorial optimization problems," in *IEEE International Solid- State Circuits Conference (ISSCC)*, vol. 64, 2021, pp. 64–66.

[3] Y. Su, H. Kim, and B. Kim, "31.2 cim-spin: A 0.5-to-1.2v scalable annealing processor using digital compute-in-memory spin operators and register-based spins for combinatorial optimization problems," in *IEEE International Solid- State Circuits Conference - (ISSCC)*, 2020, pp. 480–482.

[4] I. Ahmed, P.-W. Chiu, and C. H. Kim, "A probabilistic self-annealing compute fabric based on 560 hexagonally coupled ring oscillators for solving combinatorial optimization problems," in *IEEE Symposium on VLSI Circuits*, 2020, pp. 1–2.

[5] S. Tsukamoto, M. Takatsu, S. Matsubara, and H. Tamura, "Accelerator architecture for combinatorial optimization problems," *FUJITSU Sci. Tech. J.*, vol. 53, no. 5, pp. 8–13, 2017.

[6] [Online]. Available: https://web.stanford.edu/~yyye/yyye/Gset/

[7] C. Cook, H. Zhao, T. Sato, M. Hiromoto, and S. X.-D. Tan, "Gpu-based ising computing for solving max-cut combinatorial optimization problems," *Integration*, vol. 69, pp. 335–344, 2019.

[8] P. L. McMahon, A. Marandi, Y. Haribara, R. Hamerly, C. Langrock, S. Tamate, T. Inagaki, H. Takesue, S. Utsunomiya, K. Aihara, R. L. Byer, M. M. Fejer, H. Mabuchi, and Y. Yamamoto, "A fully programmable 100-spin coherent ising machine with all-to-all connections," *Science*, vol. 354, pp. 614–617, 2016.

2022 IEEE International Conference on
Integrated Circuits, Technologies and Applications

A 77-101GHz 6-Bit Vector-Modulated Phase Shifter with Low RMS Error in 65nm SOI CMOS

Qingzhe Zhang, Keping Wang

Tianjin University

Abstract—This paper presents a 77-101 GHz active 6-bit phase shifter based on a vector-modulated technique in 65nm SOI CMOS technology for W-band phased-array systems. Optimizations of the impedance-invariant variable gain amplifier (VGA) are performed to reduce the phase error and gain error among the phase states. In addition, a quadrature signal generator composed of a 90-degree hybrid and a pair of Marchand balun is exploited for wideband applications. The proposed phase shifter achieves 5.6° phase step, < 2.0° RMS phase error and < 0.8dB RMS gain error over 77-101 GHz. The total power consumption is 31mW and the core area of the phase shifter is only 0.2mm^2.

Keywords—W-band, phase shifter, vector-modulated, SOI CMOS

I. INTRODUCTION

W-band (75-110 GHz) provides a promising opportunity for implementation in highly directive wireless communications and high-resolution imaging systems. Therefore, as an important part of phased-array systems, the W-band phase shifter have been an active research field in recent years [1]-[4].

Phase shifter is component in phased-array systems to provide phase shifting, which can be divided into passive form and active form. Passive phase shifters, mainly including reflective-type phase shifter (RTPS) and switched-type phase shifter (STPS), typically exhibit high linearity performance, but also high losses, high noise figure (NF) and large chip area. For active phase shifters, such as vector modulation phase shifters, compared with passive phase shifters, typically exhibit higher gain, smaller chip area, higher phase shift resolution and full phase shifting range [2].

This paper presents a vector modulation phase shifter with a wideband quadrature signal generator composed of a 90° hybrid and a pair of Marchand balun and an impedance-invariant variable gain amplifier (VGA) in 65nm SOI CMOS technology. This design enables large bandwidth, low RMS phase error, low RMS gain error, and compact chip design.

II. PHASE SHIFTER DESIGN

Fig. 1 shows an architecture of a vector modulation active phase shifter. The vector modulator phase shifter is based on an IQ signal Generator, two impedance-invariant VGAs and an output combiner.

A. IQ Signal Generator

As we know, the phase error and amplitude imbalance of the IQ signal generator have a critical effect on the overall performance of the phase shifter. Thus, as a crucial part of the IQ signal generator, the 90° Hybrid and Balun designs with good phase and amplitude characteristics are necessary.

Lange coupler has symmetrical layout and ultra-wideband characteristics, and is more suitable for use in high frequency environment, such as W-band, than spiral coupler [3]. The structure of the Lange coupler is shown in Fig. 2(a). Fig. 2(c) shows the EM simulation results of the Lange coupler. It can be found that the Lange coupler has a flat intersection area

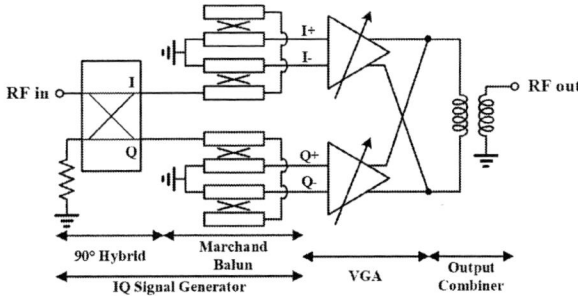

Fig. 1. Architecture of vector-modulated phase shifter

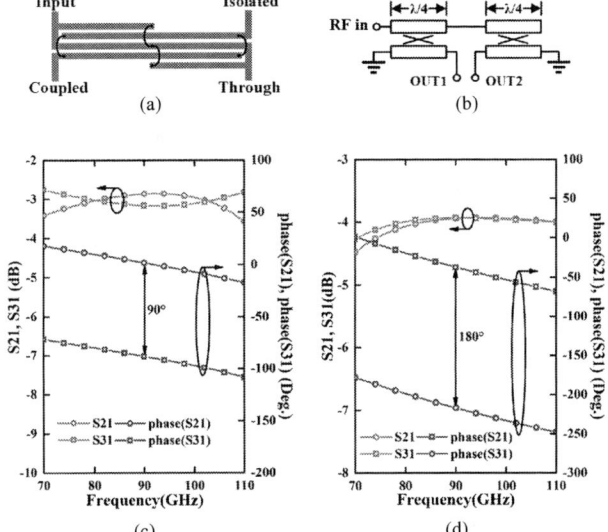

Fig. 2. (a) Lange Coupler. (b) Marchand Balun. (c) Simulation amplitude and phase of S21 and S31 of (a). (d) Simulation amplitude and phase of S21 and S31 of (b)

between the I path and the Q path, which greatly improves the bandwidth, and also has good phase characteristics in the frequency range of 70-110 GHz.

Marchand balun has the advantage of bandwidth compared with the transformer structure in high frequency environment [5]. The structure of Marchand Balun is shown in Fig. 2(b). Fig. 2(d) is the EM simulation results of Marchand balun, in the frequency range of 70-110 GHz, its amplitude imbalance and phase error are less than ±0.2 dB and ±1.4° respectively.

B. Impedance-Invariant VGA

In [6], a one shack impedance-invariant VGA to reduce phase and amplitude error of phase shifter is presented. For this architecture, a large number of capacitors are used to isolate DC control signals and RF signals, which greatly increases the layout area and is difficult to apply in high-resolution phase shifters.

To further overcome the aforementioned issues and improve the performance of phase shifters, an impedance-invariant VGA based on cascode structure is proposed, as shown in Fig. 3. First, to keep the input impedance and the output impedance of the VGA remain constant for all gain states, digital signals are used to control the size of transistors. The details of its work are as follows. A large transistor is divided into N unit transistors to form a transistor array, so that the equivalent transistor size of the access circuit can be adjusted by turning on and off the number of unit transistors. The signal path of VGA is divided

Corresponding Author: Keping Wang, email: kpwang@tju.edu.cn

978-1-6654-9270-6/22 $31.00 © 2022 IEEE

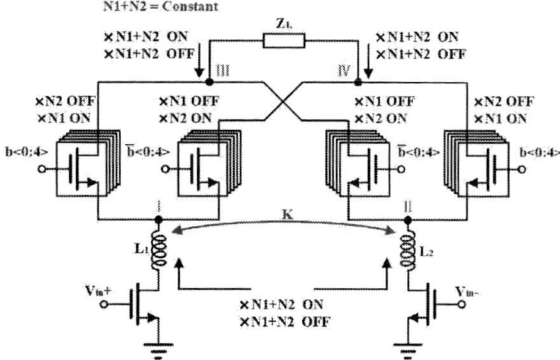

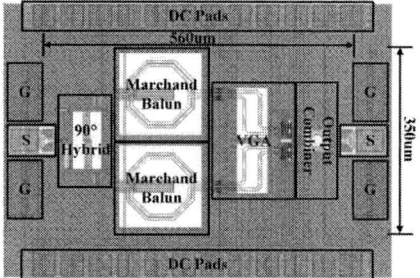

Fig. 3. Schematic of Impedance-Invariant VGA

Fig. 4. Layout of vector-modulated phase shifter

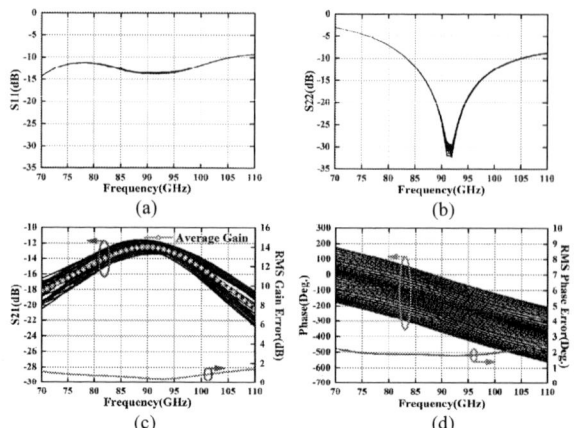

Fig.5. Simulated (a) S11, (b) S22, (c) S21 and RMS gain error, (d) phase shift and RMS phase error

TABLE I. PREFORMANCE COMPARISON

	[1]	[2]	[4]	This Work
Process	0.13um SiGe BiCMOS	28 nm FDSOI	40 nm CMOS	**65 nm SOI**
Architecture	Vector Modulator	Vector Modulator	Vector Modulator	**Vector Modulator**
Resolution (bits)	5	4	6	**6**
Freq. (GHz)	92-100	78.8-92.8	70-90	**77-101**
RMS Phase Error (deg)	<10	<11.9	<2.4	**<2**
RMS Gain Error (deg)	<1.8	<2	--	**<0.8**
DC Power (mW)	37	21.6	0	**31**
Core Area (mm²)	1.04	0.12	0.15	**0.2**

into two paths, to ensure that one path has N1 transistors with on-state and N2 transistors with off-state at the same time, while the other path has N2 transistors with on-state and N1 transistors with off-state. By making N1+N2 to be a constant (N1+N2 = N), there will always be N transistors turned on and N transistors turned off at nodes I and II. Similarly, nodes III and IV also maintain N on-state transistors and N off-state transistors. Second, the DC control signals and RF signals use different paths to drive the common-gate transistors and common-source transistors respectively, eliminating the dependence on isolation capacitors. Third, L_{1-2} can improve the gain of the VGA [2], but inductors will occupy a large chip area. Therefore, the introduction of weakly coupling (K=0.2) inductors allows for a more compact design.

III. SIMULATION RESULTS

The proposed 77-101 GHz 6bit vector-modulated phase shifter is implemented in a 65nm SOI CMOS technology. Fig. 4 shows the layout and the core area of the phase shifter. It occupies an area of 0.56×0.3mm² excluding pads. Fig. 5 is the simulation results of the phase shifter with its passive part evaluated by the full layout EM simulation. Fig. 5(a), (b) are the simulation results of S11 and S22 in all states, respectively. S11 remains below than -10dB from 70 to 104 GHz. S22 is below than -10dB from 83 to 104.5 GHz. S21 for all states and RMS gain error are shown in Fig. 5(c). The proposed phase shifter exhibits the 3-dB bandwidth from 77 to 101 GHz. The RMS gain error amounts to 0.4dB at 92GHz and is less than 0.8dB from 77 to 101 GHz. The simulation results of phase shifter for the 64 phase states are shown in Fig. 6. The phase shifter provides a phase turning range of 360° with 6-bit resolution. The RMS phase error is 1.8° at 92 GHz and remain below 2° from 77 to 101 GHz.

IV. CONCLUSION

This paper presents a 77-101GHz 6-bit vector-modulated phase shifter in 65nm SOI CMOS technology. Based on impedance-invariant VGA technique and wideband quadrature signal generator, the proposed phase shifter achieves 5.6° phase step, a low RMS phase error of < 2.0° and a low RMS gain error of <0.8dB over the bandwidth.

REFERENCES

[1] S. Afroz and K. Koh, "W-Band (92–100 GHz) Phased-Array Receive Channel With Quadrature-Hybrid-Based Vector Modulator," in IEEE Transactions on Circuits and Systems I: Regular Papers, vol. 65, no. 7, pp. 2070-2082, July 2018.

[2] D. Pepe and D. Zito, "Two mm-Wave Vector Modulator Active Phase Shifters With Novel IQ Generator in 28 nm FDSOI CMOS," in IEEE Journal of Solid-State Circuits, vol. 52, no. 2, pp. 344-356, Feb. 2017.

[3] H. Li, J. Chen, D. Hou and W. Hong, "A W-Band 6-Bit Phase Shifter With 7 dB Gain and 1.35° RMS Phase Error in 130 nm SiGe BiCMOS," in IEEE Transactions on Circuits and Systems II: Express Briefs, vol. 67, no. 10, pp. 1839-1843, Oct. 2020.

[4] P. Gu, D. Zhao and X. You, "Analysis and Design of a CMOS Bidirectional Passive Vector-Modulated Phase Shifter," in IEEE Transactions on Circuits and Systems I: Regular Papers, vol. 68, no. 4, pp. 1398-1408, April 2021.

[5] J. Yu et al., "A 300-GHz Transmitter Front End With −4.1-dBm Peak Output Power for Sub-THz Communication Using 130-nm SiGe BiCMOS Technology," in IEEE Transactions on Microwave Theory and Techniques, vol. 69, no. 11, pp. 4925-4936, Nov. 2021.

[6] G. H. Park, C. W. Byeon and C. S. Park, "A 60-GHz Low-Power Active Phase Shifter with Impedance-Invariant Vector Modulation in 65-nm CMOS," in IEEE Transactions on Microwave Theory and Techniques, vol. 68, no. 12, pp. 5395-5407, Dec. 2020.

A Fast Sampling Open-circuit Voltage Algorithm for Piezoelectric Energy Harvesting

Ying Yu[1], Xufeng Liao[1,2], Lianxi Liu[1,2]

[1] School of Microelectronics, Xidian University, Xi'an, 710071, China
[2] Xidian University Chongqing IC Innovation Institute, Research Building 1, Xiyong Microelectronics Park, Chongqing, 401331, China

Abstract—This paper presents a fast sampling open-circuit voltage algorithm for piezoelectric energy harvesting , which directly samples the output of the piezoelectric transducer and limits the sampling time to one piezoelectric vibration period. Therefore, it greatly reduces the power loss caused by the open-circuit sampling and improves the efficiency of the piezoelectric energy harvesting. The proposed circuit is implemented in 0.18 μm standard CMOS process, and the core circuit area is 670×536 μm². The circuit can sample the open-circuit voltage in less than one piezoelectric vibration cycle and output the maximum power point voltage, and the power loss rate is less than 0.79%.

Keywords—*piezoelectric energy harvesting (PEH), parallel synchronized switch harvesting on capacitor (P-SSHC), fast sampling, low power loss.*

I. INTRODUCTION

The parallel synchronized switch harvesting on capacitor (P-SSHC) can reduce the charge-discharge loss caused by the parasitic capacitor of the piezoelectric transducers (PZT) without the use of an inductor. Besides, the piezoelectric energy varies greatly with time, so the maximum power point tracking (MPPT) technology is needed to harvest more energy for the rectifier using the P-SSHC technology. The fractional open circuit voltage (FOCV) algorithm is one of the most popular MPPT algorithms with its advantages of simple structure and low power consumption [1]. However, the energy cannot be transferred to the power stage during the open-circuit sampling process when the traditional FOCV is used, which increases the power loss. Therefore, the sampling time should be as short as possible. In order to reduce the sampling time, a single-cycle peak detector using another small sampling capacitor was proposed, but it still needs to go through a charging process [2]. A complex differential circuit sampling 0.5 open-circuit voltage was proposed to shorten the sampling time, but the improvement is limited [3]. A 1.5-cycle fast sampling algorithm was proposed to shorten the sampling time to 1.5-cycle, but the algorithm can still be improved [1].

This paper proposed a fast sampling open-circuit voltage algorithm, which samples the open-circuit voltage from the output of the PZT, reducing the sampling time to less than one vibration cycle and improving the energy harvesting efficiency.

II. SYSTEM ARCHITECTURE

Fig. 1 shows the architecture of the proposed fast sampling algorithm. The P-SSHC uses the flip capacitor C_{FLIP} to flip the V_{PN} to reduce power loss caused by the parasitic capacitor C_P of the PZT. The two-stage rectifier realizes AC-DC conversion. The P-SSHC controller generates the sequence signal V_{SEQ} and the pulse signal Ø[2:0] to control P-SSHC. The sampling signal generator generates the sampling signal SAMPLE, the shielding signal SHIELD and the enable signal EN_{SAMPLE}. The sample and hold circuit (S&H) samples the open-circuit voltage V_{OC} of the PZT and outputs the maximum power point voltage V_{MPP}.

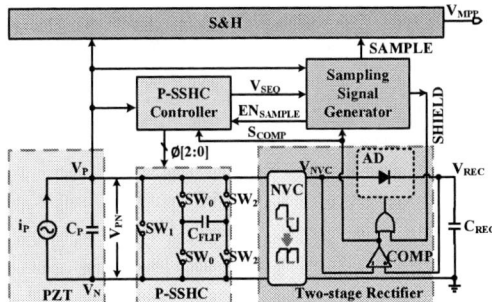

Fig. 1. Proposed fast sampling system architecture.

For the PEH circuit using the P-SSHC, the V_{MPP} [1] is

$$V_{MPP} = V_{POC} / (1 - k_{FLIP}) \qquad (1)$$

where k_{FLIP} is the flip coefficient and V_{POC} is the peak of the V_{OC}. The peak-to-peak value V_{PP} of the V_{OC} is

$$V_{PP} = \frac{1}{C_P} \int_0^{T_P/2} I_{PEAK} \sin(2\pi f_p t) dt = \frac{I_{PEAK}}{\pi C_P f_P} = 2V_{POC} \qquad (2)$$

where T_P is the vibration cycle, f_P is the vibration frequency of the PZT, and I_{PEAK} is the peak current of AC current source i_P. In addition, the average output power reduction caused by the open-circuit sampling is reflected by the power loss rate φ [1].

$$\varphi \approx (T_S / T) \times 100\% \qquad (3)$$

where T_S is the sampling time and T is the sampling period.

Due to the limitation of the process, $k_{FLIP} = 0.4$ is selected in this paper. According to (1), the V_{MPP} is $5/3 V_{POC}$. As shown in (2), the V_{PP} is $2V_{POC}$. Therefore, the proposed scheme shown in Fig. 1 can obtain the V_{MPP} by sampling the V_{PP} at the output of the PZT, which limits the sampling time to less than one vibration cycle and achieves a power loss rate φ of less than 0.79% when T is 128 vibration cycles. The proposed scheme reduces power loss rate from 24.9% in [4] by 96.8% with the same sampling period.

III. CIRCUIT IMPLEMENTATION

A. P-SSHC Controller

Fig. 2 is the P-SSHC controller. The D flip-flop generates the V_{SEQ}. The pulse sequencing circuit sequences the Ø[2:0]$_{ORI}$ generated by the pulse generator to obtain the correct Ø[2:0]. During the sampling process, the EN_{SAMPLE} jumps to low level so that Ø0 is also at low level. Therefore, V_{PN} cannot be flipped from 0 to V_{RECT} when i_P changes from negative half cycle to positive half cycle, ensuring that i_P can freely charge C_P from 0.

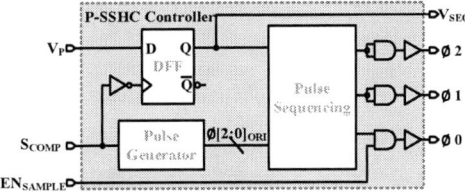

Fig. 2. Diagram of the P-SSHC controller.

B. Sampling Signal Generator

Fig. 3 (a) is the proposed sampling signal generator and Fig. 3 (b) is its key signal waveforms. The divider divides the NAND result of S_{COMP} and V_{SEQ} by 128 to obtain DIV. When

DIV jumps from low level to high level on the falling edge of the S_{COMP}, the peak detection circuit detects a peak and V_{PD} generates a rising edge, which causes the EN_{SAMPLE} to jump to low level to prepare to sample the V_{OC}. Until the next zero-crossing point of i_P, the S_{COMP} is at high level so the two-stage rectifier stops working. Therefore, the PZT is equivalent to the open-circuit. Meanwhile, the EN_{SAMPLE} is at low level, making Ø0 is also at low level, so the P-SSHC flips V_{PN} from $-V_{RECT}$ to 0 but does not to V_{RECT}. So i_P can freely charges C_P from 0, and V_{PN} increases gradually from 0 to the peak value of $2V_{POC}$ in half a vibration cycle. The peak detection circuit detects $2V_{POC}$, and the V_{PD} generates a rising edge again. The EN_{SAMPLE} jumps back to high level, and then the SAMPLE immediately generates a low-level pulse, which is the sampling signal required to control the S&H to sample the open-circuit voltage and generate the V_{MPP}. When i_P discharges C_P from $2V_{POC}$ to V_{RECT}, the sampling is completed. Then rectifier starts working normally until the next sampling.

In addition, the S_{COMP} may jump to low level during the sampling so that the sampling will be interrupted. So, two D flip-flops are used to generate a high-level signal SHIELD to shield S_{COMP} at low level during the sampling to make i_P can freely charge and discharge C_P, ensuring normal sampling.

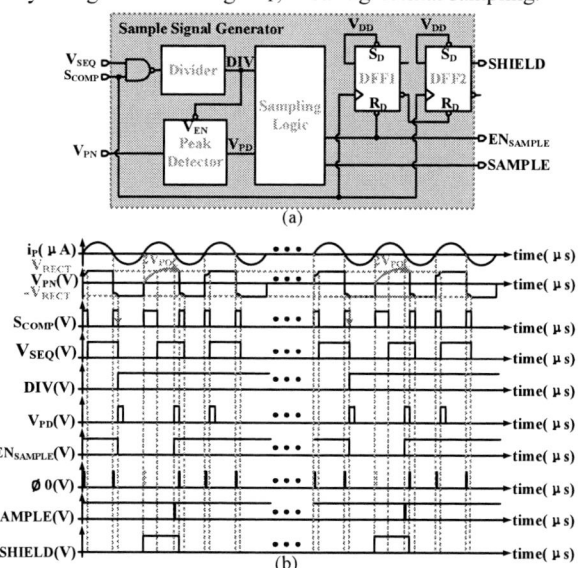

Fig. 3. (a) Diagram, (b) Key signal waveforms of the sampling signal generator.

IV. SIMULATION RESULTS

The proposed circuit is implemented by 0.18μm CMOS process. Fig. 4 (a) is the layout of the proposed fast sampling scheme, and the core area is 670×536 μm². This paper uses an AC current source with the amplitude of 140μA and the frequency of 443Hz in parallel with a capacitor of 50nF to simulate a PZT to verify the function of the proposed circuit.

Fig. 4 (b) shows the key signal simulation waveforms of the proposed circuit. V_{PN} rises to its peak value of $2V_{POC}$ of 2.00V in half a vibration cycle, which close to the theoretical value of 2.01V. At the same time, The SAMPLE generates a low-level pulse at the peak voltage of V_{PN}, and the V_{MPP} of 1.66V is obtained, which is close to the theoretical value of 1.67V. It shows that the proposed fast sampling algorithm can limit the sampling time T_S to less than one piezoelectric vibration cycle and obtain the required V_{MPP} accurately.

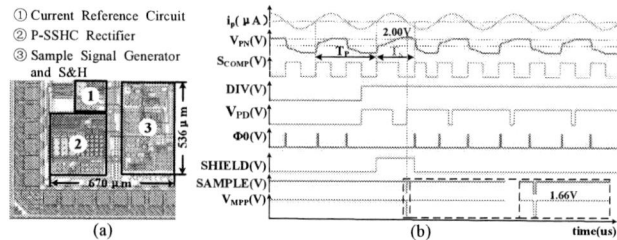

Fig. 4. (a) Layout, (b) Signal simulation waveforms of the proposed circuit.

TABLE I shows the performance comparison of the proposed fast sampling algorithm with other advanced technologies. Compared with other works, the power loss rate φ in this work is optimal. This is mainly attributed to the proposed fast sampling algorithm.

TABLE I.   PERFORMANCE COMPARISON TABLE

Parameter	References			
	[1][1]	[2][1]	[4][1]	This work[2]
Process(μm)	0.18	0.35	0.25	0.18
Type	P-SSHC	-	P-SSHI	P-SSHC
Inductor(mH)	NO	-	0.22	NO
MPPT algorithm	FOCV	FOCV	FOCV	FOCV
T_S[3]	1.5	1	-	<1
φ[4]	1.2%	2%	24.9%	<0.79%

[1]Measured [2]Simulated [3]vibration cycles [4]$\varphi=(T_S/T) \times 100\%$.

V. CONCLUSION

This paper proposed a fast sampling open-circuit voltage algorithm for piezoelectric energy harvesting, which uses the output of the PZT as the sampling point. The open-circuit voltage is sampled and the maximum power point voltage is obtained in less than one piezoelectric vibration cycle, which reduces the power loss rate caused by the open-circuit sampling to less than 0.79%, improving the efficiency of piezoelectric energy harvesting significantly.

ACKNOWLEDGMENT

This work was supported by National Natural Science Foundation of China (62131010), Natural Science Basic Research Program of Shaanxi (2022JC-39), Natural Science Foundation of Chongqing (cstc2021jcyj-msxm3684) and Chongqing Talents Program (CQYC20210301367). The authors would like to thank the Xidian University Chongqing Integrated Circuits Innovation Institute for their technical support.

REFERENCES

[1] L. Liu, J. Ma, X. Liao, Y. Ou, Y. Xie and Z. Zhu, "A 1.5-Cycle Fast Sampling P-SSHC Piezoelectric Energy Harvesting Interface," IEEE Transactions on Circuits and Systems II: Express Briefs, pp. 1-1, Jun, 2022.

[2] M. Shim, J. Kim, J. Jeong, S. Park and C. Kim, "Self-Powered 30 μW to 10 mW Piezoelectric Energy Harvesting System With 9.09 ms/V Maximum Power Point Tracking Time," IEEE Journal of Solid-State Circuits, vol. 50, no. 10, pp. 2367-2379, Oct. 2015.

[3] Z. J. Chew and M. Zhu, "Adaptive Maximum Power Point Finding Using Direct VOC/2 Tracking Method With Microwatt Power Consumption for Energy Harvesting," IEEE Transactions on Power Electronics, vol. 33, no. 9, pp. 8164-8173, Sept. 2018.

[4] L. Wu and D. S. Ha, "A Self-Powered Piezoelectric Energy Harvesting Circuit with an Optimal Flipping Time SSHI and Maximum Power Point Tracking," IEEE Transactions on Circuits and Systems II: Express Briefs, vol. 66, no. 10, pp. 1758-1762, Oct. 2019.

2022 IEEE International Conference on
Integrated Circuits, Technologies and Applications

A Wideband High-linearity Input Buffer Based on Cascade Complementary Source Follower

Tian Feng, Dengquan Li*, Jiale Ding, Shubin Liu, Yi Shen, Zhangming Zhu

School of Microelectronics, Xidian University, Xi'an, China

Abstract—The input buffer is widely used in the analog-to-digital converter (ADC) to isolate input signal from the internal sample-and-hold network and package. In this work, we propose a wide-band and high-linearity input buffer which is based on cascade complementary source follower (CCSF) structure. It is consisted of two-stage PMOS source follower (SF) and NMOS SF with improved linearity. Designed in 65-nm CMOS under 2.5-V supply, the post-layout simulation result shows that the differential input buffer achieves a Nyquist SFDR of 78.3 dB at 4 GS/s sampling rate and consumes 21.14 mW.

Keywords—Input Buffer, Source Follower, High Linearity

I. INTRODUCTION

Driven by wireless communication, high performance analog-to-digital converters (ADCs) are becoming more and more important. For high-speed high-resolution ADCs, the sampling problem becomes increasingly challenging [1]. The parasitic in package influences the linearity of the input signal seriously. An input buffer is often integrated in front of the ADC to isolate the kickback noise from the internal sampling network and the package inductance. The input buffer can also provide strong drivability with a relatively high input impedance and low output impedance. However, the performance of backend ADC can be limited by the input buffer due to distortion. Many techniques, such as flipped source follower (FSF) [2], super source follower (SSF) [1], and source follower with feed-forward compensation (FFC) [2], have been employed to suppress the non-linearity. Due to the DC coupled feedback, the signal range of the FSF is reduced, which limits its speed. For the SSF, there is a trade-off between output swing and minimum supply voltage. The FFC structure can only compensate the nonlinear current flowing through the sampling capacitor.

This paper proposes an input buffer with cascade complementary source follower (CCSF) structure to achieve wide bandwidth and high linearity. The two stages are connected by current amplifier to isolate the interference. Besides, auxiliary operational amplifiers (opamps), compensated capacitors, and bootstrapped capacitors are used to further improve linearity.

II. OPERATION PRINCIPLE

A. Compensated Capacitor

The current variation through the follower device which caused by the signal current of the sampling capacitance can lead to nonlinear distortion [3]. To suppress the non-linear current, the FFC inserts a compensated capacitor [2]. It is shown in Fig. 1(a) where a compensated capacitance C_C connecting between V_{in} and node B. The value of C_C is equal to the load capacitance C_S. Due to the low impedance of M2 source, the node B can be approximately regarded as virtual

This work was supported by National Natural Science Foundation of China under Grant 62021004, Grant 62090040, Grant 61961160703, Grant 62174047.
*Corresponding author: Dengquan Li (dqli@xidian.edu.cn).

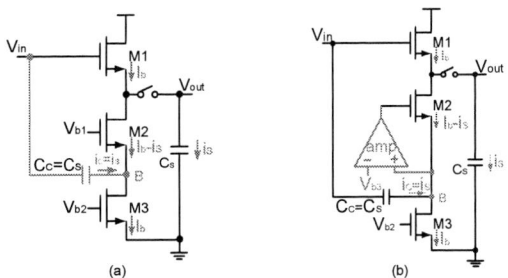

Fig. 1. Schematic of linearization technique based on (a) feed-forward compensation and (b) auxiliary operation amplifier.

ground, hence the AC current i_c flowing through M2 is equal to AC current i_s flowing through C_S. As the low source impedance of M2 and high drain impedance of M3, almost all the current i_c flows through M2 to C_S. This greatly reduces the current variation in M1 and hence improves i_s related linearity significantly. In addition, because the reduction in the current variation reduces the variation of the input impedance, this technique can also improve the distortion at the input of the buffer. However, the large C_C increases the load to the input signal.

B. Auxiliary Operation Amplifier

In Fig. 1(a), the compensated capacitor C_C cannot ensure a constant current flowing through M1. For the current source M3, the current I_b can be expressed as:

$$I_b = k(V_{gs} - V_{th})^2(1 + \lambda V_{ds}) \qquad (1)$$

where k is the conductivity factor, V_{gs} is the gate source voltage, V_{th} is the threshold voltage, and λ is the channel modulation factor. From (1) we can conclude that I_b is not only determined by V_{gs}, but also affected by V_{ds}. Since the input signal is coupled to the drain of M3 by C_C, the current will change accordingly. To keep the voltage as a constant value, we use a cascode current source and an auxiliary opamp in the source of M2, as shown in Fig. 1(b). Because the separation of the cascode structure and the clamping of the opamp, the voltage of the node B is equal to the voltage at the negative input of the opamp, which is V_{b3}. The current I_b can be independent of the input signal, which further improves its linearity.

III. CIRCUIT IMPLEMENTATION

The proposed CCSF input buffer is shown in Fig. 2, which is composed of a two-stage cascaded PMOS source follower (PSF) and NMOS source follower (NSF). The PSF is used in the first stage to realize low input common-mode voltage and high linearity. The NSF is arranged in the second stage because it has higher drive ability and bandwidth. With the two-stage input buffer structure, the kickback noise from the backend sampling switches is suppressed greatly. In addition, the input signal of the second stage is not connected directly with the bonding wire, which alleviates the feedback distortion of the sampling network. As analyzed in Section II-B, the I_b is implemented with cascode structure with a clamping opamp to attain a constant bias current. To achieve superior linearity and stability, a simple Miller compensated two-stage stage opamp with about 70-dB gain is used.

As shown in Fig. 2, to suppress input bandwidth limitation due to the increased input source load, we apply a compensated capacitor C_C with equal capacitance of the load capacitor C_L ($C_C = C_L$) in the second stage, which ensures that the current flowing through the input MOS ($MN_{1(5)}$) is almost constant and

978-1-6654-9270-6/22 $31.00 © 2022 IEEE

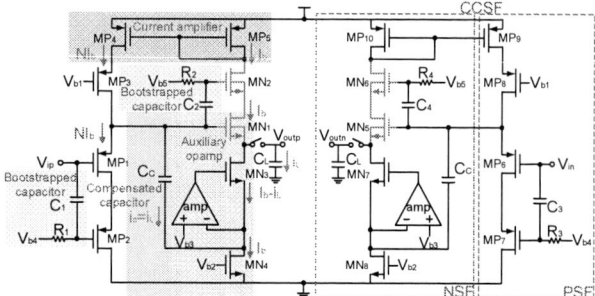

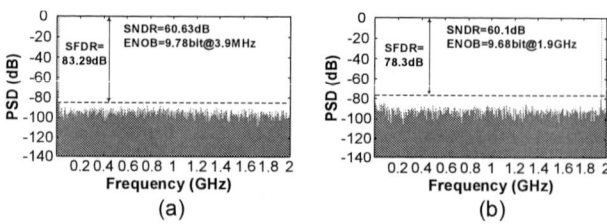

Fig. 4. Simulated output spectrum with (a) low and (b) high input frequency.

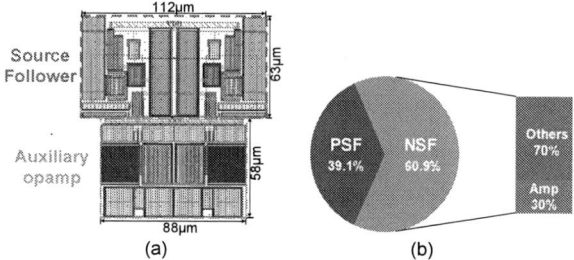

Fig. 2. The proposed input buffer based on CCSF.

Fig. 3. Input buffer (a) layout and (b) power breakdown.

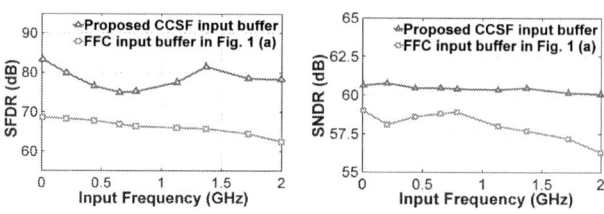

Fig. 5. Simulated SFDR and SNDR versus input frequency.

TABLE I. PERFORMANCE COMPARISON

	[4]*	[5]**	[6]*	[7]*	This work*
Process (nm)	65	28	28	28	65
Architecture	SF	FSF	SF	CAFFC	CCSF
F_S (GS/s)	0.1	8	4	6	4
Supply (V)	2.5	2.5	3	2.5	2.5
Load (pF)	5.6	0.6	1.6	-	0.6
SFDR (dB)	106.8	74	73.1	74	78.3
Input range (V_{PP})	1.8	1.4	-	1.4	1.2
Power (mW)	53.2	137	123	150	21.14
Area (mm²)	0.034	0.156	0.0391	0.0088	0.012

* Simulation results ** Measurement results

improves the linearity. With the help of the current amplifier ($MP_{4(9)}$ and $MP_{5(10)}$), the constant current of the second stage is injected into the first stage. In this way, the current of the two stages are both constant, and the non-linearity induced by current variation can be suppressed significantly.

As analyzed in [3], the variation in the V_{ds} of the input MOS can deteriorate the linearity. To solve this problem, a cascode transistor MP_2 (MN_2) is introduced on the drain of MP_1 (MN_1), and the input signal is coupled onto its gate via a bootstrapped capacitor. Since the constant current ensures V_{gs} of MP_2 (MN_2) maintaining a constant value, the V_{ds} of input device MP_1 (MN_1) keeps a constant value. Instead of the switch-capacitor structure in [3], in our design the gate of the input transistor MP_1 (MN_1) and signal is level-shifted using large capacitor C_1 (C_2) and large resistor R_1 (R_2), which can eliminate potential cross-talking and memory effect [4]. Obviously, increasing the values of R and C can reduce the signal attenuation of the AC coupling.

IV. LAYOUT AND SIMULATION RESULTS

The input buffer is designed in a 65-nm CMOS and its layout is shown in Fig. 3(a) with the area of 0.012 mm². The input buffer operates under 2.5-V supply voltage, 1.2-V_{PP} input and 4-GS/S sampling rate. Its power breakdown is shown in Fig. 3(b), and the total power consumption is 21.14 mW. The load of the buffer is 0.6 pF.

The simulated spectrum of the proposed input buffer with 3.9 MHz input frequency is shown in Fig. 4(a), illustrating the spurious free dynamic range (SFDR) and signal to noise and distortion ratio (SNDR) are 83.29 dB and 60.63 dB, respectively. Fig. 4(b) shows that with 1.9 GHz input frequency, the SFDR and SNDR are 78.3 dB and 60.1 dB, respectively. Fig. 5 compares the SNDR/SFDR of the proposed input buffer with the FFC input buffer in Fig. 1(a) versus input frequency at the same simulation condition, showing that this work has better linearity from DC to high frequency.

The detailed performance of the input buffer is summarized in Table I with comparison to other recently published high-speed input buffers, from which we can see that this work achieves high linearity with lower power at GS/s sampling rate.

V. CONCLUSION

In this paper, we propose an input buffer using CCSF to achieve wide band and high linearity. The post-layout simulation result shows that the input buffer achieves 78.3 dB SFDR and 60.1 dB SNDR at 4 GS/s with Nyquist input, consuming 21.14 mW. The technique presented in this work shows considerable improvement in linearity and is suitable for high-speed and high-resolution ADCs.

REFERENCES

[1] D. Li *et al.*, "Radio frequency analog-to-digital converters: Systems and circuits review," *Microelectron. J.*, vol. 119, no. 105331, 2021.

[2] A. M. A. Ali, Dinc, H., et al., "A 14 Bit 1 GS/s RF Sampling Pipelined ADC With Background Calibration," *IEEE J. Solid-State Circuit*, vol. 49, no. 12, pp. 2857-2867, 2014.

[3] Z. Wu, C. Wang, Y. Ding, F. Li and Z. Wang, "An ADC input buffer with optimized linearity," *IEEE International Conference on Solid-State and Integrated Circuit Technology (ICSICT)*, pp. 1-3, 2018.

[4] Y. Cao *et al.*, "An operational amplifier assisted input buffer and an improved bootstrapped switch for high-speed and high-resolution ADCs," *IEEE International Symposium on Circuits and Systems (ISCAS)*, pp. 1-5, 2018.

[5] Z. Huang *et al.*, "A 6-GHz bandwidth input buffer based on AC-coupled flipped source follower for 12-bit 8-GS/s ADC in 28-nm CMOS," *IEEE Transactions on Circuits and Systems II: Express Briefs*, 2022, doi: 10.1109.

[6] L. Zhang *et al.*, "An input buffer for 4 GS/s 14-b time-interleaved ADC," *IEEE 14th International Conference on ASIC (ASICON)*, pp. 1-4, 2021.

[7] Z. Huang *et al.*, "Low-voltage high-linearity differential input buffer with current amplifier feed-forward compensation for high-speed ADCs," *IEEE International Conference on Integrated Circuits, Technologies and Applications (ICTA)*, pp. 41-42, 2020.

2022 IEEE International Conference on
Integrated Circuits, Technologies and Applications

Low-Latency FPGA Design and Implementation of Hermitian Matrix Inversion Based on Partitioned Systolic Array for Massive MIMO

Ke Han, Daokun Li

School of Electronic Engineering, Beijing University of Posts and Telecommunications, Beijing 100876, China

Abstract—Large-scale matrix inversion is widely used in massive Multiple Input Multiple Output (MIMO) beamforming systems, but matrix inversion is very complicated in hardware implementation. In this paper, Hermitian matrix decomposition method based on partitioned systolic array is proposed, and the computing structure of the algorithm is improved flexibly by utilizing the partitioned characteristics of large-scale matrix. We compare our method with existing FPGA-based technologies on Xilinx ZCU102 FPGA. The results of the experiment show that our method has better performance than existing techniques in resource utilization, device delay and maximum working frequency when the size of Hermitian matrix is 32×32, which is a typical size for MIMO applications.

Keywords—beamforming, FPGA, matrix inversion, systolic array, partitioned matix

I. INTRODUCTION

With the increase of the number of antennas in the massive Multiple Input Multiple Output (MIMO) system, the transmission efficiency of beamforming technology is improved, but it also brings more complex numerical operations [1][2]. Matrix inversion has the highest computational complexity in the beamforming algorithm[3][4].It consumes lots of hardware resources, and complex operations will lead to instability of numerical calculation [5]. Therefore, low latency and robustness are the two most important requirements of matrix inversion for the real-time and stability of beamforming system.

In this paper, we propose an large-scale Hermitian matrix inversion method based on partitioned systolic array. The main contributions of the article include:

(1) The simplified structure of the systolic array based on the element relationship of the upper triangular matrix and the inverse matrix is designed to reduce redundant calculation.
(2) The partitioned computing characteristics of LDL decomposition and upper triangular matrix inversion are exploited to improve the computing structure of matrix multiplication.

II. SYSTEM MODEL

We assume that in a massive MIMO system, the base station has M antennas, and the base station can communicate with K single antenna users in the same frequency band, where $M \gg K$. The vector relationship between transmission and reception can be characterized as

$$\mathbf{y} = \mathbf{Hx} + \mathbf{n} \tag{1}$$

where, $\mathbf{y} \in \mathbb{C}^M$ is the received signal vector, $\mathbf{x} \in \mathbb{C}^K$ is the transmitted signal vector, $\mathbf{H} \in \mathbb{C}^{M \times K}$ is the channel matrix, $\mathbf{n} \in \mathbb{C}^M$ is the Gaussian white noise vector with zero mean and σ^2 variance. The task of MIMO detector is to detect the data transmitted signal $\mathbf{x}$ from the channel matrix $\mathbf{H}$ and the received signal vector $\mathbf{y}$. The two most widely used linear detection schemes are Zero-Forcing (ZF) and Minimum Mean Square Error (MMSE), which can be respectively expressed as

$$\begin{cases} \widetilde{\mathbf{x}}_{\mathrm{ZF}} &= (\mathbf{H}^{\mathbf{H}}\mathbf{H})^{-1}\mathbf{H}^{\mathbf{H}}\mathbf{y} \\ \widetilde{\mathbf{x}}_{\mathrm{MMSE}} &= (\mathbf{H}^{\mathbf{H}}\mathbf{H} + \sigma^2\mathbf{I}_{\mathbf{K}})^{-1}\mathbf{H}^{\mathbf{H}}\mathbf{y} \end{cases} \tag{2}$$

Where, the ^{H} denotes complex conjugation, $\mathbf{I_K}$ is the $K \times K$ identity matrix. It can be seen from the calculation results that both ZF and MMSE need to inverse the Hermitian matrix. Therefore, large-scale Hermitian matrix inversion will be the focus of the next part of this paper.

III. PROPOSED HARDWARE STRUCTURE

Hermitian matrix G can be expressed as

$$\mathbf{G} = \mathbf{L}^{\mathbf{H}}\mathbf{DL} \tag{3}$$

where $\mathbf{L}$ is the lower triangular matrix with diagonal element 1, and $\mathbf{D}$ is the diagonal matrix. Then the inverse of matrix $\mathbf{G}$ can be expressed as the product of the inverse of the decomposition matrix, as follows:

$$\mathbf{G}^{-1} = (\mathbf{L}^{\mathbf{H}}\mathbf{DL})^{-1} = \mathbf{L}^{-1}\mathbf{D}^{-1}(\mathbf{L}^{-1})^{\mathbf{H}} \tag{4}$$

A. Iterative Inverse Structure of Upper Triangular Matrix

For an upper triangular matrix, its partitioned matrix form can be expressed as:

$$\mathbf{L} = \begin{bmatrix} \mathbf{U} & \mathbf{V} \\ \mathbf{O} & \mathbf{F} \end{bmatrix} \tag{5}$$

where, $\mathbf{L} \in \mathbb{C}^{(m+n) \times (m+n)}$, $\mathbf{U} \in \mathbb{C}^{m \times m}$, $\mathbf{V} \in \mathbb{C}^{m \times n}$, $\mathbf{F} \in \mathbb{C}^{n \times n}$, $\mathbf{O} \in \mathbb{C}^{n \times m}$. Based on the definition of inverse of partitioned matrix, the following formula can be obtained:

$$\mathbf{L}^{-1} = \begin{bmatrix} \mathbf{U}^{-1} & -\mathbf{U}^{-1}\mathbf{VF}^{-1} \\ \mathbf{O} & \mathbf{F}^{-1} \end{bmatrix} \tag{6}$$

Where, $\mathbf{U}^{-1}$ is the result of the last iteration calculation, $\mathbf{F}^{-1}$ can be calculated by the systolic array structure. Fig. 1 shows the systolic array structure we designed to solve the inverse matrix of the partitioned submatrix.

In Fig. 1, the systolic array structure of partitioned submatrix inversion is composed of PE array, array controller and I/O interface. It can be seen that our computing structure is only composed of multiplication and addition, which avoids the division operation with high computational complexity. In addition, the structure of the systolic array makes full use of the calculation data before and after the iterative calculation process to reduce redundant calculation, significantly improving the speed of partitioned submatrix inversion.

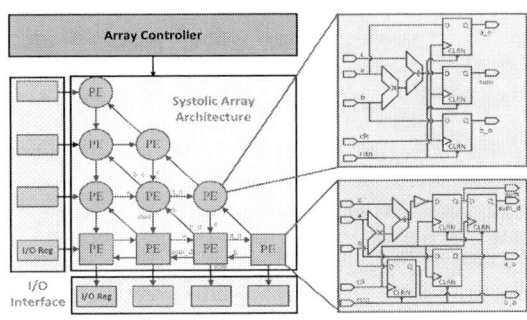

Fig. 1: Systolic array architecture

978-1-6654-9270-6/22 $31.00 © 2022 IEEE

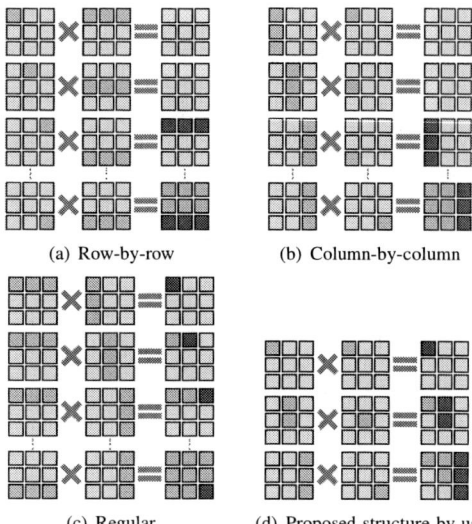

(a) Row-by-row   (b) Column-by-column

(c) Regular   (d) Proposed structure by us

Fig. 2: Three types of matrix multiplication and matrix multiplication proposed by us. Orange, dark blue, light blue and green represent matrix element results generated in previous, current iteration, partially and being calculated, respectively.

B. Partitioned Matrix Multiplication

The implementation of matrix multiplication can be divided into three types: row-by-row, column-by-column and regular matrix multiplication. However, none of them can effectively implement the matrix multiplication based on the sequence of elements generated by the partitioned matrix inversion algorithm, because they will traverse a whole row of elements of the multiplied matrix. The calculation structure of this paper determines that only with the end of the iteration, the row elements can be fully calculated. Therefore, the above three methods, if adopted, will generate waiting time between modules. We use Fig. 2 to illustrate the operation of matrix elements when applying the existing three matrix multiplication.

Therefore, we designed the partitioned matrix multiplication structure shown in Fig. 2(d), which converts the product of the upper triangular matrix and the diagonal matrix into that a series of scalar dot multiplied by column vectors. In Fig. 2(d), when any part of the elements of the upper triangular matrix are calculated, they can be immediately used by matrix multiplication, which greatly reduces the calculation delay. However, this kind of parallelism of data flow does not exist in the inverse method of matrix which is not partitioned.

IV. EXPERIMENTAL RESULTS

In this part, we choose three methods of matrix inversion to verify performance, and use Verilog language to program on Xilinx ZCU102, where, the size of Hermitian matrix is 32×32, and the size of submatrix is 5×5 in [5] and our algorithm. In order to make the comparison results more credible, the numerical calculation of the three methods adopts single precision floating-point operator (IEEE Std. 754-1985). Table I lists the resource utilization of the three methods. As can be seen form Table I, our algorithm utilizes the least resources to achieve the highest frequency and the lowest device latency.

The last experiment is to verify the numerical stability of our algorithm. We use matrix 2-norm errors, $\mathbf{c} = \left\| \mathbf{I} - \mathbf{A}\mathbf{A}^{-1} \right\|_2$,

TABLE I. UTILIZATION OF THE THREE METHODS

Resource	[3]	[5]	Our Algorithm
LUT	221483	210348	200145
FF	228191	212542	204673
BRAM	267	220	210
DSP48E	1359	1208	1054
Device Latency(us)	503.55	364.68	313.47
f_{max}(MHz)	200	250	275

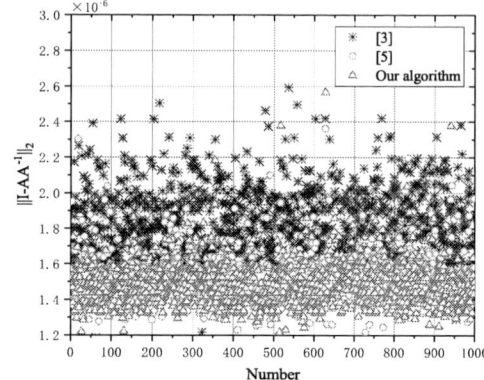

Fig. 3: Matrix 2-norm errors of the three methods. The stars, circles, triangles represent proposed methods in [3] and [5], and our algorithm, respectively.

to illustrate the performances of the three methods, where $\| \ \|_2$ represents matrix 2-norm, and $\mathbf{I}$ represents identity matrix. In this experiment, 1000 Hermitian matrices of size 32×32 generated by MATLAB are used as test data, and the experimental results processed by FPGA are shown in Fig. 3. From Fig. 3, our algorithm has a lower matrix 2-norm errors than [3], and has a similar matrix 2-norm errors to [5].

V. CONCLUSION

In this paper, we propose a large-scale Hermitian matrix inversion algorithm based on partitioned systolic array for massive MIMO. When the size of Hermitian matrix is 32×32, compared with the latest technology in [3] and [5], our method saves nearly 5% of resources, but increases the frequency by 10%, reduces the device delay by nearly 14%, and achieves the same or even better numerical stability, which shows that our method has better performance. Although this work focuses on the complex Hermitian positive definite matrix in massive MIMO detectors, the same results apply to any Hermitian positive definite matrix inversion. For a larger matrix, the inverse of the larger matrix can be obtained by more iterations.

REFERENCES

[1] X. Zhang et al., "Low Complexity Implicit Detection for Massive MIMO Using Neumann Series," in *IEEE Transactions on Vehicular Technology*, May 2022.

[2] S. Shahabuddin et al., "FPGA Implementation of Stair Matrix based Massive MIMO Detection," *2021 IEEE 12th Latin America Symposium on Circuits and System (LASCAS)*, 2021, pp. 1-4.

[3] X. -W. Zhang et al., "High-Throughput FPGA Implementation of Matrix Inversion for Control Systems," in *IEEE Transactions on Industrial Electronics*, vol. 68, no. 7, pp. 6205-6216, July 2021.

[4] H. Wang et al., "An Efficient Detector for Massive MIMO Based on Improved Matrix Partition," in *IEEE Transactions on Signal Processing*, vol. 69, pp. 2971-2986, 2021.

[5] D. Yan et al., "Revisiting the Adjoint Matrix for FPGA Calculating the Triangular Matrix Inversion," in *IEEE Transactions on Circuits and Systems II: Express Briefs*, vol. 68, no. 6, pp. 2127-2131, June 2021.

Accurate 3DIC thermal simulation for BEOL influence study

Hao Yang[1], Bin Yan[1], Jianjun Sun[1], Jian Pang[1],
Guangyao Li[1],Ouyang Keqing[2], Shuqiang Zhang[3]
[1] Department of Packaging and Testing, ZTE Corporation
[2] State Key Laboratory of Mobile Network and Mobile Multimedia Technology
[3] Ansys

Abstract—**This present report mainly covers 2.5/3DIC system efficient and accurate thermal analysis. This report not only conducts the thermal coupling among dies, PCB, package at system level, but also performs accurate and detailed thermal analysis on specific hotspot regions. More importantly, the BEOL/RDL layer has a signicant impact on the juction temperature of the die, especially the hotspot region, which is often ignored in both industrial and academic area. Thus, the BEOL/RDL layer is incorporated into the thermal simulation by simplifing the BEOL/RDL layer as the metal density. According to the simulation results, under steady-state conditions, the BEOL layer has a greater impact on the junction temperature, and under transient conditions, it has less impact on the transient temperature rise rate.**

Keywords—3D IC, finite element analysis, BEOL

I. INTRODUCTION

With the development of semiconductor technology, while the performance and functions of chips increase, the contradiction between chip area, yield and complex process is difficult to reconcile, and a new technology named 3D interconnection technology, emerges as the times require. 3D IC greatly reduces the length of the longest interconnection line on the integrated circuit, which greatly improves the performance and package size of the chip, but also brings great challenges to heat dissipation. The temperature of the chip has a huge impact on the reliability of the chip, and 55% of the reasons for chip failure are related to temperature. The temperature gradient within the chip can also cause severe thermal stress deformation, further increasing the risk of physical chip failure. Therefore, it is very important to study chip heat dissipation, especially in 3D IC[1-2].

Compared with traditional 2D IC, the main difference of 3D IC is the stacking of chips and TSV. On the one hand, the stacked structure will lead to the superposition of chip power consumption, and the power density will increase significantly. On the other hand, the heat of the lower die must use the BEOL (Back end of line) layer to transmit upwards. The BEOL layer is mainly composed of Cu and SiO2, of which Cu is a good thermal conductivity material with a thermal conductivity of 380W/mK, while the thermal conductivity of SiO2 is very poor, which is 0.27W/mK, and most of the BEOL layer is SiO2, so the overall thermal conductivity of BEOL Performance is poor[3].

The simulation software used in this article is Icepak 19.2, and the CFD simulation method is used to analyze the heat dissipation effect of the BEOL layer on the 3D IC[4]. The evaluation is carried out from two aspects of steady state and transient state. The steady state simulation analyzes the influence of the BEOL layer on the thermal resistance of the chip, and the transient simulation evaluates the influence of the BEOL layer on the transient temperature rise of the chip.

II. SIMULATION MODEL AND METHOD

This paper takes a 3D packaged CPU chip as a research case, as shown in Fig.1, the package size is 33mm*33mm, and the die size is 10mm*10mm. See Table 1 for more core package size information. The chip uses a 3D package structure (Fig.2), the bottom die is the CPU, the top die is the SRAM, and the F2B (Face to back) package type is used. The pins passing through the SRAM will be interconnected with the CPU die below through TSV. In the entire stack structure, the components from bottom to top are bump, CPU BEOL, CPU Si, SRAM BEOL, and SRAM Si. The heat source of the CPU will be attached to the bottom of the CPU Si, and the heat source of the SRAM die will be attached to the bottom of the SRAM Si.

The heat source of the chip adopts the structure of powermap, as shown in Figure 3. The powermap of the CPU mainly considers PCIe, Core, ddr, and Cache. The specific power consumption information is shown in Table 4.

For different ICs, there are large differences in the BEOL layer, and there is less information about the BEOL layer in the current literature. This paper compares three different BEOL layers, and the specific information is shown in Table 3. The three BEOL layers have 32 layers, 16 Cu layers, and 16 insulating layers. The proportion of Cu in the metal layer is 20%, and the proportion of Cu in the insulating layer is 0.5%. The difference between BEOL 1 and BEOL 2 is the thermal conductivity of Cu, BEOL 1 uses 380 W/mK of pure copper and BEOL 2 uses 190 W/mK. The difference between BEOL 3 and BEOL 1 is the thickness of each layer. In BEOL 3, the thickness of the Cu layer is 0.035um, and the thickness of the insulating layer is 1.5 um. In BEOL 1 and BEOL 2, the thickness of the Cu layer and the insulating layer are both 0.5 um. In this paper, the effects of different BEOLs on heat dissipation are analyzed in steady-state and transient conditions.

In the simulation of this paper, the simplified heat dissipation conditions are used, and the convective heat transfer coefficient of 1000 W/m2K is fixed for the top of the chip. Considering that most of the heat of the chip is dissipated from the top, the surrounding and bottom of the chip are set as adiabatic conditions, and radiation heat dissipation is ignored.

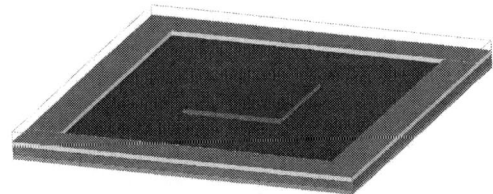

Fig. 1. Chip 3D structure diagram

Fig. 2. Schematic diagram of 3D stacking

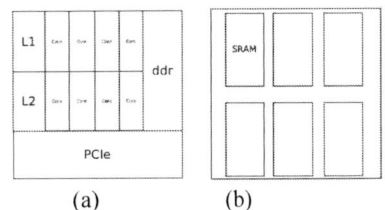

(a)   (b)

Fig. 3. (a)Powermap ofCPU (b)Powermap of SRAM

978-1-6654-9270-6/22 $31.00 © 2022 IEEE

TABLE I. PACKAGE INFORMATION

Item	Size/mm	Thickness/mm	Material
Substrate	35*35	0.919	E705G、GZ41、Cu
C4 bump	10*10	0.06	Cu
CPU BEOL	10*10	0.016	SiO2、Cu
CPU Si	10*10	0.1	Si
SRAM BEOL	10*10	0.016/0.024	SiO2、Cu
SRAM Si	10*10	0.1	Si
TIM	10*10	0.06	Resin
Lid	35*35	1	Cu

TABLE II. POWERMAP INFORMATION

Power block	Power
Core	2.0W*8
Cache	1.2W*2
ddr	2.6W
PCIE	12.0W
SRAM	2.0W*6
Total power	45.0W

TABLE III. BEOL INFORMATION

	BEOL 1	BEOL 2	BEOL 3
Total layer num	32	32	32
Total thickness	16 um	16 um	24.56 um
Metal layer num	16	16	16
Metal layer thickness	0.5 um	0.5 um	0.035 um
Metal layer Cu ratio	20%	20%	20%
Insulation layer num	16	16	16
Insulation layer thickness	0.5 um	0.5 um	1.5 um
Insultation layer Cu ratio	0.5%	0.5%	0.5%
Cu thermal conductivity	380 W/mK	190 W/mK	380 W/mk
SiO2 thermal conductivity	0.27 W/mK	0.27 W/mK	0.27 W/mK

III. SIMULATION RESULTS AND DISCUSSION

Four cases are analyzed under the steady-state condition, namely BEOL 1, BEOL 2, BEOL3 and no BEOL. The detailed simulation results are shown in Table 4 and Fig.4. According to the simulation results, the BEOL layer has a greater impact on the junction temperature of the CPU, and has less impact on the temperature of SRAM and lid.

In a 3D IC thermal simulation, without considering BEOL, the die temperature is lower than when BEOL is considered. When there is no BEOL, the junction temperature of the CPU is 80.8 ℃. While the influence of BEOL is considered, the junction temperature is 82.3 ℃, 84.0 ℃, and 87.7 ℃. The difference in junction temperature is mainly due to the SRAM BEOL layer. The heat of the CPU die needs to be conducted upwards. There is an SRAM BEOL layer with poor thermal conductivity in the path. The worse the thermal conductivity of the BEOL layer, the greater the thermal resistance of the path and the higher the CPU junction temperature. BEOL 2 has a lower thermal conductivity of metal compared to BEOL 1 (Cu 190W/mK in BEOL 2, Cu 380W/mK in BEOL1), so the overall thermal conductivity of BEOL 2 is worse. Compared with BEOL 1 and BEOL 2, BEOL 3 has thinner metal layers (0.035um in BEOL3, 0.5um in BEOL 1 and BEOL 2) and thicker insulating layers (1.5um in BEOL 3, 0.5um in BEOL 1), the overall thermal conductivity is the worst among the 3 BEOLs. In the simulation in this paper, BEOL has at least 1.5 ℃ -6.9 ℃ influence on the junction temperature, so for

thermal simulation of 3D IC, the influence of BEOL layer cannot be ignored.

The BEOL layer has little effect on the SRAM and lid temperature. For the four different BEOL cases, since the heat is transferred from the bottom to the top, after passing through the SRAM BEOL, the structure of each layer is the same, the thermal resistance is basically the same, and the temperature gradient is basically stable after the SRAM BEOL. So in the steady state simulation, the SRAM temperature and lid temperature are basically same for different BEOLs.

In transient state condition, this report analyse 4 cases which use four different BEOLs (similar to cases 1-4). The CPU die will load 45W of power consumption from a steady state of 0W and 45 ℃. Through the CFD simulation method, the CPU junction temperature growth is calculated.

The worse the thermal conductivity of the BEOL, the faster the junction temperature rises. Table 5 records the junction temperature values of the CPU at different times, Figure 6 shows the temperature cloud diagram on the CPU under different cases, and Fig.7 shows the junction temperature changes of the CPU under different cases. According to the simulation results, without considering the BEOL layer, the junction temperature increases the slowest, with a 13.1 ℃ rise at 0.1s and a 35.3 ℃ rise at 3s. Considering the BEOL layer, the worse the thermal conductivity of the BEOL layer, the faster the junction temperature increases. For BEOL 1, BEOL 2, and BEOL 3, the junction temperature increases by 13.8℃, 14.3℃, and 14.4 ℃ respectively at 0.1s; the junction temperature increased by 36.7℃, 38.2℃, and 38.7℃ respectively at 3s.

It can be seen from Fig.7, that the difference in the growth rate of the CPU junction temperature under the four BEOL cases is relatively small, indicating that the BEOL layer has relatively little influence on the transient temperature rise of the IC.

TABLE IV. STATISTICS OF STEADY STATE SIMULATION RESULTS

	BEOL type	Tj_cpu				
		0.000s	0.002s	0.100s	1.000s	3.000s
case 5	BEOL 1	45℃	47.8℃	58.8℃	71.9℃	81.7℃
case 6	BEOL 2	45℃	47.9℃	59.3℃	73.4℃	83.2℃
case 7	BEOL 3	45℃	48.8℃	59.4℃	73.8℃	83.9℃
case 8	No BEOL	45℃	47.4℃	57.1℃	70.7℃	80.3℃

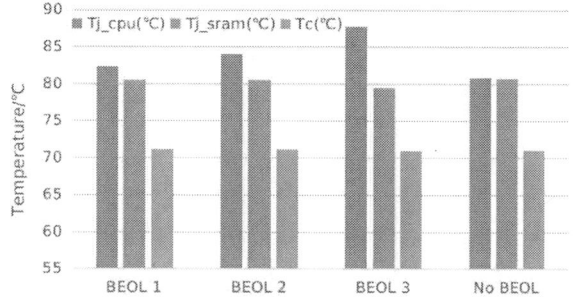

Fig. 4. Steady-state simulation results

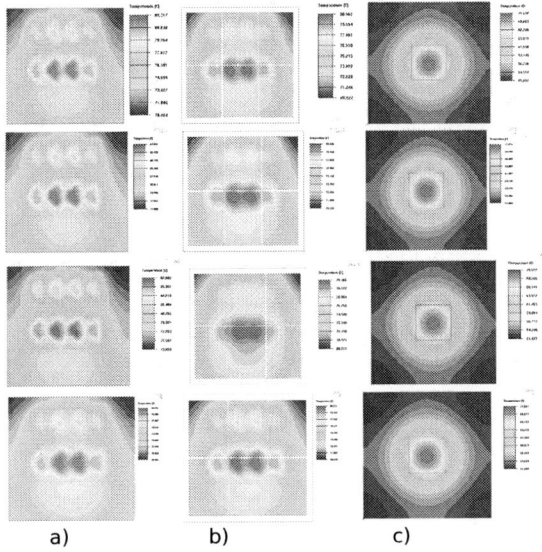

Fig. 5. Temperature cloud map for case 1-4 (a)CPU(b)SRAM(c)Lid top

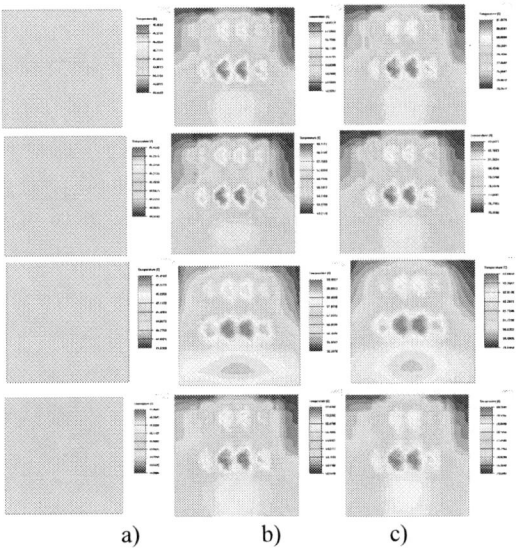

Fig. 6. Temperature cloud map for case 5-6 (a)0s(b)0.1s(c)3s

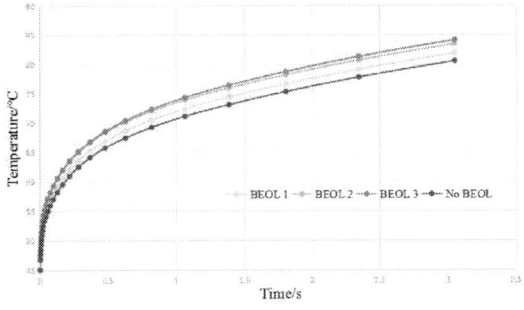

Fig. 7. CPU junction temperature rise for case 5-8

IV. CONCLUSION

This paper uses the method of CFD simulation to carry out thermal simulation analysis of the 3D packaged CPU chip. The paper mainly discusses the influence of the BEOL layer on the heat dissipation of the chip. For steady-state and transient scenarios, the conclusions are as follows:

1)   In the steady state condition, the BEOL layer has a greater impact on the junction temperature. The more SiO2 in the BEOL layer, the worse the thermal conductivity of the BEOL, resulting in an increase in the junction temperature of the chip. Therefore, the effects of BEOL must be considered in 3D IC thermal simulation.

2)   In the transient condition, the BEOL layer leads to a faster transient temperature rise. Due to the poor thermal conductivity of the BEOL, the heat of the CPU die is more difficult to dissipate, resulting in a faster increase in the junction temperature of the chip. However, according to the simulation results, different BEOLs have limited influence on the transient temperature rise, and the chip temperature rise speed will not be greatly increased due to the BEOL layer.

REFERENCES

[1]Pan, S. H. , N. Chang , and Z. Ji . "IC-Package Thermal Co-Analysis in 3D IC Environment." ASME 2011 Pacific Rim Technical Conference and Exhibition on Packaging and Integration of Electronic and Photonic Systems, MEMS and NEMS: Volume 2 American Society of Mechanical Engineers Digital Collection, 2011.

[2]Pan, S. H. , N. Chang , and Z. Ji . "A new methodology for IC-package thermal co-analysis in 3D IC environment." IEEE Electrical Design of Advanced Package & Systems Symposium IEEE, 2010.

[3]Helou, A. E. , et al. "Thermal Modeling and Experimental Validation of Heat Sink Design for Passive Cooling of BEOL IC Structures." International Workshop on Thermal Investigations of ICs and Systems 0.

[4]Im, S. , and K. Banerjee . "Full chip thermal analysis of planar (2-D) and vertically integrated (3-D) high performance ICs." International Electron Devices Meeting 2000. Technical Digest. IEDM (Cat. No.00CH37138) IEEE, 2002.[1]. Pan, S. H. , N. Chang , and Z. Ji . "IC-Package Thermal Co-Analysis in 3D IC Environment." ASME 2011 Pacific Rim Technical Conference and Exhibition on Packaging and Integration of Electronic and Photonic Systems, MEMS and NEMS: Volume 2 American Society of Mechanical Engineers Digital Collection, 2011.

Formation Mechanism of high Ni content $(Cu,Ni)_6Sn_5$ in Cu/Sn/Ni microbump for solid state aging

Haiyang Yu[1], C.R.Kao[2]

[1]Beijing Institute of Radio Measurement, Beijing, China

[2]Department of Materials Science and Engineering, National Taiwan University, Taipei, China

Abstract— **Due to its low cost, the Cu/Sn/Ni microbump is the most widely used structure in electronic packaging. Recent studies have characterized the evolution of the microstructure and phase formation in this system, and a unique $(Cu,Ni)_6Sn_5$ phase has been discovered with a high Ni content. However, there has been debate over the formation mechanism of this phase. This study builds a model of the formation mechanism of $(Cu,Ni)_6Sn_5$ and provides direct proof.**

Keywords—**$(Cu,Ni)_6Sn_5$, microbump, formation mechanism**

I. INTRODUCTION

Due to the efficient utilization of space, the three-dimensional integrated circuit integration method has the advantage of lower packaging costs and significantly more I/Os. The Cu/Sn/Cu or Cu/Sn/Ni structures are frequently used in microbump designs, and the bump size is only a few micrometers to tens of microns. Cu/Sn solders have undergone substantial research over the past few decades and are becoming a great alternative for use in electronic packaging.

There has been numerous research on microbump recently. Yang[1] stated that because of the small aspect ratio, surface diffusion phenomenon is observed in both Ni/Sn/Ni solid-state aging and Cu/Sn/Ni solid-liquid reactions. When the Cu/Sn/Ni microbump was aged at 200°C, Wang[2] reported the phenomena of Cu_3Sn growing along with the Cu_6Sn_5's grain boundary. According to Mo[3], even when the sample's other components transform into Cu_3Sn, the ultra-high Ni content $(Cu,Ni)_6Sn_5$ phase in the middle of the microbump continues to persist under certain reflow circumstances, which is consistent with earlier findings[4]. However, further investigation is required into the procedure and mechanism of phase transition. The goal of the current research, as will be demonstrated in this work, is to confirm the formation mechanism of $(Cu,Ni)_6Sn_5$ in Cu/Sn/Ni microbump by solid state aging.

II. Experimental Procedure

Cu/Sn/Ni microbumps were fabricated by electroplating on 4 inch Si wafer, the structure of which is shown in figure 1. Scanning electron microscopy (SEM) was used to characterize the evolution of morphological and structural features. The element distribution was analyzed using electron probe microscopy analysis (EMPA). Furthermore, phase identification and high-angle annular dark field (HAADF) analyses were carried out using transmission electron microscopy (TEM).

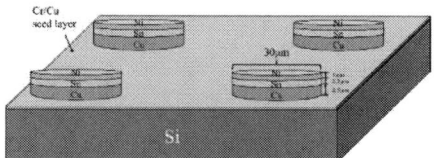

Fig. 1 Schematic illustration of Cu/Sn/Ni microbumps

III. Result and Discussion

A. Ni distribution in Cu/Sn/Ni microbump

In this work, element distribution mapping is carried out using EPMA. Figure 2 illustrates how the $(Cu,Ni)_6Sn_5$ has already been localized at the Cu side after 8 hours of aging. The same phase is present at the center line after 24 hours, as seen in figure 3. And, figure 4 displays the outcome after extensive aging. Contrary to earlier research, which believed that the $(Cu,Ni)_6Sn_5$ is formed in the center of the Cu/Sn/Ni microbump and the location of this phase does not change[3], the mapping results confirmed that the growth of Cu_6Sn_5 causes a shift in the location of $(Cu,Ni)_6Sn_5$.

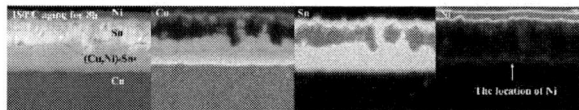

Fig. 2 EPMA result of the sample after 8h aging at 200°C

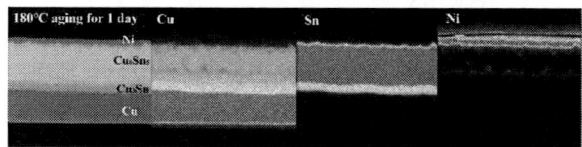

Fig. 3 EPMA result of the sample after 24h aging at 200°C

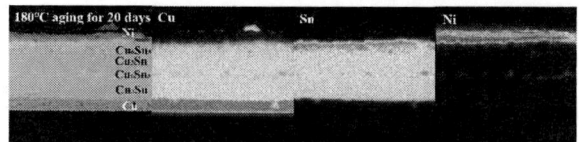

Fig. 4 EPMA result of the sample after 20 days of aging at 200°C

B. TEM result for $(Cu,Ni)_6Sn_5$ near the middle line

Due to the existence of both Cu_6Sn_5 and $(Cu,Ni)_6Sn_5$ at the center line, the focus ion beam (FIB) is used to analyze the TEM sample of the microbump fabricated after 24 hours of aging. The diffraction patterns of grains A and B, which are

978-1-6654-9270-6/22 $31.00 © 2022 IEEE

situated in the center line, are illustrated in Figure 5. Additionally, grain A is discovered to be in the monoclinic phase (η'), and according to earlier research[4], the Ni content in this region is lower than in grain B. [4 Superlattice structure, which is regarded as the crucial character to confirm the η-(Cu,Ni)6Sn5 phase and η'-(Cu,Ni)6Sn5 phase, could be seen in grain B, which has a hexagonal structure (η). Otherwise, the Ni distribution is verified using HAADF and EELS. Ni concentration is high close to the grain border of the η phase, as depicted in figure 6. And, the grain size of η phase is smaller than that of the η' phase.

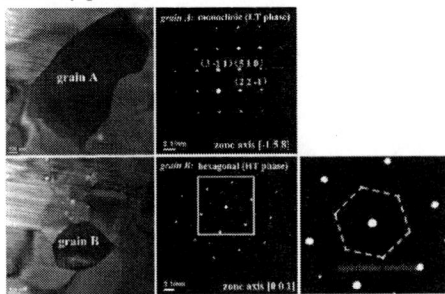

Fig. 5   TEM image and diffraction pattern of sample

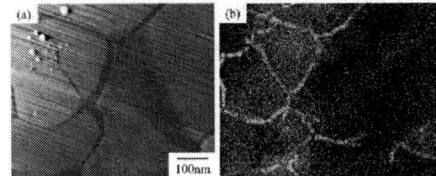

Fig. 6 (a)HAADF image of sample, and(b)EELS mapping result

C.   Formation mechanism of (Cu,Ni)₆Sn₅ in this study

The model of the microstructure evolution process is illustrated in figure 7 based on the results above. There are five stages in the evolution process.

Step 1: At this preliminary stage, a scallop (Cu, Ni)6Sn5 is formed, and through diffusion, a significant amount of Ni atom eventually diffuses into the scallop's "valley."

Step 2: Cu3Sn presents in the early stages of the reaction, and because the amount of dissolved Ni in Cu3Sn is much lower than that of Cu6Sn5, some Ni atoms are located at the interface between the scallop (Cu, Ni)6Sn5 layer and the Cu3Sn layer. However, with the accumulation of Ni atom in the valley in step 1, the Ni content quickly becomes saturated, resulting in the formation of Ni-rich (Cu, Ni)6Sn5. At the same time, there is still excess Ni as a segregation phase near the grain boundary of this layer (Cu, Ni)6Sn5. The presence of Ni segregation makes it hard for (Cu, Ni)6Sn5 to develop properly, therefore, the grain size of η phase is smaller than the η' phase.

Step 3: The (Cu, Ni)6Sn5 in the saturated state transforms into the Ni diffusion barrier due to the presence of the Ni

segregation phase, whereas the small grain (Cu, Ni)6Sn5 has no impact on the normal diffusion of Sn and Cu Atoms. In the experiment, it is still possible to see the generation and growth of new Cu6Sn5, which exists as a layer-state beneath the (Cu,Ni)6Sn5 layer with a high Ni content.

Step 4: The position of the tiny grain (Cu, Ni)6Sn5 varies as the freshly formed Cu6Sn5 gradually expands. Since the small grains of (Cu, Ni)6Sn5 are surrounded by large grains of (Cu, Ni)6Sn5 and Cu6Sn5, the diffusion path is constrained at this point, making it difficult for the Ni atoms to diffuse to the other area. At this point, the segregation phase of Ni diffuses in the system using the method of grain boundary diffusion. The sample is further brought to a relatively stable state.

Step 5: The Cu6Sn5 will then continue to react with Cu to form Cu3Sn. Since η-(Cu,Ni)6Sn5 with the small grain size is a stable phase and does not react with Cu, the Cu atoms must diffuse through the η-(Cu, Ni)6Sn5 by grain boundary diffusion and react with the upper (Cu, Ni)6Sn5 to form Cu3Sn. Finally, the (Cu, Ni)6Sn5 with high Ni content exists in the structure near the Ni side and the center line, while in the other area almost all IMCs are Cu3Sn.

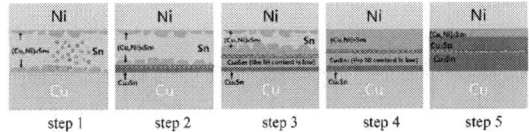

Fig. 7 Formation mechanism of (Cu,Ni)₆Sn₅

IV.   CONCLUSION

After short-term solid state aging in this study, both η'-(Cu,Ni)6Sn5 and η-(Cu,Ni)6Sn5 exist in the microbump, but the Ni content in the latter phase is higher. Ni distribution is monitored using EPMA, and the evolution of the Cu/Sn/Ni microbump's microstructure is modeled using the EPMA and TEM results. Additionally, the model describes the five steps formation mechanism of (Cu,Ni)6Sn5. Furthermore, the model of microstructure evolution can be used to explain the formation mechanism of (Cu,Ni)6Sn5 with a high Ni content.

REFERENCES

[1] Yang T H,, et al. Effects of aspect ratio on microstructural evolution of Ni/Sn/Ni microjoints[J]. Journal of Electronic Materials, 2019, 48(1): 9-16

[2]Wang, Y.W, et al. Reaction Within Ni/Sn/Cu Microjoints for Chip-Stacking Applications. J. Electron. Mater. 48, 25–31 (2019).

[3] Mo L, et al. Microstructural evolution of Cu–Sn–Ni compounds in full intermetallic micro-joint and in situ micro-bending test[J]. Journal of Materials Science: Materials in Electronics, 2018, 29(14): 11920-11929.

[4]Yu, H.Y., Yang, et al. Surface Diffusion and the Interfacial Reaction in Cu/Sn/Ni Micro-Pillars. J. Electron. Mater. 49, 88–95 (2020).

LIPFD-NPU: Low-overhead Instruction-driven Permanent Fault Detection for Neural Processing Unit

Pengfei Wu[1,2], Zheng Wang[2], Zhiming Pan[1], Weilun Wang[2]
[1] College of Electronic and Information Engineering, Shenzhen University, Shenzhen, China
[2] Shenzhen Institute of Advanced Technology, Chinese Academy of Science, Shenzhen, China

Abstract—**Neural Processing Unit (NPU) has become the state-of-the-art solution for accelerating artificial neural networks and is increasingly integrated on the System-on-Chip (SoC) of edge devices such as smartphones and cameras. However, adopting NPU in mission-critical systems, such as aerospace aircraft and autonomous driving demands high reliability, which is currently less explored on industrial NPUs. In this work, we target one of the critical reliability issues - permanent fault - for modern NPUs and provide an instruction-driven fault detection method named LIPFD-NPU. The approach executes dedicated network instructions in a self-testing fashion and generates fine-grained information on the potential fault's location, type and level of impact. An FPGA-based fault emulation framework is used to verify LIPFD-NPU. The results indicate that LIPFD-NPU effectively detects faults with tiny overheads of 0.2% in silicon area and 0.5% in power consumption.**

Index Terms—**Neural Processing Unit; Permanent fault detection; Micro-architectural reliability**

I. INTRODUCTION

Although NPUs have been widely adopted in applications such as computer vision and natural language processing, research on their architectural and circuit-level reliability techniques is still sporadic, which can be attributed to the fact that modern neural networks (NNs) exhibit levels of intrinsic fault tolerance. According to the quantitative fault injection experiments in [1], the previous claim holds only for transient fault, where a single bit of logic or storage element has its value temporarily flipped due to radiation. However, damage-induced permanent faults where logic values are stuck at zero or one lead to significant misbehavior of NPU. For instance, six stuck-at-1 faults cause an inference error of 60%+ compared to only 3% for six bit-flips. Therefore, post-fabrication micro-architecture techniques to encounter the effects of permanent faults are of prime importance during the design of future NPUs with high reliability.

A few techniques to detect permanent faults in NPU have appeared recently. Online testing techniques in [2], [3] optimize the issue of long testing time through extensive modification of micro-architectures, which results in either large silicon overhead [2] or drop of inference precision [3]. To relieve such deficits, an application-level self-test framework is proposed in [4] with zero hardware overhead. However, it fails in detecting fault locations. Consequently, fault detector in NPU simultaneously demands accuracy in analyzed fault statistics, low execution & physical overheads, as well as maintenance of inference precision.

In contrast to previous works of coarse-grained fault detection, we introduce an instruction-level fault detection approach named LIPFD-NPU. It adopts a frequency-based analysis to measure the health factor of individual processing elements (PEs). Specifically, LIPFD contains one testing instruction *N_BIST* to compare each PE's commit results with golden outputs and estimate fault statistics including location, type and impact level through hardware scoreboards and fault tables. The fault type for each PE is inferred by evaluating mismatch frequency and comparing it with a configurable threshold value in the instruction. Since fault propagation and diffusion are eliminated in instruction-level testing, LIPFD eases fault localization. Furthermore, the augmented NPU introduces negligible hardware overheads.

This work was funded by Key-Area Research and Development Program of Guangdong Province (Grant No. 2019B010155003), Guangdong Basic and Applied Basic Research Foundation (Grant No. 2020B1515120044). Zheng Wang is corresponding author (zheng.wang@siat.ac.cn).

The remaining work is organized as follows. Section II introduces the design methodology. Section III illustrate the fault emulator and experiments. Section IV concludes this work.

II. DETECTION FLOW AND ARCHITECTURE SUPPORTS

1) Baseline architecture

The baseline architecture is discussed in [5], which consists of a host interface controller, a memory access module, and a computing fabric. The interface controller communicates with desktops for loading network instructions, weights and activations and retrieving computing results. The memory access module is optimized for fast distributing activations and weights from DRAM into on-chip buffers. The computing fabric adopts output stationary dataflow [6]. The PEs are organized in a 2D array, where weights in the same row of PEs are shared and activations in the same column of PEs are shared.

2) Fault detection flow

The operation flow of LIPFD-NPU is illustrated in Fig.1. We expand the baseline NPU into two operating modes: regular and test modes. In regular mode, NPU processes network instructions such as convolution and pooling. Test mode gets triggered when *N_BIST* instruction is decoded. In the first phase of test mode, a predefined input test vector is loaded from host, forwarded to the computing fabric for processing and the output vector under test is stored into DRAM. In parallel, the predefined golden output vector from host is also loaded. The second phase starts when both output vectors are loaded into an on-chip buffer and verified with each other to detect mismatches, therefore PEs exhibiting incorrect behavior can be identified. *N_BIST* instruction typically compares PE mismatches of hundred rounds to get statistically meaningful fault information. Hardware supports such as slicing comparator and mismatch scoreboard are introduced to perform fault diagnosis. Consequently, information on fault location, type and level of impact is stored as a fault table and retrieved from host interface controller. NPU can switch to regular mode once conventional network instructions are decoded.

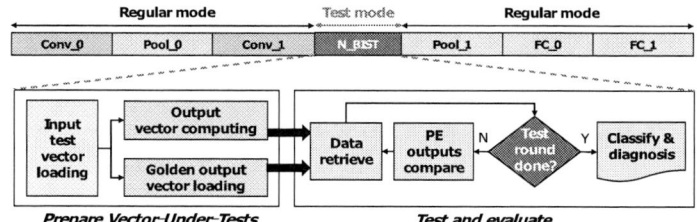

Fig. 1. Fault detection flow in LIPFD-NPU

3) Architecture supports

The augmentation to baseline architecture is illustrated in Fig.3. Additional modules are built around the output activation buffer (commit buffer) where the computing fabric, memory access and host interfaces all remain intact. The proposed architecture supports are easily ported into generic NN accelerators.

- **Commit Buffer (CB):** multiplexes the 128KB commit buffer in regular mode to cache both types of data in test mode. Therefore, no extra SRAM blocks are involved in the fault detection flow.
- **Slicing Comparator (SC):** fetches both actual and golden data from commit buffer and performs efficient comparison and identifies mismatches. Instead of comparing the outcome of each multiplication, we only compare the committed partial-sum values to significantly save fault detection logic in PE.
- **Mismatch Scoreboard (MS):** allocates a score counter for each PE to record the number of mismatches across multiple test rounds. A thread pointer register is used to update corresponding elements in MS.
- **Fault Analyzer (FA):** finally compares the values in the MS with the threshold value specified in the instruction

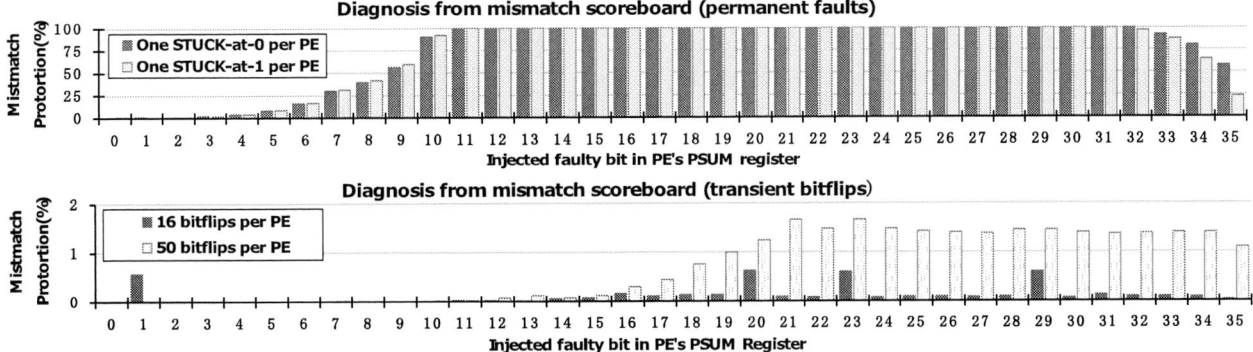

Fig. 2. Effect of *N_BIST* instruction: impact of four types of single-bit faults versus range of faulty locations in PE

to determine impact levels. Four impact levels of healthy, benign, error, and fatal can be assigned to each PE.

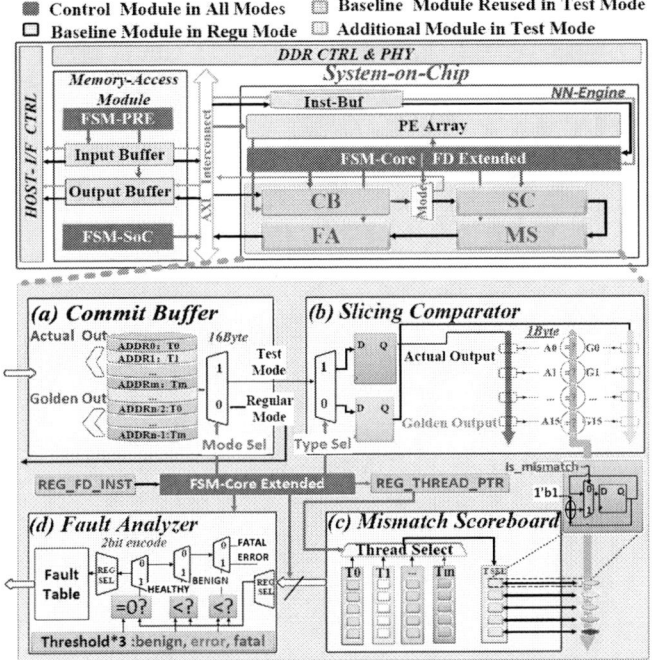

Fig. 3. Architecture augmentation to baseline accelerator

III. EXPERIMENTAL RESULTS

The architecture baseline and extension are described by Verilog HDL. Functionality and performance are evaluated on Xilinx Kintex-7 FPGA, while physical overheads are estimated through logic synthesis under SMIC 40nm standard cell library.

1) FPGA-based fault emulation

To accelerate prototyping of fault detection designs, one FPGA-based architecture emulator is developed in-house which supports precise and configurable fault injection based on minimal RTL modification and host API in python. The software API supports semi-automated generation of fault configuration including the time, type and target PE location. Scheduled faults are on-the-fly injected into instantiated design on FPGA. Currently, three types of faults e.g. stuck-at-0, stuck-at-1 and bitflip can be specified.

2) Effectiveness of fault detection

We configure *N_BIST* to verify outputs of the PE array for 500 rounds. Mixed types of faults are injected into the PSUM register on a 32 × 8 PE array. Specifically, one stuck-at fault per PE in the first 4 columns and a certain number of bitflips randomly through time per PE are injected in the rest 4 columns. We tune the faulty bit position in the PSUM register (from 0 to 35) and collect the proportion of mismatches. The results in Fig.2 show

that permanent faults have significantly larger impact compared to transient faults, where faulty bits with higher significance tend to create larger error impact. Quantitatively, it is observed that a threshold value of 2% can be used to differentiate permanent and transient faults. Faults on LSBs (0 to 3) for both categories create negligible impacts during detection.

3) Physical and latency overheads

Table I presents the area and power estimates for baseline NPU and extended designs. Both overheads are below 0.5%. The latency overhead of *N_BIST* instruction is estimated to be 700 μs at 300 MHz operating frequency. We benchmark LIPFD-NPU with relevant fault tolerant NPU designs in Table II, it is shown that our design incurs few physical overheads and detecting latency without degradation in inference precision. Regarding retrieved fault information, not only location and type can be detected, but also the level of faulty impact can be quantified.

TABLE I. Physical estimates of baseline and augmented NPUs

SMIC 40nm	Area (μm^2)	Power (mW)	Frequency (MHz)
Baseline-NPU	3775762	34.2264	300
LIPFD-NPU	3783130	34.4515	300

TABLE II. Benchmark with relevant fault detection techniques

	[2]	[3]	[4]	This work
Precision degrade	NO	YES	NO	**NO**
Extra area	34%	8%	0%	**0.2%**
Extra Power	30%	N/A	0%	**0.5%**
Detection latency	N/A	N/A	30 frames	single NN layer
Fault Information	L & T	L	T & I	**L & T & I**

∗ L: Location, T: Type, I: Impact level

IV. CONCLUSION

In this paper we present LIPFD-NPU, a low overhead fault detection solution for state-of-the-art NPUs. Our design is based on instruction-level extension to trigger test mode periodically and verify correctness of PE elements. FPGA-based fault emulator is used to demonstrate the proposed approach.

REFERENCES

[1] B. Salami, O. S. Unsal, and A. C. Kestelman, "On the resilience of rtl nn accelerators: Fault characterization and mitigation," in *2018 30th International Symposium on Computer Architecture and High Performance Computing (SBAC-PAD)*, pp. 322–329, IEEE, 2018.

[2] M. R. Roshanshah, K. Basharkhah, and Z. Navabi, "Online testing of a row-stationary convolution accelerator," in *2021 IEEE European Test Symposium (ETS)*, pp. 1–2, IEEE, 2021.

[3] S. Burel, A. Evans, and L. Anghel, "Mozart+: Masking outputs with zeros for improved architectural robustness and testing of dnn accelerators," *IEEE Transactions on Device and Materials Reliability*, 2022.

[4] F. Meng, F. S. Hosseini, and C. Yang, "A self-test framework for detecting fault-induced accuracy drop in neural network accelerators," in *Proceedings of the 26th Asia and South Pacific Design Automation Conference*, pp. 722–727, 2021.

[5] Z. Wang *et al.*, "Accelerating hybrid and compact neural networks targeting perception and control domains with coarse-grained dataflow reconfiguration," *Journal of Semiconductors*, no. 2, pp. 29–41, 2020.

[6] V. Sze *et al.*, "Efficient processing of deep neural networks: A tutorial and survey," *Proceedings of the IEEE*, vol. 105, no. 12, pp. 2295–2329, 2017.

Large Suppression to Lateral Charge Migration (LCM) Related Error Bits in Charge-Trap TLC 3D NAND Flash

Kenie Xie[1], Peng Guo[2], Fei Chen[1], Binglu Chen[1], Xiaotong Fang[1], Jixuan Wu[1], Xuepeng Zhan[1], Jiezhi Chen[1, *]

[1]School of Information Science and Engineering, Shandong University, P. R. China, *email: chen.jiezhi@sdu.edu.cn
[2]Shandong Sinochip Semiconductors Co., Ltd, P.R. China

Abstract—We present a study to suppress error bits from lateral charge migration (LCM) in charge-trap (CT) 3D NAND flash memory. For the first time, a new Baking-and-Pre-read (BPR) method is proposed with combined long-time charge diffusion by baking and short-time stabilizing by Pre-read. By characterizing 96-layer Triple-level-cell (TLC) 3D NAND chips by the raw NAND chip tester, the storage stabilities, including data retention (DR) and read disturb (RD), are studied and it is found that DR/RD error bits can be reduced up to >70%, which could be explained by the large effects of suppression to LCM-related threshold voltage (V_{th}) down-shifts.

Keywords—*3D NAND, Read Disturb, Data Retention, Cold Data, Hot Data, Lateral charge migration.*

I. INTRODUCTION

Vertically stacked 3D NAND flash memory has been widely used in digital archives, smartphones, PC, data-center, and even artificial intelligence (AI) applications [1]. As for data-centric applications, data can be mainly divided into two different types, hot data that needs frequently read operations, and cold data that needs to be stored for a long time [2]. For hot data storage, read disturb (RD) is the main concern as it occurs during read operations; while for cold data storage, data retention is important to keep data secure and stable. For the flexibility of actual use, the data type also could be changed in special circumstances. In 3D NAND flash, although it has shown good performance and reliability, bit error rate (BER) has significant degradation during DR even at room temperature (RT), which could be partially recovered by a few read cycles [3]. These specific characteristics are strongly correlated to the lateral-charge-migration (LCM) mechanism in storage-shared charge-trap (CT) 3D NAND flash.

In this work, to design 3D NAND flash memory with robust reliabilities, we present a study to suppress error characteristics from LCM-related threshold voltage (V_{th}) shift, and a novel method is proposed by combining long-time diffusion (baking at RT) and short-time stabilizing (Pre-read). Further analyses are done to understand the dominant underlying mechanisms.

II. EXPERIMENT SETUP AND CHARACTERIZATIONS

As shown in Fig.1 with a typical 3D NAND flash structure, cells in the same NAND string are connected in series to form a bit-line (BL), and the cells in a word-line (WL) share the same memory layer. The typical V_{th} distribution in Triple-level-cell (TLC) 3D NAND includes seven program states from A to G levels, and the error bits happen if the distributions of neighbor states are overlapped. Thereby, these error bits can be classified into two different categories, V_{th} down-shift errors, and V_{th} up-

This work was supported by National Natural Science Foundation of China (Nos. 62034006, 91964105, 61874068), Natural Science Foundation of Shandong Province (No. ZR2020JQ28, ZR2020KF016), the Joint fund for Intelligent Computing of Shandong Natural Science Foundation (ZR2019LZH009), Program of Qilu Young Scholars of Shandong University.

shift errors. In storage-shared CT 3D NAND, there are two main reasons for V_{th} down-shifts. One is vertical-charge-loss (VCL), the stored charges escape to the channel via the tunneling oxide, which is serious in high V_{th} states; the other one is LCM, stored charges in target cells diffuse to the space region between neighbor cells.

In this work, we measured TLC 3D NAND by using the FPGA-based raw NAND chip tester. As shown in Fig. 2, we compared two different data, hot data (Read after Program) and cold data (Read after 12 hours DR). Obviously, more error bits are generated during the retention. However, RD properties are quite different. For further analysis, Fig. 3 illustrates the errors from down-shift and up-shift. Down-shift errors are the main source of error bits for both fresh blocks and 2K PE cycled blocks. The observed error bits' recovery can be explained by the down-shift errors' lowering during read cycling, which has been studied in our previous work [5].

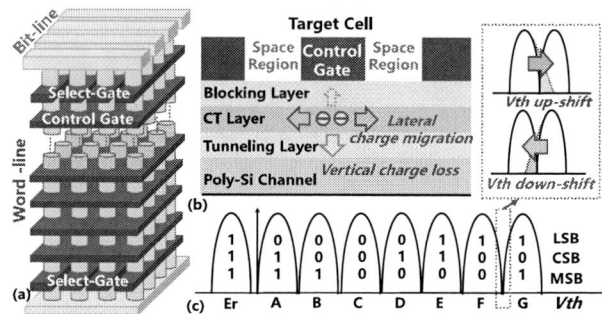

Fig. 1. (a) The structure of 3D CT NAND flash, (b) a depiction of the charge redistribution mechanisms, (c) the typical V_{th} distribution and coding method.

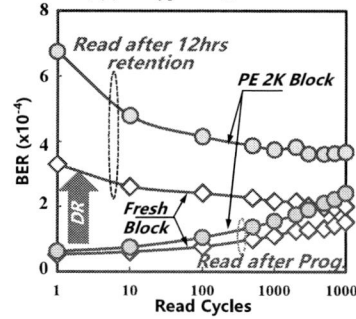

Fig. 2. Measured bit-error-rate (BER) in read cycles for hot data (read after program) and cold data (read after 12 hours' retention).

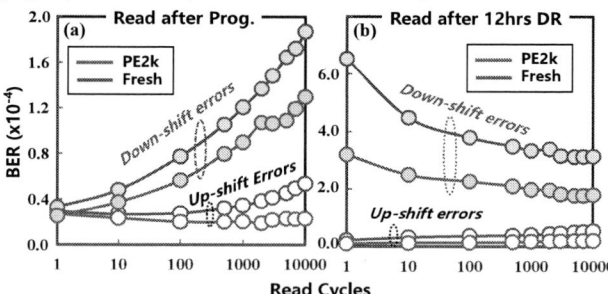

Fig. 3. Down-shift and up-shift errors in (a) hot data and (b) cold data.

III. PROPOSED SCHEMES

In this section, on the basis of previous studies [4][5], we compared the effects with different methods in 96-layer CT TLC 3D NAND flash memory, as shown in Fig.4, wherein a new method named as Baking-and-Pre-read (BPR) method is proposed with combined long-time diffusion (baking) and short-time stabilizing (Pre-read). The studied four schemes addict different pre-processing methods ahead of the RD testing. Mode-1, PE (program random data and erase); Mode-2, PPE (program random data twice continuously and then erase); Mode-3, PBK (Program random data, hold 6 hours as baking and then erase); Mode-4, BPR (program random data, hold 6 hours as baking, read data and then erase).

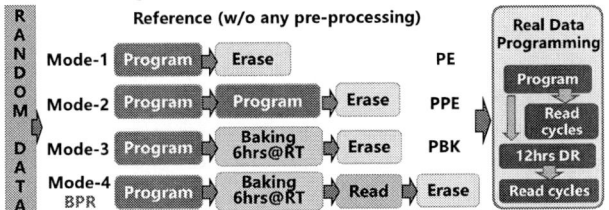

Fig. 4. The experimental processes of four different pre-processing modes, among which "mode 4" is newly proposed.

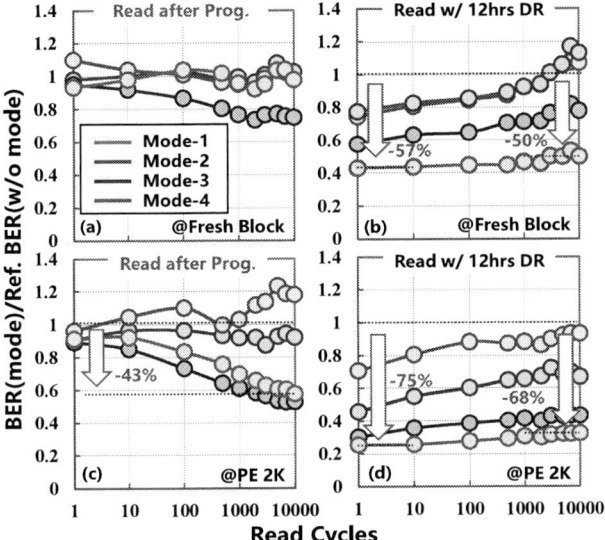

Fig. 5. Effects of four modes on BER: cold data in (a) fresh blocks and (c) 2K PE cycled blocks, hot data in (b) fresh blocks and (d) 2K PE cycled blocks.

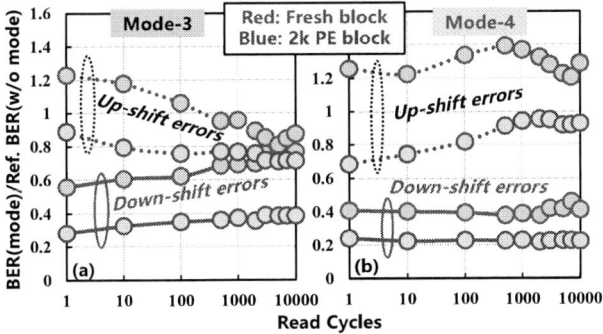

Fig. 6. Separated error bits from V_{th} down-shift and up-shift by adopting (a) mode-3 and (b) mode-4.

As shown in Fig.5, four modes are compared with the reference data without any pre-processing. For hot data (read after Prog.), mode-1 and mode-2 could cause worse BER, while mode-3 and mode-4 are effective for 2K PE cycled blocks with >40% BER reductions (@10K read cycles). For cold data (read after 12 hours' retention), all four modes help to obtain a lower BER in the early stage. However, in the fresh blocks, the BER of mode-1 and mode-2 turns out to be worse and even higher than the reference data with read cycles, while the effect of mode-3 is obvious (~40%/20% BER reduction at the initial state and 10K read cycles) and that of mode-4 is the best (~60%/50% BER reduction at the initial state and 10K read cycles). For 2K PE cycled blocks, mode-3 can achieve

~70%/60% BER reductions at the initial state and 10K read cycles, and mode-4 can have ~75%/68% BER reductions, respectively.

Then, we separate error bits to V_{th} down-/up-shift error bits respectively, as shown in Fig. 6. Obviously, large effects from mode-3 and mode-4 attribute to the great effects to suppress V_{th} down-shift, but some degradation can be observed for error bits from V_{th} up-shift. To further investigate the mechanism, we explored the effects on V_{th} down-shift error bits of each state in Fig. 7. Both two modes have better effects for bits in the higher program states. About the better effects in mode-4, it is considered that the pre-stored electrons in mode-3 can diffuse to the space region to suppress LCM, while additional read operation in mode-4 can stabilize the charges in the space since the read electric field could assist trap re-distribution [3] that can remove unstable charges and make following program process more stable. However, the pre-stored charges in the space region could diffuse to cells with low-V_{th} states and then degrade the up-shift errors. This side-effect can be observed in fresh chips due to the limited shallow traps at the fresh state.

Fig. 7. V_{th} down-shift error bits concerning cold data of each state at the first read and 10K read cycles in mode-3 and mode-4.

IV. CONCLUSIONS

A novel BPR method is proposed to suppress LCM effects in CT 3D NAND flash memory to achieve lower error bits. On the basis of raw NAND chip characterizations, we separated the error bits into V_{th} down-/up- shift errors and it is found that BPR can largely minimize the V_{th} down-shift error bits from the LCM mechanism. Especially, ~60%/70% error bits' reductions are demonstrated for the data (read after 12 hours' retention) in the fresh blocks and 2k PE cycled blocks, respectively.

REFERENCE

[1] S. -M. Jung et al., "Three Dimensionally Stacked NAND Flash Memory Technology Using Stacking Single Crystal Si Layers on ILD and TANOS Structure for Beyond 30nm Node," 2006 International Electron Devices Meeting (IEDM), 2006, pp. 1-4, Doi: 10.1109/IEDM.2006.346902.

[2] K. Mizoguchi, K. Maeda and K. Takeuchi, "Automatic Data Repair Overwrite Pulse for 3D-TLC NAND Flash Memories with 38x Data-Retention Lifetime Extension," 2019 IEEE International Reliability Physics Symposium (IRPS), 2019, pp. 1-5, Doi: 10.1109/IRPS.2019.8720420.

[3] Wang F, Cao R, Kong Y, et al. Lateral charge migration induced abnormal read disturb in 3D charge-trapping NAND flash memory[J]. Applied Physics Express, 2020, 13(5): 054002.

[4] R. Cao, J. Wu, W. Yang, J. Chen and X. Jiang, "Program/Erase Cycling Enhanced Lateral Charge Diffusion in Triple-Level Cell Charge-Trapping 3D NAND Flash Memory," 2019 IEEE International Reliability Physics Symposium (IRPS), 2019, pp. 1-4, Doi: 10.1109/IRPS.2019.8720412.

[5] Y. Kong, M. Zhang, X. Zhan, R. Cao and J. Chen, "Retention Correlated Read Disturb Errors in 3-D Charge Trap NAND Flash Memory: Observations, Analysis, and Solutions," in IEEE Transactions on Computer-Aided Design of Integrated Circuits and Systems, vol. 39, no. 11, pp. 4042-4051, Nov. 2020, Doi: 10.1109/TCAD.2020.3025514.

A Q-Band Low-Noise Amplifier in 40-nm CMOS for Q/V-band satellite communications

Qin Tian[1,2], Dixian Zhao[1,2]

[1] National Mobile Communication Research Laboratory, Southeast University, Nanjing, China
[2] Purple Mountain Laboratories, Nanjing, China

Abstract—This paper proposes a Q-band single-ended LNA fabricated in 40-nm CMOS technology, where two-stage cascode topology and inductive source degeneration technique are employed. The proposed LNA has achieved a lowest NF of 4.3 dB, and a maximum power gain of 24.2 dB with a 3-dB bandwidth of 11 GHz, consuming 17.85 mW dc power from a 1.5 V supply.

Keywords—LNA, inductive source degeneration, two-stage cascode

I. INTRODUCTION

The rapid development of satellite communication (SATCOM) has greatly expanded the coverage area of communication systems. It plays an important role in 6G networks which will integrate both terrestrial and non-terrestrial terminals [1]. In the future High Throughput Satellite (HTS) systems, higher frequency bands such as Q/V band (i.e., 37.5-42.5 GHz/ 47.2-51.4 GHz) are utilized to support Terabit connectivity, where a larger bandwidth is available for high data rate transmission [2]. Phased arrays have been increasingly adopted in SATCOM for beam-forming and beam-steering in recent years. In phased array receivers, LNA serves as the first module after antenna, and dominates the overall noise performance.

This paper presents a Q-band LNA with two-stage cascode topology and inductive source degeneration technique, achieving a lowest NF of 4.3 dB and a maximum power gain of 24.2 dB with a 3-dB bandwidth of 11 GHz. The paper is organized as follows. Section II describes the inductive source degeneration technique and the matching network of the proposed LNA. In Section III, measurement and simulation results are demonstrated. Finally, a conclusion is drawn in Section IV.

II. DESIGN OF LOW NOISE AMPLIFIER

Schematic of the proposed LNA is shown in Fig.1. Two stages of amplifiers (i.e., M_1-M_2 and M_3-M_4) are designed to satisfy the power gain demand. Cascode topology is used to reduce the effect of overlap capacitance C_{gd}, and obtain a relatively high isolation at output. The degenerated inductor L_s is added at the first stage to achieve noise and input matching simultaneously. T-match and π-match are implemented to realize a relatively flat power gain. For all the transistors (i.e., M_1, M_2, M_3, and M_4) in Fig.1, the gate width and channel length are chosen as 32 μm and 40 nm respectively. The current density of the first and second stage is 0.19 mA/μm.

A. Inductive Source Degeneration

Inductive source degeneration is a widely adopted feedback technique for narrow-band LNA to adjust the input impedance Z_{in} and the optimum noise impedance Z_{opt} [3]. With the degenerated inductor L_s, the value of Z_{in} can be shifted to the conjugate optimum noise impedance Z_{opt}^*. Then, both can be

This work was supported in part by the National Key R&D Program of China under Grant 2019YFB1803000 and the Major Key Project of PCL(PCL2021A01-2). (Corresponding Author: Dixian Zhao, email: dixian.zhao@seu.edu.cn).

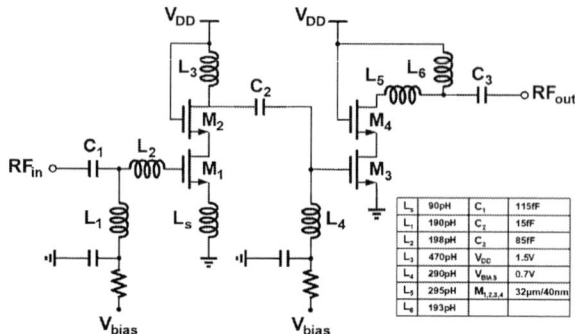

Fig.1. Schematic of the proposed LNA.

L_6	90pH	C_1	115fF
L_1	190pH	C_2	15fF
L_2	198pH	C_3	85fF
L_3	470pH	V_{DD}	1.5V
L_4	290pH	V_{Bias}	0.7V
L_5	295pH	$M_{1,2,3,4}$	32μm/40nm
L_6	193pH		

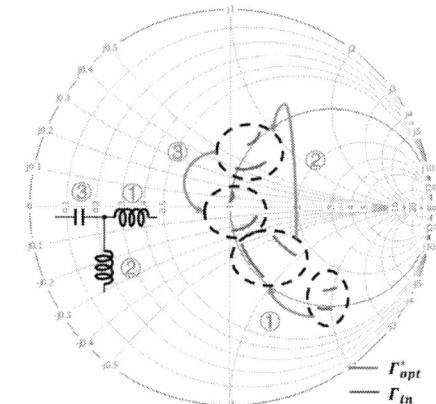

Fig.2. Simulated Γ_{in} and Γ_{opt}^* traces on Smith chart.

transformed to match the source impedance Z_s with the same input matching network, thus realizing maximum power transfer and minimum noise figure simultaneously.

As shown in Fig.2, The designing principle above can be visualized on Smith Chart, where Γ_{in} is the input reflection and Γ_{opt}^* is the conjugate optimal noise impedance point. By increasing L_s, Γ_{in} moves closer to Γ_{opt}^* [4]. Thus, simultaneous noise and input matching can be achieved by choosing appropriate L_s. The trace of the input matching network is also displayed in Fig.2. By adding series inductor L_1, shunt inductor L_2, and series capacitor C_1, both Γ_{in} and Γ_{opt}^* can be shifted to the center of Smith Chart.

B. Inter-stage Matching Network

In the proposed design, π-match is adopted as inter-stage matching network. Compared with the two-component L-match, the three-component π-match is credited with another degree of designing freedom. With an additional inductor, the quality factor Q of the matching network is tunable, which will enable the adjustment of bandwidth. As shown in Fig.3(a)(b), the Q of L-match is fixed at around 3.8, while the Q of π-match can be tuned to around 2. Since matching network's bandwidth is inversely proportional to its Q, the extension of bandwidth is possible by choosing π-match.

Due to the existence of parasitic capacitor, the intrinsic gain of the transistor declines with the increase of frequency. To achieve a relatively flat power gain over the required bandwidth, the resonant frequency of the inter-stage matching network (i.e., 42 GHz) is deliberately shifted above the center frequency (i.e., 40 GHz). In this way, the insertion loss at higher frequency will be smaller, while the loss at lower frequency will be larger, so

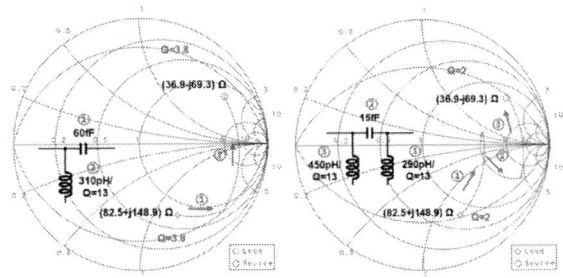

Fig.3. (a) L-match; (b) π-match.

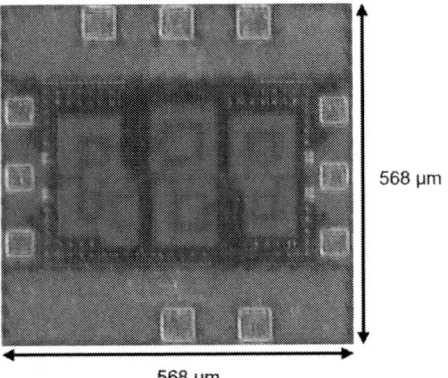

Fig.4. Microphotograph of the proposed LNA.

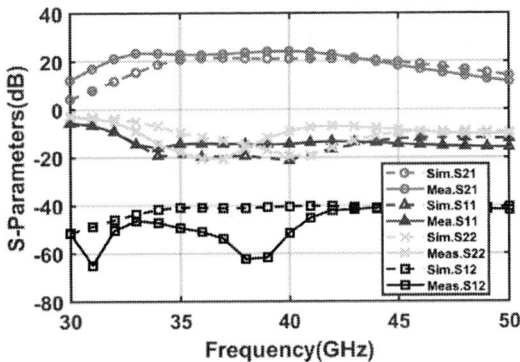

Fig.5. Measured and simulated S-parameters.

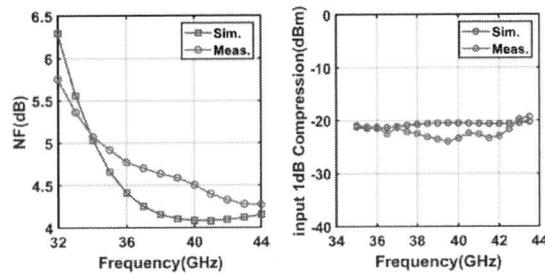

Fig.6. Measured and simulated NF and input P_{1dB}.

that the decline of transistor gain caused by increasing frequency can be compensated.

III. MEASUREMENT RESULT

Fig.4 presents the microphotograph of the proposed LNA, which occupies a 0.323 mm² chip area. Fig.5 shows the simulated and measured S parameters. A measured peak gain of 24.2 dB is achieved at 40 GHz, with a 1-dB bandwidth of 4.3 GHz (i.e., from 37.4 to 41.7 GHz), a 3-dB bandwidth of 11 GHz (i.e., from 32.1 to 43.1 GHz), and a variation of less than 2.1 dB within Q-band for SATCOM (i.e., from 37.5 to 42.5 GHz). A measured 13-14 dB input return loss is realized, as well as high reverse isolation (S12<-40 dB). Fig.6 presents the simulated and measured NF. The measured NF is 4.3-4.7 dB from 37 to 43 GHz, and the lowest NF is achieved at 43 GHz. The total power consumption is 17.85 mW under a 1.5 V supply. Fig.7 shows the simulated and measured input 1 dB compression point $P_{1dB, in}$. The measured $P_{1dB, in}$ is more than -24.6 dBm from 35 to 45 GHz.

Performance of the proposed LNA is compared with prior-art designs in Table I.

IV. CONCLUSION

A Q-band LNA fabricated in 40-nm CMOS is presented in this paper. Inductive source degeneration is employed at the first stage to achieve simultaneous noise and input matching. Two-stage cascode topology and LC matching networks are implemented to obtain a relatively high and flat power gain. The proposed LNA has achieved a lowest NF of 4.3 dB, and a maximum power gain of 24.2 dB with a 3-dB bandwidth of 11 GHz, with 17.85 mW dc power consumption.

REFERENCES

[1] Z. Zhang *et al.*, "6G Wireless Networks: Vision, Requirements, Architecture, and Key Technologies," *IEEE Veh. Technol. Mag.*, vol. 14, no. 3, pp. 28-41, Sept. 2019.

[2] G. Codispoti *et al.*, "Validation of ground technologies for future Q/V band satellite systems: The QV-LIFT project," in *IEEE Aerospace Conf.*, 2018, pp. 1-9.

[3] T.-K. Nguyen, C.-H. Kim, G.-J. Ihm, *et al.*, "CMOS low-noise amplifier design optimization techniques," *IEEE Trans. Microw. Theory Techn.*, vol. 52, no. 5, pp. 1433-1442, May 2004.

[4] J. Zhang and D. Zhao, "A Broadband 1-dB Noise Figure GaAs Low-Noise Amplifier for Millimeter-Wave 5G Base-Stations," in *Int. Conf. Microw. Millim. Wave Technol. (ICMMT)*, 2018, pp. 1-3.

[5] S. Kong, H. -D. Lee, S. Jang, et al., "A 28-GHz CMOS LNA with Stability-Enhanced Gm-Boosting Technique Using Transformers," in *IEEE Radio Freq. Integr. Circuits Symp. (RFIC)*, 2019, pp. 7-10.

[6] Z. Chen, H. Gao, D. Leenaerts, D. Milosevic and P. Baltus, "A 29-37 GHz BiCMOS Low-Noise Amplifier with 28.5 dB Peak Gain and 3.1-4.1 dB NF," in *IEEE Radio Freq. Integr. Circuits Symp. (RFIC)*, 2018, pp. 288-291.

[7] V. Chauhan and B. Floyd, "A 24-44 GHz UWB LNA for 5G Cellular Frequency Bands," in *11th Global Symp. Millim. Waves (GSMM)*, 2018, pp. 1-3.

[8] J. Zhang, D. Zhao and X. You, "A 20-GHz 1.9-mW LNA Using gm-Boost and Current-Reuse Techniques in 65-nm CMOS for Satellite Communications," *IEEE J. Solid-State Circuits*, vol. 55, no. 10, pp. 2714-2723, Oct. 2020.

TABLE I.   PERFORMANCE COMPARISON

Reference	[5]	[6]	[7]	This work
3-dB BW (GHz)	24.9-32.5	29-37	24-44	**32.1-43.1**
NF (dB)	3.25	3.1	4.2	**4.3**
Peak Gain(dB)	18.33	28.5	20	**24.2**
Power (mW)	20.52	80	58	**17.85**
FoM[a]	2.7	2.6	2.1	**5.9**
Area(mm²)	0.38	0.21	0.2	**0.322**
Technology	65nm CMOS	0.25μm SiGe	45nm CMOS	**40nm CMOS**

[a]. FoM= $\frac{|S_{21}| \times BW[GHz]}{(F-1) \times P_{DC}[mW]}$ [8]

A V-band Power Amplifier for Satellite Communications in 40-nm CMOS

Hengzhi Wan[1,2], Dixian Zhao[1,2]

[1] National Mobile Communication Research Laboratory, Southeast University, Nanjing, China
[2] Purple Mountain Laboratories, Nanjing, China

Abstract—This paper presents a V-band power amplifier (PA) with two stages. Each stage consists of a neutralized common-source amplifier pair, coupled with each other using a transformer. Implemented in 40-nm CMOS technology, the proposed two-stage PA achieves a measured small-signal gain of 16.8 dB at 47 GHz with a 3-dB bandwidth more than 6.5 GHz. The maximum 1-dB compressed output power (P_{1dB}) is 12.2 dBm with a power-added efficiency at P_{1dB} (PAE_{1dB}) of 18% measured at 47 GHz; from 45 to 51 GHz, the measured P_{1dB} is above 10 dBm under a 1.1-V supply voltage.

Keywords—CMOS, power amplifier, V-band, neutralization, transformer-coupling

I. INTRODUCTION

In recent years, satellite communications (SATCOM) using millimeter-wave (mm-Wave) bands have attracted wide attention, the traditional bands of which are Ku-band and Ka-band [1]. As the available spectrum becomes over-utilized and demand for larger bandwidth is increasing globally, future systems in higher frequency range becomes indispensable. Under the circumstances, Q-band (i.e., 33-50 GHz) and V-band (i.e., 50-75 GHz) are promising frequency bands for next generation SATCOM. Among varies building blocks of the mm-Wave wireless transceiver systems, the power amplifier (PA) plays a critical role in providing enough output power to overcome the path loss and transmitting signals. However, for the advanced CMOS technology, the output power is limited by the low supply voltage, while the low transconductance restricts transistor gain. One of the methods to increase power gain is using neutralizing capacitors in common-source amplifier pair [2]. Compared with the one without neutralization, the differential pair with neutralization can achieve higher gain, reverse isolation and stability in differential mode.

Another challenge for PA design in mm-Wave is the chip area. Since phased array systems are widely applied to compensate for the high path loss of mm-Wave communication, PAs can occupy nearly half the area of a transmitter front end. To minimize the area of matching networks, which are the largest parts in PA, transformers become a preferred choice at mm-Wave frequencies. Transformers can help realize stage matching networks and output baluns in a compact form. Additionally, transformers can isolate the DC path and providing a virtual ground at the center tap for power supply [3].

This work presents a V-band two-stage PA. Neutralized common-source amplifier pairs are employed to boost the power gain and transformers are used to implement the area-efficient matching networks. The PA is fabricated in 40-nm CMOS technology. Measured results show that it achieves a power gain of 19 dB at 44 GHz and 1-dB compressed output power (P_{1dB}) of > 10 dBm across the 47-51 GHz, with a power-added efficiency at P_{1dB} (PAE_{1dB}) higher than 10%. The chip size of the proposed PA is 623×573 mm^2.

This work was supported in part by the National Key R&D Program of China under Grant 2019YFB1803000 and the Major Key Project of PCL(PCL2021A01-2). (Corresponding author: Dixian Zhao, email: dixian.zhao@seu.edu.cn)

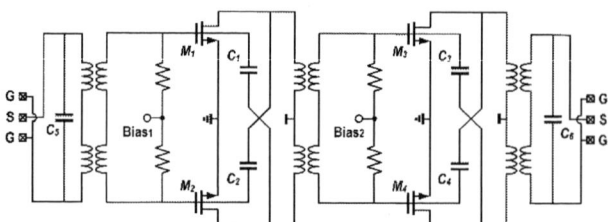

Fig. 1. Schematic of the two-stage PA.

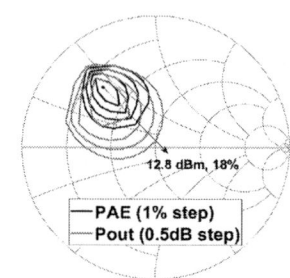

Fig. 2. Simulated P_{out} and PAE contours of PA output stage.

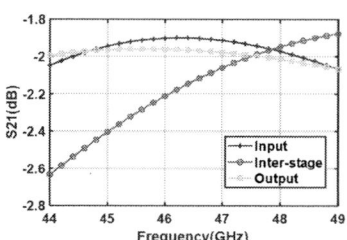

Fig. 3. Simulated S_{21} of input, inter-stage and output matching networks.

II. CIRCUIT DESIGN

The schematic of the two-stage PA is shown in Fig. 1. The input and output baluns are realized in the form of a transformer with a shunt capacitor, and the matching network between two stages is a transformer only.

A. Common-source Amplifier Pair with Neutralization

Traditional common-source amplifier pairs suffer from the Miller effect introduced by the parasitic capacitance between transistor's gate and drain, which can cause instability or oscillation. Additionally, the gain of the differential pair is also reduced due to the current feedback through this parasitic capacitance. Thus, neutralization topology is considered in this work, for it can boost gain and stability simultaneously. Using a capacitor to connect the transistor's drain of one side and the gate of the other side like C_1 or C_2 in Fig. 1, whose capacitance equals the parasitic one between the gate and drain, a compensating current can be created to counteract the unwanted feedback caused by Miller effect. In this work, the neutralizing capacitors are 26 fF and 40 fF for the first and second stage respectively.

Apart from the neutralization topology, the size and the bias of transistors are carefully designed. The second stage consists five transistors in one side, whose gate width is 32 μm, to provide enough output power, while the first stage consists three transistors to obtain a balance between gain and driving ability. The first stage is biased in class-A, so that it can provide enough gain and enjoy a decent linearity, and the second stage is biased in class-AB to achieve high P_{1dB}. The cut-off frequencies of transistors at two stages are 250 GHz and 96 GHz respectively.

978-1-6654-9270-6/22 $31.00 © 2022 IEEE

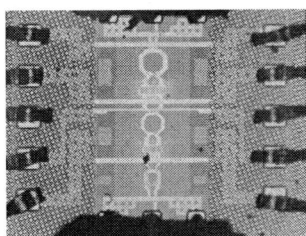

Fig. 4. Chip micrograph.

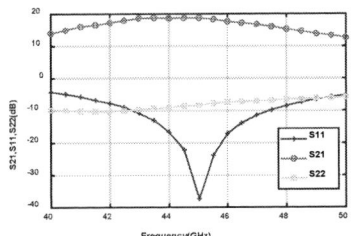

Fig. 5. Measurement results of S-parameters.

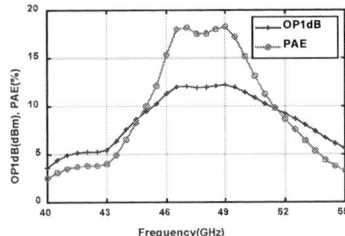

Fig. 6. Measurement results of large signal parameters.

B. Optimization of Matching Networks

The input and output matching networks both consist of a transformer and a capacitor which is in parallel with the primary and secondary coil respectively, while the one between two PA stages does not contain the capacitor. The two coils of three transformers are all designed to overlap each other in order to achieve high mutual induction and low transmission loss.

Additionally, load-pull simulation is applied in designing the output balun, for the S-parameters of PA in large signal are different from those in small signal. As shown in the Fig. 2, PAE and output power (P_{out}) have peak values of 18% and 13 dBm respectively. The PAE contours are drawn in the step of 1% and the P_{out} contours are drawn in the step of 0.5 dB. Considering that PAE contours are denser, an optimum load impedance $(15+ j15\Omega)$ is chosen for the output stage, offering an output power of 12.8 dBm and PAE of 18%.

After optimization, a two-turn transformer with 11.75- and 13.27-μm inner radius realizing 153 pH and 48 pH at 47 GHz is employed as input balun, while a center-tapped one-turn transformer with 24.21- and 31.21-μm inner radius realizing 75.86 pH and 85.32 pH at 47 GHz is employed as output balun, and the transformer between two stages consists of two one-turn coils with 21.8- and 22.8-μm respectively, realizing 78.45 pH and 95.3 pH at 47 GHz. The simulated S_{21} of three matching networks are shown in Fig. 3. Input and output matching networks are designed to have a flat S_{21} to realize a decent S_{11} and S_{22} for PA, while the S_{21} of inter-stage one rises along with frequency to compensate for the gain curves of two PA stages.

III. MEASUREMENT RESULTS

Fabricated in 40-nm CMOS, the power amplifier occupies a core chip area of 623 × 573 mm², including G-S-G pads (see Fig. 4). The overall circuit consumes 78.5 mW from 1.1-V supply. In Fig. 5, the measured S-parameters are plotted. As depicted in the figure, the measured S_{21} reaches a peak of 19 dB at 44 GHz and is higher than 17.5 dB from 42.2 to 46.4 GHz. Meanwhile, the measured S_{11} and S_{22} are lower than -7.5 dB across the same frequency band.

The large signal characteristics of the circuit are also measured. The measured P_{1dB} and PAE_{1dB} are depicted across 40-55 GHz in Fig. 6. The P_{1dB} is above 10 dBm from 45-51 GHz, peaking at 47 GHz (12.2 dBm). Besides, the PAE_{1dB} is higher than 10% across 47-51 GHz, whose peek can be found at 47 GHz (18%).

In Table I, the performance of the proposed power amplifier is summarized and compared with the prior-art designs. With a relatively low power dissipation and chip area, this work achieves sufficient output power and power gain.

IV. CONCLUSION

A V-band power amplifier with two stages is presented in this work. Utilizing two differential neutralized common-source amplifier pairs and transformer-based matching networks, the

TABLE I. PERFORMANCE COMPARISON

Reference	WPTC 2017 [4]	TMTT 2019 [5]	IMS 2021 [6]	This Work
Technology	90nm CMOS	65nm CMOS	45nm CMOS SOI	40nm CMOS
Output Frequency (GHz)	60	60	55	47
Gain (dB)	6	29.7	18.8/16.4	16.8
P_{1dB} (dBm)	7.3	19.9	13.5/15	12.2
PAE_{1dB} (%)	1.81	11.1	28.9/35.2	18
V_{DD} (V)	1.2	2.4/1.2	1.1	1.1
P_{DC} (mW)	215	816	N/A	78.5
Core Area (mm²)	0.56	0.653	0.18	0.357

gain is increased and the chip area is reduced. Implemented in 40-nm CMOS, the V-band PA achieves a measured small-signal of 16.8 dB at 47 GHz and shows a maximum P_{1dB} of 12.2 dBm at 47 GHz with a maximum PAE_{1dB} of 18%. The overall circuit consumes 78.5 mW, occupying 623 × 573 mm² silicon area.

REFERENCES

[1] L. R. Shet et al., "Challenges and configuration of ISRO's future Q/V band satellite," in 2016 International Conference on Wireless Communications, Signal Processing and Networking (WiSPNET), 2016, pp. 674-679.

[2] D. Zhao and P. Reynaert, "A 60-GHz Dual-Mode Class AB Power Amplifier in 40-nm CMOS," IEEE J. Solid-State Circuits, vol. 48, no. 10, pp. 2323-2337, Oct. 2013.

[3] J. Zhong, D. Zhao and X. You, "A Ku-Band CMOS Power Amplifier With Series-Shunt LC Notch Filter for Satellite Communications," IEEE Trans. Circuits Syst. I, Reg. Papers, vol. 68, no. 5, pp. 1869-1880, May 2021.

[4] J. Chen, T. Chang and Y. Chiang, "A V-band power amplifier using Marchand balun for power combining in 90n-nm CMOS process," in 2017 IEEE Wireless Power Transfer Conference (WPTC), 2017, pp. 1-3.

[5] Y. Chang, Y. Wang, C. -N. Chen, Y. -C. Wu and H. Wang, "A V-Band Power Amplifier With 23.7-dBm Output Power, 22.1% PAE, and 29.7-dB Gain in 65-nm CMOS Technology," IEEE Trans. Microw. Theory Techn., vol. 67, no. 11, pp. 4418-4426, Nov. 2019.

[6] T. -W. Li et al., "A V-Band Doubly Hybrid NMOS/PMOS Four-Way Distributed-Active-Transformer Power Amplifier for Nonlinearity Cancellation and Joint Linearity/Efficiency Optimization," in 2021 IEEE MTT-S International Microwave Symposium (IMS), 2021, pp. 382-385

First Demonstration of High PAE Performance Using InGaN Channel HEMT for 5G RF Applications

Hao Lu[1], Likun Zhou[2], Longge Deng[1], Ling Yang[1], Bin Hou[1], Xiaohua Ma[1], Yue Hao[1]

[1] State Key Discipline Laboratory of Wide Band-gap Semiconductor Technology, School of Microelectronics, Xidian University, Xi'an 710071, China

[2] School of Advanced Materials and Nanotechnology, Xidian University, Xi'an 710071, China

Abstract—The conventional GaN channel HEMT will suffer the significant short channel effect and high-temperature degradation due to its weak channel confinement. Although the InGaN channel double-heterostructure HEMT (DH-HEMT) has been reported to address this issue well due to the strong quantum confinement, the efficiency of the InGaN channel lacks investigation. In this work, a high PAE performance of the InGaN channel heterostructure has been reported for the first time. The fabricated InGaN channel device with a gate length of 200-nm achieved a high f_T/f_{max} of 36.7 and 97 GHz, respectively. 3.6 GHz continuous-wave load-pull measurements gain a high power-added efficiency (PAE) of 59.4 %, and an associated output power density (P_{out}) of 2.14 W/mm at V_{DS} = 20 V. It is the first time for the InGaN channel achieved so high PAE performance, the results presented here are benchmarked against the state-of-the-art (SOA) InGaN channel. This work illustrated that the InGaN channel with reasonable design can boost the 5G base-station applications.

Keywords—Group III-Nitrides, GaN, HEMT, InGaN channel, recess etching, power amplifier.

I. INTRODUCTION

5G has been widely infiltrated in every aspect of people's lives. GaN-based HEMT has attracted much attention in 5G applications with high frequency and high output power performance due to the superior natural material properties [1]. When targeting the high-frequency band, the conventional GaN HEMT will suffer the short channel effect (SCE), and will happen electrons spill over into the GaN buffer under high temperature/field stress [2]. These phenomena are due to the weak channel quantum confinement deriving from the similar material between the channel and buffer. Historically, the InGaN channel double-heterostructure HEMT (DH-HEMT) has been reported to address this issue well due to the strong quantum confinement due to the inverse piezoelectric field on the GaN buffer [3]. Moreover, attributing to the small electron mass and high saturation velocity, the InGaN is desired channel material among the group III-Nitrides.

In terms of the InGaN channel as a power amplifier for base-station applications, there are several reports on the output power characteristics. Adivarahan et al. reported a high output power performance using the AlGaN/InGaN/GaN MOSFET [4]. Sohel *et al.* employed the InGaN sub-channel to achieve X-band linearity performance with OIP3/P_{DC} of 9.7 dB [5]. In our early work, we also proposed a GaN/InGaN coupling channel architecture to gain DC and RF gain linearity and ultralow subthreshold slope characteristics [6], [7]. However, considering the giant energy consumption, the efficiency of the power amplifier needs to be deeply improved. To the best of the authors' knowledge, there are no reports on the high PAE of the InGaN channel.

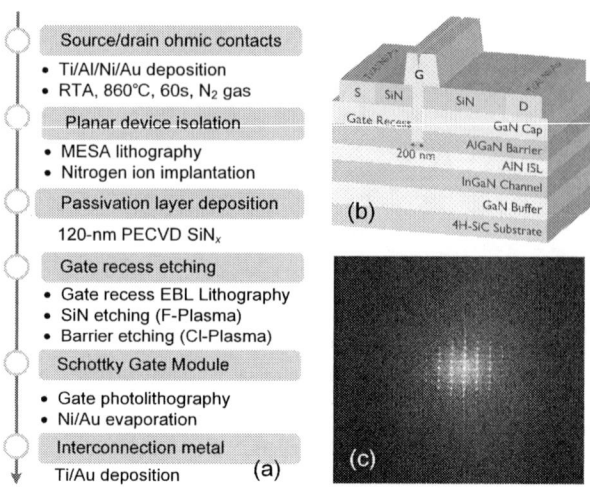

Fig.1 (a) Fabrication flow of the InGaN-channel DH-HEMT with gate-recess etching (b). Schematic diagram of the designed InGaN-channel device. (c) Fast-fourier transformation (FFT) of the local epitaxy area for the InGaN Channel heterostrucutre.

In this work, the high PAE performance of the GaN-based HEMT was achieved by the strong-confinement InGaN channel double-heterostructure high electron mobility transistor (DH-HEMT). Attributed to the strong quantum confinement of the InGaN channel, the fabricated device shows a high peak g_m of 0.27 S/mm. Moreover, the InGaN channel device with a gate length of 200-nm achieved a high current gain cutoff frequency f_T and maximum oscillation frequency f_{max} of 36.7 and 97 GHz, respectively. 3.6 GHz continuous-wave large signal measurements gain a high power-added efficiency (PAE) of 59.4 %, and an associated output power density (P_{out}) of 2.14 W/mm at V_{DS} = 20 V. Moreover, the transistor delivers a linearly f_{max} and high constant PAE over a wide range of drain bias variations. These excellent results have demonstrated the InGaN channel with careful device design could be an attractive platform to facilitate the GaN-based HEMTs for 5G wireless communication base stations.

II. GROWTH AND DEVICE FABRICATION

The proposed AlGaN/InGaN/GaN DH-HEMTs were fabricated based on a 4H-SiC substrate. The Epilayers are composed of an AlN nucleation layer, GaN buffer layer, unintentionally doped (UID) GaN layer, 5nm InGaN channel, AlN insert layer (ISL),20nm AlGaN barrier layer, and 2-nm GaN cap layer from down to top. The full width at half maximum (FWHM) of the epitaxial stacks for (002) and (102) measured by HR-XRD are 226 and 487 arcsec, respectively. As shown in Fig. 1(a), the overall fabrication flow started with the Ti/Al/Ni/Au source/drain ohmic contact, followed by annealing at 860 °C for 60 s in N_2 ambient. After the MESA lithography, the planar electrical isolation was achieved by nitrogen ion implantation. The two-terminal isolation test to investigate buffer leakage. The isolation leakage current depicted an ultra-low isolation leakage current density of < 1 µA/mm at V_{DS}=200 V for the source-drain spacing of 2 μm, indicating an excellent isolation effect and epitaxial growth quality. Subsequently, the device was deposited with 60 nm PECVD SiN_x, The gate foot defined by electron-beam lithography (EBL) was opened using CF_4-based and Cl-based plasma etching to enhance gate electrostatic control capability. Then the device was evaporated with Ni/Au metal stacks for

978-1-6654-9270-6/22 $31.00 © 2022 IEEE

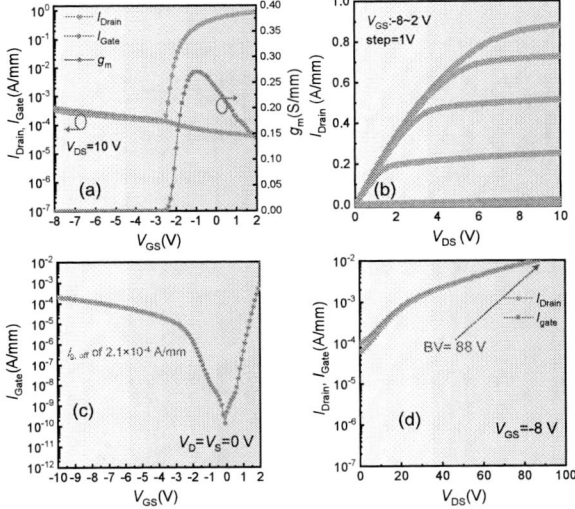

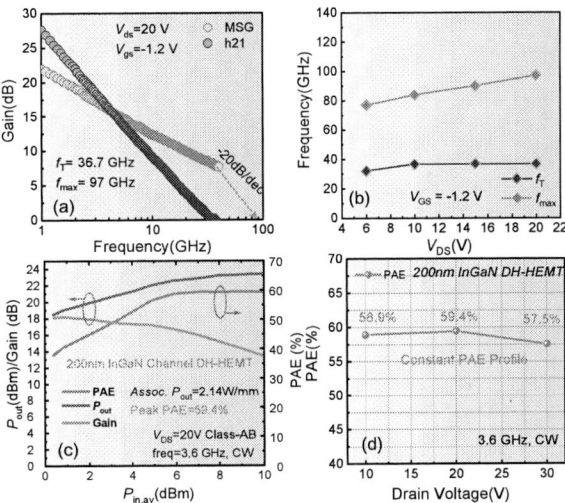

Fig. 2. The DC IV characteristics for the 200 nm InGaN channel DH-HEMT. (a) Transfer I-V, (b) output IV, (c) Schottky gate leakage, and (d) three-terminal off-state breakdown characteristics of the InGaN channel DH-HEMT.

Fig. 3. (a) RF small signal characteristics of the InGaN-channel DH-HEMT at V_{DS}=20V. (b) f_T and f_{max} as functions of drain voltage for the InGaN-channel DH-HEMT. (c) RF small signal characteristics of the InGaN-channel DH-HEMT at V_{DS}=20V. (d) PAE as a function of drain voltage for the 200-nm InGaN-channel DH-HEMT performed at 3.6 GHz.

the gate and Ti/Au metal stacks for the interconnection to accomplish the RF test. The gate length (L_g), gate-drain length (L_{gd}), and gate width (W_g) of the InGaN DH-HEMT are 0.2, 3.7, and 2 × 50 μm, respectively. The device after the overall process flow was illustrated in Fig. 1(b). Fig. 1(c) suggests the Fast-fourier transformation (FFT) image of the local epitaxy area for the proposed heterostructure. The well-defined crystalline structure illustrates the high growth quality of the epitaxial stacks.

III. RESULTS AND DISCUSSION

The static device performance was evaluated by Keysight B1500. As can be seen in Fig. 2(a), the transfer characteristics at V_{DS} = 10 V deliver a low leakage current of 3.7×10^{-4} A/mm and a high peak g_m of 0.27 S/mm. As shown in Fig. 2(b), the output I_D-V_{DS} characteristics of the fabricated devices show a reasonable saturation drain current of ~0.9 A/mm, and no obvious short channel effect was observed. Fig. 2(c) illustrates the Schottky leakage current performance, which indicates the main source of the off-state leakage current is the gate leakage current. As demonstrated in Fig. 2(d), the three-terminal off-state breakdown results suggest that the BV of the transistor is 88 V.

As demonstrated in Fig. 3(a), small-signal RF performances of the InGaN channel were evaluated at V_{DS} = 20 V and V_{GS} = -1.2 V. The InGaN channel HEMTs deliver an f_T of 36.7 GHz and f_{max} of 97 GHz, respectively. The undesired cut-off frequency characteristics of the InGaN channel DH-HEMT can be improved by shrinking the device dimension [8] in future work. As shown in Fig. 3(b), the f_T and f_{max} as functions of drain voltage for the InGaN-channel DH-HEMT. It can be seen that f_{max} linearly increased with the drain voltage increased, and f_T remain flat over a wide range of drain voltage variations.

A large signal load-pull measurement was performed to evaluate the power performance of the InGaN channel DH-HEMT with PAE-tune at 3.6 GHz. As illustrated in Fig. 3(c), peak PAE of 59.4% and associated P_{out} of 2.14 W/mm were measured for the InGaN DH-HEMTs at V_{DS} = 20 V. The excellent power performance is attributed to the suppressed gate leakage current and superior dispersion control. Fig. 3(d) illustrates the drain bias dependence of the PAE of the InGaN

channel DH-HEMTs. The InGaN channel DH-HEMTs show a high PAE plain with the variation of drain voltage, which means that the InGaN channel can operate at dynamic bias operation RF front-end-module (RF-FEM).

IV. CONCLUSION

In this paper, we proposed an InGaN channel double-heterostructure HEMT with high PAE performance of 59.4%. Attributed to the strong quantum confinement for the InGaN channel, the fabricated InGaN channel device with a gate length of 200-nm achieved a high f_T/f_{max} of 36.7 and 97 GHz, respectively. 3.6 GHz continuous-wave large signal measurements gain a high PAE of 59.4 %, and an associated output power density of 2.14 W/mm at V_{DS} = 20 V. It is the first time for the InGaN channel achieved so high PAE performance, the results presented here are benchmarked against the state-of-the-art (SOA) reports for the InGaN channel. This work illustrated that the InGaN channel with reasonable design can boost the 5G base-station applications.

REFERENCES

[1] H. Lu et al., "High RF Performance GaN-on-Si HEMTs With Passivation Implanted Termination," *IEEE Electron Device Letters*, vol. 43, no. 2, pp. 188-191, 2022.

[2] R. Gaska, et al., "Electron mobility in modulation-doped AlGaN–GaN heterostructures", *Appl. Phys. Lett.* 74, 287-289, 1999.

[3] T. Palacios, et al., "AlGaN/GaN high electron mobility transistors with InGaN back-barriers," in IEEE Electron Device Letters, vol. 27, no. 1, pp. 13-15, Jan. 2006.

[4] V. Adivarahan, et al., "Selectively Doped High-Power AlGaN/InGaN/GaN MOS-DHFET," *IEEE Electron Device Letters*, vol. 28, no. 3, pp. 192-194, 2007.

[5] S. H. Sohel, et al., "Polarization engineering of AlGaN/GaN HEMT with graded InGaN sub-channel for high-linearity X-band applications," *IEEE Electron Device Letters*, vol. 40, no. 4, pp. 522-525, 2019.

[6] H. Lu et al., "AlN/GaN/InGaN Coupling-Channel HEMTs for Improved g_m and Gain Linearity," *IEEE Transactions on Electron Devices*, vol. 68, no. 7, pp. 3308-3313, 2021.

[7] H. Lu, et al., "AlN/GaN/InGaN coupling-channel HEMTs with steep subthreshold swing of sub-60 mV/decade," *Appl. Phys. Lett.* 120, 173502, 2022.

[8] H. Lu et al., "Improved RF Power Performance of AlGaN/GaN HEMT Using by Ti/Au/Al/Ni/Au Shallow Trench Etching Ohmic Contact," *IEEE Transactions on Electron Devices*, vol. 68, no. 10, pp. 4842-4846, 2021.

2022 IEEE International Conference on
Integrated Circuits, Technologies and Applications

A 41-GHz 19.4-dBm P_{SAT} CMOS Doherty Power Amplifier for 5G NR Applications

Zheng Li, Zixin Chen, Qiaoyu Wang, Jian Pang,
Atsushi Shirane, Kenichi Okada

Tokyo Institute of Technology

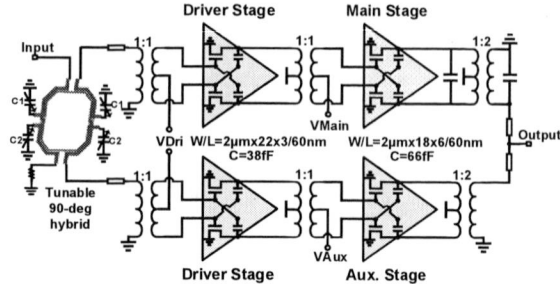

Fig.1. The block diagram of the proposed Doherty PA.

Abstract—In this paper, a 41-GHz Doherty power amplifier (PA) in a standard 65nm CMOS technology is presented for 5G New Radio (NR) applications. The PA implements transformer-based parallel-combined Doherty structure to enhance the power-added efficiency (PAE). And the tunable 90° hybrid is proposed for output phase compensation. This work achieves a saturated output power (P_{SAT}) of 19.4dBm and an OP1dB of 18.6dBm at 41.5GHz under 1-V power supply. The peak PAE and the PAE at 6-dB output power back-off (PBO) are 30.4% and 19.2%, respectively. The core chip area is 0.22 mm^2 with a static power consumption of 76mW.

Keywords—41GHz, Doherty, power amplifier, 5G New Radio, CMOS.

I. INTRODUCTION

The 5G New Radio has drawn a lot of attention due to its plenty of band resources and higher data rate compared with the sub-6GHz cellular network. As the recently licensed frequency band, 5G FR2 n259 band (39.5GHz~43.5GHz) has 4-GHz available bandwidth, which is adjacent to the n260 band (37GHz~40GHz). In mm-wave base station, beam-forming technique is necessary to achieve an adequate effective isotropically radiated power (EIRP) to compensate the increased free-space path loss (FSPL). Meanwhile, the high peak-to-average power ratio (PAPR) of 5G OFDMA-mode modulated signal limits the transmitter to operate at 6-dB or even more back-off point [1-2]. As the key component of the phased-array transmitter, the power-efficient PA with sufficient output power and good linearity performance is required. This work focuses on the n259 band PA design.

Considering the basic PA structure to improve the output power, the differential PA is better than 2-way single-ended in-phase-combined PA for additional common-mode rejection. And the biases are usually optimized at class-AB to obtain a good linearity feature. In order to further enhance the output power and PAE at deep PBO region, several solutions are available for mm-wave PA design, such as mixed-signal PAs, out-phasing PAs and Doherty PAs. The mixed-signal PAs need external digital controls, while the out-phasing PAs require additional signal separation and suffer from nonlinearity. As for Doherty PAs, instead of the bulky and lossy $\lambda/4$ transmission line, the transformer-based passive networks are selected together with the MOM capacitors to realize both impedance matching and load modulation.

In this work, a 41-GHz CMOS transformer-based parallel-combined Doherty power amplifier is introduced. With the tunable 90° hybrid, the output power combining phase mismatch can be compensated to enhance both P_{SAT} and PAE performance. Measured at 1-V power supply, a P_{SAT} of 19.4dBm and an OP1dB of 18.6dBm at 41.5GHz are obtained. The peak PAE and the PAE at 6-dB PBO are 30.4% and 19.2%, respectively.

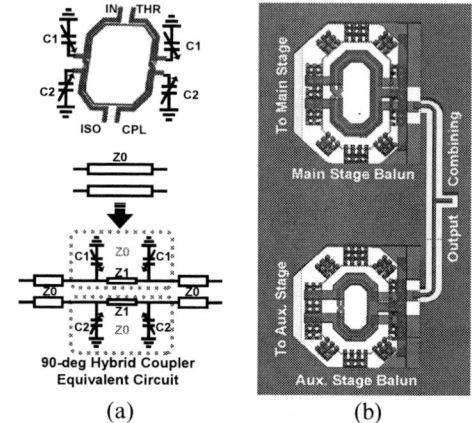

Fig.2. Passive components modeling: (a) tunable 90° hybrid and equivalent circuit; (b) output combining network.

II. PROPOSED 41GHz DOHERTY PA

The block diagram of the 41-GHz power amplifier is shown in Fig. 1. The amplifier consists of a tunable 90° hybrid and two symmetric amplifier paths. Each path is composed of a driver stage (W/L=2μm×22×3/60nm) and an amplifier stage (W/L=2μm×18×6/60nm), both realized by differential capacitive-neutralized common-source pair. The driver stages are biased at class-B for better AM-AM performance. In order to conduct the load modulation, the amplifier stages of the main path and the auxiliary path are biased at class-A and class-C mode, respectively.

The input signal is split by a tunable 90° hybrid as shown in Fig. 2(a). Differ from the conventional 90° hybrid, the tunable 90° hybrid has 3 sections for each coil to realize characteristic impedance of Z0, Z1 and Z0, respectively. 2 pairs of varactors are connected in shunt with the centre coil section to perform a Z0 together. In this case, the control bias VC1 and VC2 are set to 0, the outputs phase difference is 90°. While tuning the value of VC1 and VC2, a phase tuning range of 82° (VC1=1, VC2=0) ~102° (VC1=0, VC2=1) can be realized.

Following the tunable 90° hybrid, the 1:1 input baluns and 1:1 inter-stage transformers are used for input and inter-stage matching, respectively. Same passive sizes are chosen for both PA main path and auxiliary path, and optimized towards the best power gain for the main path. Considering the output matching network, 1:2 baluns are selected for impedance up-scaling. For main stage, the 1:2 balun is designed together with MOM capacitors to not only match the load pull Z_{MAIN} to 2 times of 50-Ohm output impedance, but also realize the load

978-1-6654-9270-6/22 $31.00 © 2022 IEEE

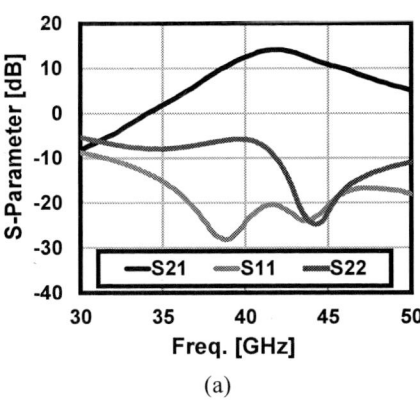

(a)

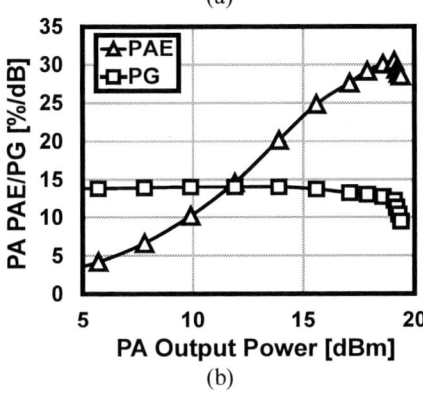

(b)

Fig.3. (a) Measured S-parameter (b) measured CW power gain and PAE vs. P$_{out}$ at 41.5GHz.

modulation. As for auxiliary stage, the 1:2 balun is also optimized for a high impedance to isolate the influence by auxiliary path when the input power is small. Both these passives are modelled by HFSS, and the model of output combining network is shown is Fig. 2(b).

III. MEASUREMENT RESULTS

The following measurements are conducted on wafer-level. At 1-V power supply, the static power consumption is 76mW. Fig. 3(a) shows the S-parameter characteristics measured by Keysight N5247A network analyser. The peak gain is 14.1dB at 41.5GHz with a -3dB bandwidth from 39GHz to 45GHz. As for large-signal measurements, the measured continuous-wave (CW) power gain and PAE versus Pout are given in Fig. 3(b) at 41.5GHz. The proposed PA achieves a 19.4-dBm P$_{SAT}$ and an 18.6-dBm OP1dB. The peak PAE reaches 30.4% and the PAE at OP1dB is 30.1%. When operating in the deep PBO region, this PA still obtains a PAE of 19.2% at 6dB PBO and 13.7% PAE at 8dB PBO, respectively.

Fig. 4 shows the die micrograph. The proposed PA is manufactured in a standard 65nm CMOS technology. The total chip size including pads is 0.89mm×0.61mm and the core chip area is 0.22mm².

TABLE I summarizes the measured performance of this work, compared with other 5G band PAs [3-5].

IV. CONCLUSIONS

This paper presents a 41-GHz Doherty PA fabricated in a standard 65nm CMOS technology. The total chip area is 0.89mm x 0.61mm with a compact core area of 0.22mm². By

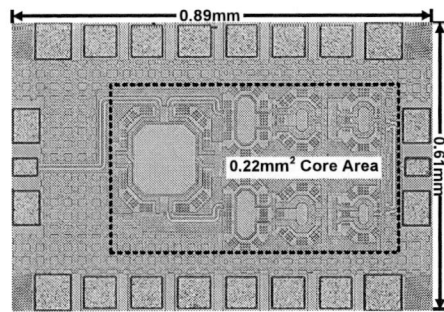

Fig.4. Die micrograph of proposed Doherty PA.

TABLE I. Performance comparison of mm-Wave PAs

	This Work	[3]	[4]	[5]
Technology	65-nm CMOS	65-nm CMOS	45-nm CMOS	130-nm SiGe
Freq. [GHz]	41	39	39	39
VDD Supply [V]	1.0	1.0	2.0	1.5
Topology	Doherty	Class-AB	Hybrid class-F/F^{-1}	Doherty
P$_{SAT}$ [dBm]	19.4	16.3	18.9	17.0
OP1dB [dBm]	18.6	14.9	17.4	15.4
Peak PAE [%]	30.4	31.2	36.0	21.4
PAE @6dB PBO [%]	19.2	16.0*	15.0*	12.6
Core Area [mm²]	0.22	0.08	0.21	0.21

* Estimated from the measurement.

taking the advantage of Doherty structure, a good back-off PAE performance with high linearity and compact area is achieved at 1-V power supply.

Acknowledgments

This work was partially supported by the Ministry of Internal Affairs and Communications in Japan (JPJ000254), JSPS, STAR, and VDEC in collaboration with Cadence Design Systems, Inc., Mentor Graphics, Inc., and Keysight Technologies Japan, Ltd.

REFERENCES

[1] Z. Li, et al., "A 39-GHz CMOS Bi-Directional Doherty Phased-Array Beamformer Using Shared-LUT DPD with Inter-Element Mismatch Compensation Technique for 5G Base-Station." *2022 IEEE Symposium on VLSI Technology and Circuits (VLSI Technology and Circuits)*, pp. 98-99, 2022.

[2] J Pang, et al, "21.1A 28GHz CMOS Phased-Array Beamforming Utilizing Neutralized Bi-Directional Technique Supporting Dual-Polarized MIMO for 5G NR," *IEEE International Solid-State Circuits Conference (ISSCC)*, pp. 344-346, Feb 2019.

[3] Y. Wang, R. Wu, and K. Okada. "A compact 39-GHz 17.2-dBm power amplifier for 5G communication in 65-nm CMOS." *2018 IEEE International Symposium on Radio-Frequency Integration Technology (RFIT)*, 2018.

[4] T-W. Li, M-Y. Huang, and H. Wang. "Millimeter-wave continuous-mode power amplifier for 5G MIMO applications," *IEEE Transactions on Microwave Theory and Techniques*, vol. 67, no.7, pp. 3088-3098, 2019.

[5] S. Hu, F. Wang, and H. Wang. "A 28-/37-/39-GHz linear Doherty power amplifier in silicon for 5G applications," *IEEE Journal of Solid-State Circuits*, vol. 54, no. 6, pp. 1586-1599, 2019.

A 120nW, 121kHz, -20∼100°C CMOS Relaxation Oscillator with Digital Current Comparator and On-Chip Voltage and Current Reference

Renwei Chen, Yifei Zhang, Chenchang Zhan
School of Microelectronics, Southern University ofScience and Technology, Shenzhen, China

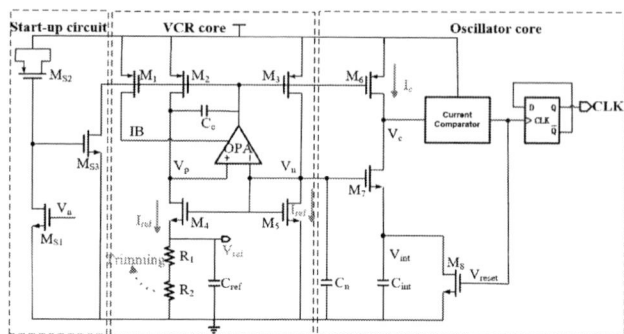

Fig. 1: Schematic of the proposed relaxation oscillator

Abstract—A relaxation oscillator(RO) composed of a voltage and current reference (VCR) and a digital current comparator is presented in this paper. By using two different types of resistors with opposite temperature coefficient (TC) in the VCR, this design successfully stabilizes the operation frequency from -20 to 100°C with a very simple structure. Designed in a standard 180nm CMOS process, the proposed RO consumes 120nW under a 0.8V supply and operates at 121kHz, with a TC as low as 36.2 ppm/°C.

Keywords—relaxation oscillator, digital current comparator, temperature coefficient(TC), low power.

I. INTRODUCTION

Relaxation oscillators have been widely used for reference clock generation in low-power Internet of Things(IoT) applications[1-3]. The potentially well-balanced performance in frequency stability, low power consumption and small area make them favorable when compared with ring oscillators or LC oscillators. The design in [1] adopts a concise structure that can work under low supply voltage using comparator delay for compensation. But it becomes less useful at high frequency range(e.g., higher than 100kHz) in which the delay of the comparator takes a larger proportion. [2] exploits the positive temperature coefficients(TC) of the RC delay and comparator delay to cancel the negative temperature coefficient(TC) of the SR-latch delay. [3] similarly uses a digital compensation loop to obtain good temperature stability. But both the TCs in [2] and [3] are relatively large. Two different reference offset compensation techniques are developed in [4] and [5], but their energy efficiency can be further improved.

In this paper, a low-power, temperature-independent RC oscillator is presented with an oscillator core based on a digital current comparator [6] and a voltage and current reference(VCR) that has two different resistors with opposite TCs. Thus, the oscillator can have a better reference voltage as well as reference current with adjustable temperature coefficient for delay compensation.

II. PROPOSED RELAXATION OSCILLATOR

Fig. 1 shows the schematic of the proposed relaxation oscillator composed of a start-up circuit, a VCR and a digital current comparator. The start-up circuit consists of M_{S1}-M_{S3}, which enables the VCR to work in normal operating state. A folded-cascode operational amplifier is inserted to provide negative feedback and let V_p be equal to V_n, whose detailed schematic is shown in Fig. 2. All transistors in VCR work in sub-threshold region to reduce quiescent current and generate the reference voltage. For a higher than ∼100mV V_{DS} of a MOSFET, the sub-threshold current I can be expressed as [2]:

$$I = \frac{W}{L} I_0 exp(\frac{V_{GS} - V_{TH}}{\eta V_T}) \quad (1)$$

This work is supported by NSFC under grant 62174080 and SZSTI under grant JCYJ20200109141225025. (Comesponding author: Chenchang Zhan.)

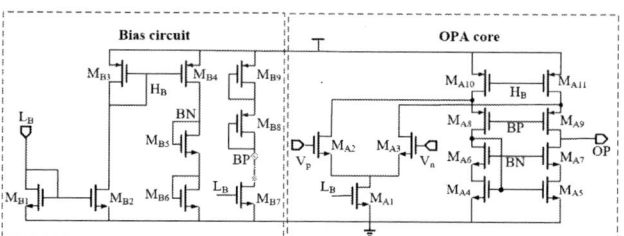

Fig. 2: Schematic of the OPA used in the VCR circuit

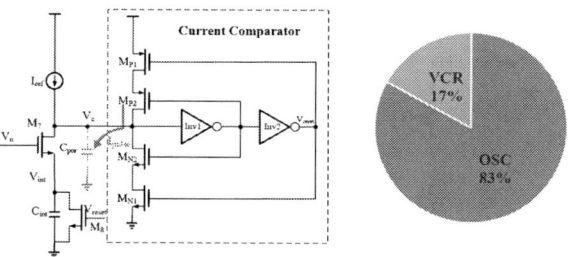

Fig. 3: Schematic of the digital current comparator [6] and power distribution of the proposed RO

where I_0 is a process-dependent parameter. η is the sub-threshold slope factor. $V_T(= k_B T/q)$ is the thermal voltage. Transistor M_4 has the same unit-cell size as M_5, and hence V_{TH4} is almost the same as V_{TH5}. Reference voltage V_{ref} can then be easily derived as:

$$V_{ref} = V_{GS5} - V_{GS4} = \frac{\eta k_B T}{q} \cdot ln[\frac{(W/L)_{M4}}{(W/L)_{M5}}] \quad (2)$$

Reference current I_{ref} then can be derived as:

$$I_{ref} = \frac{1}{R_0[1 + \alpha(T - T_0)]} \frac{\eta k_B T}{q} \cdot ln[\frac{(W/L)_{M4}}{(W/L)_{M5}}] \quad (3)$$

where R_0 is the sum resistance of R_1 and R_2. Its temperature coefficient α can be trimmed to provide suitable compensation, based on the different resistance ratio of R_1 and R_2 which have opposite TCs. The multiplier ratio of M_2, M_3 and M_6 is 1:1:4 for copying I_{ref}. Thus, the charging current I_c that pass through M_6 is equal to $4*I_{ref}$.

Fig. 3 shows the schematic of the digital current comparator [6]. Initially, V_{int} and M_7's drain voltage V_c are reset to zero after previous cycle and no dc consumption exists in the comparator. As I_{ref} charges C_{int}, the current pass through M_7 will decrease exponentially according to Eq (1). Thus

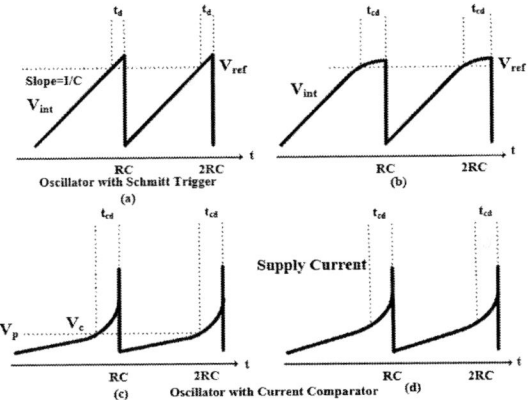

Fig. 4: Voltage and current waveform of coventional oscillator with schmitt trigger and proposed current comparator

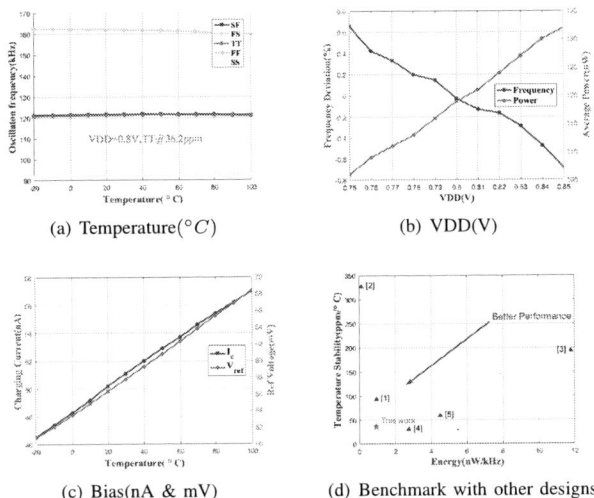

(a) Temperature($^\circ C$)

(b) VDD(V)

(c) Bias(nA & mV)

(d) Benchmark with other designs

Fig. 5: Simulation results

the current difference (I_{ref}-$I_{DS,M7}$) increases to charge the parasitic capacitance C_{par} and raises V_c exponentially.The output of Inv1 then changes from high to low and turns on M_{P2}. Due to the small delay of Inv2, M_{P1} and M_{P2} are on and charge V_c rapidly. Fig 4 compares the charging waveform of the coventional oscillator with normal OPA and Schmitt trigger in (a) and the proposed RO with VCR and digital current comparator in (b)(c)(d). Compared with the conventional design, the proposed RO not only consumes less switching losses and avoids dc loss from OPA, but also reduces comparator delay [6]. M_4 over M_7 has the same size ratio as M_2 over M_6 (both are 1:4) to ensure V_c rises close to V_P when V_{int} is charged to V_{ref} and will then flip the current comparator. The comparator delay and reset delay change with temperature, and their delay time's TC can be sensitive to power supply and process variations. Hence, in the proposed RO, resistors R_1 and R_2 with opposite TCs can be trimmed to make sure that the comparator delays' TC is compensated by the RC charging delay. After suitable compensation, the output clock frequency can be given by [6]:

$$F_{CLK} \simeq \frac{I_c}{2C_{int}V_{ref}} \simeq \frac{2}{R_0 C_{int}} \qquad (4)$$

III. SIMULATION RESULTS

The proposed relaxation oscillator is designed and simulated in a standard 180nm CMOS process. In Figure 5 (a), the output frequency is 121.2kHz with 36.2 ppm/$^\circ$C temperature coefficient from -20 - 100 $^\circ$C at TT corner and the total power consumption is 119.6nW under 0.8V supply. The worst temperature coefficient is simulated to be 148.2ppm/$^\circ$C at FF corner. Frequency deviation reaches near $0.7\pm$% when VDD changes from 0.75V to 0.85V in (b), which could be alleviated using a low-dropout regulator. (c) shows the PTAT trend of reference voltage and average bias current I_c versus temperature for frequency stabilization. (d) compares this work with other state-of-the-art designs with respect to energy efficiency and temperature stability.

Table I compares this relaxation oscillator with other state-of-art designs. The proposed design obtains a high energy efficiency and favorable temperature stability. It satisfied the requirement of low power consumption over a wide temperature range and with good frequency stability.

TABLE I. PERFORMANCE COMPARISON AND SUMMARY

	*This work	[1]	[2]	[3]	[4]	[5]
Process(nm)	180	180	180	180	180	180
VDD(V)	0.8	0.4	0.6	1.2	0.8	1.1
Frequency(kHz)	121	1.22	122	13400	128	1020
Temperature Range($^\circ$C)	-20-100	-20-70	-20-100	-20-100	-20-100	-40-125
T Stability(ppm/$^\circ$C)	**36.2**	94	327	193	30.6	59
V Stability (%/V)	12.8	17.2	6	N/A	7.7	3.3
Power(μW)	**0.120**	0.001	0.014	157.8	0.344	5.72
FOM(nW/kHz)	**0.99**	0.93	0.12	11.78	2.76	4.5

* Simulation Results

IV. CONCLUSION

A relaxation oscillator composed of a VCR and a digital current comparator is proposed in this work. The V_{ref} and I_{ref} generated by the VCR are PTAT with adjustable temperature coefficient. Combined with the digital current comparator, this design achieves a temperature stability of 36.2ppm/$^\circ$C with an energy efficiency of 0.99nW/kHz. This simple structure with good temperature stability proves to be suitable for the application in low cost SoC designs.

REFERENCES

[1] H. Jiang, P. P. Wang, P. P. Mercier and D. A. Hall, K. Huang and D. D. Wentzloff, *A 0.4-V 0.93-nW/kHz Relaxation Oscillator Exploiting Comparator Temperature-Dependent Delay to Achieve 94-ppm/$^\circ$C Stability,"* in *IEEE Journal of Solid-State Circuits*, vol. 53, no. 10, pp. 3004-3011, Oct. 2018, doi: 10.1109/JSSC.2018.2859834.

[2] Dai, Shanshan and Rosenstein, Jacob K, "*A 14.4nW 122KHz dual-phase current-mode relaxation oscillator for near-zero-power sensors,"* in *2015 IEEE Custom Integrated Circuits Conference (CICC)*, pp. 1-4, 2015, doi: 10.1109/CICC.2015.7338396.

[3] Chang, Yi-An and Liu, Shen-Iuan, "*A 13.4-MHz Relaxation Oscillator With Temperature Compensation,"* in *IEEE Transactions on Very Large Scale Integration (VLSI) Systems*, vol. 27, no. 7, pp. 1725–1729, 2019. doi: 10.1109/TVLSI.2019.2908204.

[4] Y. Liu et al., "*A Low-Power RC Oscillator with Offset and Path Delay Cancellation,"* in *2021 IEEE International Conference on Integrated Circuits, Technologies and Applications (ICTA)*, pp. 12–13, 2021, doi: 10.1109/ICTA53157.2021.9662022.

[5] C. Chen, C. Zhan, Z. Zhang, N. Zhang and L. Wang, "*A 5.72-W, 1.02-MHz, -40 125°C, 1s-Startup Time Relaxation Oscillation with Fully-on-Chip Voltage Reference and LDO Regulator,"* in *2021 IEEE International Conference on Integrated Circuits, Technologies and Applications (ICTA)*, 2021, pp. 237-238, doi: 10.1109/ICTA53157.2021.9661823.

[6] R. Magod, B. Bakkaloglu and S. Manandhar, "*A 1.24 μA Quiescent Current NMOS Low Dropout Regulator With Integrated Low-Power Oscillator-Driven Charge-Pump and Switched-Capacitor Pole Tracking Compensation,"* in *IEEE Journal of Solid-State Circuits*, vol. 53, no. 8, pp. 2356-2367, Aug. 2018, doi: 10.1109/JSSC.2018.2820708.

A 4.2-to-5.6 GHz Transformer-Based PMOS-only Stacked-g_m VCO in 28-nm CMOS

Mingkang Zhang, Zihao Zhu, Yueduo Liu, Zehao Zhang, Rongxin Bao, Jiahui Lin, Haoyu Zhuang, Jiaxin Liu, Xiong Zhou, Shiheng Yang, Qiang Li

Institute of Integrated Circuit and System, University of Electronic Science and Technology of China

Abstract—This paper presents a low power, low phase noise, and small chip area transformer-based PMOS-only stacked-g_m LC VCO in a 28-nm process. Both the top and bottom use PMOS cross-coupled pairs to provide negative resistances with suppressing the flicker noise. An interleaved transformer is used to get a higher coupling coefficient to achieve small chip area and high passive voltage gain. The VCO exhibits a wide turning range from 4.2 to 5.6 GHz, and low phase noise from -120 dBc/Hz to -113 dBc/Hz at 1MHz frequency offset, respectively. It consumes minimum 678 μW and maximum 680 μW from a 0.8V supply voltage which does not require internal LDO or DC-DC converter to reduce the supply voltage.

Keywords—Voltage Controlled Oscillator, PMOS-only stacked-g_m, transformer, low power, low phase noise, small chip area.

I. INTRODUCTION

With the development of short-distance communication technologies such as Internet of Things technology (IoT). It demands for low-power communication systems where the voltage-controlled oscillators (VCOs) play a crucial role. Besides the power requirements, the trade-off with other metrics such as the phase noise, tuning range, and small chip area in this high integration IC era renders the ultra-low-power VCOs a continuous research direction in the latest CMOS technologies [2].

This paper proposes a low power low phase noise and small chip area VCO with PMOS-only stacked-g_m transistors and a transformer with a high coupling coefficient to minimize the phase noise and power consumption at a standard supply voltage and the area of inductors. In addition, a 4-bit binary-sized switched capacitor array (SCA) is implemented to enlarge the frequency tuning range and a pair of varactors is used to realize fine tuning.

II. CIRCUIT DESIGN

A. Analysis of VCO

The traditional CMOS VCO with drain resistors topology is advantageous for obtaining a large loop gain in the current-limited region [3]. To achieve a low power consumption under a high supply voltage, large transconductance is required. However, the parasitic capacitance from the NMOS and PMOS transistor degrades the Q-factor of the VCO tank and the drain resistance limited the voltage swing. Thus, it is difficult to achieve a good performance with power and phase noise. An NMOS-only transformer-based stacked-g_m VCO is a promising topology for Ultra-low-power performance due to its large-g_m provided by both the top and bottom cross-coupled NMOS pair [1]. However, the 2:4 transformer-based inductor occupies a large area and the tail current source contributes a large noise factor to the circuit to reduce the phase noise performance.

In this paper, a PMOS-only stacked-g_m transformer-based VCO is proposed. As shown in the Fig. 1. The two pairs of PMOS cross-coupled transistors are stacked to form a "cascode-type" positive feedback which have a high loop gain to sustain

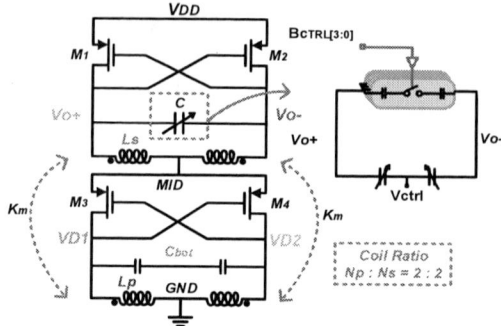

Fig. 1. The proposed transformer-based PMOS-only stacked-g_m LC VCO

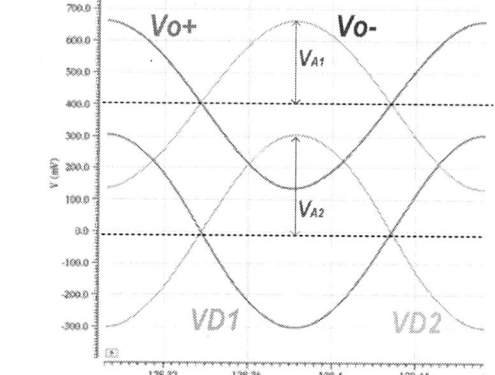

Fig. 2. Simulated transient voltage

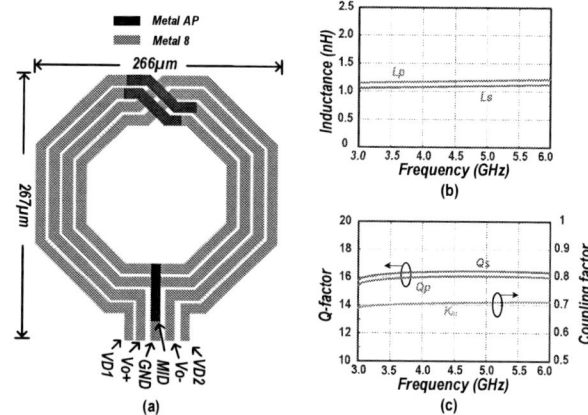

Fig. 3. (a) 2:2 transformer layout. (b) Transformer EM simulation results.

the oscillation. Compare to the way the resonators are paralleled which can lower phase noise [6], the series connection of the resonator enables lower power consumption. The transistor sizes in the top cross-coupled pair are 4.5 μm/30 nm and 4.2 μm/30 nm for the bottom cross-coupled PMOS pair. Compare with the ultra-low-power NMOS-only stacked-g_m VCO in [1], a smaller transistor is adopted here, a higher oscillation frequency can be achieved.

As shown in the Fig. 2, because of the same dc current in the circuit and the use of 2:2 transformer, it generates two equal voltage swings under a standard supply voltage. As derived in [4], the main oscillation frequency ω_{osc} of the tank can be expressed as

$$\omega_{osc}^2 = \frac{1}{L_p C_p (1 + |k_m|)}$$

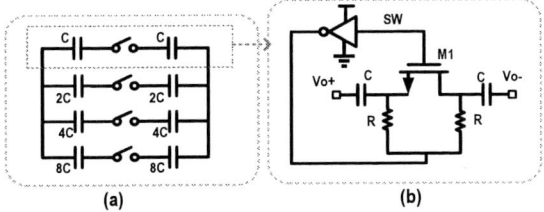

Fig. 4. (a) 4-bit switched capacitor bank. (b) Unit of switched capacitor bank.

Where C_p represent the bottom capacitance. As analyzed in [7], the main advantage of interleaved transformer is that we can achieve a high coupling coefficient and approximately one-half of the chip is required to realize the same self-inductance. Therefore, a small area transformer is designed which is smaller than the inductor in [2]. And the transformer-based resonator can provide a voltage gain of k_m, then we will get a larger passive voltage gain which can improve the phase noise and power consumption by decreasing the size of transistors. Overall, the figure of merit (FoM) of the VCO can be improved.

Compare to the NMOS-only stacked-g_m VCO in [1], the use of PMOS-only relaxes the noise requirement. PMOS has more flexibility and less flicker noise than NMOS. In addition, it eliminated the use of tail current source here, a lower noise factor contributes to the circuit can be achieved.

B. Design of Transformer

As shown in Fig. 3(a), a 2:2 transformer structure is adopted [7], there are only three cross sections exists in the transformer used to connect metals. In the EM simulation shown in the Fig. 3(b), There are two windings of inductances L_P and L_S, the primary winding (L_P) and the secondary winding (L_S) are 1.2 and 1.1 nH at 4.9 GHz, respectively. The magnetic coupling coefficient k_m is 0.76. The Q-factors of the primary winding and the secondary winding are 16.8 and 15.2 respectively. Both of the two coils realize a stable high self-inductance value and a high Q-factor over the full frequency tuning range. In addition, the transformer occupies $266\mu m \times 267\mu m$ area, a smaller area is achieved.

The transformer is implemented in 28-nm 1P8 CMOS process with ultrathick metal layer. The 2:2 transformer is constructed with the $3.3\mu m$ top ultrathick (M8 layer) copper metal.

C. Frequency Tuning

The switched capacitor array combined with a pair of varactors is used to tune the oscillator to achieve a wide tuning range without phase noise deterioration. Fig. 4(a) depicts the structure of the capacitor array. It consists of 4-bit binary weighted resistor biased switched capacitor array, which contributes to a coarse frequency tuning. Fig. 4(b) depicts the switched capacitor unit cell. The capacitor and switch sizes in the unit cell are 104 fF and 40 μm/30 nm, respectively. High-Q MOM capacitors are also used to compose of varactors to tune the frequency continuously.

III. SIMULATION RESULTS AND CONCLUSIONS

The proposed LC VCO are designed in 28-nm CMOS process. Fig. 5(a) depicts the frequency tuning range versus control voltage. Fig. 5(b) shows the simulated phase noise at f_{max} (f_{min}) with V_{DD} = 0.8 V. Power consumption is 678 μW (680 μW). The phase noise reaches -120 to -113 dBc/Hz at 1 MHz offsets, and the FOM is 195 to 190 dBc/Hz, respectively.

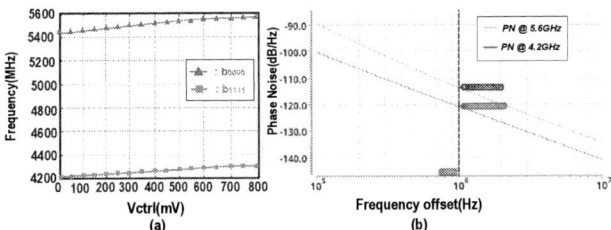

Fig. 5. Simulation results (a) oscillation frequency versus control voltage and (b) phase noise of VCO versus frequency at 4.2GHz and 5.6GHz.

TABLE I.   PERFORMANCE SUMMARY AND COMPARISON

Parameters	[2]	[3]	[1]	[5]	**This Work (simulation)**
Technology	65nm	65nm	65nm	180nm	**28nm**
Supply (V)	0.8	1.2	0.45	1.8	**0.8**
Frequency (GHz)	4.2-5.1	3.3	2.46	6.09-7.50	**4.2-5.6**
Tuning Range (%)	19	18	38.5	10.4	**29**
Power (uW)	349-297	720	107	2200	**680-678**
PN (dBc/Hz) @1MHz	-106.4/-110.1	-114.4	-107	-120^1	**-120/-113**
FOM^2 (dBc/Hz) @1MHz	183.4/189.4	187	184.5	189-191	**195/190**

^{1}PN@2MHz; ^{2} FoM $= -\text{PN} + 20log_{10}\left(\frac{f}{\Delta f}\right) - 10log_{10}(\frac{P_{dc}}{1mW})$

In conclusion, A VCO with low power, low phase noise, small chip area and wide frequency tuning range at a standard high supply voltage is presented. The VCO with PMOS-only stacked-g_m transistors is able to achieve low power and low phase noise. A wide tuning range is achieved with the help of a 4-bit binary weighted switched capacitor array and varactors. By using a 2:2 interleaved high-Q transformer, smaller area and higher loop gain can be achieved. The results of simulations and Table I show that this VCO achieves a competitive FoM better to that of other VCO designs operating at standard supply voltage.

ACKNOWLEDGMENT

This work is supported in part by the National Natural Science Foundation of China under Grant 62004028 and in part by the Department of Human Resources and Social Security of Sichuan Province under Grant 2021LXHGKJHD22.

REFERENCES.

[1] H. Liu *et al.*, "A 265-μw fractional-*N* digital PLL with seamless automatic switching sub-sampling/sampling feedback path and duty cycled frequency-locked loop in 65-nm CMOS," *IEEE J. Solid-State Circuits*, vol. 54, no. 12, pp. 3478–3492, Dec. 2019.

[2] R. Martins *et al.*, "Design of a 4.2-to-5.1 GHz Ultralow-Power Complementary Class-B/C Hybrid-Mode VCO in 65-nm CMOS Fully Supported by EDA Tools," in *IEEE Transactions on Circuits and Systems I: Regular Papers*, vol. 67, no. 11, pp. 3965-3977, Nov. 2020.

[3] F. Pepe, A. Bonfanti, S. Levantino, C. Samori, and A. L. Lacaita, "Suppression of flicker noise up-conversion in a 65-nm CMOS VCO in the 3.0-to-3.6 GHz band," *IEEE J. Solid-State Circuits*, vol. 48, no. 10, pp. 2375–2389, Oct. 2013.

[4] A. Mazzanti and A. Bevilacqua, "Second-order equivalent circuits for the design of doubly-tuned transformer matching networks," *IEEE Trans. Circuits Syst. I, Reg. Papers*, vol. 65, no. 12, pp. 4157–4168, Dec. 2018.

[5] A. Mazzanti and P. Andreani, "A Push–Pull Class-C CMOS VCO," in *IEEE Journal of Solid-State Circuits*, vol. 48, no. 3, pp. 724-732, March 2013.

[6] M. Tohidian, S. A. Reza Ahmadi Mehr and R. B. Staszewski, "Dual-core high-swing class-C oscillator with ultra-low phase noise," 2013 *IEEE Radio Frequency Integrated Circuits Symposium (RFIC)*, 2013, pp. 243-246.

[7] J. R. Long, "Monolithic transformers for silicon RFIC design," *IEEE J. Solid-State Circuits*, vol. 35, no. 9, pp. 1368–1382, Sep. 2000.

A Multiplying Delay-Locked Loop design with low jitter and high linearity

Jiahao Hu[1,2], Zhongxian Huang[1,2], Baoxing Duan[1], Qing Li[2], Ziqi Song[2], Dian He[2]

[1] College of Microelectronics, Xidian University
[2] Xidian-wuhu Research Institute

Abstract—In this paper, a Multiplying Delay-Locked Loop (MDLL) for high-precision Time to Digital Converter(TDC) is proposed, which has low jitter and high delay linearity. In order to reduce the phase noise, an internally compensated charge pump(CP) is used to achieve better current matching between charging and discharging.The improved reverse differential delay cell structure is used to improve the resolution of multi-phase clock. An MDLL with an output frequency of 80-240MHz and an area of 0.08mm² is realized by using 0.18um CMOS process. The test results show that the total power consumption under 1.8V power supply is 11.52mW@240MHz , RMS jitter is 10ps@240MHz。

Keywords—Multiplying Delay-Locked Loop(MDLL), Time to Digital Converter(TDC), low jitter, high linearity

I. INTRODUCE

In order to obtain better jitter performance in the clock generation circuit based on TDC application, lower static phase error is required. However, static phase error is usually limited by Phase Detector(PD) detection range and CP current mismatch. By expanding the pulse width to amplify the error, further compressing the static phase difference can effectively improve the linearity of multi clock phase [1]. Another way to solve this problem is to use the charge pump calibration technology to compensate the charge discharge mismatch current of CP [2].

In this work, the influence of non ideal factors is reduced by the methods of layout matching and key signal isolation and protection. The charge pump can reduce the loop noise by calibrating and compensating the mismatch caused by non ideal factors in the system. MDLL noise includes two kinds, one is the noise outside the loop, that is, the noise of the reference clock, and the other is the noise inside the loop, including CP, Loop Filter(LPF) and Voltage-Controlled Delay Line(VCDL). These two kinds of noise have a great impact on the final output jitter.

At the same time, to obtain lower jitter in the multi-phase clock generation circuit based on TDC application, a smaller loop bandwidth is required to reduce the current noise of the charge pump and offset the jitter of each phase clock of the DLL. Therefore, considering the layout area and loop response speed, the larger the filter capacitance, the better. In the second section, the overall architecture and specific circuit modules of the proposed MDLL are introduced. The third section analyzes the noise optimization in circuit design. The fourth section gives the test results and related analysis. The fifth section summarizes the full text.

II. MDLL ARCHITECTURE

A. The proposed MDLL with low jitter and high linearity

Figure 1 shows the architecture of the proposed MDLL. In each cycle, VCDL avoids MDLL jitter accumulation by multiplexing the jitter free reference edge of delay. Through logic control, the input channel of the first delay unit is changed

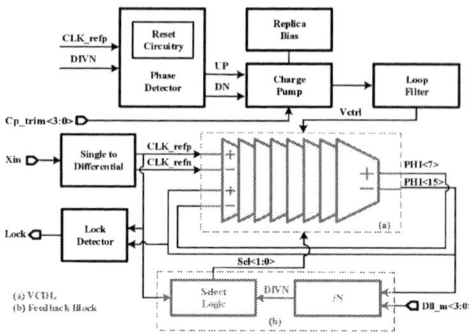

Figure 1. The system block of the proposed MDLL

to realize signal switching. Through the reverse differential delay unit structure of VCDL, the propagation delay of the signal is smaller than that of the buffer delay element, and the resolution is improved. At the same time, by minimizing the loop bandwidth, the charge pump can achieve very low in band phase noise, generate Vctrl and control VCDL until the system is locked.

B. Voltage-Controlled Delay Line

Figure 2 shows the schematic diagram of VCDL and vdelay cell. The delay unit is composed of a differential delay adjustable multiplexer unit with two input channels, and VCDL is composed of eight voltage controlled delay units, with a total of 16 phase outputs. Through the multi-channel selector MUX control, the external reference edge without jitter is gated into the loop before the start of each clock cycle, and the result of the delay chain is output in multiple phases.

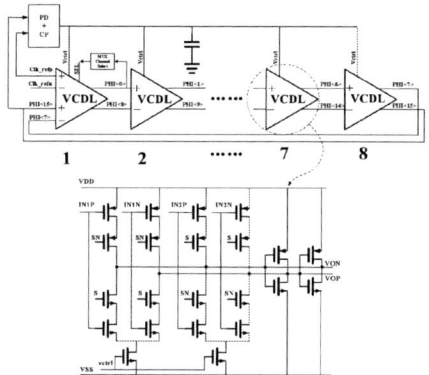

Figure 2. Schematics of the VCDL and delay cell

C. charge pump

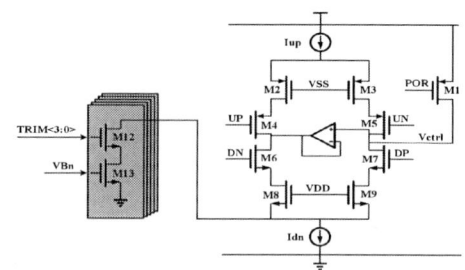

Figure 3. Schematic of CP

978-1-6654-9270-6/22 $31.00 © 2022 IEEE

Figure 3 is the circuit schematic of the CP. The right side is the basic charge pump unit structure, and the left side is the pull-down current calibration array. The charge pump unit is a current steering charge pump, which adopts an operational amplifier structure between the charge pump switch and the bootstrap switch, which can effectively suppress the charge sharing effect. The current calibration array can calibrate the current of the charge pump unit through the current source switch according to the trim<3:0> control signal, eliminate the charge and discharge mismatch of the charge pump unit, and realize the current calibration of -15uA~15uA.

III. MDLL CIRCUIT DESIGN AND NOISE ANALYSIS

A. Noise and bandwidth analysis of MDLL

DLL noise includes noise outside the loop and noise inside the loop. The noise of the reference clock outside the loop has all pass characteristics, and the setting of the loop bandwidth has no effect on it. The noise transfer function of CP in the loop is low-pass, and that of LPF and VCDL is high pass. For DLL applications that only use the last phase, the loop bandwidth should be a compromise between output jitter, tracking performance, and loop stability. For TDC applications, multi-phase clocks are used for output. Even if the bandwidth is increased, all the multi-phase clock noise of VCDL cannot be suppressed, so DLL only needs a small bandwidth to reduce CP current noise to reduce the jitter of each multi-phase. However, as the bandwidth decreases, the ability of the loop to track phase jitter will deteriorate.

Therefore, under the condition of ensuring the phase tracking ability and the layout area, the filter capacitance is increased as much as possible in the design to obtain a lower loop bandwidth and suppress loop noise.

B. Relationship between frequency locking range and noise

In VCDL, the delay of each stage delay unit is expressed in τ, the total delay of n-stage delay unit is $N\tau$, and the gain of VCDL is proportional to $\Delta N\tau/\Delta V_{ctrl}$.

Therefore, K_{VCDL} is proportional to VCDL series N, and the frequency locking range is inversely proportional to N. The loop bandwidth of DLL is proportional to $I_{CP}K_{VCDL}/C$. therefore, when I_{CP}/C is constant, the larger the frequency locking range, the smaller the loop gain, and the smaller the bandwidth, the better the noise performance. Therefore, the frequency locking range of VCDL should be increased as much as possible in order to obtain better noise performance.

IV. TEST RESULTS AND DISCUSSION

Figure 4 shows the layout and micrograph of the MDLL.The proposed MDLL is realized by 180nm CMOS process, with an area of 0.08mm². At 1.8V power supply, the power consumption is 11.52mW@240MHz , RMS jitter is 10ps@240MHz 。

Figure 5 shows the proposed 3D diagram of MDLL jitter and phase separation uniformity test, showing the influence of

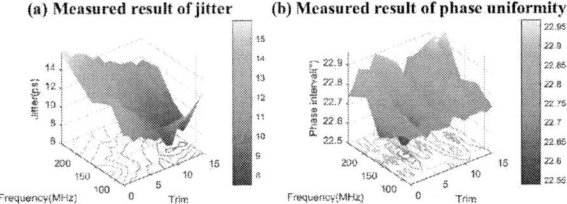

Figure 5. 3D diagram of MDLL jitter and phase separation uniformity test

charge pump trim on jitter and clock phase linearity.

Figure 5 (a) distribution shows that the trough of jitter distribution is the optimal value for CP mismatch calibration, while for higher frequency bands, there is no trough, and the jitter decreases monotonically with the increase of trim, indicating that the mismatch current of CP is not fully compensated due to the limitation of calibration range.

Figure 5 (b) shows that the phase separation uniformity distribution of DLL does not change significantly with the change of trim. That is, while suppressing the phase noise, the phase characteristics of DLL are not deteriorated, and the multi-phase clock still maintains high linearity.

Table 1 gives a detailed comparison with previous advanced DLLs [3][4][5].

TABLE I.   PERFORMANCE COMPARATION

	[3]	[4]	[5]	**This work**
Technology	65nm	180nm	65nm	180nm
Input frequency (Hz)	N/A	170M	87.5M	40M
Output frequency (Hz)	0.7-2G	0.4-1.2G	1.4175G	80-240M
RMS jitter (ps)	2.859@2 GHz	2.56@1.1 9GHz	2.8@1.41 75GHz	10@240M Hz
RMS jitter / Tout	0.0057	0.003	0.0039	0.0024
Current (mA)	3.31	12.4	6.7	6.4
Core area (mm²)	0.019	0.06	0.054	0.08

V. SUMMARY

This paper presents a MDLL with low jitter and high linearity.The analysis indicates that the output jitter is related to the loop bandwidth and the current mismatch of CP. The chip test results show that at normal temperature and 1.8V supply voltage, the output jitter is better than 16ps in the range of 80-240MHz, and the current mismatch of CP can be effectively reduced through CP calibration, so that the output jitter is less than 10ps.Compared with previous work, The MDLL can achieve better relative noise (jitter/T) and higher linearity. Therefore, the proposed MDLL is very suitable for the application of high-precision and high-resolution TDC.

REFERENCES

[1] Wu, Jin , et al. "Low-jitter DLL applied for two-segment TDC." *Iet Circuits Devices & Systems* 12.1(2018):17-24.

[2] X. Fu, K. El-Sankary and Y. Yadong, "A Pulse injection background calibration technique for charge pump PLLs," *2020 18th IEEE International New Circuits and Systems Conference (NEWCAS)*, 2020, pp. 98-101.

[3] T. A. Ali, A. A. Hafez, R. Drost, R. Ho and C. -K. K. Yang, "A 4.6GHz MDLL with −46dBc reference spur and aperture position tuning," *2011 IEEE International Solid-State Circuits Conference*, 2011, pp. 466-468.

[4] C. -S. Hwang, T. -L. Chu and W. -C. Chen, "A clock generator based on multiplying delay-locked loop," *2014 27th IEEE International System-on-Chip Conference (SOCC)*, 2014, pp. 98-102.

[5] S. Kundu, B. Kim and C. H. Kim, "19.2 A 0.2-to-1.45GHz subsampling fractional-N all-digital MDLL with zero-offset aperture PD-based spur cancellation and in-situ timing mismatch detection," *2016 IEEE International Solid-State Circuits Conference (ISSCC)*, 2016, pp. 326-327.

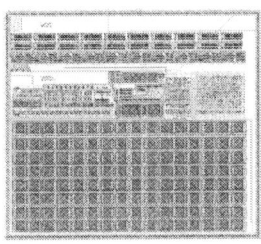

Figure 4. Layout and micrograph of proposed MDLL

Cooperative surface-activation strategy for low-temperature Cu/SiO₂ hybrid bonding

Qiushi Kang, Ge Li, Fanfan Niu, Chenxi Wang*

Harbin Institute of Technology

Abstract—Cu/SiO₂ hybrid bonding is a potent tool to effectively mitigate data-movement issues within von Neumann architecture due to the shortening of the distance between the processor and the memory unit. To protect stacked chip performance, the realization of hybrid bonding at low temperatures (<260 °C) is paramount. The essence of low-temperature hybrid bonding lies in the construction of desirable chemical structures on Cu and SiO₂ surfaces. Therefore, this paper presents two types of feasible surface-activation strategies to achieve selective/non-selective hydrophilization of the Cu/SiO₂ surface. Regardless of activation strategy, the Cu-Cu interface with sufficient grain growth and seamless amorphous SiO₂-SiO₂ interface structure were obtained at 200 °C. Moreover, the non-selective hydrophilization of Cu/SiO₂ surface based on Ar/O₂→NH₄OH activation realized interfacial layer-free SiO₂-SiO₂ interface, which can provide more reliable mechanical support for next-generation data-centric applications.

Keywords—hybrid bonding, low-temperature, surface activation, interface structure

I. INTRODUCTION

Currently, advanced computing systems and consumer electronics have an increasing demand for high-density and multi-functional integration between chips. Meantime, the energy consumption caused by massive data movement between the processor and the memory cannot be ignored. Thereby, three-dimensional (3D) integration based on microbumps and underfill came into being, which facilitates the development of Micron's Hybrid Memory Cube (HMC) and Hynix's High Bandwidth Memory (HBM) [1]. However, only a minimum interconnection pitch of 10 um can be achieved based on the conventional combination of microbumps and underfill. In this scenario, a Cu/SiO₂ hybrid bonding technology is proposed by Ziptronix Inc., which can scale the pitch down to <1 um by replacing microbumps and underfill with Cu-Cu and SiO₂-SiO₂ direct bonding [2]. Nowadays, Samsung, TSMC, and CEA Leti are dedicated to integrating chiplets via hybrid bonding, providing unprecedented 3D stacking flexibility [3].

To maintain optimal chip performance, the realization of Cu/SiO₂ hybrid bonding at low temperatures (<260 °C) is necessary [4]. The essence of low-temperature hybrid bonding lies in the construction of desirable chemical structures on Cu and SiO₂ surfaces. Regarding the SiO₂-SiO₂ bonding, surfaces terminated by hydrophilic functional groups are mandatory. Contrastingly, the atomically clean surfaces are suitable for Cu-Cu direct bonding due to the removing of diffusion barrier. Meantime, He et al [5] proved that the hydrophilic Cu surface also has potential for low temperature bonding. Therefore, this paper presents two feasible surface-activation strategies to selectively/non-selectively hydrophilize the Cu/SiO₂ surface. Regardless of activation strategy, the Cu-Cu interface with sufficient grain growth and seamless amorphous SiO₂-SiO₂ interface structure were obtained at 200 °C. Based on chemical affinity measurement and X-ray photoelectron spectroscopy (XPS) test, the mechanism of selectively/non-selectively establishment of -OH groups on surfaces was illustrated.

Corresponding Author: Chenxi Wang. Email: wangchenxi@hit.edu.cn

II. EXPERIMENTS

A. Materials

The standard damascene process was adopted to fabricate Cu/SiO₂ hybrid bonding samples. The 500-nm-thick Cu connections were defined via electroplating and lithography on SiO₂ surfaces. Eventually, the Cu connections and SiO₂ dielectrics were planarized via chemical mechanical polishing.

B. Methods

Regardless of activation strategy (as shown in Fig. 1), the activation approaches were based on the Ar/O₂ (0.2%) plasma activation for 120 s followed by solution immersion. To selectively hydrophilize hybrid surface, the plasma-activated surfaces were immersed in formic acid (FA, 50%) for 20 min (denoted as Ar/O₂→FA). In contrast, the Ar/O₂ plasma-activated surfaces were treated by NH₄OH (26.5±1.5%) for only 120 s to construct -OH groups non-selectively (denoted as Ar/O₂→NH₄OH). After activation, the samples were bonded at a pressure of 5 MPa for 30 min under atmosphere conditions, and a strengthened process was followed at 200 °C for 2 h.

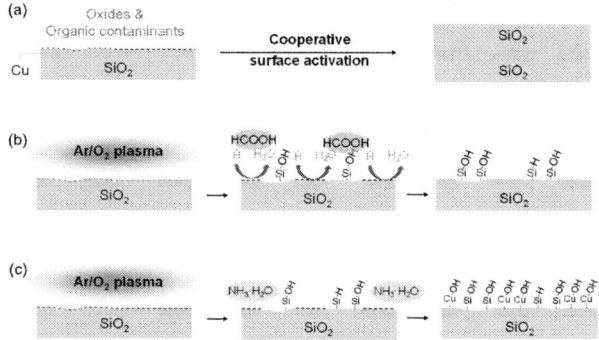

Fig. 1. Schematic diagram of (a) hybrid bonding and surface-activation processes of (b) Ar/O₂→FA and (c) Ar/O₂→NH₄OH.

III. RESULTS AND DISCUSSION

A. Surface energy characterization

The surface terminated with functional groups will result in the variation of surface energy, which can be evaluated by the water contact angle (CA) test. As listed in Table I, the CA of the Cu surface activated by Ar/O₂→FA decreased from 86.5° to 46.8°, which was close to that of the oxide-free Cu surface (~45°) [6]. Similarly, the surface energy of the SiO₂ substrate was improved after Ar/O₂→FA activation. It turns out that oxide-free Cu surface and hydrophilic SiO₂ surface were achieved simultaneously by Ar/O₂→FA activation. Nevertheless, the CA of the Cu surface further decreased to 19.6° after Ar/O₂→NH₄OH activation, indicating the formation of hydrophilic structures (such as CuO or Cu(OH)₂). Meantime, the CA of the SiO₂ surface activated by Ar/O₂→NH₄OH dropped to 2°, which illustrated that hydrophilic chemical structures were formed on both Cu and SiO₂ surfaces.

TABLE I.   WATER CONTACT ANGLE

Substrate	Cooperative surface activation		
	Bare	*Ar/O₂→FA*	*Ar/O₂→NH₄OH*
Cu	86.5°	46.8°	19.6°
SiO₂	59.1°	35.2°	<2°

B. Chemical composition investigation

Although the surface chemical affinity can be demonstrated via the CA test initially, it is necessary to confirm the chemical composition in detail. Therefore, the O 1s spectra of Cu and SiO_2 obtained via XPS are displayed in Fig. 2. The bare Cu surface was covered by Cu-O, Cu-OH and organic contaminants. Regardless of the activation method, oxides were always presented on the Cu surface due to the short-term storage in ambient air before XPS test. However, the $C-O/H_2O$ peak could still be detected after $Ar/O_2{\rightarrow}FA$ activation due to the organic acid solution. In addition, the relative content ratio of Cu-OH increased from 67.9% to 78.1% after $Ar/O_2{\rightarrow}NH_4OH$ activation, which is consistent with the increase in surface energy. The SiO_2 surface consisted of stable Si-O-Si and dangling Si-O bonds. It should be noted that unstable Si-O bonds will spontaneously aggregate into Si-O-Si under a high-vacuum environment of the XPS test. Therefore, the higher the proportion of Si-O-Si bonds on the surface, the higher the density of Si-O dangling bonds in essence. On the results shown in Fig. 2 (b), two activation approaches could break Si-O-Si bonds effectively, showing the potential for high-quality bonding. Notably, $Ar/O_2{\rightarrow}NH_4OH$ activation could introduce more hydrophilic functional groups on the surface, which was in very good accordance with surface energy results.

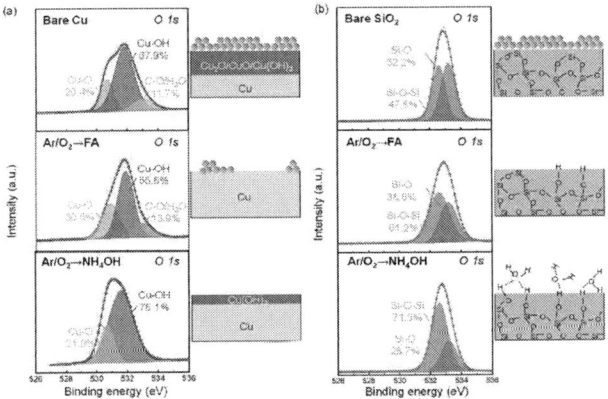

Fig. 2. XPS O 1s core-level spectra of (a) Cu and (b) SiO_2.

C. Interface structure and performance

The above results have proved that $Ar/O_2{\rightarrow}FA$ and $Ar/O_2{\rightarrow}NH_4OH$ activations realized selectively/non-selectively hydrophilization of Cu/SiO_2 hybrid surface, respectively. Herein, SEM and TEM observations were performed to evaluate interfacial structures.

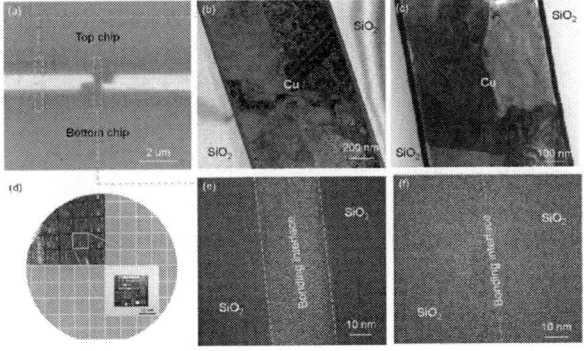

Fig. 3. (a) Cross-sectional image of Cu/SiO_2 hybrid bonding interface observed by SEM and (d) image of hybrid bonding samples. TEM observations of homogeneous bonding interface activated by (b)(e) $Ar/O_2{\rightarrow}FA$ and (c)(f) $Ar/O_2{\rightarrow}NH_4OH$ are also shown.

As displayed in Fig. 3, the intimate hybrid bonding interfaces were attained based on two activation strategies, and sufficient grain growth occurred at Cu-Cu interfaces. It is noteworthy that seamless SiO_2-SiO_2 bonding structures were achieved, but an amorphous interfacial layer was formed at $Ar/O_2{\rightarrow}FA$-activated interface. The composition of the interfacial layer was examined as the enrichment of carbon, as presented in Fig. 4. Albeit the tensile strength of SiO_2-SiO_2 was sufficient to provide mechanical support (~ 4 MPa), the enriched carbon nanolayer might have a negative effect on interfacial performance.

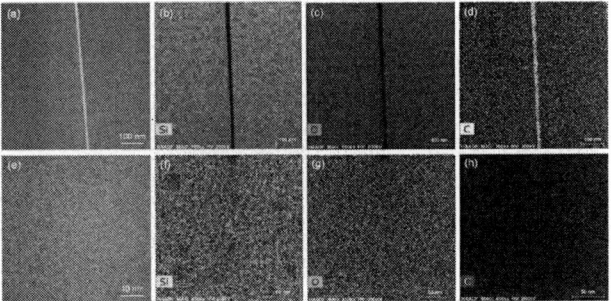

Fig. 4. The elemental analysis across the SiO_2-SiO_2 bonding interface obtained by (a-d) $Ar/O_2{\rightarrow}FA$ and (e-h) $Ar/O_2{\rightarrow}NH_4OH$.

IV. CONCLUSION

This paper proposes two feasible activation strategies to selectively/non-selectively hydrophilize Cu/SiO_2 hybrid surface, and the high-quality hybrid bonding was successfully achieved at 200 °C. The Cu-Cu interface with sufficient grain growth and seamless amorphous SiO_2-SiO_2 interface structure were observed by TEM. In addition, an enriched carbon nanolayer was formed at the selectively hydrophilized SiO_2-SiO_2 interface, which might have a negative effect on interfacial performance. Therefore, the non-selectively hydrophilization strategy can provide a more reliable Cu/SiO_2 hybrid bonding structure for next-generation data-centric applications.

ACKNOWLEDGMENT

This work was supported by the National Natural Science Foundation of China (Grant No. 92164105 and 51975151), the Heilongjiang Provincial Natural Science Foundation of China under grant LH2019E041, and the Heilongjiang Touyan Innovation Team Program (HITTY-20190013).

REFERENCES

[1] Singh G, et al. "Near-memory computing: Past, present, and future," *Microprocessors and Microsystems*, pp. 102868, 2019.

[2] Enquist P, et al. "Low cost of ownership scalable copper direct bond interconnect 3D IC technology for three-dimensional integrated circuit applications," *IEEE International Conference on 3D System Integration*, IEEE, 2009.

[3] Jouve A, et al. "1μm Pitch direct hybrid bonding with< 300nm wafer-to-wafer overlay accuracy," *IEEE SOI-3D-Subthreshold Microelectronics Technology Unified Conference (S3S)*, IEEE, 2017.

[4] Ko, C. T.; Chen, K. N. "Low Temperature Bonding Technology for 3D Integration," *Microelectron. Reliab.* vol. 52, no. 2, pp. 302–311, 2012.

[5] He R, et al. "Combined Surface-Activated Bonding Technique for Low-Temperature Cu/SiO_2 Hybrid Bonding,". *ECS Transactions*, vol. 69, no. 6, pp. 79-88, 2015.

[6] Tu S H, et al. "Time-varying wetting behavior on copper wafer treated by wet-etching," *Appl. Surf. Sci*, vol. 341, pp. 37-42, 2015.

A DLL-Based Offset Calibration Loop Technology for Wake-Up Receivers

Yuhang Xie[1], Xufeng Liao[1,2], Xincai Liu[1], Lianxi Liu[1,2]

[1] School of Microelectronics, Xidian University, Xi'an, 710071, China
[2] Xidian University Chongqing IC Innovation Institute, Research Building 1, Xiyong Microelectronics Park, Chongqing, 401331, China.

Abstract—A DLL-based offset calibration loop (OCL) is proposed to eliminate the DC offset and low-frequency flicker noise of the two differential paths to optimize the input signal-to-noise ratio before signal demodulation. The loop technology that can effectively calibrate the offset reduces the false alarm rate of the wake-up receiver (WuRX), and improves the sensitivity and robustness. This design uses 65nm LP CMOS process for layout design and simulation verification. With a supply voltage of 0.4V, the DC offset voltage on the signal path is reduced from an initial 5mV to a calibrated 39µV, resulting in a total system power consumption of 7.4nW.

Keywords—offset calibration, robustness, delay-locked loop, wake-up receiver (WuRX)

I. INTRODUCTION

To reduce the power consumption of the wireless sensor network, WuRX is a commonly used solution. WuRX's structure with an envelope detector (ED) as the first stage is popular because it does not require circuit modules with high power consumption such as RF amplifiers. The single-ended ED has the problem that when the input signal is small, the existence of offset may cause the signal to be overwhelmed. If a differential ED is used, although the sensitivity and conversion gain can be improved, the common-mode voltage between the output ports of the ED is different because the forward-biased and reverse-biased diodes have different channel resistances. The offset voltage is larger than the minimum detectable signal at the ED output. It increases the false alarm rate and decrease the sensitivity. In view of the above problems, this paper proposes an offset calibration loop based on the delay locked loop (DLL) to eliminate the DC offset and low-frequency flicker noise of the two differential paths, which can effectively reduce the DC offset voltage and optimize the input signal-noise-ratio before signal demodulation.

II. WURX CIRCUIT OVERVIEW

Fig.1 shows the overall framework for WuRX. The RF signal is amplified by the baseband amplifier after envelope detection. The DC offset of the two differential paths is eliminated by the offset elimination loop. After that, the demodulated signal is oversampled and digitized by the regenerative comparator time domain comparator (TDC). The output of the comparator is fed to the digital correlator, which is compared with the reference codebook on the node to decide whether to output the wake-up enable signal.

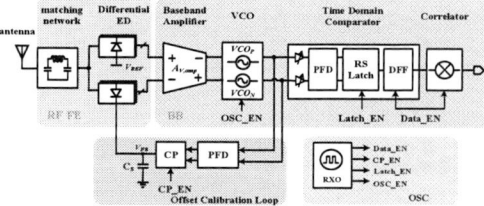

Fig.1. The overall scheme of the WuRX designed in this paper

III. DLL-BASED OFFSET CALIBRATION LOOP

A. Envelope Detector Offset

Fig.2 shows the N-stage differential Dickson ED. This architecture is sensitive to the DC offset of the ED and baseband amplifiers. This makes WuRX insensitive without special design considerations. Figure 2 illustrates this problem, where a forward-biased diode is placed in series with the input resistance R_{in} of the baseband amplifier. The negative branch consists of a reverse biased diode in series with the input resistor. Forward and reverse biased diodes have different channel resistances, which result in a non-zero ED offset voltage between $V_{ED,p}$ and $V_{ED,n}$. The diode resistance is a function of bulk bias voltage V_{bulk} and temperature, as shown in Figure 2(b) and (c), respectively. According to the simulation, the offset voltage varies between 10 and 500 µV, which is significantly larger than the smallest detectable signal at the ED output. It increase the false alarm rate and decrease the sensitivity

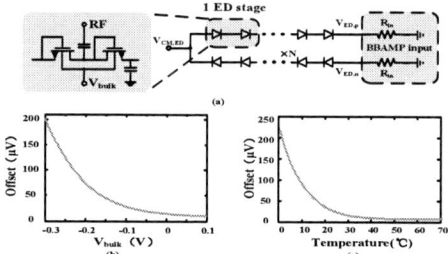

Fig.2. (a) Circuit model of the offset problem of the ED. Simulation results of offset between (b) substrate voltage (c) temperature

B. Loop Offset Cancellation Principle

In order to solve the problem, a DLL-based offset calibration loop and replica ED structure are proposed, as shown in Fig.3. We connect all ED outputs to the input of the baseband amplifier and design the same replica ED as the signal path ED, therefore, they should produce the same DC output voltage, i.e. $V_{ED,p/n}=V_{ED,p/n,replica}$. The signal path ED is connected to the RF input through the AC coupling capacitor, while the replica ED is not connected to the RF input. Therefore, unique RF input signal information exists at $V_{ED,p}$ and $V_{ED,n}$. The effect of the DC offset is equivalent to the positive input of a fully differential baseband amplifier. The DLL works as follows: During the WuRX startup, the pre-charged module charges the storage capacitor C_S until its voltage reaches V_{REF}; during the calibration phase, S_1 and S_3 are turned off, while S_0 and S_2 are turned on. The DC offset V_{OS} is amplified by the baseband amplifier, and then processed by the voltage controlled oscillator (VCO) and phase/frequency detector (PFD) in the time domain. If $V_{OS}>0$, charge pump (CP) charges C_S, and the voltage of C_S rises to $V_{REF}+V_{OS}$ in several sampling cycles. The offset voltage is stored in C_S; during the comparison phase, S_1 and S_3 are turned on, and S_0 and S_2 are turned off. Because the DC voltages at the positive and negative inputs of the fully differential baseband amplifier are equal, the offset is calibrated.

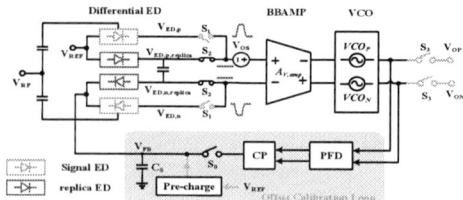

Fig. 3. Circuit structure of replica ED and delay locked loop

C. Stability Analysis of Loop

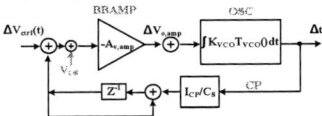

Fig.4. Multi-domain model of the DLL-based offset calibration loop

The multi-domain model of the DLL-based offset calibration loop is shown in Fig.4. At the k-th sampling period, the output of the fully differential baseband amplifier is:

$$\Delta V_{o,amp}(t) = -A_{v,amp}\left\{\Delta V_{CTRL}(t) + V_{OS}[(k-1)T_S]\right\} \quad (1)$$

$V_{OS}[(k-1)T_S]$ is the offset voltage of the (k-1)-th sampling period, and $\Delta V_{CTRL}(t)$ is the input voltage difference of baseband amplifier .

At the end of each integration period, the offset calibration loop is turned on for 2/8 T_S, during which time the pulse width of the two VCDL outputs is

$$\Delta t_l[kT_S] = \int_{(k-2/8)T_S}^{kT_S} K_{VCO}T_{VCO}\Delta V_{o,amp}(t)dt \quad (2)$$

A separate frequency and phase detector is used to detect $\Delta t_l[kT_S]$, and a charge pump with load capacitance C_S is used to feed back to the common-mode reference node of the differential passive ED, offset calibration The loop sets $\Delta t_l = 0$ at DC, thus forming a delay locked loop, which sets the amplifier input DC potential to V_{REF} while also biasing baseband amplifier. The output of the charge pump can be expressed as:

$$V_{OS}[kT_S] = V_{OS}[(k-1)T_S] + \frac{I_{CP}}{C_S}\Delta t_l[kT_S] \quad (3)$$

I_{CP} is the charge and discharge current of the charge pump.

The z-domain transfer function from sig(z) to $\Delta t_l[z]$ is:

$$\frac{\Delta t_l(z)}{sig(z)} = \frac{1-z^{-1}}{1-(1-\beta)z^{-1}} \quad (4)$$

where β is the loop gain, and its value is:

$$\beta = \frac{1}{4}A_{v,amp}K_{VCO}T_{VCO}T_S\frac{I_{CP}}{C_S} \quad (5)$$

The function region of convergence is:

$$|z| > |(1-\beta)| \quad (6)$$

The refion of Convergence (ROC) must contain the unit circle |z|=1 for stability, so 0<β<1. Accordingly, the current of the charge pump and the size of the load capacitance can be set in our design. The charge pump charge and discharge current is 60 pA, and the offset capacitor C_S is 30 pF.

IV. SIMULATION RESULTS AND DISCUSSION

The proposed WuRX is designed in 65nm CMOS process. Fig 5 shows the layout of the proposed WuRX with core area of 0.82×0.36 mm².

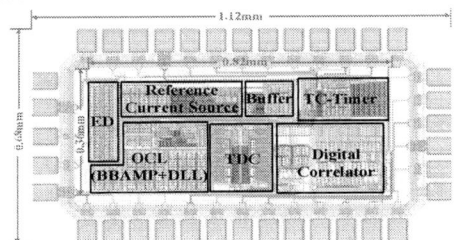

Fig.5. Overall layout of the designed WuRX chip

A 5mV DC offset voltage is superimposed on the positive input terminal of the baseband amplifier, and the simulation of

the DLL loop is performed. The DLL working process is shown in Fig.6. The result shows that after 12 sampling periods, the voltage difference at the input of the fully differential baseband amplifier decreases from an initial 5mV to 39μV.

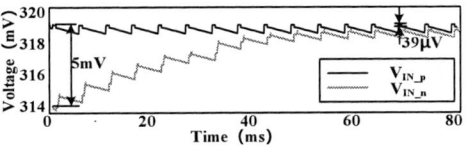

Fig.6. Transient simulation results of DLL working process

Table I shows the performance tested by the proposed WuRX and state-of-the-art works. This design has better FOM[1][2]. The WuRX sensitivity performance is improved[1][2][3] thanks to an offset calibration loop with DC offset cancellation and low frequency flicker noise suppression.

TABLE I. PERFORMANCE SUMMARY AND COMPARISON OF WuRXs

References	This work	[1] JSSC'19	[2] JSSC'20	[3] ISSCC'20
Process(nm)	65	130	65/180	65
Frequency (MHz)	915	433	9600	9600
Supply voltage(V)	0.4	1/0.6	0.4	N/A
Transfer rate(bps)	100	200	33	20
Correlator	16 bit	8 bit	36 bit	63 bit
Power (nW)	3.6	7.6	7.3	N/A
Delay (ms)	80	82.5	540	3150
P_{sen} (dBm)	-77.2	-71	-64	-65
$P_{sen,norm}$ (dBm)	-82.6	-76.4	-65.3	-62.5
FOM(dB)	-166.8	-157.6	-146.7	N/A
Area (mm²)	0.3	1.95	0.14	3

FoM (dB) $= P_{sen,norm} + 10\log(P_{DC}/1\text{W})$.

V. CONCLUSION

Aiming at the problem of the output offset of the ED, a DLL-based offset calibration loop is proposed to eliminate the offset of the ED. This paper uses 65nm process to realize the circuit and layout of the proposed WuRX, and completes the post-simulation verification. The results show that in the DLL, the PFD output pulses are fed back through the charge pump to the input reference node of the ED, eliminating the DC signal due to any DC offset introduced by the baseband signal processing circuitry at the receiver input, which reduces false alarm rate and improves sensitivity and robustness.

ACKNOWLEDGMENT

This work was supported by National Natural Science Foundation of China (62131010), Natural Science Basic Research Program of Shaanxi (2022JC-39), Natural Science Foundation of Chongqing (cstc2021jcyj-msxm3684) and Chongqing Talents Program (CQYC20210301367). The authors would like to thank the Xidian University Chongqing Integrated Circuits Innovation Institute for their technical support.

REFERENCES

[1] J. Moody, P. Bassirian, A. Roy et al. Interference robust detector-first near-zero power wake-up receiver [J]. IEEE Journal of Solid-State Circuits, 2019, 54(8): 2149–2162.

[2] H. Jiang, et al. A 22.3-nW, 4.55 cm² Temperature-Robust Wake-Up Receiver Achieving a Sensitivity of -69.5 dBm at 9 GHz [J]. IEEE Journal of Solid-State Circuits, 2020, 55(6):1530-1541.

[3] P. Bassirian, D. Duvvuri, D. S. Truesdell et al. 30.1 A Temperature-Robust 27.6nW −65dBm Wakeup Receiver at 9.6GHz X-Band [C]. IEEE International Solid- State Circuits Conference (ISSCC), 2020:460-462.

978-1-6654-9270-6/22 $31.00 © 2022 IEEE

An Efficient FPGA Design for Fixed-point Exponential Calculation

Weiyi Zhang, Chun Zhang, Liting Niu, Fasih Ud Din Farrukh, Hanjun Jiang*

School of Integrated Circuits, Tsinghua University

Abstract—**Exponential calculation is widely used in different algorithms, such as the activation functions of artificial neural networks. However, it is hard to implement on FPGA, consuming much time and resources. In this work, a novel exponential calculation module for fixed-point number is proposed based on the theory of Fast InvSqrt. The proposed exponential unit achieves at most 3.7x throughput while the resource utilization is largely reduced compared with previous works. The efficiency and accuracy are suitable for different applications.**

Keywords—FPGA, exponential, fast inverse square root

I. INTRODUCTION

Mathematical operations such as addition and multiplication are hardware-friendly, making it possible for various algorithms to be accelerated by FPGA. For example, the acceleration for neural networks has been widely researched, bringing great potential to the real usage on lightweight devices [1][2]. However, there are operations hard to implement in FPGA such as exponential calculation. The exponential calculation is used in areas such as activation functions of neural networks. The main stream methods for exponential are based on CORDIC, which get the result by iterations [3][4]. However, iterative computing is time and resource consuming. A novel exponential calculation module is proposed based on the theory of Fast InvSqrt[5] in this work. The throughput improved at most to 3.7x compared with related works while the resource utilization is largely reduced.

II. BACKGROUND KNOWLEDGE

Fast InvSqrt leverages the storage format of floating-point numbers as shown in Fig.1. The actual value of a floating-point number can be calculated as (1), where $e_x = E_x - B$ represents the number of shifting bits and m_x is the actual value of the significand part ranging from 0 to 1 when considered as a fixed-point number with no integer bit and 23 fraction bits. The value of the bias B is set as 127 in IEEE 754 standard. When m_x is relatively small, the logarithm of x can be estimated by (2), where σ denotes the estimation error. Denoting the value of the significand part as M_x when considered as an unsigned integer, we have $m_x = M_x/L$, where $L = 2^{23}$. In this work, we denote I_γ ($\gamma = x, y, ...$) as the value of the 32-bit floating-point number γ when considered as an unsigned integer. Thus, I_γ and γ actually share the same storage space but represent different value. The relationship between I_x and $\log(x)$ can be derived as (3). Finally, the estimation of $\log(x)$ is calculated as (4). Given $y_1 = 1/\sqrt{x}$, (5) is derived. According to (4) and (5), the first estimation of $y_1 = 1/\sqrt{x}$ can be calculated as (6). The estimation error σ is set as 0.0450466 [5]. Finally, the second estimation can be calculated by Newton method as (7).

$$x = 2^{e_x}(1 + m_x) \tag{1}$$

$$\begin{aligned}\log(x) &= e_x + \log(1 + m_x) \\ &= e_x + m_x + \sigma\end{aligned} \tag{2}$$

$$\begin{aligned}I_x &= E_x L + M_x \\ &\approx L \log(x) + L(B - \sigma)\end{aligned} \tag{3}$$

Supported by the National Natural Science Foundation of China (No.U20A20220).

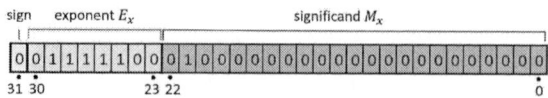

Fig.1. Storage format of floating-point numbers

$$\log(x) \approx \frac{I_x}{L} - (B - \sigma) \tag{4}$$

$$\log(y_1) = -\frac{1}{2}\log(x) \tag{5}$$

$$I_{y_1} = -\frac{1}{2}I_x + \frac{3}{2}L(B - \sigma) \tag{6}$$

$$y_2 = y_1 * (1.5 - 0.5 * x * y_1 * y_1) \tag{7}$$

III. PROPOSED DESIGN

To simplify the task, only the input ranging from 0 to 1 is calculated because given a rational number x as input, it can always be decomposed as the sum of one integer and one fraction. The integer part can be calculated by multiplication directly. For cases where the exponential base equals to 2^N, the multiplication can be further simplified into logic shifting.

A. Theoretical derivation

The proposed design comprises of two stages: first estimation based on storage format of floating-point number, and the second estimation based on Newton method. For the first stage, (8) can be derived directly given $y = 2^x$. According to (3), I_{y_1} can be approximated by (9). By replacing $\log(y)$ with x, the relationship between I_{y_1} and x can be represented as (10), where y_1 is the first estimation of y. Then Newton method is executed based on y_1. Firstly, the calculation is rewritten as (11), where the zero point of $f(y)$ is the result of 2^x. The calculation of y_2 is derived as shown in (12). The logarithm calculation of $\ln(y_1)$ is hard to implement directly. Noting that the range of x is restricted to [0,1] and y_1 is limited to [1,2], $\ln(y_1)$ can be approximated by a piecewise linear function $g(x)$. The design of the piecewise function trades off between the accuracy and the implementation efficiency (speed and resources). There are two factors influencing the performance of the linear function: the number and position of the breaking points. Increasing the number of breaking points will improve the accuracy, while consume more hardware resources and clocks to execute. The selection of the breaking points will also influence the accuracy. Given N breaking points, the optimal positions of breaking points can be calculated as (13). In this work, N is set as 1 to achieve the highest throughput and least resource utilization. One-norm is used to measure the difference between $\ln(x)$ and $g(x)$. The calculation of the breaking point x_1 is shown as (14), and the optimal position is $x_1 = 1.431$. Thus $g(x)$ is derived as (15).

$$\log(y) = x \tag{8}$$

$$I_{y_1} \approx L \log(y_1) + L(B - \sigma) \tag{9}$$

$$I_{y_1} = (x + B - \sigma) * L \tag{10}$$

$$f(y) = \log(y) - x \tag{11}$$

$$y_2 = y_1 - \frac{f(y_1)}{f'(y_1)} = y_1[1 - \ln(y_1) + x\ln(2)] \tag{12}$$

$$x_1, x_2, ..., x_N = \underset{x_1, x_2, ..., x_N}{\mathrm{argmin}}\left(\mathrm{norm}\big(\ln x - g(x)\big)\right) \tag{13}$$

$$x_1 = \underset{x_1}{\mathrm{argmin}}\left(\int_1^2 (|\ln x - g(x)|)\,\mathrm{d}x\right) \tag{14}$$

$$g(x) = \begin{cases} 0.831x - 0.831 & (1 \le x \le 1.431) \\ 0.588x - 0.483 & (1.431 < x \le 2) \end{cases} \tag{15}$$

B. Theoretical generalization

The proposed design is originally targeted for the base 2, while cases for other bases can be derived. The calculation of $y = e^x$ is designed as an example of generalization. Given $y = e^x$, (16) can be derived. Thus, the first and second estimation can be derived as shown in (17) and (18). Because y_1 ranges from 1 to e, the piecewise linear function is extended into (19). To ensure the accuracy, when input x is larger than 0.8, the second estimation will execute the Newton method twice.

$$\log(y) = x * \log(e) \qquad (16)$$

$$I_{y_1} = [x * \log(e) + B - \sigma] * L \qquad (17)$$

$$y_2 = y_1 * [1 - \ln(y_1) + x] \qquad (18)$$

$$g(x) = \begin{cases} 0.831x - 0.831 \, (1 \le x \le 1.431) \\ 0.588x - 0.483 \, (1.431 < x \le 2) \\ 0.427x - 0.162 \, (2 < x \le e) \end{cases} \qquad (19)$$

C. Hardware implementation

The proposed design for 2^x is shown in Fig.2. The fixed-point input x will firstly be converted into a fixed-point number with 9 bits for integer and 23 bits for fraction (fixed <32,9>), which is denoted as I_x. The conversion ensures the fraction bit number of I_x equals to the length of significand of floating-point number, so that it can be operated as a floating-point number. The bias B operates on bits [30,23] while the error σ operates on bits [22,0], which is in accordance with Fast InvSqrt. Meanwhile, the early type conversion minimizes the accuracy drop resulted from the truncated error of σ. Then I_x goes through a fixed <32,9> adder ADD1 and is added by constant $(B - \sigma)$. The value of $(B - \sigma)$ is 126.9549534 as shown in Fig.2(a). The result of ADD1 is y_1 in floating-point format. Then y_1 is converted into fixed <32,16> format for the second estimation. The design of second estimation is shown as Fig.2(b). Firstly, y_1 is input into logarithm sub-unit for $\ln(y_1)$ which is shown in Fig.2(c). The comparator CMP decides the interval of the y_1. Different k and b are selected by multiplexers according to the interval. And the fixed <32,16> result $\ln(y_1) = ky_1 + b$ is calculated by the multiplier MUL2 and adder ADD3. Meanwhile, $[1 + x\ln(2)]$ is calculated by the fixed <32,16> multiplier MUL1 and <32,16> adder ADD2. The subtractor SUB calculates the difference of $[1 + x\ln(2)]$ and $\ln(y_1)$. The result from SUB is multiplied by y_1 in multiplier MUL2 to get the second estimation y_2.

The proposed design for e^x is shown in Fig.3. The result of type conversion, I_x, is multiplied by $\log(e)$ as shown in Fig.3(a). The multiplier in the second estimation is reduced according to (18). In addition, the logarithm sub-unit is extended to support 3

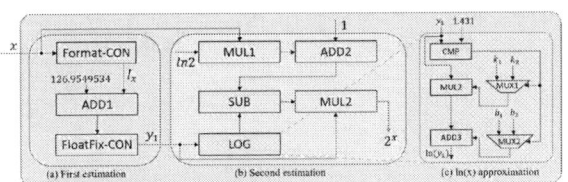

Fig.2. Proposed hardware design for 2^x

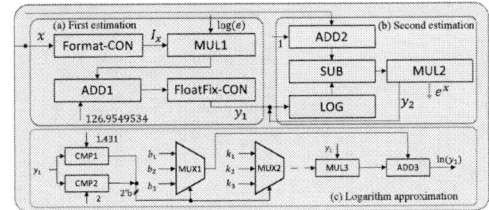

Fig.3. Proposed hardware design for e^x

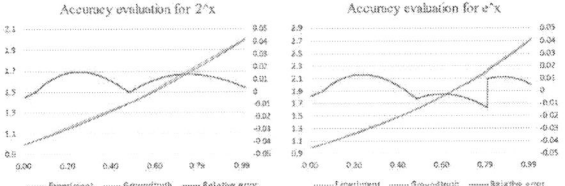

Fig.4. The accuracy of proposed design

TABLE I. EFFICIENCY AND RESOURCE OF PROPOSED DESIGN

	proposed 2^x	proposed e^x	Ref [3]
DSP48E	12	20	6
FF	319	535	5562
LUT	1227	1810	5507
Clock cycles	3	4~5	13
Frequency (MHz)	151.2	151.2	177.8
Throughput (exp/s)	50.4M	30.2~37.8M	13.7M

intervals. The result y_2 is input to the second estimation to perform the Newton method again for when x is larger than 0.8.

IV. EXPERIMENT RESULTS

The proposed design is synthesized by Vivado High Level (HLS) Synthesis 2018.3 and implemented on Ultra96-v2 platform at 151.2MHz. The format conversion of fixed-point number, and the conversion from floating point number to fixed point number are achieved by Vivado HLS APIs. Figure.4 shows the accuracy of the proposed design, where the calculated value, the ground truth value, and the relative error are illustrated. The design for 2^x has a relative error within 1.6%, while the design for e^x has a relative error within 1.3%. The accuracy of both modules is capable of general usage. The relative error of 2^x has two continuous intervals, while e^x has three, which is in accordance with the number of intervals of the piecewise function. Table.1 shows the resource utilization and efficiency. The efficiency is measured by exponential calculation per second, which is calculated as ratio of frequency and execution cycle. The proposed design for $y = 2^x$ achieves 50.4M exp/s, which is 3.7x of Ref [3]. The module for $y = e^x$ achieves 30.2~37.8M exp/s, because when the input ranges from 0.8 to 1.0, one more Newton iteration is used for higher accuracy. The throughput is at most 2.76x of Ref [3]. The utilization of flip-flops (FFs) and look-up tables (LUTs) is also largely reduced.

V. CONCLUSION

In this work, a novel exponential calculation module is proposed base on theory of Fast InvSqrt. The throughput improves to at most 3.7x compared with reference works while the resource utilization is reduced. The accuracy of the design ensures the application for different areas, and the design theory also inspires for modules of other functions.

REFERENCES

[1] Nguyen, Duy Thanh, et al. "A high-throughput and power-efficient FPGA implementation of YOLO CNN for object detection." IEEE Transactions on Very Large Scale Integration (VLSI) Systems 27.8 (2019): 1861-1873.

[2] Zhang, Weiyi, et al. "A portable accelerator of proximal policy optimization for robots." 2021 IEEE International Conference on Integrated Circuits, Technologies and Applications (ICTA). IEEE, 2021.

[3] Rekha, Ramesh, and Karunakara P. Menon. "FPGA implementation of exponential function using cordic IP core for extended input range." 2018 3rd IEEE International Conference on Recent Trends in Electronics, Information & Communication Technology (RTEICT). IEEE, 2018.

[4] Sudha, J., et al. "A novel method for computing exponential function using cordic algorithm." Procedia Engineering 30 (2012): 519-528.

[5] Lomont, C. (2003). Fast inverse square root. Tech-315 nical Report, 32.

2022 IEEE International Conference on Integrated Circuits, Technologies and Applications

A 6-18GHz Low-Noise Amplifier Using Noise Canceling Technique in 130-nm CMOS PD-SOI

Jialong Xue[1], Tenghao Zou[1], Hao Xu[1], Tingting Han[2], Mi Tian[2], Weiqiang Zhu[2], Zhijian Li[2], Na Yan[1,3]
[1]State Key Laboratory of ASIC & System, Fudan University, Shanghai 201203, China
[2]Nanjing Electronic Equipment Institute (NEEI), Nanjing 211103, China
[3]National Integrated Circuits Innovation Center, 201210, China

Abstract—This paper presents the design of a 6-18GHz low-noise amplifier (LNA) utilizing noise canceling technique to achieve large bandwidth and low noise figure (NF) simultaneously. The LNA is composed of three stages, resistive shunt feedback cascode topology is adopted for the first one, which is convenient for wideband input impedance matching. Besides, the second and third stage are designed for noise canceling and gain compensation respectively. Inductive peaking technique is employed to broaden the bandwidth. Implemented in 130-nm CMOS PD-SOI technology, the proposed LNA achieves maximum 15.44dB gain and minimum 2.42dB NF with flatness of ±1.44dB and 0.109dB/GHz respectively across 6-18GHz, whose fractional bandwidth is as large as 100%.

Keywords—noise canceling, low-noise amplifier (LNA), inductive peaking, silicon on insulator (SOI)

I. INTRODUCTION

Satellite communication has the advantages of large communication range, high reliability and multiple access application, the most commonly used frequency bands for which are C-band (4-8GHz), X-band (8-12GHz) and Ku-band (12-18GHz), Ka-band (27-40GHz) has also been widely adopted in recent years. A fully integrated wideband receiver covering several frequency bands is appealing for academia and industry.

As the first active block of the receiver, the performance of LNA determines the whole system's noise, sensitivity and dynamic range. How to design a wideband, high gain, low noise, high linearity LNA with given power consumption is the key issue.

A number of topologies have been proposed to trade off between these requirements. Common gate (CG) amplifier is available for wideband input matching due to its input impedance of $1/(g_m + sC_{gs})$ [1]. However, the gain and noise figure of CG amplifier are limited by its transconductance g_m. Inductive source degeneration is one of the most commonly used LNA topology, which can achieve excellent proximity of noise matching and power matching [2]. There are several techniques for broadening the bandwidth of this narrowband LNA, such as filter-type input matching network [3] and additional feedback [4], both at the price of extra noise introduced by passive devices. Noise canceling is a distinctive technology which is known for breaking the tradeoff between input reflection coefficient S_{11} and NF [5].

II. CIRCUIT IMPLEMENTATION

The schematic of the proposed noise canceling LNA is illustrated in Fig. 1(the bias circuit is omitted). Three stages work together to compensate for a good gain flatness. Several techniques are utilized for bandwidth extension.

This work was supported by the project fund (2019-JCJQ-ZD-232-00) of Nanjing Electronic Equipment Institute (NEEI).
(Corresponding Author: Na Yan, email: yanna@fudan.edu.cn)

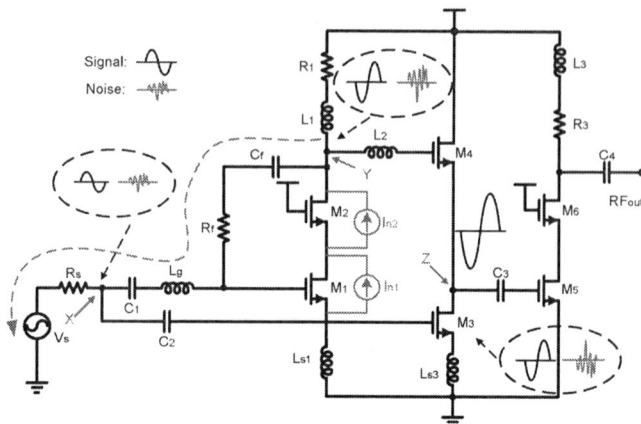

Fig. 1: The schematic of the proposed LNA

A. Noise Canceling Technique

As shown in Fig. 1, the thermal noise current $I_{ni}(i = 1, 2)$ of M_1 and M_2 flows from drain of M_2 to gate of M_1 and eventually signal source V_s, generating noise voltage of same polarity at node X and Y. On the other hand, owing to the amplification characteristic of cascode, the polarity of signal voltage at node X is opposite to that at node Y. As a result, if voltages of signal and noise at X are inversely amplified by M_3, meanwhile, voltages at Y are converted to currents by source follower (SF) M_4, eventually the currents of M_3 and M_4 are added together at node Z, it can be illustrated that signal is superimposed and noise is cancelled.

It should be noted that the perfect noise canceling condition occurs at a single frequency point, and the auxilliary transistors M_3 and M_4 itself contribute extra noise. To suppress NF at the whole working frequency band, in this work, the optimum noise canceling is implemented at the highest frequency of 18GHz, which is shown in Fig. 4(b) on the next page.

B. Bandwidth Extension Techniques

The calculated fractional bandwidth of this 6-18GHz LNA is 100%, which is too large to implement using traditional topologies. As a result, several bandwidth extension techniques are utilized in this work.

Source degenerative inductors L_{s1} and L_{s3} make the point of noise matching and power matching as close as possible. L_g is constantly used in narrowband LNA for impedance matching. To broaden the working frequency band, feedback resistor R_f is added between drain of M_2 and gate of M_1 for reducing quality factor of the input matching network, whose value is $1k\Omega$ to trade off between matching and noise.

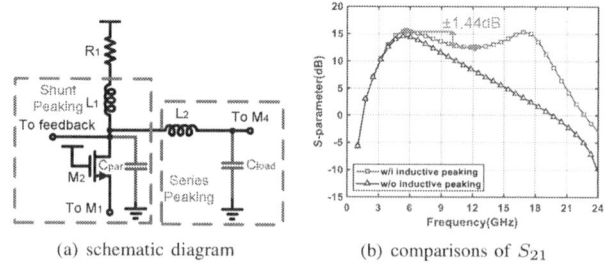

(a) schematic diagram

(b) comparisons of S_{21}

Fig. 2: Inductive peaking technique

TABLE I. PERFORMANCE COMPARISON

Reference	Technology	Freq.(GHz)	FBW(%)[1]	Max. Gain(dB)	Min. NF(dB)	IP_{1dB}(dBm)	P_{DC}(mW)	$FoM_1{}^2$	$FoM_2{}^3$
This Work	130nm CMOS PD-SOI	6-18	100	15.44	2.42	-14.11	55.44	0.39	563
[1]	65nm CMOS	17-22	24.5	14.9	3.3	-24	1.9	0.07	31
[3]	55nm CMOS	6.5-12	59.4	20.7	3.26	-12	75	0.29	343
[6]	150nm pHEMT	3.7-10.5	95.7	11	1.8	-12	45	0.22	160
[7]	45nm CMOS SOI	26.5-30	12.4	23	2.7	-19.2	28	0.043	100

^{1} $FBW = (BW/f_c) \cdot 100\%$, where BW is the abbreviation of bandwidth and f_c is the center frequency.
^{2} $FoM_1[GHz] = (Gain_{lin} \cdot FBW \cdot BW_{[GHz]} \cdot IP_{1dB[mW]})/((F-1) \cdot P_{dc[mW]})$
^{3} $FoM_2[GHz] = (Gain_{lin} \cdot FBW \cdot BW_{[GHz]})/(F-1)$, where the F is the noise factor.

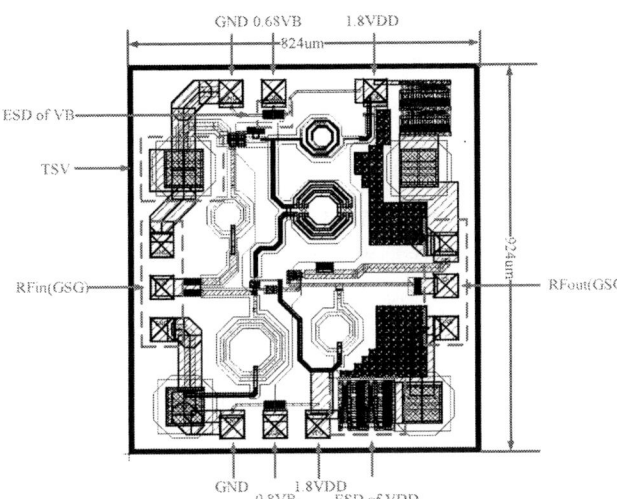

Fig. 3: Layout of the LNA

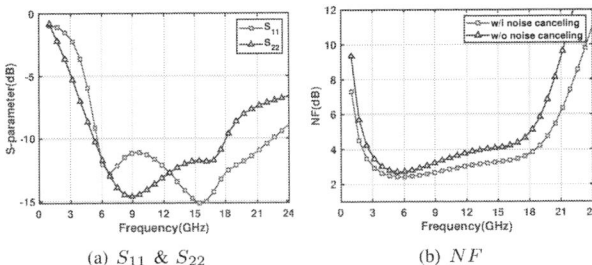

(a) S_{11} & S_{22} (b) NF

Fig. 4: Post-simulation results

As shown in Fig. 2(a), series and shunt inductive peaking techniques are employed to implement bandwidth extension as well. L_1 and parasitic capacitor at the drain of M_2 (denoted by C_{par}) are connected in parallel to form a low Q load, thus expanding the output bandwidth of the first stage. With the increase of working frequency, the effect of gain roll-down caused by load capacitance C_{load} (representing the parasitic capacitor of M_4) becomes more and more significant, thus interstage series inductor L_2 is added to boost the gain at high frequency. The resonant frequency generated by L_1 and L_2 are 6GHz and 18GHz respectively.

The effect of inductive peaking for bandwidth extension is shown in Fig. 2(b) by post-simulation result of S_{21}. Thanks to this technique, the maximum frequency of 3dB bandwidth is extended from 8GHz to larger than 18GHz.

III. POST-SIMULATION RESULTS

Implemented in 130-nm CMOS PD-SOI technology, the proposed LNA occupies an area of $0.9 \times 0.8 mm^2$, including pads, TSV and ESD, etc. (see Fig. 3). The overall circuit consumes 55.44mW from 1.8V supply. Simulated S-parameters and NF are plotted in Fig. 2(b) and Fig. 4. For the design frequency band of 6 to 18GHz, the power gain S_{21} of the LNA is 12.55dB to 15.44dB, with a gain ripple of ±1.44dB, input and output reflection coefficient S_{11} and S_{22} are both less than -10dB for good return loss. The NF varies from the minimum of 2.42dB at 6GHz to 3.73dB at 18GHz. Linearity is a critical parameter for wideband LNA, simulated IP1dB and IIP3 are -14.11dBm and -5.31dBm respectively(not plotted here), which is sufficient for the application of this LNA.

In Table I, the performance of the proposed LNA is summarized and compared with other state-of-the-art wideband LNA fabricated in different technologies, including CMOS, pHEMT and SOI. Two FoMs [3] are utilized to compare them fairly. As can be seen, our LNA achieves wideband, high gain and low noise simultaneously thanks to the proposed noise canceling and bandwidth extension techniques.

IV. CONCLUSION

A 6-18GHz LNA utilizing noise canceling and bandwidth extension techniques is implemented in 130-nm CMOS PD-SOI. Post-simulation results show that the LNA has a maximum 15.44dB gain and minimum 2.42dB NF. The overall circuit consumes 55.44mW from 1.8V power supply, occupying $0.72mm^2$ silicon area.

REFERENCES

[1] J. Zhang, D. Zhao, and X. You, "A 20-GHz 1.9-mW LNA Using gm-Boost and Current-Reuse Techniques in 65-nm CMOS for Satellite Communications," *IEEE J. of Solid-State Circuits*, vol. 55, no. 10, pp. 2714-2723, Oct. 2020.

[2] B. Cui and J. R. Long, "A 1.7-dB Minimum NF, 22–32-GHz Low-Noise Feedback Amplifier With Multistage Noise Matching in 22-nm FD-SOI CMOS," *IEEE J. of Solid-State Circuits*, vol. 55, no. 5, pp. 1239-1248, May 2020.

[3] H. Gao et al., "A 6.5–12-GHz Balanced Variable-Gain Low-Noise Amplifier With Frequency-Selective Gain Equalization Technique," *IEEE Trans. Microw. Theory Techn.*, vol. 69, no. 1, pp. 732-744, Jan. 2021.

[4] H. Chen, H. Zhu, L. Wu, W. Che and Q. Xue, "A Wideband CMOS LNA Using Transformer-Based Input Matching and Pole-Tuning Technique," *IEEE Trans. Microw. Theory Techn.*, vol. 69, no. 7, pp. 3335-3347, July 2021.

[5] H. Yu, Y. Chen, C. C. Boon, P. -I. Mak and R. P. Martins, "A 0.096-mm^2 1 –20-GHz Triple-Path Noise- Canceling Common-Gate Common-Source LNA With Dual Complementary pMOS–nMOS Configuration," *IEEE Trans. Microw. Theory Techn.*, vol. 68, no. 1, pp. 144-159, Jan. 2020.

[6] Y. Hsiao, C. Meng, and M. C. Li, "Analysis and design of broadband LC-ladder FET LNAs using noise match network," *IEEE Trans. Microw. Theory Techn.*, vol. 66, no. 2, pp. 987–1001, Feb. 2018.

[7] S. Li et al., "A Millimeter-Wave LNA in 45nm CMOS SOI with Over 23dB Peak Gain and Sub-3dB NF for Different 5G Operating Bands and Improved Dynamic Range," *2021 IEEE Radio Frequency Integrated Circuits Symposium (RFIC)*, 2021, pp. 31-34.

A CGP-based Efficient Approximate Multiplier with Error Compensation

Qiao Shen[1], Renyuan Zhang[1], Hao Zhang[3], Hao Cai[2], Bo Liu[2], Jian Xiao[4]*
[1]School of Integrated Circuits, Southeast University, Wuxi 211189, China
[2]School of Electronic Science and Engineering, Southeast University, Nanjing 210096, China
[3]Nanjing Research Institute of Electronics Technology, Nanjing 210012, China
[4]Nanjing University of Posts and Telecommunications, Nanjing 210042, China
*Email: xiaoj@njupt.edu.cn

Abstract—As one of the most promising energy-efficient paradigms in deploying Neural Network (NN) on hardware, approximate computing (AxC) has recently gained great traction to replace exact computing. This paper proposes an efficient approximate multiplier design method, which combines the Cartesian Genetic Programming (CGP)-based automatic design method and manual design method. Besides, an error compensation scheme based on the traversal search of truth table is proposed for higher-order multiplier construction. Experiments show that compared to exact multiplier, the proposed approximate multiplier can reduce the area, power consumption, and delay by 54.9%, 55.7%, and 36.86%, respectively. It also shows superiority to the state-of-the-art approximate multiplier. In addition, when deployed in LeNet-5 for MINIST datasets, the proposed multipliers show higher efficiency than exact multiplier with comparable recognition accuracy.

Keywords—approximate computing, multiplier, Cartesian Genetic Programming

I. INTRODUCTION

Due to the fault tolerance of NN, exact computing is not always a must in NN based applications including speech recognition and image classification. Therefore, approximate computing has received great attention as an emerging computing paradigm to help NN deployment achieve higher power and area efficiency with less accuracy cost [1]. Apart from the conventionally utilized deliberate design method based on the simplification of the Boolean functions, a CGP-based automatic design method was proposed by Sekanina et al. [2] to generate approximate computing circuits, which can achieve a well-balanced performance of power consumption, and area. However, unlike most conventional methods with configurable approximate bit width, CGP-based method shows disadvantage in controlling errors with respect to the construction of higher order multiplier, and may lead to non-convergence because of the large numbers of nodes. These factors conflict with the requirement of multi-bit multiplication in NN. [3] proposed the Constrained Cartesian Genetic Programming (CCGP) along with a partitioning methodology for higher-order multiplier construction to address the above problems. However, few works on approximate multipliers before have considered the combination of approximate adders and multipliers, or the combination of automatic method and manual method.

This paper proposes a CGP-based approximate multiplier with error compensation, which proves to be power and area efficient. The multi-object CGP based lower-order approximate multiplier is applied in the less significant bits for energy and area saving while OR adders are adopted in the accumulation stage to reduce the critical path delay. When constructing the higher-order approximate multiplier, the error generated by the LSBs multiplier and OR adders is compensated by an

---

Algorithm 1: CGP

Input: CGP parameters, fitness function.
Output: Individual p.

1    $Q \longleftarrow$ given parent p and its λ offspring;
2   **while** *the termination condition is not satisfied* **do**
3   FitnessFunction(Q);
4      $a \longleftarrow$ lowest-fitness-scored individual in Q;
5   **if** *fitness(a)<fitness(p)* **then**
6        $p \longleftarrow a$;
7   **else**
8        $p \longleftarrow p$;
9   **end**
10      $Q \longleftarrow$ generated λ offspring from p by point mutation;
11 **end**

---

elaborately customed multiplier in the MSBs, which can be obtained by the rectification of the truth table.

II. CGP-BASED APPROXIMATE MULTIPLIERS WITH ERROR COMPENSATION

A. Automatic Generation of Approximate Circuits Based on CGP

The main idea of CGP is to model the candidate circuit as a two-dimensional array of programmable nodes which are comprised of NOT, AND, XOR, OR, NOR, XNOR, XAND. The population starts from a given circuit and is iteratively updated by point mutation and selection by fitness function (multi-object or single object in different bits multipliers as is described in Section II. B). The pseudo-code of CGP in this work is given in Algorithm 1.

B. Higher-Order Multiplier Construction with Error Compensation

To construct a higher-order $2n \times 2n$ multiplier with $n \times n$ multiplier, the $2n$ bits multiplier A and multiplicand B should be divided into high n bits A_H, B_H and low n bits A_L, B_L, respectively. In the partial product generation stage, multipliers generated by different CGP configurations are adopted. As is shown in the step 1 in Fig. 1, in lower and middle bits, the multipliers are obtained by multi-object CGP considering area, power and delay. OR adders are adopted in the final accumulation phase, which effectively simplify the circuit structure and reduce the area, power and delay. Now that the structures of lower bits multipliers and OR adders are determined, to eliminate the error produced by them, we need to find out the truth table of the higher bits multiplier which can best compensate for the current structure. This can be achieved by rectifying the truth table of the exact multiplication through a traversal search. As is shown on the right of the step 2 in Fig. 1, an n bits array R is added to the output of the exact multiplier truth table. In order to determine the most suitable one, all possible values of R should be traversed and evaluated in accuracy. On finishing the search, we can obtain the best compensation scheme of R for current structure. Therefore, the circuit structure closest to the rectified truth table can be obtained by single-object CGP considering only accuracy.

Take the construction of the unsigned 8×8 approximate multiplier shown in Fig. 2 for example. The 8 bits multiplier A and multiplicand B are divided into high 4 bits and low 4 bits.

978-1-6654-9270-6/22 $31.00 © 2022 IEEE

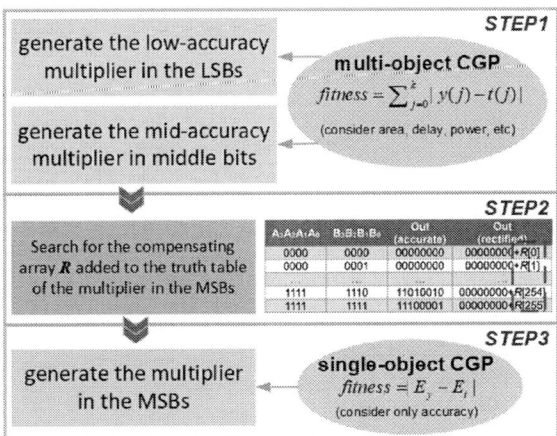

Fig. 1: The design flow of CGP-based higher order multiplier with error compensation

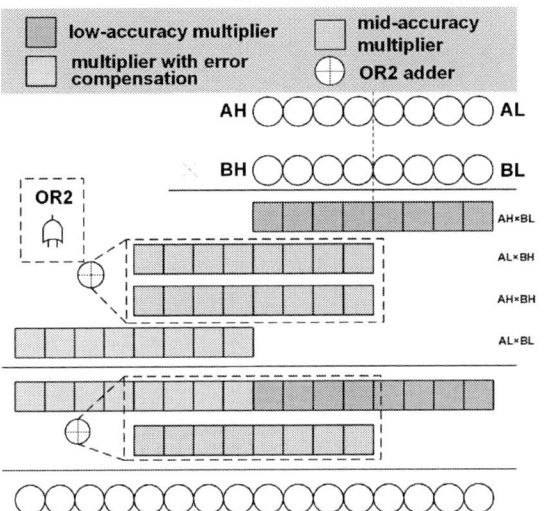

Fig. 2: 8×8 unsigned approximate multiplier constructed with the proposed scheme

In order to increase the area efficiency and power efficiency while minimizing the accuracy loss, the least significant 4 bits multiplication $A_L \times B_L$ uses a multiplier generated by multi-object CGP whose accuracy is relatively low but with high area and power efficiency; the middle bits multiplications $A_H \times B_L$ and $A_L \times B_L$ use a mid-accuracy approximate multiplier generated by multi-object CGP; the most significant 4×4 multiplication uses a customed approximate multiplier generated by single-object CGP to compensate the error.

III. EVALUATION ANG ANALYSIS

In this section, the accuracy, power consumption and area of the approximate multipliers constructed under the proposed scheme are evaluated. To compute the recognition accuracy, we also applied them to the LeNet-5 network for the classification of MNIST dataset. Error rate (ER), normalized mean error distance (NMED) and PRED, which is defined as the probabilities that relative error distance (RED) is lower than 5%, are used in this paper to evaluate the accuracy. The critical path delay and the power consumption were evaluated on TSMC 22nm ULL process technology with the logic supply voltage of 0.6V and

TABLE I. COMPARISON BETWEEN THE PROPOSED 8×8 APPROXIMATE MULTIPLIERS, THE EXACT MULTIPLIER AND THE STATE-OF-THE-ART MULTIPLIER

	Exact	mul8_350 [4]	CGP_COM_1	CGP_COM_2
ER	/	98.98%	96.73%	97.88%
NMED	/	0.96%	0.76%	1.12%
PRED	/	55.2%	34.2%	27.8%
Power (nW)	35.2	17.3	18.0	15.6
Power impro.	/	2.03×	1.96×	2.26×
Area (um^2)	111.52	52.04	58.60	50.27
Area impro.	/	2.14×	1.90×	2.22×
Delay(ns)	2.55	1.67	1.31	1.61
Delay impro.	/	1.53×	1.95×	1.58×
Testing accuracy	98.17%	97.85%	98.12%	97.37%

the clock frequency 250KHz.

Table I presents the comparisons of the proposed unsigned 8×8 approximate multipliers, exact multipliers and the state-of-the-art work. Compared to the exact multiplier, our design CGP_COM_2 achieves a power saving and area saving of 55.7% and 54.9%, respectively, which performs the best; CGP_COM_1 achieves a power saving of 48.9% and an area saving of 52.9%. Both of them significantly reduce the critical path delay. Compared to the state-of-the-art design mul8_350, CGP_COM_1 reduces the delay by 18.6% and increases the precision by 21.3% with a 12.6% increase in area and comparable power. CGP_COM_2 is superior to mul8_350 in each of power, area and delay with error compensation.

To evaluate the performance of the proposed approximate multiplier on NN based applications, we deployed them in the retraining process of the LeNet-5 for MNIST dataset with all multipliers in CONV layers replaced with our approximate multiplier. The testing accuracy after 5 epochs are shown in Table I. It can be learnt from the table that the proposed approximate multipliers have very little impact on the recognition accuracy of this NN application.

IV. SUMMARY

This paper proposes an efficient approximate multiplier based on CGP method with error compensation. Experiments and comparison with other multipliers illustrate that the proposed method can reach a significant trade-off between power consumption, area and delay with little loss of accuracy, and the proposed multiplier can perform well in NN based tasks.

ACKNOWLEDGMENT

This work was supported by the National Key R&D Program of China (Grant No. 2018YFB2202102).

REFERENCES

[1] H. Jiang, et al., "Approximate arithmetic circuits: A survey, characterization, and recent applications," in *Proceedings of the IEEE*, Dec 2020, vol. 108, no. 12, pp. 2108-2135.
[2] L. Sekanina and Z. Vasicek, "Approximate circuit design by means of evolvable hardware," in *ICES*, 2013, pp. 21-28.
[3] K. K. Senthilkumar, et al., "Approximate multipliers using bio-inspired algorithm", in *JEET*, 2020, vol. 16, no. 1, pp. 559-568.
[4] M. S. Ansari, et al., "Improving the accuracy and hardware efficiency of neural networks using approximate multipliers," in *IEEE TVLSI*, 2020, vol. 28, no. 2, pp. 317-328.

978-1-6654-9270-6/22 $31.00 © 2022 IEEE

Photon-Memristive Device for Neuromorphic Computing

Yuqing Fang[1], Qingxuan Li[1], Tianyu Wang[1,2], Jialin Meng[1]*, QingQing Sun[1,2], David Wei Zhang[1,2], Lin Chen[1,2]*

[1] State Key Laboratory of ASIC and System, School of Microelectronics, Fudan University, Shanghai 200433, China
[2] Zhangjiang Fudan International Innovation Center, Shanghai 201203, China
* E-mail: linchen@fudan.edu.cn; jlmeng@fudan.edu.cn

Abstract—With the development of artificial intelligence technology, the usage of perception, storage and computing integrated devices for neuromorphic computing has become a research hotspot. We have successfully fabricated a photon-memristive device with optical sensing ability, which could simulate the synaptic behaviors in the human brain. The device could receive and respond to ultraviolet light, and realize synaptic functions, such as short-term plasticity, long-term potentiation and paired-pulse facilitation. In addition, it was able to be applied in the image identification field, the recognition rate could reach 92.2%. The device still had high stability and low operating current after light pulses excitation. The two-terminal photonic memristive unit is expected to be integrated into a 3D system and have broad application prospects in the field of artificial intelligence.

Keywords—memristor, nickel oxide, neuromorphic computing, photonic synapse

I. INTRODUCTION

Synapses are the physical structures that link neurons and transmit impulses in the human brain [1]. In 1971, Chua proposed the fourth basic circuit element, memristor [2]. After that, significant achievements have been made in simulating the synaptic functions by memristors. With the development and popularization of the Internet of Things and 5G technology, the research on devices integrating sensing, storage and computing has attracted widespread attention [3-5]. Memristors with optical sensing capabilities are not limited by bandwidth. They are able to solve the memory wall caused by the physical separation of memory and CPU, and break the von Neumann bottleneck [6]. Compared with three-terminal devices, two-terminal memristors are more likely to be applied into practical process production due to simple structure and easy preparation [7].

In this work, we reported a two-terminal $Si/SiO_2/Pt/NiO/ITO$ photon-memristive device. Based on the photosensitive characteristics of the dielectric layer NiO, the device can respond to the light pulses and realize synaptic functions like short-term plasticity (STP), long-term potentiation (LTP) and paired-pulse facilitation (PPF). The device has superior memristive capability, high stability and high resistance, and can be used in the three-layer artificial neural networks (ANNs) to recognize images with high accuracy, which have great development potential in neuromorphic computing, laying a foundation for the photonic sensing, memory and computing systems.

II. RESULTS

A. Device Structure and Electrical Characteristics

Fig. 1 shows a schematic structure and the process flow of device. Firstly, 10 nm Ti was grown on a Si/SiO_2 substrate by DC magnetron sputtering as the adhesive layer. Next, 100 nm metal Pt was deposited as the bottom electrode. Then, 100 nm NiO was grown at 10% O_2/Ar ratio. After annealing at a high temperature, photolithography was carried out to define the shape of the top electrode as a rectangle with side length of 80 μm. Finally, 100 nm ITO top electrode was deposited by DC magnetron sputtering. A part of ITO film was removed by soaking in a degumming solution to obtain the device. The I-V curve of device was shown in Fig. 2. Under ultraviolet of 350 nm, its resistance decreased due to the appearance of photocarriers, which indicated that the resistance of the device could be controlled by ultraviolet light.

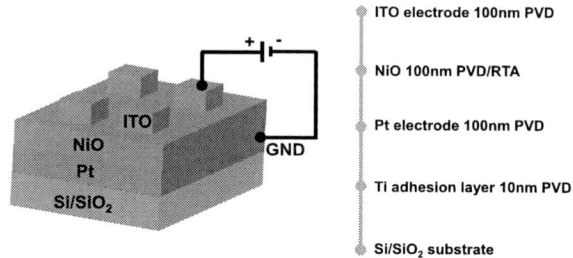

Fig. 1. The structure schematic and the process flow of the device.

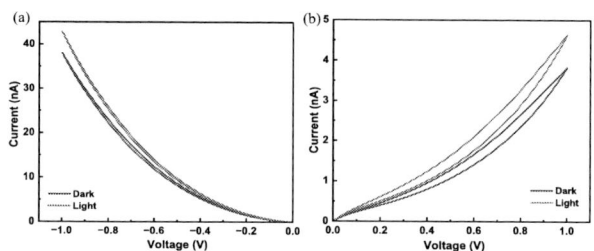

Fig. 2. The I-V characteristic curves in the dark and under UV light under (a) 0~–1 V, and (b) 0~+1 V voltage sweep.

B. Post-Synaptic Current Modulated by Light

In order to explore the modulation ability of light pulses on the device, we changed the pulse width and pulse number and measured the post-synaptic current (PSC). The light wavelength was 350 nm, the intensity was 24μW/cm², and the DC test voltage was 100 mV. Fig. 3 (a) presents the relationship between PSC variation (△PSC) and pulse width. When the pulse width increased gradually, △PSC rose due to the increase of photo-induced carrier concentration.

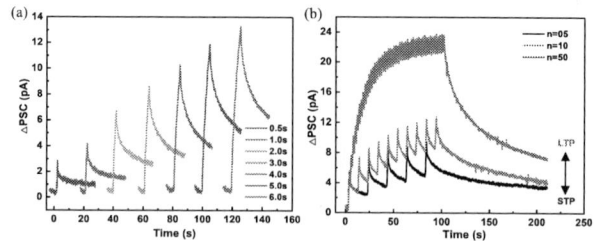

Fig. 3. (a) △PSC under light pulse widths of 0.5, 1.0, 2.0, 3.0, 4.0, 5.0 and 6.0 s. (b) △PSC under light pulse numbers of 5, 10 and 50 within 100 s.

A single optical pulse was kept for 1 s, 5, 10 and 50 optical pulses were applied continuously within 100 s (the period is 20 s, 10 s and 2 s respectively), as shown in Fig. 3 (b). The resistance of the device reduced as it was stimulated by continuous light pulses. After removing the light, the PSC gradually decreased toward the initial state because of the STP. However, when the total numbers of pulses rose, the device showed the function of LTP. While the light was turned off, the PSC did not return to the initial state, demonstrating a nonvolatile memristive ability. It suggests that the device can pick up and respond to ultraviolet light, which is promising to mimic synaptic functions in the human brain for neuromorphic computing.

C. Photon-Memristive Properties

PPF is one of the important functions of synapses [8]. It means that when the time interval ($\triangle$ t) between two consecutive pulses increases within a certain range, the second amplitude of rise will decline. Fig. 4 (a) shows the current of the device when the time interval was 500 ms (light pulse width was 1 s, test voltage was 100 mV), where A_1 and A_2 was the first and second $\triangle$PSC, respectively.

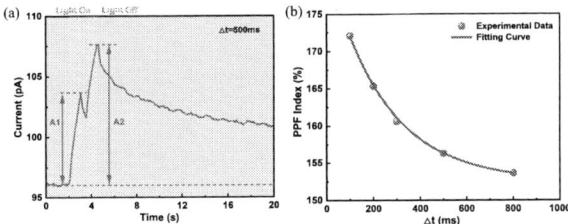

Fig. 4. (a) The current triggered by applying two optical pulses with a time interval of 500 ms. (b) Functional relationship between PPF index and Δt.

The PPF index is $A_2/A_1 \times 100\%$. PPF index can be fitted by a double exponential function:

$$PPF = A_1 exp(-\triangle t/\tau_1) + A_2 exp(-\triangle t/\tau_2) + A_0 \qquad (1)$$

Where $\triangle$t is the time interval between two pulses. The functional relationship between the PPF index and $\triangle$t was shown in Fig. 4 (b). In this work, τ_1 was 221.05 ms and τ_2 was 270.17 ms. As shown in Fig. 5, the memristive phenomenon still existed when continue to increase $\triangle$t between two light pulses to 19 s. The memristive ability decreased until $\triangle$t was 79 s. Even under light illumination, the device had relatively high stability and low operating current, which was conducive to the application of low-power systems.

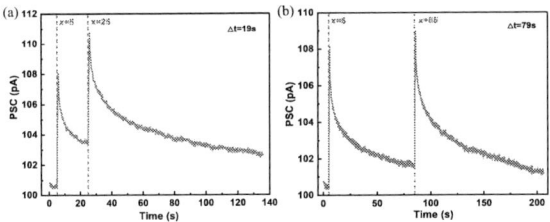

Fig. 5. The PSC triggered by applying two optical pulses with time interval of (a) 19 s, and (b) 79 s.

D. Image Recognition Function

Based on the multi-state conductance and memory ability under the ultraviolet light, the device was applied to a three-layer artificial neural network (ANN) to recognize handwritten digits "0-9" (64×64 pixels) of MNIST. The backpropagation algorithm was used to train the three-layer ANN (the number of input, hidden and output layer was 1024, 64, and 10, respectively), and the weight was replaced by 50 conductance. Then, we obtained the relationship between recognition rate and training epochs, as shown in Fig. 6 (a). After 600 training epochs, the recognition rate reached 92.2%. In Fig. 6(b), the distribution of conductance after 10, 50, 100, and 600 training epochs were fitted by gaussian curves. It could be seen that after 600 training epochs, the distribution became dispersed and the image recognition ability was improved.

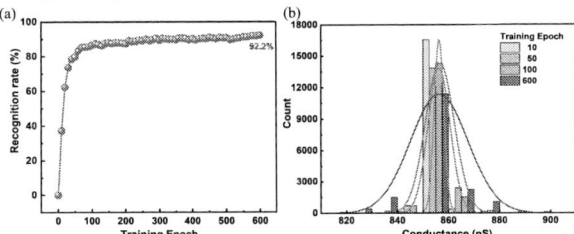

Fig. 6. (a) The relationship between recognition and training epochs. (b) Distribution of conductance after 10, 50, 100, and 600 training epochs.

III. CONCLUSION

In summary, we proposed a two-terminal photon-memristive device with good optical response capability based on the structure of Si/SiO₂/Pt/NiO/ITO. Physical vapor deposition method was used to prepare electrodes and NiO dielectric layer with photoconductivity, which made the process simpler. This kind of optical memristive device realized the function of synapses with high memristive ability, high stability, and low working current, and identified handwritten digit images with high recognition rates, providing a new way for an integrated system of light sensing, storage and computing.

ACKNOWLEDGMENT

This work was supported by the National Key R&D Program of China (2021YFA1202600), NSFC (92064009, 61904033, 62004044), Shanghai Rising-Star Program (19QA1400600), the Program of Shanghai Subject Chief Scientist (18XD1402800), China Postdoctoral Science Foundation (Grant 2022TQ0068, BX2021070, 2021M700026), the Zhejiang Lab's International Talent Fund for Young Professionals, and the Young Scientist Project of MOE Innovation Platform.

REFERENCES

[1] V. M. Ho, J.-A. Lee, K. C. Martin, "The cell biology of synaptic plasticity," Science, 334 (6056), pp. 623-628, 2011.

[2] L. Chua, "Memristor-The missing circuit element," IEEE Transactions on Circuit Theory, vol. 18, (5), pp. 507-519, 1971.

[3] C. Ríos, et al, "In-memory computing on a photonic platform," Sci. Adv. 5, eaau5759, 2018.

[4] H. W. Tan, Y. F. Zhou, Q. Z. Tao, J. Rosen, S. V. Dijken, "Bioinspired multisensory neural network with crossmodal integration and recognition," Nat. Commun. 12, 1120, 2021.

[5] J. L. Meng, et al, "Integrated In-Sensor Computing Optoelectronic Device for Environment-Adaptable Artificial Retina Perception Application," Nano Lett. 22, 1, 81–89, 2022.

[6] J. S. Tang, et al, "Bridging biological and artificial neural networks with emerging neuromorphic devices: fundamentals, progress, and challenges," Adv. Mater. 31, 1902761, 2019.

[7] G. C. Liu, et al, "Ultralow-power and multisensory artificial synapse based on electrolyte-gated vertical organic transistors," Adv. Funct. Mater. 32, 2200959, 2022.

[8] L. Yin, et al, "Synaptic silicon-nanocrystal phototransistors for neuromorphic computing," Nano Energy 63, 103859, 2019.

An Input Buffer with 85dB SFDR for High-Speed Pipeline ADC

Cece Huang[1,2], Yuanfu Zhao[2], Yafei Ji[2], Xin Yang[2], Tieliang Zhang[2], Weixin Gai[1]

[1] School of Integrated Circuits, Peking University, Beijing, China
[2] Beijing Microelectronics Technology Institute, Beijing, China

Abstract—**This paper presents a high linearity input buffer with proposed two-level bootstrapping scheme for high-speed pipeline ADC. In high input frequency, the parasitic capacitance of active devices and inductance of packaging are the main sources of non-linearity. In order to improve the linearity, the proposed input buffer drives the bootstrapping block by the output signal instead of the input, which prevents the non-linear sink current from flowing through the inductance, and thus the linearity is improved. The input buffer was designed together with a 14-bit 500MSPS pipeline ADC in a 28nm CMOS technology. The measured results show that the SFDR achieves 85dB at 2nd Nyquist frequency, which is 8dB larger than the conventional one.**

Keywords— Input buffer, High SFDR, Linearity, ADC

I. INTRODUCTION

Analog-to-Digital Converter (ADC) is one of the key building blocks in modern electronic systems. ADCs with both high sample rate and resolution are widely used in radar systems, image processing and wireless communication [1-3]. Linearity, usually evaluated using SFDR, is very important in high-speed ADCs. For example, higher SFDR can bear larger block signals in communication systems. However, SFDR is strongly affected by the sampling network that is located at the right front-end of an ADC. Fig. 1 is the commonly used pipeline architecture for high-speed and high-resolution ADCs [4]. It is usually composed of sampling network, two or more sub-stages and digital back-ends. The input signal is sampled by the sampling network, and quantized by the following sub-stages until the lowest LSBs are generated. In other words, the linearity of the entire ADC is limited by the sampling network.

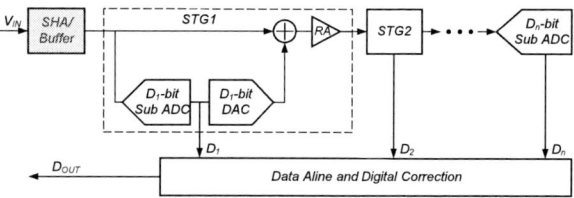

Fig. 1. Structure of the pipeline ADC.

Sample and Hold Amplifier (SHA) and input buffer are two commonly used blocks in sampling [4,5]. Although the input buffer may bring aperture error, it is preferred in high-speed ADCs. Because it can provide better isolation, lower current drawn from the source and lower noise contribution. However, as mentioned above, the overall distortion of ADC may be limited by the linearity of the buffer itself, and the improvement in the linearity may consume significant power, so the design of high linearity input buffers is challenging. In order to solve the problem, this paper introduces an input buffer that can improve the linearity of sampling network without increasing the power dissipation.

This work was supported in part by the National Key R&D Program of China under Grant 2018YFB2202301.

II. ANALYSIS OF LINEARITY

The equivalent circuits of sampling network are shown in Fig. 2, which includes anti-alias filter, package routing, series resistance R_{IN}, ESD protection, input buffer and the sampling capacitance C_S. When the input frequency reaches hundreds of megahertz, the parasitic parameters gradually become the bottleneck to improve SFDR. These parasitic parameters mainly include the parasitic inductance L_P of the package's bond wires, parasitic capacitance of active devices, such as C_{P1} and C_{P2} that related to ESD and input buffer respectively.

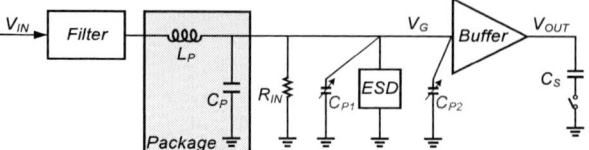

Fig. 2. The equivalent circuits of front-end network including parasitic parameters.

The value of C_{P2} is relatively larger than C_{P1} and varies with V_G, resulting in non-linear. The value of C_{P2} is shown in (1).

$$C_{P2} = C_{P2\_0}\left(1 + a_0 V_G + a_1 V_G^2 + a_3 V_G^3 + \cdots\right) \quad (1)$$

As a result, the charging current of C_{P2} (I_{IN}) is also non-linear. As shown in (2), when I_{IN} flows through L_P, the voltage across L_P (ΔV_L) also has non-linearity.

$$\Delta V_L = L_P \cdot \frac{dI_{IN}}{dt} = L_P \frac{dV_{IN}}{dt^2} C_{P2\_0}\left(1 + a_0 V_G + a_1 V_G^2 + a_3 V_G^3 + \cdots\right) \quad (2)$$

That's to say, the input signal already has non-linearity before reaching the sampling capacitance, which determines the upper level of SFDR. We should minimize C_{P2} to increase linearity.

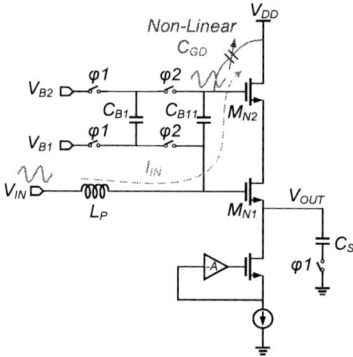

Fig. 3. The commonly used input buffer structure in high-speed ADCs.

The frequently used input buffer utilizing source follower with gain-boosting cascode current source and bootstrapping between the input and the drain is shown in Fig. 3 [4,6], and C_{P2} mainly comes from C_{GD}. Because the gate voltage of M_{N2} varies with input voltage, while the drain voltage is fixed at supply voltage, C_{GD} changes with the input voltage. As a result, the charging current of C_{GD} introduces strong non-linearity.

III. PROPOSED BUFFER

To reduce the negative effect of C_{GD} on linearity, this paper proposed an input buffer as shown in Fig. 4. Instead of driving the bootstrapping capacitance (C_B) by V_{IN}, V_{OUT} is used.
The non-linearity of C_{GD} still exits, but the charging current doesn't flow through L_P, the non-linear voltage drop also disappears. The charging current is provided by V_{OUT} instead, but V_{OUT} is a low-resistance node, the current doesn't affect the

linearity. Besides, the proposed input buffer utilizes a two-level bootstrapping to further improve the linearity of the buffer itself [6]. In addition, driving the bootstrapping circuits by V_{OUT} reduces the parasitic capacitance of input node, thus increases the overall bandwidth.

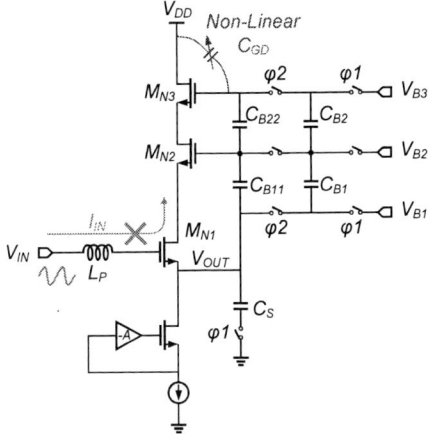

Fig. 4. The proposed input buffer that eliminates the negative effect of C_{GD}.

IV. EXPERIMENTAL RESULTS

The proposed input buffer is verified with a 14-bit 500MSPS pipeline ADC in 28nm CMOS process. The die photograph of the proposed input buffer is shown in Fig. 5.

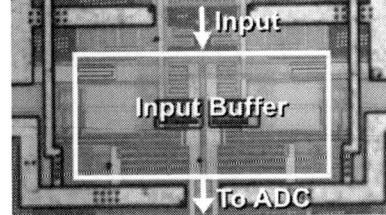

Fig. 5. Die photo graph of the proposed input buffer.

Fig. 7. provides the measured DNL and INL, the DNL is within ±1LSB, and INL is within ±1.2LSB。Fig. 7 shows the output spectrum at 450MHz (2nd Nyquist) input. The measured SFDR with proposed input buffer is 85.2dB, which is 8dB larger than the conventional one. The measured SFDR versus sampling rates is shown in Fig. 8.

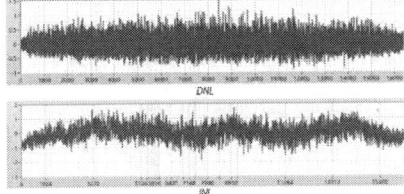

Fig. 6. Measured DNL and INL.

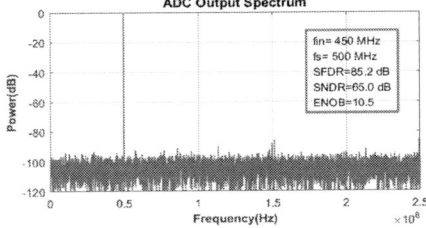

Fig. 7. Output spectrum of the ADC with proposed input buffer.

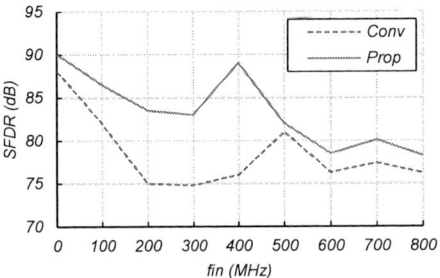

Fig. 8. Measured SFDR versus input frequency.

The comparison with other works is shown in TABLE I. This design has obvious advantages in linearity.

TABLE I. COMPARISON WITH OTHER WORKS

Parameter	ISSCC 2017[7]	ISSCC 2019[8]	VLSI 2020[9]	This work
Architecture	Pi-SAR	Pi-SAR	Pipeline	Pipeline
Technology	65nm	28nm	16nm	28nm
Resolution (bits)	12	12	11	14
Sample Rate (MSPS)	330	1000	1000	500
SFDR@2nd Nyq. (dB)	75	70	76	85
SNDR@Nyq. (dB)	63.5	60.0	59.5	65.0
FoMWalden (fJ/Conv-step)	15.4	9.28	14.1	48.8
FoMSchreier (dB)	167.8	168.2	166.1	160.0

V. CONCLUSION

This paper presents a high linear input buffer for high-speed pipeline ADC. The proposed scheme removes the non-linear capacitance of input buffer from the signal path, thus reduces the negative effect of parasitic inductance and non-linear capacitance on linearity in high input frequency. By utilizing this proposed input buffer, the SFDR and bandwidth are improved at the same time. Experimental results illustrate that the 14-bit 500MSPS ADC with proposed input buffer achieves over 85dB SFDR at 2nd Nyquist frequency.

REFERENCES

[1] M. Kashmiri, et al., "A 4GS/s 80dB DR Current-Domain Analog Front-End for Phase-Coded Pulse-Compression Direct Time-of-Flight Automotive LiDAR," ISSCC 2020.

[2] R. Garg et al., "A 28-GHz Beam-Space MIMO RX With Spatial Filtering and Frequency-Division Multiplexing-Based Single-Wire IF Interface," JSSC, vol. 56, no. 8, pp. 2295-2307, 2021.

[3] H.-J. Kim, "11-bit Column-Parallel Single-Slope ADC With First-Step Half-Reference Ramping Scheme for High-Speed CMOS Image Sensors," JSSC, vol. 56, no. 7, pp. 2132-2141, 2021.

[4] A.M.A. Ali et al., "A 14-bit 125 MS/s IF/RF sampling pipelined ADC with 100 dB SFDR and 50 fs jitter," JSSC, vol. 41, no. 8, pp. 1846-1855, 2006.

[5] B. P. Brandt and J. Lutsky, "A 75-mW, 10-b, 20-MSPS CMOS subranging ADC with 9.5 effective bits at Nyquist," JSSC, vol. 34, no. 12, pp. 1788-1795, 1999.

[6] A. M. A. Ali et al., "A 14-bit 2.5GS/s and 5GS/s RF Sampling ADC with Background Calibration and Dither," VLSI-CIRCUITS 2016

[7] H. Huang, et al., "A 12b 330MS/s pipelined-SAR ADC with PVT stabilized dynamic amplifier achieving <1dB SNDR variation," ISSCC 2017.

[8] Wenning Jiang et al., "A 7.6mW 1GS/s 60dB SNDR Single-Channel SAR-Assisted Pipelined ADC with Temperature-Compensated Dynamic Gm-R-Based Amplifier", ISSCC 2019.

[9] B. Hershberg et al., "A 1MS/s to 1GS/s Ringamp-Based Pipelined ADC with Fully Dynamic Reference Regulation and Stochastic Scope-on-Chip Background Monitoring in 16nm", VLSI 2020.

TSV Defects Classification with Machine Learning Approaches

Haitao He, Changhao Luo, Junchen Dong, Yudi Zhao*, Min Miao, Kai Zhao

Key Laboratory of Information and Communication Systems, Ministry of Information Industry,

Beijing Information Science and Technology University, Beijing, China.

E-mail: zhaoyd@bistu.edu.cn

Abstract—The S parameter amplitude, latency, resistance, and inductance of TSV-RDL structures with the presence of five kinds of defects are simulated as feature vectors for defect detection and classification. Three nondestructive defect classification schemes for the TSV-RDL structure in advanced packaging are evaluated. Feedforward neural network with rectified linear unit activation function for the backpropagation algorithm is superior for defect classification and may play an important role in design for test and build-in self-repair circuit design.

Keywords—TSV, Defect Classification, Machine Learning

I. INTRODUCTION

High-density advanced packaging (HDAP), including 2.5D and 3D integration, is a promising solution to achieve higher interconnect bandwidth, higher integration density, and lower latency [1]. As one of the key technologies of HDAP to implement vertical stacking of multi-dies, the manufacturing of through silicon via (TSV) is a costly process in which various defects may be introduced, such as TSV voids, pinch-off, pinholes defect in oxide, voids, cracks in micro-bumps, short and open defects, electromigration defects, etc. [2]. The aforementioned defects may cause glitches and delays, and therefore must be taken into account in the design for test process. In this paper, we present the simulation results of five major defects and the comparison of defects classification with different machine-learning (ML) techniques.

II. CHARACTERISTICS OF TSV WITH DEFECTS

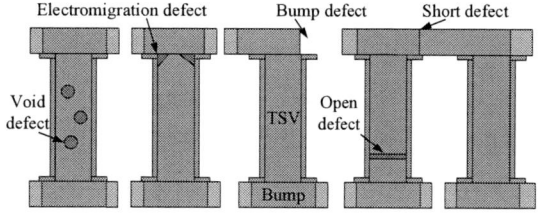

Fig. 1. Cross-sectional diagram of defected TSVs: void, electromigration, bump, open, and short defects.

As illustrated in Fig. 1, defects including TSV voids, electromigration (EM) defects, bump defects, open and short defects are simulated in a ground-signal-ground (GSG) structure by ANSYS HFSS™ to generate their S-parameters, resistance and inductance. The TSV and bump diameter/height are 6/10μm and 8/1μm, respectively. The TSV pitch is 7μm. The liner thickness is 0.1μm.

Contrary to previous work [3,4], we not only study the S parameter amplitude but also focus on more measurable quantities including latency [5], resistance, and inductance. The amplitude of S11 and S21, resistance and inductance with

This research was funded by the National Natural Science Foundation of China (Grant No. 62074017, 62004005), and the Science and Technology Project of the Beijing Municipal Education Commission (Grant No. KM202111232016).

central void defects, electromigration defects and bump defects are shown in Fig. 2-4, respectively. For short and open defects, resistance and inductance are rather distinguishing, so we choose the S parameter phase as a more testable quantity. The amplitude and phase of S11 and S21 with short and open defects are demonstrated in Fig. 5 and 6.

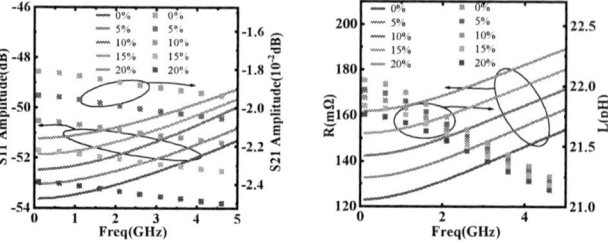

Fig. 2. The S parameter amplitudes, resistance, and inductance of TSVs with void ratios vary from 0% to 20%.

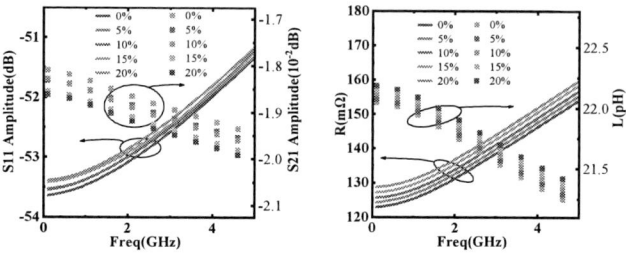

Fig. 3. The S parameter amplitudes, resistance, and inductance of TSVs with EM vacancy ratio vary from 0% to 20%.

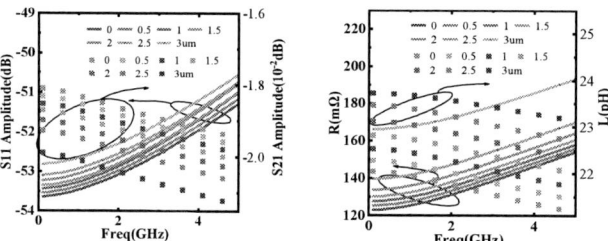

Fig. 4. The S parameter amplitudes, resistance, and inductance of TSVs with bump misalignment vary from 0 to 3μm.

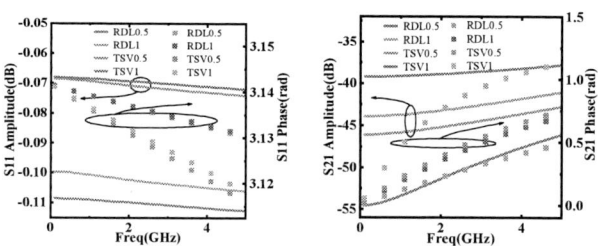

Fig. 5. The S parameter amplitudes and inductance of TSVs with RDL or TSV short contact length vary from 0.5 to 1μm.

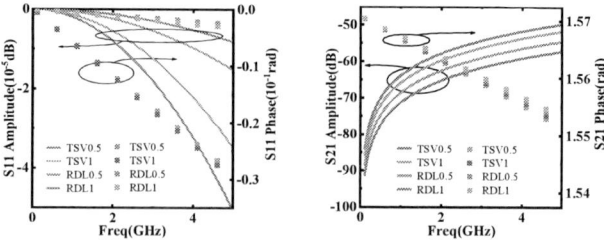

Fig. 6. The S parameter amplitudes and inductance of TSVs with RDL or TSV open distance vary from 0.5 to 1μm.

978-1-6654-9270-6/22 $31.00 © 2022 IEEE

III. CLASSIFICATION OF TSV DEFECTS

The HFSS simulation results of the five kinds of defects are divided into training and testing data sets for the application of supervised ML algorithms, including K-Nearest Neighbors (KNN), Support-Vector Machine (SVM) and Feedforward Neural Network (FNN) approaches. The training data set is to train the ML algorithms with labeled defects, and the testing data set is for the validation of the classification accuracy.

The correct rate of the KNN algorithm is shown in Fig. 7. It reaches its highest value of 97.05% for weighted average voting when k=4. The correct rate with weighted average voting is better than that with equally weight voting, but both show overfittings although their correct rates are satisfying. Besides, the choice of K value is relatively random, and it is hard to determine it by algorithms.

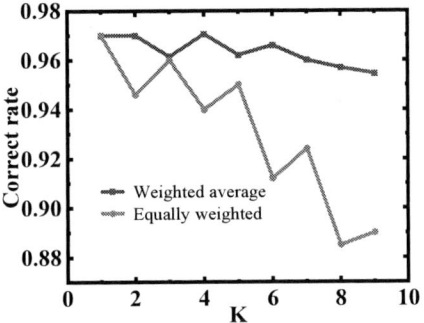

Fig. 7. The correct rate of the KNN approach is a function of the K value.

The correct rates of the SVM algorithm with Gaussian, polynomial and linear kernel functions are shown in Fig. 8. As the penalty parameter C increases to about 25, the correct rates of both Gaussian and linear kernel function converge rapidly, while that of polynomial kernel function increase much slower. Gaussian kernel function presents better correct rates. The transmission characteristics for RDL-TSVs with EM or bump defects are somehow analogous, which makes them linearly inseparable, and cause troubles here.

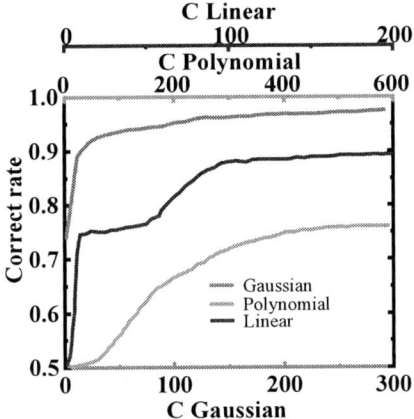

Fig. 8. The correct rate of SVM approach with Gaussian, polynomial, and linear kernel function, respectively.

The FNN method with the sigmoid, hyperbolic tangent, and Rectified Linear Unit (ReLU) activation functions are applied for the defect classification. The correct rate and loss function of their training and testing data sets are shown in Fig. 9. The tanh activation function shows better correct rates than the sigmoid function, however, they both exhibit discrepancies between the training and testing sets. Their training correct rates are higher than the testing rate, which means overfitting both occurs. The ReLU activation function demonstrates the highest correct rates. In the meantime, the agreements between training and testing data sets are the smallest, which means no overfitting and underfitting happens. The loss function with tanh and ReLU activation function drop rapidly in a similar manner as their iteration number increases. However, the discrepancies between training and testing sets of ReLU are better than that of tanh activation function. Moreover, the convergence speed of ReLU activation function is satisfying. It starts to converge, and reaches 90% correct rate after 100 iterations, while the sigmoid activation function takes more than 400 iterations. To summarize, employing ReLU as the activation function for the backpropagation algorithm of the FNN learning exhibits fast convergence behavior, high correct rates without overfitting or underfitting.

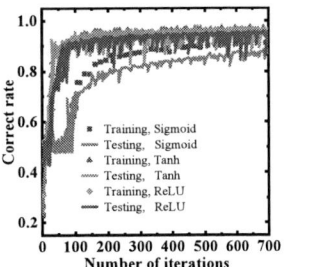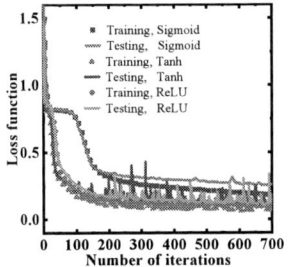

Fig. 9. The correct rate (left) and loss function (right) of NNC approach for both training and testing sets with sigmoid, hyperbolic tangent (tanh), Rectified Linear Unit (ReLU) activation function, respectively.

IV. CONCLUSIONS

This paper presents electromagnetic field simulations of a GSG TSV-RDL structure with the presence of five kinds of defects. Defect classification with KNN, SVM, and FNN algorithms is evaluated based on the simulation results, including S parameters amplitude, phase, resistance, and inductance. The KNN classification approach shows certain randomness in the selection of the K value although the correct rates are high. SVM approach with a Gaussian kernel function exhibits superior correct rates compared to that with polynomial and linear kernel functions. Overfitting can be observed in KNN, SVM, and FNN with sigmoid and tanh activation functions. FNN with the ReLU activation function for the backpropagation algorithm is the best way for the defect classification for its quick convergence, high correct rate, and better agreement between training and testing sets.

REFERENCES

[1] J. H. Lau, "Recent Advances and Trends in Advanced Packaging," *Trans. Comp. Packag. Technol.*, vol. 12, no. 2, pp. 228-252.

[2] K. Chakrabarty, S. Deutsch, H. Thapliyal and F. Ye, "TSV defects and TSV-induced circuit failures: The third dimension in test and design-for-test," *2012 IEEE International Reliability Physics Symposium (IRPS)*, 2012, pp. 5F.1.1-5F.1.12.

[3] H. Liu, R. Fang, M. Miao, Y. Yang and Y. Jin, "Defect Detection for the TSV Transmission Channel Using Machine Learning Approach," *2019 IEEE 69th Electronic Components and Technology Conference (ECTC)*, 2019, pp. 2168-2172.

[4] Y. Huang, C. Pan, S. Lin and M. Guo, "Machine-Learning Approach in Detection and Classification for Defects in TSV-Based 3-D IC," *Trans. Comp. Packag. Technol.*, vol. 8, no. 4, pp. 699-706.

[5] C. Luo, K. Zhao, X. Sun, M. Miao and Z. Li, "Detection and Classification of Typical Defects in TSV and RDL," *2019 20th International Conference on Electronic Packaging Technology (ICEPT)*, 2019, pp. 1-4

SAUST: A Scheme for Acceleration of Unstructured Sparse Transformer

Yifan Song, Shunpeng Zhao, Song Chen, Yi Kang

University of Science and Technology of China, China

Abstract—Transformer achieves impressive results on many AI tasks. However, it also introduces a huge amount of computation. Pruning is a promising method to reduce the computation load by generating sparse transformer models. To avoid load imbalance caused by computing involved in zero elements, previous works explore structured pruning combined with hardware acceleration. However, tight constraints in structured pruning usually make training much harder and reach a lower sparsity level in the end. This paper proposes SAUST, a scheme that exploits the high sparsity level of unstructured pruning and addresses the load imbalance problem using both hardware and software methods. FPGA implementation shows that SAUST can achieve 3.35x and 2.76x execution time speedup compared to two state-of-the-art references on hardware accelerators.

Keywords—Transformer, unstructured pruning, sparse-dense matrix multiplication

I. INTRODUCTION

Transformer [1] achieves excellent results on many natural language processing and computer vision tasks at the expense of intensive computation. To reduce the computation load, pruning is a promising solution by removing many redundant weights and generating sparse transformer models. Then, a large number of ineffectual multiplications with zero weights can be skipped in designated hardware. Pruning is divided into two categories, namely unstructured pruning and structured pruning. Unstructured pruning usually generates a highly irregular zero distribution in weight matrices, leading to load imbalance when skipping zero weights and processing nonzero (NZ) weights in parallel. Thus, structured pruning is adopted by most of the previous works [2][3][4]. By generating regular sparsity patterns in weight matrices, processing elements (PE) can be carefully arranged both in hardware and software to work out in a balanced way. However, structured pruning may be harder to train and lead to a lower sparsity level because structuring weights may limit the degrees of freedom during the training. Because the structured pruned model has a lower sparsity level and more NZ weights, it requires a larger amount of computation compared to the unstructured pruned model.

To exploit the high sparsity level of the unstructured pruned transformer model in the meantime address the load imbalance problem, this paper proposes a scheme called SAUST (Scheme for Acceleration of Unstructured Sparse Transformer). The accelerator contains several AttenCores. Each AttenCore is a basic computing unit including multiple PEs that are connected by a common bus. Thus, Transformer model is computed in two-granularity parallel, a single SDMM (Sparse-Dense Matrix Multiplications) computed in parallel on PEs in a single Atten-Core and multiple SDMMs in multi-head attention/feed forward network (FFN) computed in parallel on multiple AttenCores. Hence, load balancing is also carried out in two types of granularities, fine-grained load balance over PEs in a single AttenCore and coarse-grained load balance over AttenCore. To achieve fine-grained load balance, we propose an outer product method for computing SDMM, in which the sparse matrices are represented in Run Length encoding. To achieve coarse-grained load balance, we propose a two-stage load balance aware pruning method. Experiments on Tiny-Transformer show that SAUST implemented on FPGA can achieve 3.35x and 2.76x

This work was supported by the National Key R&D Program of China under grant No. 2019YFB2204800

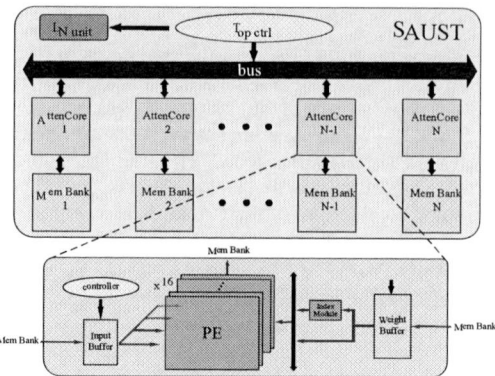

Fig. 1. An overview of SAUST with AttenCores.

execution time speedup compared to two state-of-the-art hardware accelerators [3] [2], respectively.

II. OUR SCHEME

A. SAUST Hardware Architecture

As shown in Fig. 1, SAUST hardware architecture consists of several AttenCores connected by a bus, an LN (Layer Normalization) unit based on lookup table and a top controller to control LN unit and the synchronization of AttenCores. Each AttenCore contains a controller, 16 PEs connected by a bus and an Index Module to locate the NZ weights. The NZ weights and input matrices are stored in Weight Buffer and Input Buffer, respectively. The controller coordinates the data movement and controls PEs. The architecture of PE is detailed in section B.

B. Fine-grained Load Balance

To address the fine-grained load imbalance problem, we propose the outer product based SDMM. The sparse weight matrices are compressed using Run Length encoding, in which run is 4 bits. Each SDMM is mapped to all PEs of each AttenCore. As illustrated in Fig. 2, each PE contains four multipliers and accumulators, a register-based Input FIFO to store input data in the row of the dense input matrix, and a Psum Buffer to store the partial result matrices.

All PEs have a three-stage pipeline so that one multiply-accumulate (MAC) operation can be processed per clock cycle. First, the Index Module in AttenCore accumulates runs and compares the sum with the number of rows of the weight matrix to determine the position of an NZ weight. The position is used for reading the corresponding partial sums from Psum Buffer and writing the results back into Psum Buffer. Second, the NZ weight is broadcasted to all PEs and multiplied by the input data in Input FIFOs and then plus the partial sums. Third, the results of MAC operations are written into Psum Buffer. As a result, all PEs work simultaneously when performing a single SDMM, and fine-grained load balance is achieved.

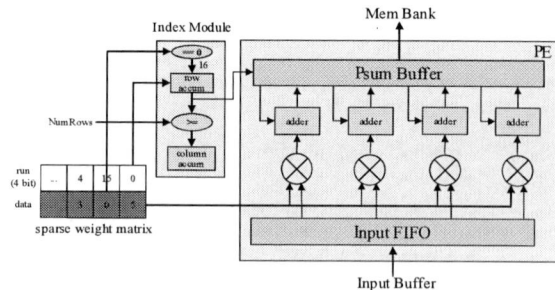

Fig. 2. PE architecture.

C. Coarse-grained Load Balance

To address the coarse-grained load imbalance problem, we propose a two-stage load balance aware pruning method. In stage 1, unstructured pruning based on magnitude is applied to a pre-trained transformer model. In stage 2, Coarse-Grained Balanced Pruning (CGBP) algorithm is employed for the model obtained from stage 1, as described in Algorithm 1.

In multi-head attention consisting of N attention heads, three weight matrices in the i-th attention head and the weight matrix in the output linear projection are denoted as W_{Qhi}, W_{Khi}, W_{Vhi} and W_O, respectively. In FFN, weight matrices are denoted as W_{F1} and W_{F2}. In sparsity vector $\boldsymbol{S} = (S_{Qh}, S_{Kh}, S_{Vh}, S_O, S_{F1}, S_{F2})$ of each encoder/decoder layer, S_{Qh}, S_{Kh} and S_{Vh} are the average sparsity ratios of $(W_{Qh1}, ..., W_{QhN})$, $(W_{Kh1}, ..., W_{KhN})$ and $(W_{Vh1}, ..., W_{VhN})$, respectively. S_O, S_{F1} and S_{F2} are the sparsity ratios of W_O, WF_1, and W_{F2}, respectively.

Algorithm 1: Coarse-grained Balanced Pruning

1	**Input:** the output model of stage 1 T_{in},
	sparsity vector $\boldsymbol{S_{in}}$ in each layer of T_{in},
	the number of attention heads N
2	**Output:** coarse-grained balanced pruned model T_{out},
	the accuracy of T_{out} A_{out}
3	initialize $T_{out} = T_{in}$
4	**for** each encoder/decoder layer in T_{out} **do**
5	initialize $\boldsymbol{S_{out}} = \boldsymbol{S_{in}}$
6	split W_O, W_{F1} and W_{F2} into N matrices $W_{O1}, ..., W_{ON}$, $W_{F11}, ..., W_{F1N}$ and $W_{F21}, ..., W_{F2N}$, respectively
7	**for** each i in range $(1, N)$ **do**
8	$W = [W_{Qhi}, W_{Khi}, W_{Vhi}, W_{Oi}, W_{F1i}, W_{F2i}]$
9	**for** each S, W in $\boldsymbol{S_{out}}$, W **do**
10	set the weights of S percent in W as zero based on the magnitude
11	**end for**
12	**end for**
13	**end for**
14	**While** (not converge) **do**
15	update weights of T_{out} in training
16	same as steps **4 - 13**
17	validate and obtain the A_{out}
18	**end**

Compared to structured pruning, such as block-wise pruning followed by vector-wise pruning in each block [2], block-balanced pruning [3] and column balanced block pruning [4], our loose pruning constraint is at the matrix level, leading to more degrees of freedom in weight updating. Thus, coarse-grained load balance is achieved with CGBP while maintaining the high sparsity ratio of the unstructured pruned model.

III. EXPERIMENT

Our scheme is tested on Tiny-Transformer using WikiText-2 dataset. TABLE I lists the key parameters of Tiny-Transformer. We run 200 training epochs and obtain the model with the best accuracy of 98.21% as the pre-trained model. The weights and activations are quantized to 8-bit fixed point, but the internal calculations in softmax are implemented with 32-bit fixed point, as in [5]. For exponential operation in softmax, e^x is converted to $2^{x/ln2}$ and a linear function is used for approximation like [6]. To fit Tiny-Transformer model, 4 AttenCores are used. Our hardware is implemented on the Xilinx Zynq ZCU102 board using Verilog HDL and utilizes 11713 logic cells (LUT), 13020 FFs, 276.5 block RAMs (BRAM) and 256 DSPs.

TABLE I.   TINY-TRANSFORMER CONFIGURATIONS

Encoder Layers	Decoder Layers	Attention Heads	Feed-forward Size	Hidden Size
2	1	4	200	800

TABLE II.   SPARSITY RATIO COMPARISON

	[2]	[3]	[4]	This work
Sparsity Ratio	84%	50%	70%	**88%**
Accuracy	97.4%	96.5%	96.0	**97.77%**

The two-stage load balance aware pruning method is applied to the pre-trained model at different sparsity ratios. The model with a sparsity ratio of 88% and an accuracy of 97.77% (less than 1% accuracy loss compared to the pre-trained model) is used for comparison. As shown in TABLE II, our two-stage load balance aware pruning method reaches a higher sparsity ratio with higher accuracy than other structured pruned models.

The effectiveness of SAUST to address the load imbalance problem is evaluated here by comparing execution time with an ideal-balanced system that evenly distributes the total computation load among all PEs. The ideal-balanced system has an execution time that equals the clock cycle times the total number of effective multiplications, i.e., all NZ weights multiply with inputs, then divided by the number of multipliers. We measure the execution time of SAUST performing Tiny-Transformer model with an 88% sparsity ratio, and it can achieve 76.4% speedup over the ideal-balanced system.

Furthermore, the system performance is compared among Intel i5-11400 (2.60 GHz) CPU, NVIDIA GeForce GT 1030 GPU and FPGA using the same model and same dataset. As shown in TABLE III, our FPGA implementation achieves 3.35x and 2.76x execution time speedup compared to [3] and [2] on FPGA, respectively.

TABLE III.   PERFORMANCE COMPARISON

	CPU	GPU	FPGA [3]	FPGA [2]	This work
Platform	Intel i5-11400	NVIDIA GeForce GT 1030	Alveo U200	Alveo U200	ZCU102
Latency (ms)	33.46	21.64	7.85	6.45	**2.34**
Throughput (GOPs/s)	16.21	25.07	69.11	84.11	**231.84**

IV. CONCLUSION

This paper proposes SAUST, a scheme to accelerate the computation of unstructured sparse transformer and address the load imbalance problem. Experiments show that SAUST reaches a higher sparsity ratio than structured pruning schemes and achieves load balance. Moreover, FPGA implementation of SAUST can achieve 3.35x and 2.76x execution time speedup compared to two state-of-the-art hardware accelerators.

REFERENCES

[1] A. Vaswani, N. Shazeer, N. Parmar, J. Uszkoreit, L. Jones, A. N. Gomez, Ł. Kaiser, and I. Polosukhin, "Attention is all you need," in *Advances in neural information processing systems*, 2017, pp. 5998–6008.

[2] Panjie Qi, Edwin Hsing-Mean Sha, Qingfeng Zhuge, Hongwu Peng, Shaoyi Huang, Zhenglun Kong, Yuhong Song, Bingbing Li, "Accelerating Framework of Transformer by Hardware Design and Model Compression Co-Optimization," in *IEEE/ACM International Conference On Computer Aided Design (ICCAD)*, 2021, pp. 1-9.

[3] Panjie Qi, Yuhong Song, Hongwu Peng, Shaoyi Huang, Qingfeng Zhuge, and Edwin Hsing-Mean Sha, "Accommodating Transformer onto FPGA: Coupling the Balanced Model Compression and FPGA-Implementation Optimization," In *Proceedings of the 2021 on Great Lakes Symposium on VLSI*, 2021, pp. 163-168.

[4] H. Peng et al., "Accelerating Transformer-based Deep Learning Models on FPGAs using Column Balanced Block Pruning," in *22nd International Symposium on Quality Electronic Design (ISQED)*, 2021, pp. 142-148.

[5] Sehoon Kim, Amir Gholami, Zhewei Yao, Michael W Mahoney, and Kurt Keutzer, "I-bert: Integer-only bert quantization," In *arXiv preprint arXiv:2101.01321*

[6] M. Wang, S. Lu, D. Zhu, J. Lin and Z. Wang, "A High-Speed and Low-Complexity Architecture for Softmax Function in Deep Learning," In *2018 IEEE Asia Pacific Conference on Circuits and Systems (APCCAS)*, 2018, pp. 223-226.

Improve the Robustness of Diffusive Memristor based True Random Number Generator via Voltage-to-Time Transformation

Haoyang Li[1], Yuyang Fu[2], Tianqing Wan[1], Yifan Lu[1], Ling Yang[1], Yi Li[1*]

[1]School of Integrated Circuits, Huazhong University of Science and Technology, Wuhan 430074, China (e-mail: liyi@hust.edu.cn).
[2] School of Microelectronics, Hubei University, Wuhan 430062, China.

Abstract—**Fluctuations in Vth distribution can pose reliability problems for true random number generators (TRNGs). Here, we propose an explanation of the mechanism for voltage-to-time transformation scheme, which improve the reliability of TRNGs by fine-grained segmentation of time. Compared with conventional schemes, test scheme in this work can tolerate 50% cycle-to-cycle (C2C) and device-to-device (D2D) distribution shift and ensure the independence of random bits. This work provides a distinct and reliable evidence for tolerating frequent parameter shifts in voltage-to-time transformation scheme, which increase Vth robustness, reduce device requirements in practical use, and avoid redundant measurements and extra calibration in the test.**

Keywords—diffusive memristor, true random number generator, threshold switching, variation, security primitive

I. INTRODUCTION

True random number generators (TRNGs) play a crucial role in various hardware systems as security primitives [1], [2]. Conventional CMOS-based TRNGs usually require post-processing circuits and entropy-tracking feedback loops [2], [3] which makes integration in lightweight systems difficult. Among the many volatile device candidates, TRNGs based on diffusive memristors have garnered much attention [4]–[7]. Due to the stochastic nature of the ion migration or vacancy formation process at the microscopic level, the position and morphology of the CFs are unpredictable [8]–[12]. These unstable CFs may vary considerably from C2C and D2D, leading to frequent shifts in the parameter distribution. These shifts will result in the disable of entropy extraction circuit, which usually use the mean value of parameter distribution to distinguish the bit "1" and "0". Therefore, how to solve the problem of poor robustness of TRNG caused by the common shift phenomenon in diffusive memristors is a key issue before realizing practical applications.

In this work, we propose an essential aspect to discuss the robustness improvement of diffusive memristors-based TRNGs via voltage-to-time transformation scheme. Unlike conventional schemes, voltage-to-time transformation scheme has multiple references to distinguish random bits. Through our experiments and mathematical analysis, voltage-to-time transformation scheme is proved to be able to operate more stable and convenient than the conventional one.

II. EXPERIMENT

The schematic structure of the Ag/HfOx/Pt via-hole device is shown in the Fig. 1(a). The bottom electrode (BE) consisting of 5 nm Ti and 100 nm Pt was sequentially deposited by magnetron sputtering. Subsequently, 100 nm SiO2 was deposited by plasma-enhanced chemical vapor deposition at 300 ℃. Then electron beam lithography and inductive coupled plasma etching are applied to yield the bottom electrode pad area and nanoscale via-holes. After defining the via-holes, 6 nm HfOx was grown at 280 ℃ by atomic layer deposition. UV lithography was then used to define the top electrode area. The top electrode (TE) of 80 nm Ag and 10 nm Au was deposited by electron beam evaporation. The DC I-V characterization was done with Keysight B1500A semiconductor analyzer. For the pulse test, a Keysight WGFMU B1530A was used to generate the sawtooth wave pulse.

III. RESULTS AND DISCUSSION

Fig. 1(b) shows DC I-V sweeping curves of 380 consecutive cycles under the compliance current (I_{CC}) of 100 μA. The device shows typical threshold switching characteristics with an on/off ratio of over 10^5. The inset in Fig. 1(b) illustrates the forming process. The I-V curves under different I_{CC} from 10 nA to 100 μA are shown in Fig. 1(c). The distribution of Vth extracted from 380 DC cycles in Fig. 1(d) shows stochasticity with a coefficient of variation ($Cv = \delta/\mu$) of 8%. Besides, it is noticed that the distribution has a significant shift after about 190 cycles, which is commonly observed in different kinds of volatile switching devices [13], [14].

To reveal the influence of the V_{th}-distribution shift in C2C on the reliability of TRNG, we compare our voltage-to-time transformation scheme (S2) with a conventional one (S1), [7], [15]. S1 is shown in Fig. 2(a). In contrast, instead of using the reference, S2 (Fig. 2(b)) utilizes the sawtooth wave to transfer the normal distributed V_{th} to ON-time to generate the random bits. The comparator transforms the ON-state of the device into a high voltage level and sends it to the T-Flipflop. Because the duty ratio of the clock is 50%, the theoretical probability of the high voltage level of Q will be 0.5. Herein we define the high voltage level as a bit "1", and low voltage level as a bit "0".

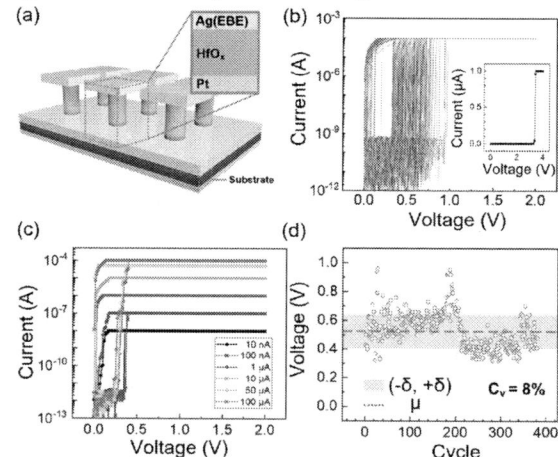

Fig. 1. (a) The structure and 3D schematic diagram of the Ag/HfOx/Pt device. (b) DC I-V sweeping characteristics of 380 cycles under 100 μA. The inset shows that the forming voltage is 3.4 V. (c) The threshold switching property under I_{CC} from 10 nA to 100 μA. (d) The stochastic distribution of threshold voltages in 380 cycles.

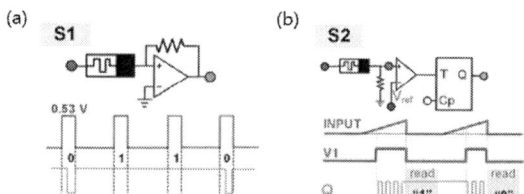

Fig. 2. (a) A conventional TRNG circuit scheme S1 consists of an

---

This work was supported by the National Natural Science Foundation of China under Grant Nos.92064012.

978-1-6654-9270-6/22 $31.00 © 2022 IEEE

amplifier and a resistor. (b) TRNG circuit scheme S2 includes a comparator, a resistor, and a T-Flipflop.

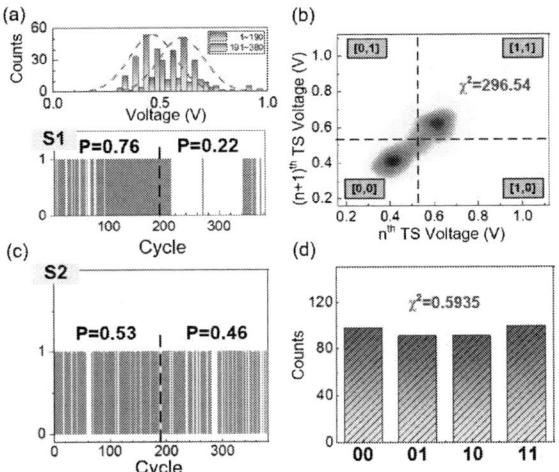

Fig. 3. (a) The upper figure shows the fluctuation of V_{th} distribution after 190 cycles. The bottom figure shows the 380 random bits generated by S1 and the probabilities of bit "1" in the first and last 190 cycles. (b) A 2-D kernel density image of the V_{th} in 380 cycles. (c) The 380 random bits generated by S2. (d) The counts of the four pairs of bits in the random steam

Fig. 3(a) shows that the distribution of the Vth shifts after 190 cycles and the probability of bit "1" varies considerably between the first and last 190 cycles due to shifting in the mean value of Vth. The correlation between the random bits is another important factor that ensures the unpredictability of the TRNG. Fig. 3(b) shows the 2-D kernel density image of the TS voltage at the nth cycle as a function of it at the $(n+1)^{th}$ cycle. There is a clear linear relationship between the TS voltage at the nth cycle and the $(n+1)^{th}$ cycle. Due to the residual metal interstitials after the CF dissolution, new CF formation is promoted at nearby sites, making Vth subject to the previous cycle. Therefore, the random bits from S1 are dependent, and the related Chi-squared value χ^2 was calculated as 296.54.

In contrast, the voltage to time transformation scheme blocks the influence from shifts of Vth distribution, the probability of generating a bit "1" is stable and close to 0.5, as shown in Fig. 3(c). In addition, the counts of "00", "01", "10", and "11" are almost equal in the generated random sequence with an associated χ^2 is 0.5935, which implies that the generated random bits are with strong independence. The above results show that the S2 scheme can effectively tolerate the Vth distribution shift of the device in C2C.

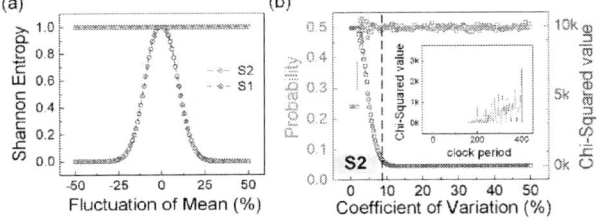

Fig.4. (a) The Shannon entropy of bits generated by the S1 and S2 schemes under the fluctuation of mean value. The reference mean value and C_v of device are set as 0.53 and 8%, respectively. (b) The probability and chi-square value under different C_v. The inset shows the influence of clock rate on the χ^2.

To elucidate the superiority of the voltage-to-time transformation scheme, further simulations were performed. Fig.

4(a) indicates that the S1 is far more stringent in requirements of device uniformity than the S2. On the other hand, of course, a certain degree of device variability is necessary to ensure the randomness of the random bits. As can be seen from the results in Fig. 4(b), the correlation between random bits is high at smaller Cv (8%), although the probability of randomly generating 0 and 1 bits is close to 0.5. In this regard, the sampling rate of the clock should be adapted to the Cv.

IV. CONCLUSION

In this study, we propose a voltage-to-time transformation scheme to mitigate the effect of Vth distribution shift and thus improve the reliability of diffusive memristor-based TRNG. We demonstrate the feasibility of the scheme with experimental data from the Ag/HfOx/Pt device, showing tolerance to the shift over 50%. The simple operation and low requirement for the device make this scheme versatile for all diffusive memristors. Our work provides a new technical solution for building a highly reliable and practical TRNG using diffusive memristors.

REFERENCES

[1] M. Ammar et al., "Internet of Things: A survey on the security of IoT frameworks," J. Inf. Secur. Appl., vol. 38, pp. 8–27, 2018, doi: 10.1016/j.jisa.2017.11.002.

[2] R. Carboni et al., "Stochastic Memory Devices for Security and Computing," Adv. Electron. Mater., vol. 5, no. 9, pp. 1–27, 2019, doi: 10.1002/aelm.201900198.

[3] S. K. Mathew et al., "2.4 Gbps, 7 mW all-digital PVT-variation tolerant true random number generator for 45 nm CMOS high-performance microprocessors," IEEE J. Solid-State Circuits, vol. 47, no. 11, pp. 2807–2821, 2012, doi: 10.1109/JSSC.2012.2217631.

[4] H. Jiang et al., "A novel true random number generator based on a stochastic diffusive memristor," Nat Commun, vol. 8, no. 882, 2017, doi: 10.1038/s41467-017-00869-x.

[5] K. S. Woo et al., "A Combination of a Volatile-Memristor-Based True Random-Number Generator and a Nonlinear-Feedback Shift Register for High-Speed Encryption," Adv. Electron. Mater., vol. 1901117, pp. 1–7, 2020, doi: 10.1002/aelm.201901117.

[6] B. Dang et al. "Physically Transient True Random Number Generators Based on Paired Threshold Switches Enabling Monte Carlo Method Applications," IEEE Electron Device Lett., vol. 40, no. 7, pp. 1096–1099, 2019, doi: 10.1109/LED.2019.2919914.

[7] Z. Chai et al. "GeSe-based Ovonic Threshold Switching Volatile True Random Number Generator," IEEE Electron Device Lett., vol. PP, no. c, p. 1, 2019, doi: 10.1109/LED.2019.2960947.

[8] Z. Wang et al., "Threshold Switching of Ag or Cu in Dielectrics: Materials, Mechanism, and Applications," Adv. Funct. Mater., vol. 28, no. 6, pp. 1-19, 2018, doi: 10.1002/adfm.201704862.

[9] K. Xue et al., "Theoretical investigation of the Ag filament morphology in conductive bridge random access memories," J. Appl. Phys., vol. 124, no. 15, 2018, doi: 10.1063/1.5042165.

[10] W. Wang et al., "Surface diffusion-limited lifetime of silver and copper nanofilaments in resistive switching devices," Nat. Commun., vol. 10, no. 1, pp. 1–9, 2019, doi: 10.1038/s41467-018-07979-0.

[11] Q. Liu et al., "Real-time observation on dynamic growth/dissolution of conductive filaments in oxide-electrolyte-based ReRAM," Adv. Mater., vol. 24, no. 14, pp. 1844–1849, 2012, doi: 10.1002/adma.201104104.

[12] H. Sun et al., "Direct Observation of Conversion Between Threshold Switching and Memory Switching Induced by Conductive Filament Morphology," Adv. Funct. Mater., vol. 24, no. 36, pp. 5679-5686, 2014, doi: 10.1002/adfm.201401304.

[13] B. Song et al., "A HfO2/SiTe based dual-layer selector device with minor threshold voltage variation," Nanomaterials, vol. 9, no. 3, 2019, doi: 10.3390/nano9030408.

[14] X. Zhao et al., "Modulating the filament rupture degree of threshold switching device for self-selective and low-current nonvolatile memory application," Nanotechnology, vol. 31, no. 14, 2020, doi: 10.1088/1361-6528/ab647d.

[15] E. Piccinini et al., "Self-Heating Phase-Change Memory-Array Demonstrator for True Random Number Generation," IEEE Trans. Electron Devices, vol. 64, no. 5, pp. 2185–2192, 2017, doi: 10.1109/TED.2017.2673867.

A statistics-based background capacitor mismatch calibration algorithm for SAR ADC

Zhiqiang Luo, Peng Wang, Fule Li, Chun Zhang, Zhihua Wang

School of Integrated Circuits, Tsinghua University, Beijing, China

Abstract—This paper presents a statistics-based background capacitor mismatch calibration algorithm for successive approximation register (SAR) analog-to-digital converter (ADC). The calibration algorithm is capable of detecting capacitor mismatch errors based on statistical principles and signal correlation is eliminated by introducing additional dummy capacitors, leading to fast convergence. This calibration increases the signal-to-noise-and-distortion ratio (SNDR) from 64.57dB to 82.03dB and achieves 29dB spurious-free dynamic range (SFDR) improvement. The simulated differential nonlinearity (DNL) and integral nonlinearity (INL) are +0.17/-0.13LSB and +0.36/-0.38LSB respectively.

Keywords—Background calibration, Signal independent, SAR ADC

I. INTRODUCTION

SAR ADC is widely used in medium-to-high resolution and low power situations due to high digitization. However, capacitor mismatch becomes a bottleneck for SAR ADC resolution beyond 12b. Increasing the dimensions of the capacitor can improve matching at the expense of higher power dissipation and drive capability which eliminates the advantages of SAR ADCs [1]. Alternatively, mismatch calibrations are typically applied for high resolution with a small area of capacitor which is only limited by thermal noise. Normally, calibrations can be divided into two categories including foreground calibration [2] and background calibration [3]. Under most circumstances, background calibration is preferred because it can be performed without interrupting the normal ADC operation where digital calibration can achieve high resolution [4] while analog calibration has lower power consumption and area cost [5-6].

In this paper, an analog correction scheme of a 14b SAR ADC with background calibration is proposed. To obtain fast convergence, additional dummy capacitors are performed in the capacitor array for the elimination of signal correlation. In this case, the mismatch error of capacitors can be calibrated within a certain range using the statistical properties of quantization error and comparator noise. Simulation results show the calibration algorithm can achieve high performance.

II. ARCHITECTURE OF CDAC AND CORRECTION METHOD

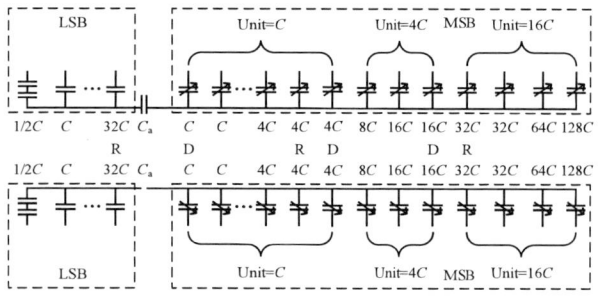

Fig. 1. Schematic diagram of 14-bit CDAC.

This work was supported by Program of Science and Technology Development No. Z201100004320003 and Beijing Smartchip Microelectronics Technology.

The capacitor array is shown in Fig. 1 which contains LSB capacitor array and MSB capacitor array. The LSB capacitor array is a 6-bit capacitive digital-to-analog converter (CDAC) where the highest bit is the redundant bit to tolerate the errors from the MSB capacitor array while the MSB capacitor array is an 8-bit CDAC including 2 redundant bits and 3 dummy capacitors which are used as reference capacitors to calibrate other capacitors. The number of dummy capacitors increases linearly with the number of bits. The MSB capacitor array can be divided into 3 parts and their unit capacitances are C, 4C, and 16C respectively.

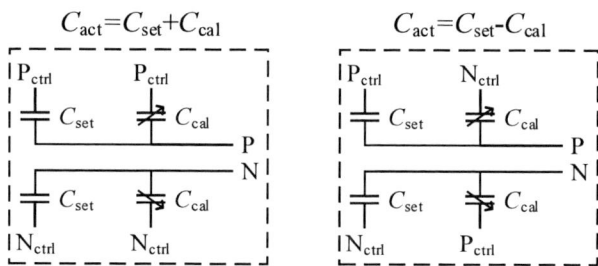

Fig. 2. Compensation principle of capacitor mismatch.

Fig.2 shows the compensation principle of capacitor mismatch for a single capacitor. A programmable capacitor array C_{cal} is in parallel with the capacitor C_{set} of ideal value. Through this analog correction method, the actual capacitance C_{set} can be adjusted a certain stepwise. Benefiting from differential circuits, swapping calibration capacitor C_{cal} on the same side or opposite side is equivalent to increasing or decreasing the capacitance [6]. Although additional calibration capacitor arrays will be introduced, the area cost is also limited due to their small capacitance and low accuracy requirements for their total value.

III. PROPOSED CALIBRATION ALGORITHM

During the sampling phase, the 4 MSB capacitors are used to sample input signals and all remaining ones are connected to common mode voltages. After the successive approximation process, the bottom plate of other capacitors is connected to the reference voltages according to the comparison result of each bit except the dummy capacitors are still connected to the common mode voltage.

To calibrate the dummy capacitor with a capacitance of C, the highest bit of the LSB capacitor array and the dummy capacitor exchange voltages of the bottom plate with each other. Since the weights of the two capacitors are the same, the actual relative size of the two capacitors can be determined by comparing the errors after the exchange is completed. It should be noted that the size of the calibrated capacitor can be judged by only detecting the sign of the error, not the magnitude. In order to reduce the influence of the residual voltage, the last comparison result is also sent to the CDAC with a capacitance of 1/2C which means that the magnitude of the quantization error is reduced by half. To further weaken the influence of the residual voltage, an additional register is adopted to average errors and store comparisons after exchanging the voltage of the bottom plate between the dummy capacitor and the calibrated capacitor. Only when the probability of 1 or 0 is close to 50%, the capacitor mismatch error is small enough. The capacitor will be calibrated well if the probability of 1 or 0 is larger than 58% based on the magnitude of comparator and quantization noise and the calibration accuracy of the capacitor. On the one hand, a better calibration effect can be obtained. On the other hand, hysteresis can be achieved. Since only one extra DA and one more comparison operation are performed and the calibration capacitor is small, the cost of power and speed is negligible. The flowchart of the calibration algorithm is shown in Fig.3.

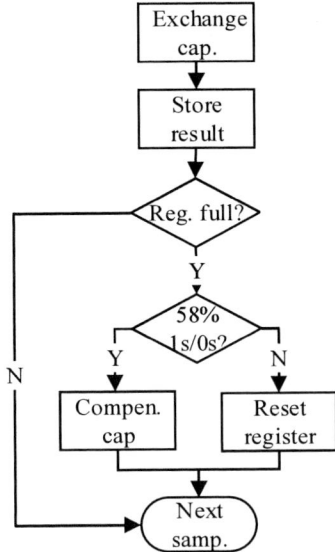

Fig. 3. Flowchart of the calibration algorithm.

When the dummy capacitor with a capacitance of C is calibrated, it will be used to calibrate other capacitors in the first part in the same way. After the dummy capacitor with a capacitance of 4C is calibrated, it will calibrate other capacitors in the second part and the dummy capacitor with a capacitance of 16C will be selected to calibrate other capacitors in the third part. Although the number of dummy capacitors required increases with the improvement of resolution, the increase is linear and will not cause much extra overhead.

IV. SIMULATION RESULTS

The proposed calibration is simulated in MATLAB to verify its effectiveness and performance. The output spectrum with 1% capacitor mismatch before and after calibration is shown in Fig. 4. The SNDR and SFDR before calibration are 64.57dB and 69.67dB, respectively. The SNDR and SFDR are improved by about 17dB and 29dB after calibration. The SFDR against simulated convergence time is illustrated in Fig. 5. It takes about 5M samples to settle the SFDR of the ADC above 90dB. The simulated DNL and INL are +0.17/-0.13LSB and +0.36/-0.38LSB respectively (see Fig.6).

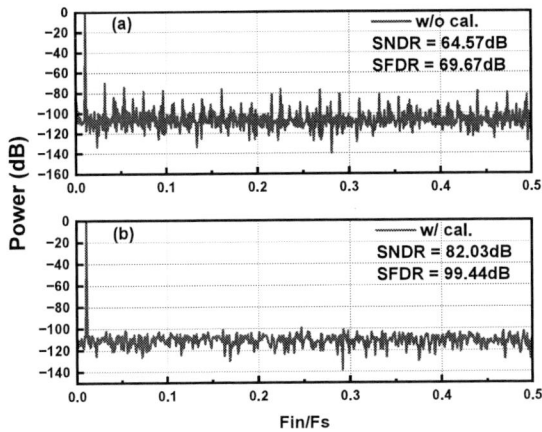

Fig. 4. Output spectrum with 1% mismatch (a) before and (b) after calibration.

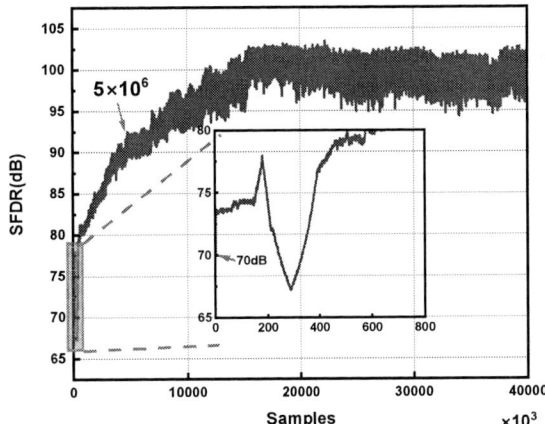

Fig. 5. Simulated SFDR against convergence time.

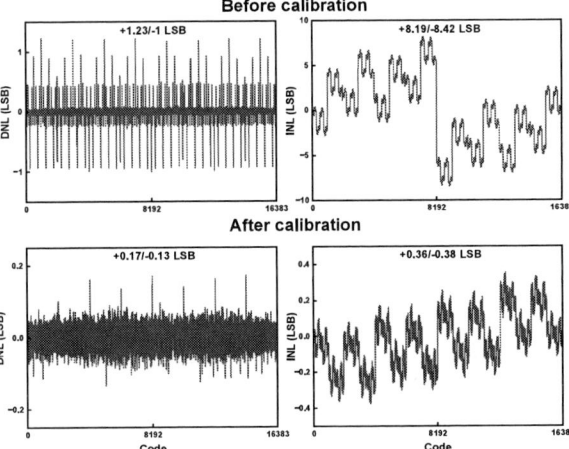

Fig. 6. Simulated DNL and INL before and after calibration.

V. CONCLUSION

A statistics-based signal-independent background capacitor mismatch calibration is presented in this paper and simulated results show that the proposed algorithm can significantly improve the performance of the SAR ADC.

ACKNOWLEDGMENT

This work was supported by Program of Science and Technology Development No. Z201100004320003 and the Laboratory Open Fund of Beijing Smartchip Microelectronics Technology Co., Ltd.

REFERENCES

[1] G. Promitzer, "12-bit low-power fully differential switched capacitor noncalibrating successive approximation ADC with 1 MS/s," in *IEEE J. Solid-State Circuits*, vol. 36, no. 7, pp. 1138-1143, July 2001.

[2] J. Shen, A. Shikata and C. W. Coln, "A 16-bit 16-MS/s SAR ADC With On-Chip Calibration in 55-nm CMOS," in *IEEE J. Solid-State Circuits*, vol. 53, no. 4, pp. 1149-1160, April 2018.

[3] H. Li, and N. Naeem, "A signal-independent background-calibrating 20b 1MS/S SAR ADC with 0.3ppm INL," in *ISSCC 2018*, pp. 242-244.

[4] J. A. McNeill, K. Y. Chan, "All-Digital Background Calibration of a SAR ADC Using the 'Split ADC' Architecture" in *IEEE Trans. Circuits Syst. I: Reg. Papers*, vol. 58, no. 10, pp. 2355-65, Oct. 2011.

[5] Z. Zhu, X. Zhou, and Q. Li, "A 14-bit 4-MS/s VCO-Based SAR ADC With Deep Metastability Facilitated Mismatch Calibration," in *IEEE J. Solid-State Circuits*, vol. 55, no. 6, pp. 1565-1576, June 2020.

[6] M. Ding, P. Harpe, and H. de Groot, "A 46 uW 13 b 6.4 MS/s SAR ADC With Background Mismatch and Offset Calibration," in *IEEE J. Solid-State Circuits*, vol. 52, no. 2, pp. 423-432, Feb. 2017.

TEDOP: A Tiny Event-Driven Neural Network Hardware Core Enabling On-Chip Spike-Driven Synaptic Plasticity

Cong Shi[1], Sihao Chen[1], Haibing Wang[1], Zhengqing Zhong[1], Ping Li[2], Junxian He[1], Tengxiao Wang[1], Jianyi Yu[1], Min Tian[1*]
[1]School of Microelectronics and Communication Engineering, Chongqing University, Chongqing 400044, China
[2]China Resources Microelectronics (Chongqing) Ltd., Chongqing 401331, China
* E-mail: tianmin@cqu.edu.cn

Abstract—For edge intelligent applications, this work proposes a tiny neuromorphic hardware core embedding high-speed on-chip synaptic plasticity, by adopting the proposed Temporal-Integrate neuron model and a simplified supervised spike-driven synaptic plasticity rule for on-chip learning. The proposed hardware core was prototyped on a very-low-cost Zybo Zynq-7010 FPGA device, and attained comparably high classification accuracies on many datasets (e.g. 90.4% on MNIST), with a learning and inference speed as high as 11,268 and 11,749 frame/s, respectively, while dissipating only 39 mW power under a 250 MHz clock frequency.

Keywords—Spike-driven synaptic plasticity (SDSP), Event-driven, Neuromorphic hardware, On-chip learning, Spiking Neural Network

I. INTRODUCTION

With the rapid increase of intelligent computing requirement in ubiquitous artificial internet-of-things (AIoT) edge devices, the neuromorphic computing paradigm based on spiking neural network (SNN) models have garnered more and more interests, due to their computational efficiency by emulating the human brain cortex that processes sensory information via the form of spatiotemporally sparse electrical pulses (i.e., spike events) [1]. However, many large-scale general-purpose neuromorphic processors demand high area and energy consumptions [1]-[2], which are not suitable for embedded or mobile applications at the edge side under stringent energy and cost budgets. To fit edge constraints, small-scale neuromorphic hardware systems specifically tailored to dedicate usage, SNN topology, neuron model or synaptic plasticity (i.e., on-chip learning method) are proposed. Among them, the FPGA-based implementations [3]-[6] permits flexible architectural investigations and enjoy faster prototyping cycles than their ASIC tape-out counterparts [7]-[8]. Yet, these designs still do not exhibit a good tradeoff among hardware cost, real-time performance and on-chip learning efficacy for extremely resource-constrained situations.

In this work, we propose a tiny event-driven neural network hardware core containing only ten neurons in a single-layer network and embedding high-speed on-chip learning capability. We propose the Temporal-Integrate (TI) neuron model, which is a simplified version of the well-known Integrate & Fire (IF) spiking neuron model, coupled with a simplified stochastic supervised Spike-Driven Synaptic Plasticity (S4DSP) rule for on-chip learning. The proposed hardware core was prototyped and consumed merely 360 (8.18%) Slices, 741 (2.1%) FFs, 4 (6.7%) BRAMs and 0 Multiplier/DSP on the low-cost Zybo Zynq-7010 FPGA chip. It attained a comparable classification accuracy of 90.4% on the MNIST dataset, with a learning and inference speed of 11,268 and 11,749 frame/s, respectively, while only dissipating 39 mW power at a 250 MHz clock rate.

II. NEURON MODEL AND ON-CHIP LEARNING METHOD

The proposed TI neuron is a behaviorally simplification from the IF model by removing the firing mechanism. In the event-driven computing manner, the membrane potential V_j of neuron j is updated upon every input spike event:

$$V_j(t) = V_j(t_{\text{prev}}) + \sum_i (w_{ij} - E_{\text{inh}})s_i(t) + T_j(t), \ \ s.t. \ \ V_j(t=0) = V_{\text{init}} \quad (1)$$

where i is the synapse index, t the algorithmic (discrete) time-step, t_{prev} the latest previous time when V_j was updated, $w_{ij} \geq 0$ the connecting synaptic weight, E_{inh} a constant inhibitory term, V_{init} the initialization value for V_j when the spike trains of a new input sample begin to present, $s_i(t) = 1$ for synapse i receiving an input spike at time t, and 0 otherwise. Unlike an IF neuron, the TI neuron has no firing threshold and never fires output spikes. The teacher signal T_j in (1) is omitted during inference, and is calculated before V_j update as below during the learning procedure: when the category labels of the training sample and the neuron match, and $V_j(t_{\text{pre}}) < V_{\text{th1}}$, then $T_j(t) = \eta^+ > 0$. If their labels do not match and $V_j(t_{\text{pre}}) > V_{\text{th2}}$, then $T_j(t) = \eta^- < 0$, where $V_{\text{th2}} < V_{\text{th1}}$. Otherwise, $T_j(t) = 0$. The parameters E_{inh}, V_{init}, η^+, η^-, V_{th1}, V_{th2} are all configurable.

Upon an input spike arriving at synapse i of neuron j, after the V_j update in (1), the proposed S4DSP learning algorithm adjusts the weight w_{ij} as:

$$w_{ij} \leftarrow \begin{cases} w_{ij} - \Delta w, \text{if } \theta_1 \leq V_j(t) < \theta_2 \text{ and } prob \geq rand \\ w_{ij} + \Delta w, \text{if } \theta_2 \leq V_j(t) < \theta_3 \text{ and } prob \geq rand \\ w_{ij}, \text{ otherwise} \end{cases} \quad (2)$$

where updating thresholds $\theta_1 < \theta_2 < \theta_3$, weight change amount $\Delta w > 0$ and learning probability $0 < prob < 1$ are configurable. The *rand* is a random variable between 0 and 1, and sampled at every weight update. Compared to the previous supervised SDSP algorithm [9], our method eliminates the allocation of the Ca^+ ion variable, which is essentially a spiking neuron's output spike count. By leveraging the approximate linear relationship $V_{j,\text{TI}}(t) \approx Ca^+_{\text{IF}}(t) \times V_{\text{th,IF}}$ between the non-firing TI neuron's V_j, and the firing threshold $V_{\text{th,IF}}$ and the output spike count Ca^+_{IF} of a reference IF neuron, the dependence of weight change on the Ca^+ condition [9] in the IF neuron can be transferred to be on the V_j condition in the TI neuron as shown in (2). Thus, computational overheads and hardware costs are reduced. Moreover, we incorporate the stochasticity into weight updates [8], which reduces the weight bit precision requirement and saves synaptic weight storage space.

III. HARDWARE DESIGN

The architecture and key block circuits of the proposed high-speed, low-cost, event-driven neuromorphic hardware core are depicted in Fig. 1(a)-(c). This core can train and run a single-layer fully-connected neural network consisting of up to $J_{\text{max}} = 10$ TI neurons, as well as up to $I_{\text{max}} = 1024$ input nodes that just passes input spikes to the neurons via fully-connected synapses w_{ij}. The core mainly contains a global scheduler, an input spike FIFO, a parameter register bank, a neural state updater, a learning engine, a decision unit, along with an $8b \times (I_{\text{max}} \times J_{\text{max}}) = 8b \times 10240$ synaptic weight (w_{ij}) memory, and a $16b \times J_{\text{max}} = 16b \times 10$ membrane potential (V_j) memory.

Input spikes are buffered in the input FIFO using the address-event representation (AER) [7] indicating the index of the input node receiving the input spike. Upon each AER event, neurons and synapses are processed sequentially by this single core in a time-multiplexing manner under the control of the global scheduler. The neural state updater performs V_j updates according to (1). The learning engine executes the S4DSP algorithm in (2). The random variable for stochastic weight updates is generated online by a linear feedback shift register (LFSR). The *prob* and *rand* in Fig. 1(c) are 8-bit unsigned fractions. After all spikes of an inference sample are handled, the decision unit serially compares each neuron's V_j, and outputs the label of the neuron with the maximum V_j as the inferred result of the sample. The decision unit and the learning engine are inactivated and skipped for training and inference samples, respectively. All computing blocks are pipelined.

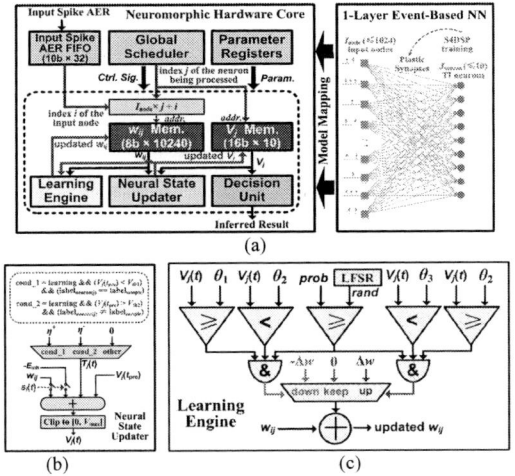

Fig. 1. (a) The hardware architecture of the proposed neuromorphic core. (b) The circuit of the Neural Sate Updater. (c) The circuit of the Learning Engine.

IV. FPGA PROTOTYPE

We implemented an FPGA prototype of the neuromophic core on a very-low-cost Zybo Zynq-7010 chip. We tested the on-chip learning accuracy of our prototype on static image datasets including MNIST (90.4%), Fashion-MNIST (80.93%) and ETH-80 (79.57%), as well as dynamic sensor captured AER streams including MNIST-DVS (74.2%), Poker-DVS (95.65%) and Posture-DVS (98.3%). For each dataset, the counts of input nodes and the TI neurons of the mapped neural network equaled to the sample spatial resolution and the number of categories, respectively. The prototype evaluation system was built following [10]. Briefly, a host PC paritioned training/inference subsets, converted static image pixels to rate-coded spike trains in 200 time-steps, and preprocessed the DVS streams, and downloaded spike AERs to the FPGA via a 1000 Mbps Ethernet link sample by sample for processing. More details about the evaluation setup can be found in [10].

Table I compares our work with other recent FPGA-based neuromorphic systems that have reported their on-chip learning accuracies on the MNIST set. Our design consumed very few resources, while still achieving a comparable recognition accuracy. Our prototype ran as fast as 11,268 and 11,749 frame/s during training and inference while dissipating only 39 mW power at a 250 MHz clock frequency, leading to an energy efficiency 3-6 orders of magnitude higher than other works.

V. SUMMARY

This work proposes a low-cost event-driven neuromorphic hardware core supporting high-speed on-chip S4DSP leaning. An FPGA prototype was implemented and realized comparable accuracies on MNIST images at 11,268 and 11,749 frame/s during training and inference, respectively, while comsuming very few FPGA resources and only 39 mW power dissipation. This demonstrate that our design is quite suitable for extremely resource-constrained real-time intelligent edge applications.

ACKNOWLEDGMENT

This work was funded in part by the National Key Research and Development Program of China (Grant No. 2019YFB2204303), in part by the National Natural Science Foundation of China (Grant No. U20A20205), in part by the Key Project of Chongqing Science and Technology Foundation (Grant. cstc2019jcyj-zdxmX0017, cstc2021ycjh-bgzxm0031), in part by the Open Research Funding from the State Key Laboratory of Computer Architecture, ICT, CAS. (Grant No. CARCH201908), and in part by the Chongqing Social Security Bureau and Human Resources Dept. (Grant No. cx2020018).

REFERENCES

[1] F. Akopyan, et al., "A million spiking-neuron integrated circuit with a scalable communication network and interface," *Science*, vol. 345, no. 6197, pp. 668-673, Aug. 2014.

[2] M. Davies, et al., "Loihi: A Neuromorphic Manycore Processor with On Chip Learning," *IEEE Micro*, vol. 38, no. 1, pp. 82-99, Jan./Feb. 2018.

[3] Q. Wang, et al., "Energy efficient parallel neuromorphic architectures with approximate arithmetic on FPGA," *Neurocomputing*, vol. 221, 2017.

[4] H. Zheng, et al., "Balancing the Cost and Performance Trade-Offs in SNN Processors," *IEEE Trans. Circuits and Systems II: Express Briefs*, vol. 68, no. 9, pp. 3172-3176, Sept. 2021.

[5] S. Li, et al., "A Fast and Energy-Efficient SNN Processor With Adaptive Clock/Event-Driven Computation Scheme and Online Learning," *IEEE Trans. Circuits and Systems I: Regular Papers*, vol. 68, no. 4, pp. 1543-1552, April. 2021.

[6] J. Wu, et al., "Efficient Design of Spiking Neural Network With STDP Learning Based on Fast CORDIC," *IEEE Trans. Circuits and Systems I: Regular Papers*, vol. 68, no. 6, pp. 2522-2534, June. 2021.

[7] C. Frenkel, et al., "A 0.086-mm^2 12.7-pJ/SOP 64k-Synapse 256-Neuron Online-Learning Digital Spiking Neuromorphic Processor in 28-nm CMOS," *IEEE Trans. Biomed. Circuits Syst.*, vol. 13, pp. 145–158, 2019.

[8] C. Frenkel, et al., "MorphIC: A 65-nm 738k Synapse/mm^2 Quad-Core Binary-Weight Digital Neuromorphic Processor With Stochastic Spike-Driven Online Learning," *IEEE Trans. Biomed. Circuits Syst.*, vol. 13, no. 5, pp. 999-1010, Oct. 2019.

[9] Y. Zhang, et al., "A Digital Liquid State Machine With Biologically Inspired Learning and Its Application to Speech Recognition," *IEEE Trans. Neural Networks and Learning Systems*, vol. 26, no. 11, pp. 2635-2649, Nov. 2015.

[10] H. Wang, et al., "TripleBrain: A Compact Neuromorphic Hardware Core with Fast On-Chip Self-Organizing and Reinforcement Spike-Timing Dependent Plasticity," *IEEE Trans. Biomed. Circuits Syst.*, 2022, doi: 10.1109/TBCAS.2022.3189240.

TABLE I WORK COMPARISON

FPGA (& Ref.)	Virtex-6 [3]	ZCU102 [4]	Virtex-7 [5]	Zynq-7035 [6]	Zynq-7045 [10]	Zynq-7010 (Ours)
slice LUTs	97,287	23,573	N/A	78,586	10,052	952 (5.4%[1])
slice reg. (flip-flops)	58,826	8713	N/A	23,498	8505	741 (2.1%)
slices	N/A	N/A	N/A	N/A	4146	360 (8.2%)
DSPs	N/A	0	N/A	433	32	0 (0%)
BRAMs	34	431.5	N/A	N/A	131	4 (6.7%)
clock freq.	120 MHz	200 MHz	100 MHz	301.8 MHz	250 MHz	250 MHz
input spike encoding	N/A	rate	rate	rate	rate	rate
SNN model	784-800	784-2304-10	784-100	784-100	784-256	784-10
on-chip learn.	STDP	TSTDP	STDP	STDP	SOM-STDP&R-STDP	S4DSP
learning speed	0.06 fps	N/A	61 fps	164 fps	1349 fps	11,268 fps
inference speed	0.12 fps	46.45 fps	317 fps	N/A	2698 fps	11,749 fps
accuracy @MNIST[2]	89.1%	90.58%	85.28%	83.4%	93.21%~95.10%	90.4%
power	80~134 mW	782 mW	1610 mW	1090 mW	938 mW	39 mW
learning energy efficiency	1340 mJ/img	N/A	26.32 mJ/img	6.64 mJ/img	0.70 mJ/img	3.46 μJ/img
inference energy efficiency	1127 mJ/img	16.83 mJ/img	5.04 mJ/img	N/A	0.35 mJ/img	3.32 μJ/img

[1] Utilization percentage of all that type of resources available on the Zybo Zynq-7010 FPGA chip.

[2] Our accuracies of our prototype on other more complicated datasets are given in the texts in Section IV.

A 92.7% Peak Efficiency 48/1V DSD Power Converter with 102mV Droop and 1.6μs Settling Time for a 1A/10ns Load Transient

Yongchao Zhang, Zhuoqi Guo*, Zhongming Xue, Zhuoneng Li, Xihao Liu, Shangzhou Zhao, Dexuan Lv, Mengqi Duan, Li Geng*

School of Microelectronics, Xi'an Jiaotong University
Corresponding Author Email: guozhuoqi2004@163.com, gengli@xjtu.edu.cn

Abstract—With the rapid growth of data centers, the power supply has shifted from 48/12/1V two-stage architecture to 48/1V single-stage. In this paper, a new two-phase sawtooth voltage mode PWM control is proposed for the double step-down (DSD) converter. In order to solve the problem of inherent cycle delay of PWM control, a fast-transient response scheme is proposed. The converter also has a precharge and soft start scheme, which is designed with a 0.18 μm BCD process. It achieves peak efficiencies of 92.7%, 90%, 87.8%, and 86% at 250 kHz, 500 kHz, 750 kHz, and 1 MHz, respectively. During a 1A/10 ns load jump, the undershoot is reduced from 200 mV to 102 mV and the setting time is reduced from 5.3 μs to 1.6 μs.

Keywords—48/1V，double step down (DSD)，two-phase sawtooth, fast transient response

I. INTRODUCTION

In recent years, with the development of data centers to ultra-large scale, more and more attention has been paid to the energy consumption of their power supply systems. As shown in Fig. 1, the data center's power supply system has gradually shifted from 48/12/1V two-stage architecture to 48/1V single-stage to reduce the conduction loss of the power grid.

Double step-down (DSD) converter is a suitable candidate due to its double duty cycle and double equivalent frequency [1,2]. However, it is still a challenge for stable control to achieve fast response when the load jumps. Because DSD converter requires two control signals, which have 180° phase shift and cannot be overlapped, hysteretic control is obviously not applicable. There is an inherent cycle delay when load-jump appears in the traditional PWM control, which leads to the risk of two-phase control overlapping. [3] proposed an adaptive on-off time (AO²T) control. A mirror circuit is designed, achieving settling time of 8.2μs.

In this paper, a new two-phase sawtooth voltage mode control scheme is proposed for the DSD converter, and a fast-transient response scheme is proposed to solve the PWM control cycle delay problem. The overall system control scheme is proposed and the key circuits are analyzed in the second part. Then the simulation results are presented, and in the last part, we conclude the paper.

II. PROPOSED CONTROL SCHEME AND FAST RESPONSE TECHNOLOGY

A. Principle of two-phase sawtooth voltage mode PWM control loop and fast transient response loop

Fig. 2 is the block diagram of the proposed system based on DSD converter, including power stage, two-phase sawtooth voltage mode control, fast transient response module, non-overlapping module, deadtime & driver module, and precharge & soft-start module. During operation, CLK&RAMP module generates two intersecting staggered sawtooth waves, RAMP1 and RAMP2, and then generates two independent control signals, PWM1 and PWM2 by comparing with the same EA's output

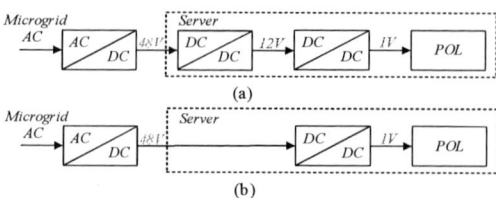

Fig. 1 (a) Two-stage power conversion and (b) single-stage power conversion.

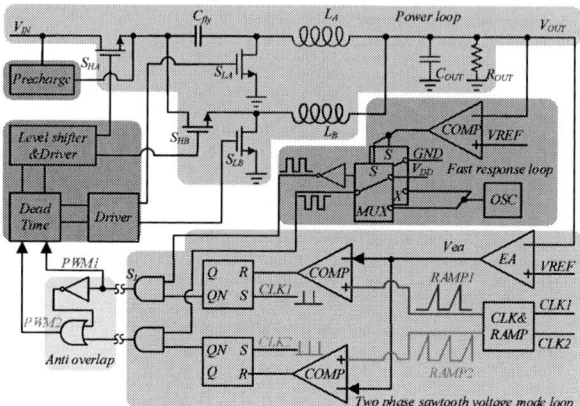

Fig. 2 Overall block diagram of the proposed DSD converter.

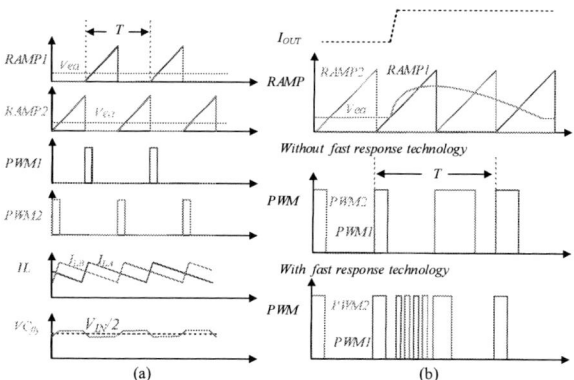

Fig. 3 (a) Steady state waveform and (b) transient waveform.

signal, V_{ea}, avoiding the risk of simultaneous conduction of two-phase high-side MOSFET in traditional PWM controller. The steady-state waveform is shown in Fig. 3 (a). If the load jump occurs after the falling edge of the PWM wave, the system needs to wait for the next PWM wave to arrive before adjusting. The waiting period is the inherent cycle delay of the PWM control. During the waiting period, the output voltage will produce a large voltage drop and the setting time will be extended. To solve this problem, this paper proposes a fast-transient response scheme. The acceleration principle is equal to that when the load jumps, the switching frequency of the system is equivalently increased, so as to quickly respond to the change of the load. When the load jumping is detected, the fast-transient response loop acts as follows. Before the PWM waveform arrives, two high frequency pulse signals with opposite phase are generated to control the two switches, the two-path inductor currents are rapidly interleaved for charging and discharging, which reduces the current difference and reduces the undershoot and setting time.

B. Design of key circuit modules

Fig. 4 is the clock and sawtooth generation circuit module. Through D trigger, the clock signal generated by the oscillator is divided into two-phase signals OSC1,2 with the duty ratio of

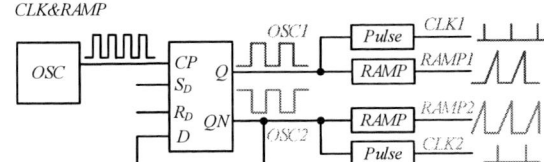

Fig. 4 Clock and sawtooth wave generation circuit.

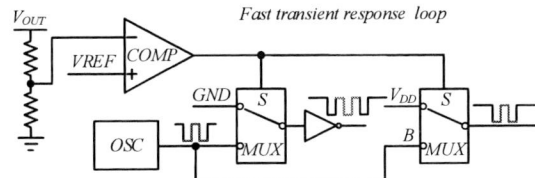

Fig. 5 Fast-transient response circuit.

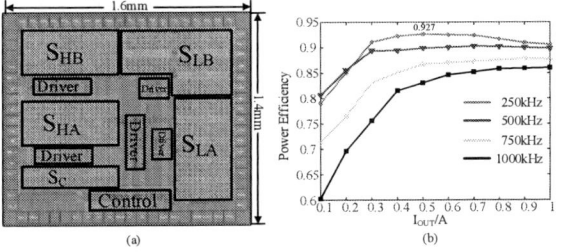

Fig. 6 Transient response during 1A load jump, with and without the fast-transient response loop.

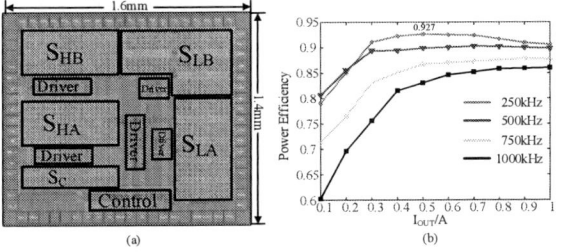

Fig. 7 (a) Layout and (b) efficiency curves of proposed DSD converter.

TABLE I. PERFORMANCE COMPARISON

	JSSC 2020[3]	ISSCC 2021 [4]	This Work
Process	0.18μm BCD	0.18μm BCD	0.18μm BCD
Topology	DSD	12-level series capacitor	DSD
Control Scheme	AO²T	VCDL	PWM
Power Switch Type	GaN	GaN & On-chip MOSFET	On-chip MOSFET
V_{IN}	48V	36~60V	48V
V_{OUT}	1V	0.5~1V	1V
Maximum I_{LOAD}	1.5A	8A	1A
Area	1.46 mm² (without power switch)	18.29 mm²	2.24 mm²
Frequency	100~2000kHz	2500kHz	250~1000kHz
Inductor	2×0.9μH	NA	2×4.7μH
C_{OUT}	22μF	4×47μF	2×4.7μF
C_{fly}	1μF	NA	1μF
Efficiency	85.4%@100kHz; 79%@250kHz; 56.8%@2000kHz	90.2%@2500kHz	92.7%@250kHz; 90%@500kHz; 87.8%@750kHz; 86%@1000kHz
Load step (rise time)	1A	5A	1A(10ns)
Undershoot	160mV	114.5mV	102mV
Settling time	8.2μs	80μs	1.6μs

signals with opposite phase to control the switches, and the two inductors' currents are alternately charged and discharged rapidly. The undershoot is reduced to 102mV and the recovery time is reduced to 1.6μs. Besides, the difference between the two inductors' currents is reduced from 0.605A to 0.125A.

Fig. 7(a) is the layout of the DSD converter. This design has been simulated and verified in a 0.18μm BCD process. The overall layout area is 2.24mm². Fig. 7(b) is the efficiency curves of proposed DSD converter, which achieved peak efficiencies of 92.7%, 90%, 87.8%, and 86% at 250 kHz, 500 kHz, 750 kHz, and 1 MHz, respectively. Table I is the performance comparison. Compared with other works. This design has higher efficiency, lower undershoot and shorter setting time.

IV. CONCLUSION

In this paper, a new two-phase sawtooth voltage mode PWM control is proposed for the DSD converter, which achieves the high reliability control of 48/1V converter with a peak efficiency of 92.7%. Meanwhile, a fast-transient response loop is proposed to improve the equivalent switching frequency of the system when the load jumps. The undershoot is reduced to 102mV, and the setting time is reduced to 1.6μs during a 1A/10ns load jump.

ACKNOWLEDGMENT

The authors thank the supports from National Key Research and Development Program of China (2019YFB2204700), Xi'an Major Scientific and Technological Innovation Platform and Local Transformation Project of Scientific and Technological Achievements (20KYPT0001-11-1), and High-end Power Management Chip Project of the Ministry of Industry and Information Technology (TC210H03T).

REFERENCES

[1] O. Kirshenboim et al, "Closed-Loop Design and Transient-Mode Control for a Series-Capacitor Buck Converter," *IEEE Trans. Power Electron.*, vol. 34, no. 2, pp. 1823-1837, Feb. 2019.

[2] P. S. Shenoy et al., "A 5 MHz, 12 V, 10 A, monolithically integrated two-phase series capacitor buck converter," *2016 IEEE Applied Power Electronics Conference and Exposition (APEC)*, 2016, pp. 66-72.

[3] D. Yan et al, "Direct 48-/1-V GaN-Based DC–DC Power Converter with Double Step-Down Architecture and Master–Slave AO2T Control," *IEEE J. Solid-State Circuits*, vol. 55, no. 4, pp. 988-998, Apr. 2020.

[4] X. Yang et al., "33.4 An 8A 998A/inch3 90.2% Peak Efficiency 48V-to-1V DC-DC Converter Adopting On-Chip Switch and GaN Hybrid Power Conversion," *2021 IEEE International Solid- State Circuits Conference (ISSCC)*, 2021, pp. 466-468.

50%. Then through the sawtooth wave generating circuit and the pulse reduction module, two-phase sawtooth waves RAMP1,2 and two-phase clock signals CLK1,2 are generated.

Fig. 5 is the fast-transient response circuit module. In the steady state, the comparator's output is low, and the output signals of the two multiplexers are low and high, respectively, which will not affect the normal PWM control. When the load jumps from 10mA to 1A, V_{OUT} will drop rapidly. As it drops to the pre-set voltage threshold, the comparator generates a high signal, which makes the multiplexers generate high frequency pulse signals. As shown in Fig. 2, the pulse signals are merged into the main loop through an AND gate, the power MOSFET can be turned on and off in advance to achieve fast response.

III. SIMULATION RESULTS

This design realizes the stable control of 48/1V converter based on DSD topology, and the output ripple is 11.4mV at the switching frequency of 500kHz. Fig. 6 is the waveforms during the load jump. The load current varies from 10mA to 1A in 10ns, the undershoot is 200mV and the setting time is 5.3μs without fast transient response loop. After adding fast transient response scheme, extra loop will generate two high frequency pulse

A 14.39ppm/kPa Stress Sensor with Low Temperature-drift and High Linearity for turbulence Stress

Lanxiang Xiao[1,2], Lei Chen[2], Fengwei An[2*]

[1] Harbin Institute of Technology, Harbin, China
[2] Southern University of Science and Technology, Shenzhen, China (Email: anfw@sustech.edu.cn)

Abstract—In the analysis and calculation of turbulent flow, the stress generated by the fluid needs to be accurately measured by sensors. In this research, a mechanical stress sensor was developed based on standard 180nm CMOS technology for real-time measurement of stress. The main contributions of this study include the following: 1) The temperature-compensated RC oscillator circuit changes mechanical stress-induced chip deformation into a frequency output. When the temperature was changed from -40°C to 120°C, the oscillator frequency only changed to 0.303Hz/°C. 2) No complicated calibration and complicated measurement equipment are required. The chip area does not exceed 0.569 mm^2. Moreover, the power consumption does not exceed 22.9 μW. 3) The average sensitivity measured by the chip is 14.39ppm/kPa, and the linear fitting curve's determination coefficient (R^2) is 0.9983.

Keywords—Turbulence, mechanical and thermal stresses, RC oscillator circuit, temperature-compensation, stability, sensitivity.

I. INTRODUCTION

Stress measurements in turbulent boundary layers are very important for fluid mechanics. Due to their large available bandwidth and small spatial size, MEMS sensors are often used for fundamental research on turbulence, flow control feedback, and aerodynamic drag measurements, as well as for validating computational turbulence models [1]. Currently, the commonly used floating element sensors have fluid damping and heat transfer problems to thermal sensors, making it difficult to

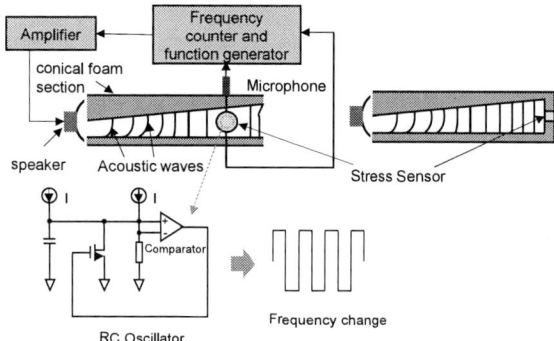

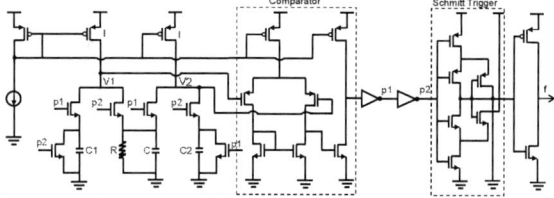

Fig.1. Schematic of plane-wave tube device used for stress measuring in air turbulence models.

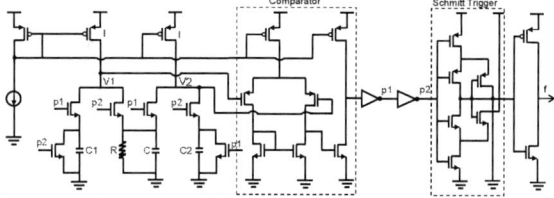

Fig.2. Architecture of the stress sensor integrated with the RC Oscillator converting the stress to a frequency.

accurately predict the frequency response function [2]. In order to improve the measurement accuracy, MEMs sensors with small size and high-temperature stability can be used. The integrated capacitive stress sensor manufactured using the standard CMOS and microelectromechanical system (MEMS) technology has been applied in pressure sensors in [3]. Piezoresistive stress sensors result in resistance changes due to the piezoresistive effect that can be measured [4]. This paper developed a mechanical stress sensor based on 180nm CMOS technology for real-time stress measuring. A sensor circuit with on-chip integration is used in order to reduce parasitic capacitance. This sensor has a good frequency and temperature stability, small area, low noise, and good linearity within the measurable pressure range.

II. PRINCIPLE AND REALIZATION OF THE STRESS SENSOR

A. Principle of the capacitive stress sensor

The RC oscillator circuit converts the stress-induced capacitance variation into frequency output, as shown in Fig. 1. To measure the stress generated by air turbulence, sound wave disturbance is adopted to excite the air turbulence effect. The sound wave generated by the speaker affects the airflow, causing the air density to change, resulting in the turbulent flow phenomenon. The strain sensor is properly installed at the bottom of the tube. The stress sensor converts the force into a frequency, which a frequency counter can record. The function generator is used to generate acoustic signals with different waveforms, and then the generated signals are amplified by amplifiers.

B. Offset-canceling oscillator

The RC oscillator in Fig. 2 has a symmetrical switching mechanism that eliminates the comparator voltage offset. The working principle of the symmetrical switching mechanism is that p1 and p2 turn on the MOS tubes in turn, and the current I passes through resistor R, while the capacitor C in parallel is charged for stabilizing the voltage. Voltage V2 at both ends of resistance R remains at I·R. When phase p1=1, capacitor C1 is charged, the voltage across C1 is V1, and it comes to V2 within the time lag of R·C. After t_{delay}, which corresponds to the delay of comparator and buffers, ph2 opens the switch for discharging C1, voltage V1 immediately drops and remains at I·R. After that, the capacitor C2 starts to charge by the current mirror until it reaches I·R. In the end, V1 and V2 fall and rise in turn within 2(R·C) with negligible comparator and buffer delays. Capacitors C1 and C2 operate in parallel so that the charging/ discharging delay of the capacitor is not included in the cycle. The frequency f0 flaps when the sensing capacitor array C1 and C2 changes caused by the warpage due to a stress loading. The temperature stability of this oscillator is mainly affected by the temperature drift of the resistance/capacitor and on-resistance/ leakage of the switches. A POLY resistor with a temperature drift of only 3.4835×10^{-6}/°C is adopted in this work.

Inter-plate capacitance is a function of overlapping plate area (A), plate spacing (d), and dielectric constant (ε), (C=εA/d). When a load is applied to the chip, the deformation of the chip results in a variation of the distance (Δd) between the capacitor plates. It is observed that the frequency of the oscillation is not related to the current and voltage, but its fluctuation only relies on Δd. In this work, we verify the sensing array with 20 cells of the MIM capacitor. The capacitor used in this sensor is square with the same width and length. When the capacitor is under pressure, it can be seen as a square sheet under forces. According to elastic theory, the relation between stress and deformation can

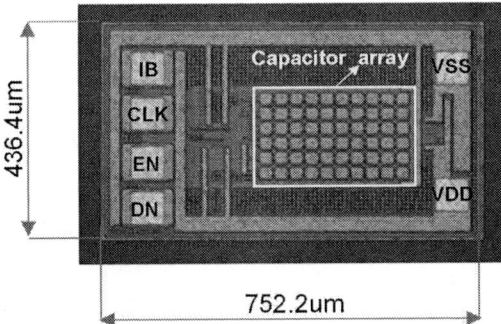

Fig.3. The oscillator in 180 nm CMOS technology.

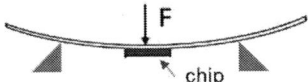

Fig.4. The experimental environment.

Index	MEMS solution [5]	This work
Temperature drift (/°C)	33.26ppm	30.34ppm
Process	0.35um CMOS	180nm CMOS
Power Consumption (μW)	/	22.65 @5V
Sensitivity (/kPa)	14.55 ppm	14.39ppm
Area (μm²)	600 × 800	436×752

Fig.5. The comparison of the temperature drift and sensitivity.

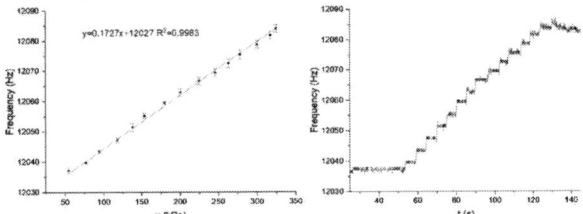

Fig.6 Frequency Measurements of oscillator vary with time and force

be obtained according to the displacement equation of equilibrium in [5].

C. Relation between the stress and frequency

When a square capacitor is stressed, it can be regarded as a square sheet under stress, which will lead to distance variation between plates, which leads to the change in capacitance and the oscillation frequency. Its relation is described in formula (1).

$$f_0=1/kRC \qquad (1)$$

Where f_0 is the frequency variation of the measured signal, R represents the oscillator resistance, k is the proportional coefficient, and C is the capacitance of the stress sensor.

III. EXPERIMENTAL RESULTS

The test chip with a standard 180 nm CMOS process demonstrated in Fig. 3 is designed and manufactured for performance testing of the stress sensor chip. A test system including a power supply (NF E36312A), a frequency counter (Keysight 53210A), and a material testing machine (INSTRON 5943) are used to apply a force, as demonstrated in Fig. 4. During the experiment, the pressure head moved downward with the constant 0.05mm displacement each time. Experiments were carried out using the three-point bending method, as shown in Fig. 4. The sensor chip was mounted on PCB with bottom filler and wire bonding. The chip is placed in the center of the bottom surface of the PCB board while the distance between the holder and the PCB center is 40mm. The pressure head can apply a known displacement or uniaxial compressive stress to the sensor chip. Based on the principle of three-point bending, the chip in the middle of the PCB will generate a deformation. This causes a change in the distance of the capacitor plates in the oscillator, which changes the oscillator frequency. Insulating glue was pasted on a three-point bending fixture to prevent leakage. The temperature stability measurement was shown in Fig. 5 over −40°C to 120°C. Furthermore, the power consumption is 22.65uW.

In Fig. 6, the output frequency is proportional to the intensity of pressure on PCB from 57 kPa to 326 kPa. The measured average sensitivity is 0.1736 Hz/kPa (14.39ppm/kPa). The linear fitting curve's determination coefficient (R^2) is 0.9983, which is better than the CMOS+MEMS solution in [6] with R^2−0.9862, indicating that the sensor has an excellent linear response.

IV. CONCLUSION

The on-chip capacitive stress sensor test chip with an integrated RC oscillator circuit was successfully realized in the 180 nm CMOS process. The sensor consumes 22.65 uW at 5V, and the sensitivity is 0.1736Hz/kPa, while the linear fitting curve's determination coefficient (R2) is 0.9983.

ACKNOWLEDGMENT

This work is supported by the Shenzhen Science and Technology Innovation Committee Fund JSGG20200102162401765.

REFERENCES

[1] Chandrasekharan V , Sells J , Meloy J , et al. A metal-on-silicon differential capacitive shear stress sensor[C]// Solid-State Sensors, Actuators and Microsystems Conference, 2009. TRANSDUCERS 2009. International. IEEE, 2009.

[2] Mark, Sheplak, et al. Dynamic Calibration of a Shear-Stress Sensor Using Stokes-Layer Excitation[J]. AIAA Journal, 2001, 39(5):819-819.

[3] Dai, C. L et al., "Capacitive Micro Pressure Sensor Integrated with a Ring Oscillator Circuit on Chip", Sensors,2009, 9(12):10158-10170.

[4] Mahsereci Y et al., "An Ultra-Thin Flexible CMOS Stress Sensor Demonstrated on an Adaptive Robotic Gripper", JSSC,2015,58(1):290-291.

[5] Rui L et al., "On the finite integral transform method for exact bending solutions of fully clamped orthotropic rectangular thin plates", Applied Mathematics Letters,2009,22(12):1821-1827.

[6] Chan W P , George A K , Narducci M S , et al. A Monolithically Integrated Pressure/Oxygen/Temperature Sensing SoC for Multimodality Intracranial Neuromonitoring[C]// Solid-state Circuits Conference. IEEE, 2014.

2022 IEEE International Conference on Integrated Circuits, Technologies and Applications

A 30W and 95% Efficiency Class-E Wireless Power Transfer Transmitter with Vector Algorithm Control

Shangzhou Zhao, Zhongming Xue*, Yuhao Xiong, Zhuoneng Li, Xihao Liu, Yongchao Zhang, Zhuoqi Guo, Li Geng*

School of Microelectronics, Xi'an Jiaotong University

Abstract—This paper proposes a vector algorithm control (VAC) method for class-E power amplifier (PA) to compensate the variation of load resistance based on an introduced mathematical model. In contrast to the conventional Class-E PA adaptive control method with impedance matching, which introduces additional power loss, the proposed VAC loop directly adjusts the core parameters of the Class-E PA, thus enhances the system efficiency. A transmitter applicable to wireless power transfer (WPT) systems is implemented to verify the proposed VAC loop. Simulation results show that the system achieves a peak output power of 30W and a very high peak efficiency of 95% with wide variable loads.

Keywords—*class-E PA, vector algorithm control, wireless power transfer, load variation capability.*

I. INTRODUCTION

Wireless power transfer (WPT) based on resonant inductive links has received a lot of attention recently to provide power to broad applications such as portable devices, medical implants, household appliances, and even electric vehicles, etc. With the increasing power levels of WPT systems, the requirement of output efficiency is higher. The class-E power amplifier (PA) is often applied in WPT systems thanks to its simple topology and soft switching property, i.e., zero-voltage-switching (ZVS) and zero-voltage-derivative-switching (ZVDS). However, the class-E PA is sensitive to the variations of load impedance. In real applications, difficulties in the control arise from various power demands and uncertainties such as the deviation in the relative position of coils and load characteristics.

Traditional tuning methods based on impedance matching and frequency tuning are showed in Fig.1. Impedance matching is the most intuitive approach, but additional components occupies a large area and increases the ohmic power loss due to the equivalent series resistances. In contrast, frequency tuning is simple, but the transmission frequency deviates from standard communication protocol. In this paper, the mathematical model of the class-E PA is introduced, and thus a novel vector algorithm control (VAC) loop is proposed, which directly adjusts the core parameters such as the parallel capacitance (C_P) and the switch on duty cycle (D) to compensate the load variations. The rest of the paper is organized as follows. Section II introduces the vector algorithm of the proposed VAC loop and the implementation of the transmitter with VAC loop. Section III illustrates the simulation results and Section IV concludes this paper.

II. PROPOSED SYSTEM

A. Vector Algorithm

As shown in Fig.2(a), when the input voltage and switching frequency are fixed, the unique C_P and D can be found under ZVS and ZVDS conditions for arbitrary R_L in power range, which means we can compensate R_L variations by utilizing C_P and D.

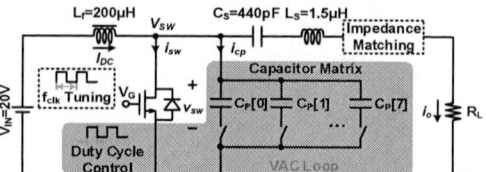

Fig. 1. Class-E PA topology and its tuning method

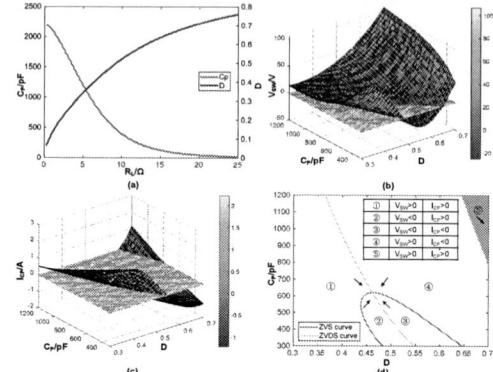

Fig. 2. (a) C_P and D versus R_L under ZVS and ZVDS conditions (b) V_{SW} versus C_P and D (R_L=8Ω) (c) I_{CP} versus C_P and D (R_L=8Ω) (d) ZVS and ZVDS curves (R_L=8Ω)

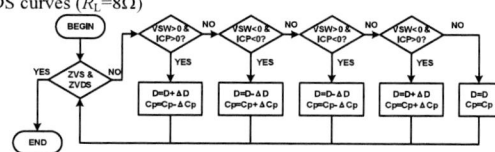

Fig. 3. Flowchart of vector algorithm

In [1], the explicit expressions for steady-state behavior of the class-E PA are derived. These equations can be used to predict the behavior of the class-E PA outside the designed conditions. With these equations, V_{SW} and I_{CP} surface can be plotted over the whole plane composed of C_P and D for a typical R_L, which are shown in Fig.2(b), (c). The switch-on duty cycle range is set to 0.3~0.7, because the power MOS is subjected to high current stress when $D<0.3$, on the contrary, the switch is subjected to high voltage stress when $D>0.7$. The parallel capacitance range is set to 300pF~1200pF because C_P is sensitive to the parasitic capacitance when it's too small, and the adjustment accuracy will be reduced when C_P is too large.

As shown in Fig.2(d), the projections of V_{SW} and I_{CP} surfaces on the zero plane are the curves that satisfy the ZVS and ZVDS conditions, respectively. The intersection of the ZVS and ZVDS curves is the optimal operating point for the current load impedance. If the operating point is shifted, it needs to be adjusted accordingly. The whole plane can be divided into five regions. When the signs of V_{SW} and I_{CP} are used as the judgment basis for regulation, there are four definite directions of adjustment, which are showed in Fig. 3. Region ① and region ⑤ correspond to the same V_{SW} and I_{CP} conditions, so the corresponding adjustment direction is also the same. The adjustment direction of region ⑤ is then opposite to the desired direction, so it is necessary to ensure that the duty cycle and shunt capacitance do not fall into region ⑤ during the adjustment process. The flow chart of the proposed vector algorithm is shown in Fig.3.

B. Implementation of the proposed VAC Loop

The circuit of the proposed VAC loop is shown in Fig.4, which consists of five parts: V_{SW} sampling, I_{CP} sampling,

978-1-6654-9270-6/22 $31.00 © 2022 IEEE

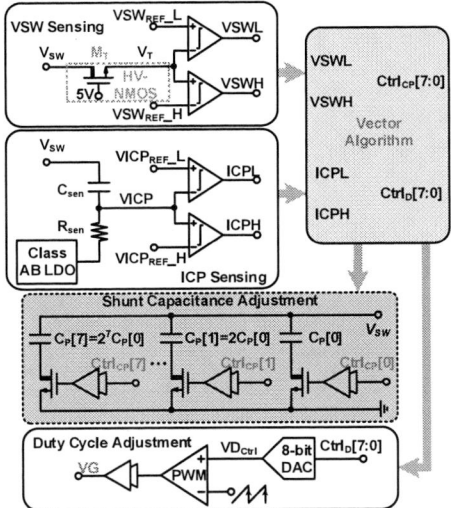

Fig. 4. The circuit of the proposed VAC loop

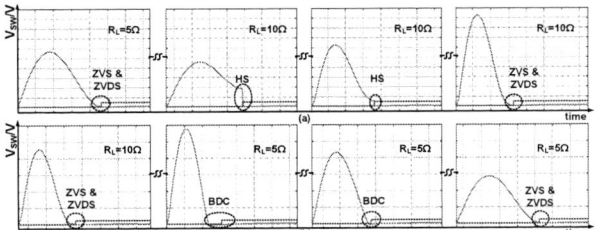

Fig. 5. V_{SW} transient waveform during load steps.

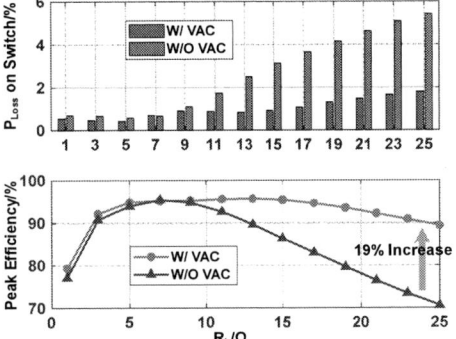

Fig. 6. P_{Loss} of Switch and peak efficiency versus load variations

vector algorithm, shunt capacitor adjustment and duty cycle adjustment. In the V_{SW} sampling module, high voltage MOS M_T is used to block the high voltage above $V_{DD}-V_{th}$ which does not satisfy the ZVS condition to protect the subsequent circuit. The window comparator generates the corresponding outputs, VSWL and VSWH, based on V_T. In the I_{CP} sampling module, a class-AB LDO is used to boost the sampling level to ensure positive V_{ICP}. Likewise, the window comparator generates the corresponding outputs ICPL and ICPH based on V_{ICP}. The VAC loop adjusts the output digital code $Ctrl_D[7:0]$ and $Ctrl_{CP}[7:0]$ accordingly to the input signal. The shunt capacitor adjustment module changes the total value of Cp according to $Ctrl_{CP}[7:0]$. Thus in the duty cycle adjustment module, a 8-bit DAC and PWM generator convert $Ctrl_D[7:0]$ to correspond the duty gate driving signal VG.

III. SIMULATION RESULTS

Parameters of the proposed transmitter of the WPT system with proposed VAC loop are showed in Fig 1.

The V_{SW} transient waveforms during load step are shown in Fig. 5. In Fig. 5(a), the ZVS and ZVDS conditions are satisfied when the initial system load is 5Ω, and the hard-switching (HS) occurs after the system load steps to 10Ω, and the HS voltage becomes smaller and smaller during the system adjustment. Finally, the ZVS and ZVDS conditions are achieved at 10Ω load. In Fig. 5(b), the ZVS and ZVDS conditions are satisfied when the initial load of the system is 10Ω, and the body diode conduction (BDC) occurs after the system load steps to 5Ω, and the BDC time is gradually eliminated after the system is enabled, thus finally realizing the ZVS and ZVDS conditions.

When the optimal C_P is within the regulation range of the capacitor array, VAC ensures both ZVS and ZVDS to minimize HS and BDC power losses, otherwise VAC minimizes HS and BDC losses. The statistic results in Fig.6 depict that the total power loss on switch is significantly reduced. 19% maximum efficiency is improved and >80% efficiency is achieved over a wide load range. Finally, peak output power of 30W and peak efficiency of 95% are achieved, which are very suitable for the WPT applications. The performance comparison with previous state-of-the-art WPT transmitters is shown in Table I.

IV. CONCLUSION

A transmitter for WPT systems based on class-E PA is developed, which utilizes a novel VAC loop to reduce the HS and BDC losses. In addition, the voltage and current sampling circuit designed to provide information for the VAC loop solves the problem of negative voltage. Simulation results verify the feasibility of the VAC. Finally, the transmitter achieves high peak efficiency over a wide load range, which meets the complex application requirement for WPT systems very well.

ACKNOWLEDGMENT

This work was supported by National Natural Science Foundation of China (62004159), National Key Research and Development Program of China (2019YFB2204700) and Xi'an Major Scientific and Technological Innovation Platform and Local Transformation Project of Scientific and Technological Achievements (20KYPT0001-11-1).

REFERENCES

[1] T. Suetsugu and M. Kazimierczuk, "Steady-state behavior of class E amplifier outside designed conditions," 2005 IEEE International Symposium on Circuits and Systems, vol. 1, pp. 708-711, 2005

[2] C. -Y. Xie et al., "15.3 A 100W and 91% GaN-Based Class-E Wireless-Power-Transfer Transmitter with Differential-Impedance-Matching Control for Charging Multiple Devices," ISSCC, pp. 242-244, 2019

[3] H. Oh et al., "6.78 MHz Wireless Power Transmitter Based on a Reconfigurable Class–E Power Amplifier for Multiple Device Charging," TPE, vol. 35, no. 6, pp. 5907-5917, June 2020

[4] X. Ma, Y. Lu and W. -H. Ki, "A 27W D2D Wireless Power Transfer System with Compact Single-Stage Regulated Class-E Architecture and Adaptive ZVS Control," ISSCC, pp. 1-3, 2022

TABLE I.   COMPARISON TABLE

	ISSCC' 19 [2]	TPE'19 [3]	ISSCC' 22 [4]	This work*
Power Switch	eGaN	GaN	eGaN	0.18μm CMOS SOI
Operating Frequency [MHz]	13.56	6.78	6.78	6.78
PA type	Class-E	Class-D	Class-E	Class-E
Tuning Method	Impedance matching, duty	Impedance matching	Duty	Parallel capacitance, Duty
Peak output power [W]	100	98	27	30
Peak Efficiency [%]	91	90.4	N/A	95

A Novel Segmented Temperature Monitor for Adaptive MRAM

Yu'ang Wu, Mingyang Zhou, Hao Cai

School of Microelectronic, Southeast University, Nanjing, 210096, China

Abstract—The influence of temperature change on device directly affects the memory performance, especially in access latency and energy consumption. Based on temperature monitor, temperature adaptive magnetic random access memory (MRAM) eliminates the impact of temperature on storage performance. However, limited by the characteristics of the magnetic tunnel junction (MTJ) device and the operating mode of the MRAM array, wider operating temperature brings challenges to the design of monitor. In this work, based on MRAM array, using the method of segmented detection, we propose a novel temperature monitor for monitoring temperature under -55~125°C. Simulation results show that the temperature monitor can detect the temperature with an accuracy of 5°C within 1.2μs.

Keywords—MRAM, segmented, temperature, monitor, adaptive

I. INTRODUCTION

With the increasing diversity of memory application scenarios, spin-transfer torque magnetic random access memory (STT-MRAM) has occupied a place in the new memory with its nonvolatile and high compatibility [1]. In the conventional MRAM, the write operation time is set to the required time at the lowest temperature, causing huge energy waste at high temperature. For another, the effect of temperature on TMR is mapped to the accuracy of read operation. A promising method to handle the temperature issue in MRAM is temperature adaptive scheme. In 2021, S. Wang proposed an adaptive MRAM with MTJ variation monitor [2]. As is shown in Fig.1, the monitor gets the information from write probability under fixed write condition and then feeds it back to writing or reading circuits. As the most important part of the adaptive scheme, the proposed monitor can detect the temperature with an accuracy of 10°C from 270 K to 370 K within 1-10μs.

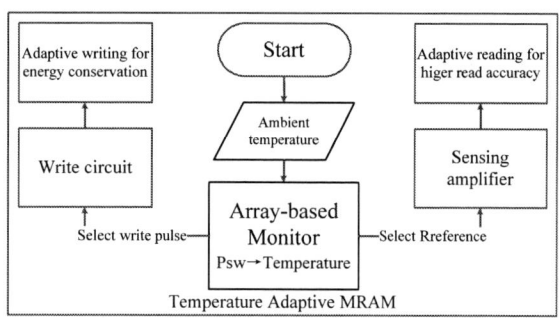

Fig. 1. Workflow diagram of temperature adaptive MRAM based on monitor

However, a single fixed write condition cannot measure a wider temperature and takes a long time to sense the high and low temperature areas. For example, it takes 1μs to measure the normal temperature but 10μs to measure the lowest or highest temperature. In fact, the limitation of this scheme is due to the existence of tails at the beginning and end of the probability-temperature curve, which will be explained in detail in the following chapters.

In this paper, we propose a novel MRAM-based temperature monitor, with a wider temperature range in a smaller delay in a segmented manner. The main contribution includes:

1) To the best of the authors knowledge, this is the first work to measure temperature in MRAM by segmented method. The segmented method makes the process of temperature measurement more efficient, and makes the theoretically measurable temperature unlimited.

2) The proposed design changes the mixture of coarse-grained and fine-grained methods in the traditional scheme into pure coarse-grained method, and reduces the temperature measurement latency in the worst case from ten times to three times that in the best case.

II. ARRAY-BASED TEMPERATURE MONITOR

A. Fundamental monitoring principle

Switching delay of STT-MTJ is intensely related to temperature. According to Landau–Lifshitz–Gilbert equation, switching probability Psw can be dined as followed [3-6]:

$$P_{sw} = \exp\left\{-4 \cdot f \cdot \zeta \cdot exp\left[-\frac{2\cdot(t_{pulse}-delay)}{\tau_{sw}}\right]\right\} \quad (1)$$

where ζ is the thermal stability factor and τ_{sw} is the average latency for state switching, which are both inversely proportional to temperature. In a word, switching probability is positively correlated with temperature, as shown in Fig.2. Thus, the fundamental principle of MRAM-based temperature is that on the premise that the writing conditions remain unchanged, get the Psw by repeated reading and writing operations of MRAM, through which temperature is sensed and monitored.

Since each write operation is a probability event, more samples are required to obtain more accurate write probability, that is, more repeated read and write operations. Especially in the high-temperature and low-temperature regions, there is little difference in the write probability of each temperature. This phenomenon is shown in Fig.2 that there are tails at the beginning and end of the curve. More read and write operations are required to accurately read the highest or lowest temperature, which means more latency and energy consumption for monitoring. What's more, this scheme with a single write condition cannot adapt to a wider temperature range. The tail is lengthened under a wider temperature, and the monitor with single write condition has no ability to sense the temperature.

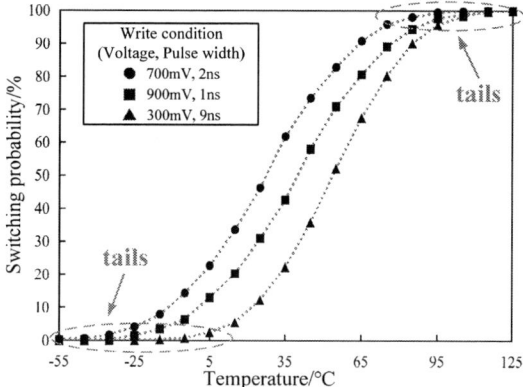

Fig. 2. Switching probability of STT-MTJ with different write conditions under temperature change

B. Segmented Temperature Monitor

Considering that the time consumption of the monitor in the high temperature and low temperature area is about 10 times greater than that in the middle temperature area, that is, the number of reads and writes required is 10 times greater, we believe that the segmented monitoring is faster and more accurate. We propose a novel segmented temperature monitor, as illustrated in Fig.3. Using a MTJ compact model [4], this monitor is based on a MRAM array with 416×1 bitcells, with the ability to read or write 104 bits at the same time.

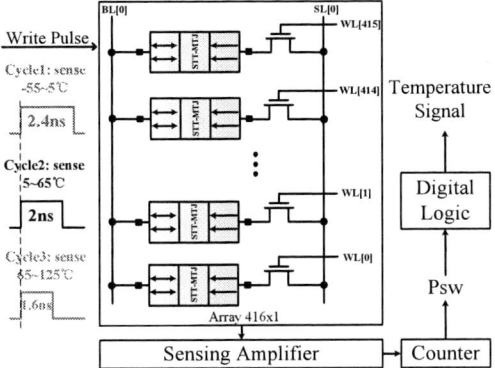

Fig. 3.   Circuit composition of array-based segmented temperature monitor

The monitor divides the temperature range of -55~125℃ into three intervals and senses the temperature in three cycles. The first cycle is used to sense -55~-5℃. If the ambient temperature is detected to be within the range of -55~-5℃, the whole monitoring will be terminated. On the contrary, the next cycle will be carried out. In the first cycle, the write pulse width of the monitor is set to 2.4ns. Finally, the write probability under this write condition is measured through 3016 repeated reading and writing operations of the MTJ, and the temperature information is obtained through the Psw. The second cycle is used to sense -5~55℃, and the write pulse width is set to 2ns. The third cycle is used to sense 55~125℃, and the write pulse width is set to 1.6ns. The reason for this setting scheme is to maximize the difference in switching probability under different temperatures, to reduce the number of repeated reads and writes required, to reduce the delay and energy consumption of the monitor.

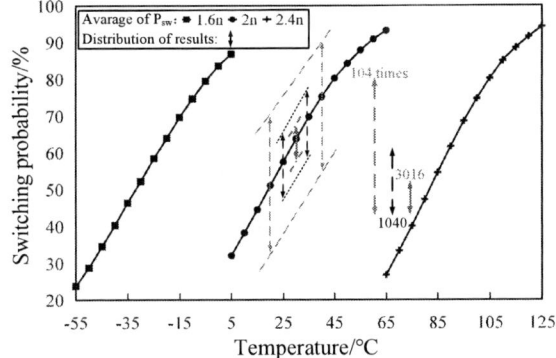

Fig. 4.   Simulation results of segmented temperature monitor with different times of total repeated writing and reding operations (Dashed lines and two-way arrows represent distribution area. 104/1040/3016 represents the total number of times to read and write one MTJ during the monitoring process.)

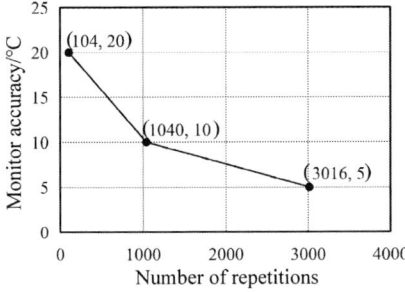

Fig. 5.   Monitor accuracy with varying number of repetitions

III.   SIMULATION RESULTS

The accuracy of the obtained Psw determines the accuracy of the final perceived temperature signal. Through MC simulation, we got the monitor accuracy with different number of repeated write and read operations, as is illustrated in Fig.4 and Fig.5. Obviously, more repetitions make the probability variance measured by the monitor smaller and the distribution more concentrated, which means more higher accuracy. Finally, we set the number of repetitions to 3016, where the error range of the monitor is within 5℃. The delay of the monitor is about 0.4µs in the fastest case and 1.2µs in the slowest case. In Table I, we compare the proposed segmented temperature monitor with the monitor in [2] and the conventional resistor-based thermal sensor in [5]. The proposed monitor has considerable advantages in monitoring range and speed. What's more, the proposed one does not need ADC required in the traditional sensor like [8].

TABLE I.   COMPARISON OF DIFFERENT THERMAL MONITORS/SENSORS

	This work	TETC2021[2]	JSSC2018 [8]
Process/Type	STT-MTJ array-based	STT/VC-MTJ array-based	65nm-CMOS resistor-based
Monitoring Range	-55~125℃	270~370K	-45~80℃
Delay	0.4~1.2µs	1~10µs	10µs
Accuracy	5℃	10℃	0.121℃
ADC	No need	No need	10-bit

IV.   DISCUSSION AND SUMMARY

As the proposed monitor starts to sense from the low temperature domain, the monitoring of the previous two cycles is objectively wasted when the actual temperature is high. If there is a simple circuit to judge the range before sensing the temperature through the array, the monitoring delay and energy consumption will be greatly reduced. For example, a low precision but high-speed simple resistor-based thermal sensor can be used as an auxiliary circuit to judge the temperature range first, and then a high-precision array-based monitor is used to measure the temperature.

In this work, we propose a novel segmented temperature monitor for temperature adaptive MRAM. Based on MRAM array, by repeating the write and read operations, the switching probability of MTJ can be obtained, which can then be used for temperature monitoring in MRAM. Also, the segmented monitoring method greatly reduces the time delay and energy consumption of the monitor, and obtains higher accuracy. Above all, it helps array-based monitors break through the limitations of device properties and can be applied to a wider temperature range.

REFERENCES

[1] S. Peng et al., Magnetic Tunnel Junctions for Spintronics: Principles and Applications, 12 2014, pp. 1– 16.

[2] S. Wang et al., "Adaptive MRAM Write and Read with MTJ Variation Monitor," in IEEE Transactions on Emerging Topics in Computing, vol. 9, no. 1, pp. 402-413, 1 Jan.-March 2021.

[3] Y. Wang et al., "Compact Model of Dielectric Breakdown in Spin-Transfer Torque Magnetic Tunnel Junction," in IEEE Transactions on Electron Devices, vol. 63, no. 4, pp. 1762-1767, April 2016.

[4] R. H. Koch et al., "Time-resolved reversal of spin-transfer switching in a nanomagnet," Phys. Rev. Lett., vol. 92, p. 088302, Feb. 2004.

[5] D. C. Worledge et al., "Spin torque switching of perpendicular Ta|CoFeB|MgO-based magnetic tunnel junctions," Appl. Phys. Lett., vol. 98, no. 2, p. 022501, 2011.

[6] H. Tomita et al., "High-speed spin-transfer switching in GMR nano-pillars with perpendicular anisotropy," IEEE Trans. Magn., vol. 47, no. 6, pp. 1599–1602, Jun. 2011.

[7] MTJ compact model. [Online]. Available: http://spinlib.com/

[8] H. Park and J. Kim, "A 0.8-V Resistor-Based Temperature Sensor in 65-nm CMOS With Supply Sensitivity of 0.28 °C/V," in IEEE Journal of Solid-State Circuits, vol. 53, no. 3, pp. 906-912, March 2018.

A High Reliability Sensing Amplifier for Hybrid MTJ/CMOS Circuits

Jiawei Fu, Pengcheng Wu, Hao Cai

School of Microelectronic, Southeast University, Nanjing, 210096, China

Abstract—Spin transfer torque magnetic tunnel junction (STT-MTJ) based MRAM shows great performance such as zero standby power, outstanding CMOS compatibility, high density and endurance. Hybrid MTJ/CMOS circuits have been extensively studied for energy efficient applications. However, MRAM sensing operations still suffer from reliability issue owing to the inevitable process variations, voltage and temperature fluctuations. This paper proposes a high reliability sensing amplifier (HRSA) for hybrid MTJ/CMOS circuits, simulation is performed based on 28-nm CMOS design kit and 40-nm MTJ model. Simulation results show that the proposed sensing circuit achieves a lower sensing error rate (SER) compared to previous works over a wide temperature range (-55°C~125°C). Proposed HRSA exhibits excellent tolerance to the temperature and process variations.

Keywords—*STT-MTJ, Sensing Amplifier, Sensing Error Rate, Reliability, Temperature*

I. INTRODUCTION

With the decline of CMOS technology, the leakage power problem has become prominent. Owing to the zero standby power consumption, unlimited endurance and excellent CMOS compatibility performance of Spin Transfer Torque Magnetic Tunnel Junction (STT-MTJ), Investigating hybrid CMOS/MTJ circuits is an effective method to reduce static power [1-2]. However, due to the inevitable process variations, voltage and temperature fluctuation, there is still a sensing reliability issue. To overcome this problem, a separated pre-charge sensing amplifier and variation immune sense amplifier have been proposed to improve the read reliability [3-4]. Moreover, Reliability-Enhanced Separated Pre-charge Sensing amplifier and Variation-Tolerant sensing circuits were designed to further reduce the Sensing Error Rate (SER) [5-7]. However, there is a lack of consideration about the influence of temperature aware sensing. MTJ and CMOS transistor are affected by temperature, thus threatening the reliability and performance of hybrid MTJ/CMOS circuits under temperature fluctuations.

In this paper, we propose a novel high reliability sensing amplifier (HRSA) for hybrid MTJ/CMOS circuits. The main contribution can be summarized as follows:

- The temperature-aware performance of MTJ and sensing circuits is simulated and evaluated. Low-VDD and low TMR sensing are studied.

- We proposed a novel HRSA for reliability-aware scenario. Reduced SER is realized over a wide temperature range.

The rest of this paper is organized as follows: Section II introduces the MTJ and the effect of temperature on it. Section III presents the proposed HRSA. section IV shows the simulation results and compares the proposed circuit with other works. Section V concludes the paper.

II. PRELIMINARIES

Fig. 1(a) illustrates the structure of the perpendicular magnetic anisotropy STT-MTJ, it consists of three layers. Two layers are separated by one oxide barrier layer. Depending on the magnetization state of two ferromagnetic layers, the MTJ device

is with parallel (P) and antiparallel (AP) state. The resistances characteristic of two states can be quantified by tunnel magnetoresistance ratio (TMR=(R_{AP}-R_P)/R_P). Depended on the direction of switching current, the state of MTJ can be switched between P and AP by a bidirectional current which is higher than the critical current.

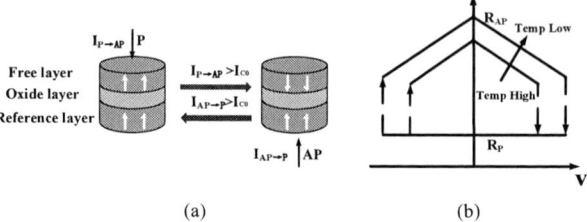

Fig. 1. (a) The structure of perpendicular magnetic anisotropy STT-MTJ (b) R-V curves under different temperature and voltage

As illustrated in Fig. 1(b), with the increase of temperature, R_P remains almost unchanged, while R_{AP} gradually decrease, leads to a reduced TMR. For the STT-MTJ, the TMR as a function of temperature can be obtained in the following equation

$$TMR(T) = \frac{(TMR_0 + 1)}{1 + 2Q \cdot \beta_{AP} \cdot ln\left(\frac{k_B T}{E_c}\right)} - 1 \quad (1)$$

where TMR_0 is the TMR at zero temperature, Q indicates the probability of magnons participating in the tunneling process and Ec represents the cutoff energy of magnon energy, β_{AP} = $Sk_B T/E_m$, S denotes the spin parameter, while E_m is a parameter associated with the Curie temperature. R_{AP} is also influenced by the bias voltage across MTJ. As the bias voltage increases, R_{AP} will decrease, resulting in a lower TMR [8]. Moreover, the drivability of the MOS transistor will reduce under high temperature environment. All of these will lead to a decline in sensing reliability at high temperature.

III. PROPOSED HIGH RELIABILITY SENSING CIRCUIT

In this section, a novel HRSA is proposed for high reliability scenario. The previous SA used to separate the sensing path by using two inverters [3-7]. However, the inevitable induction errors caused by the mismatch of the two inverters reduce sensing reliability. To mitigate sensing error, we remove both two inverters and a strong-arm current latched sensing amplifier has been used, as shown in Fig. 2.

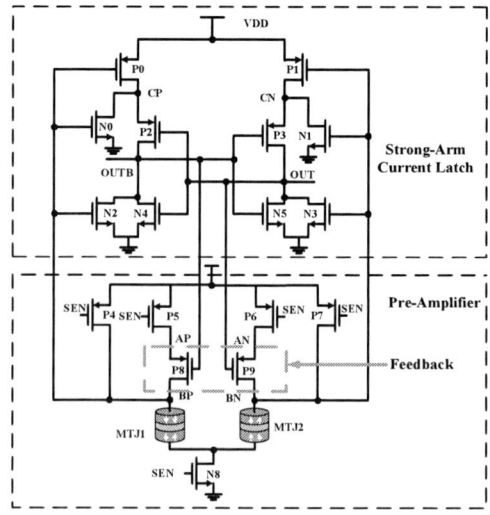

Fig. 2. (a) The structure of proposed high reliability sensing amplifier

978-1-6654-9270-6/22 $31.00 © 2022 IEEE

During the reset phase, SEN signal is low, turning M4~M7 on. This will charge the nodes AP, AN and BP, BN to VDD. While the nodes OUT, OUTB are discharged to Gnd. As soon as the signal SEN rises to VDD, N8 turns on, the node BP, BN begin to discharge with a different speed according to the resistance of two paths. Then the charge stored at node AP and AN starts to flow through P8 and P9 to nodes BP, BN. The gate voltages of P8 and P9 are controlled by the voltages of output nodes OUTB and OUT as feedback to improve the voltage difference between the node BP and BN. When the voltages at nodes BP and BN become low enough, N2 and N3 will turn off, while P0 and P1 will turn on and nodes OUT and OUTB will start charging. Finally, one of the nodes OUTB and OUT will be pull up to VDD and the other will discharge to Gnd.

Owning to the charging of nodes BP and BN to nodes AP and AN, the HRSA has a larger sensing current in the discharging paths than the previous designs, the feedback in the charging path further increases the voltage difference between AP and AN node. HRSA can represents high reliability under various conditions.

IV. SIMULATION RESULTS AND COMPARISON

The simulation is performed with a 28-nm CMOS design kit and a 40 nm MTJ compact model [8]. Table I lists the basic parameters of STT-MTJ.

TABLE I.   PARAMETERS OF STT-MTJ

Parameters	Description	value
α	Gilbert Damping Coefficient	0.027
Area	MTJ layout surface	40nm × 40nm × π/4
t_{sl}	Free Layer Thickness	1.3nm
t_{ox}	Oxide Barrier Thickness	0.85nm
RA	Resistance Area Product	5
TMR	Tunnel Magnetoresistance Ratio	150%

Hybrid process variation-aware Monte Carlo (MC) simulation is executed to analysis the SER of the proposed sensing circuits. Table II presents the SER results of 3k Monte-Carlo simulations at -55, 27 and 125°C. A more intuitive result is shown in Fig. 3, the SER and delay of different circuits are analyzed under process variation and temperature fluctuations. Although the proposed HRSA presents a lager delay, it always shows the lowest error rate compared to other sensing circuits when temperature varies from -55 to 125°C.

TABLE II.   PERFORMANCE COMPARISON WITH OTHER DESIGNS

		This work	[3]	[5]	[6]	[7]
Number of transistor		17	15	18	17	19
SER (%)	-55°C	6.3	14.8	13	8.3	8.3
	27°C	7.6	17.9	15.6	12.1	11.8
	125°C	11.1	23	18.4	18.4	17.2

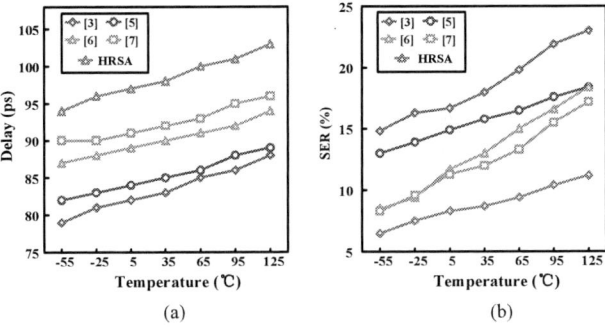

Fig. 3.   Performance under wide temperature range (a) sensing delay, (b) SER.

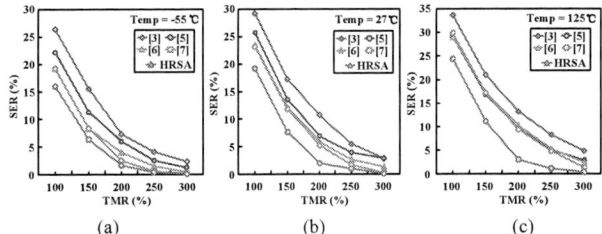

Fig. 4.   SER versus TMR at (a) -55°C (b) 27°C (c) 125°C temperature

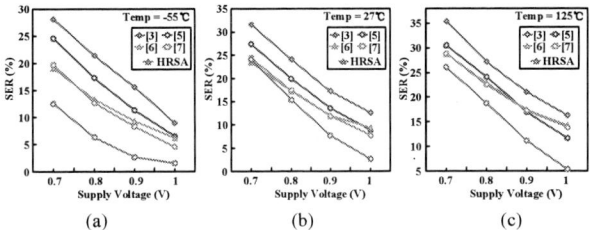

Fig. 5.   SER versus supply voltage at (a) -55°C (b) 27°C (c) 125°C temperature

As the temperature rises, the SER of the proposed SA increases slighter compared to others [3-7]. The proposed HRSA shows a more pronounced enhancement of SER at high temperatures. The sensing operation of HRSA presents less sensitive to temperature. Fig. 4 shows the SER across different TMR, the HRSA always remains the lowest when TMR varying from 100% to 300%. Moreover, SER at different supply voltages shown in fig. 5 also proves that the proposed HRSA has outstanding sensing reliability.

V. CONCLUSION AND DISCUSSION

Taking into account the influence of temperature, we propose HRSA for hybrid MTJ/CMOS circuits in this paper. Simulation result shows HRSA achieves a lower sensing error rate within the temperature range of -55°C to 125°C. In the case of low TMR and supply voltage, HRSA still presents high sensing reliability. Compared to other previous work. HRSA shows a particularly significant improvement in SER at high temperatures. In order to improve the sensing reliability further, the operating voltage can be increased at high temperatures, with extra layout area and energy consumption overhead.

REFERENCES

[1] H. Cai et al., "Exploring Hybrid STT-MTJ/CMOS Energy Solution in Near-/Sub-Threshold Regime for IoT Applications," in IEEE Transactions on Magnetics, vol. 54, no. 2, 2018, pp. 1-9.

[2] W. Zhao et al., "High Speed, High Stability and Low Power Sensing Amplifier for MTJ/CMOS Hybrid Logic Circuits," in IEEE Transactions on Magnetics, vol. 45, no. 10, 2009, pp. 3784-3787.

[3] W. Kang et al., "Separated Precharge Sensing Amplifier for Deep Submicrometer MTJ/CMOS Hybrid Logic Circuits," in IEEE Transactions on Magnetics, vol. 50, no. 6, 2014, pp. 1-5.

[4] S. Salehi and R. F. DeMara, "Process Variation Immune and Energy Aware Sense Amplifiers for Resistive Non-volatile Memories," ISCAS, 2017, pp. 1-4.

[5] S. Salehi et al., "Mitigating Process Variability for Non-Volatile Cache Resilience and Yield," in IEEE Transactions on Emerging Topics in Computing, vol. 8, no. 3, 2020, pp. 724-737.

[6] D. Zhang et al., "Reliability-Enhanced Separated Pre-Charge Sensing Amplifier for Hybrid CMOS/MTJ Logic Circuits," in IEEE Transactions on Magnetics, vol. 53, no. 9, 2017, pp. 1-5.

[7] J. Kim and J. Park, "Variation-Tolerant Separated Pre-Charge Sense Amplifier for Resistive Non-Volatile logic circuit," ISOCC, 2020, pp. 147-148.

[8] Y. Wang et al., "Compact Thermal Modeling of Spin Transfer Torque Magnetic Tunnel Junction," Microelectronics Reliability, Volume 55, Issues 9–10, 2015, pp 1649-1653.

A 0.7-2.5GHz NB-IoT/GNSS/BLE Hybrid PLL with PA Pulling Mitigation and Out-of-Band Phase Noise Reduction

Jiahao Zhao, Xuansheng Ji, Su Han, Ziwei Wang, Woogeun Rhee, Zhihua Wang

School of Integrated Circuits, Tsinghua University, China

Abstract—**A 0.7-2.5GHz NB-IoT/GNSS/BLE hybrid PLL with a single D/VCO is implemented in 28nm CMOS. With careful frequency planning, the PA pulling effect is mitigated by using a multi-mode divider chain. A divider-by-2.5 relaxes the tuning range requirement of the D/VCO and mitigates the PA pulling for NB-IoT HB band, while a divider-by-6 is designed for NB-IoT LB. With an 8-tap FIR filtering method, a wideband fractional-N PLL is designed without increasing the out-of-band phase noise. The proposed PLL consumes the maximum 4.7mW with 0.9V supply. Experimental results show that the PLL meets the phase noise and spur requirements of the NB-IoT/GNSS/BLE standards.**

Keywords—*Hybrid PLL, NB-IoT, GNSS, BLE, pulling effect, fractional-N*

I. INTRODUCTION

There is an increasing demand for the PLL to cover a wide frequency band to support multiple communication protocols in low-power mobile connectivity. The use of multiple oscillators for multiband frequency generation occupies a large area [1], and the mixer-based structure suffers from spur generation [2]. NB-IoT demands stricter blocker performance than BLE. For example, 30dB higher blocking is required at 10-60MHz offset frequency, while the minimum bandwidth (200kHz) is 5 times narrower. Therefore, out-of-band phase noise and spur performances are important [3]. In addition, the VCO pulling by a PA must be carefully considered. Using a single divider-by-2 (DIV2) makes the VCO pulled by a second harmonic of the PA. Three methods to reduce the PA pulling effect are addressed in [4]; i) reduce the interference amplitude, ii) split oscillator frequency and PA output harmonic frequency as far as possible, iii) increase the PLL bandwidth. In this work, following design aspects are considered in the PLL design to cover NB-IoT, GNSS and BLE frequency bands with a single oscillator and mitigate the PA pulling. Firstly, a careful frequency plan is made by using a multi-mode frequency divider chain. Secondly, a hybrid PLL architecture is chosen to achieve linear loop dynamics, low spur, and compact area. Thirdly, an out-of-band phase noise reduction method is employed for the PLL to have a wide loop bandwidth, thus mitigating the VCO pulling.

II. PROPOSED PLL STRUCTURE

Fig. 1 shows a frequency plan to cover NB-IoT low band (LB), NB-IoT high band (HB), GNSS and BLE standards. Using a divider-by-2.5 (DIV25) for NB-IoT HB not only relaxes the tuning range requirement of the VCO but also increase the frequency interval between the harmonic frequencies of the PA (e.g. 2GHz and 4GHz) and the fundamental frequency of the VCO (e.g. 5GHz). A divider-by-6 (DIV6) for NB-IoT LB reduces the interference to the VCO

This work was supported by National Key Research and Development Program of China under Contract #2020YFB2205602.

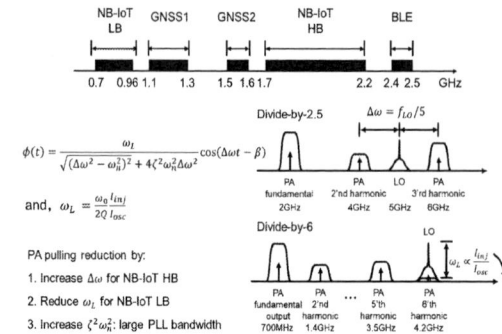

Fig. 1. PLL frequency plan with PA pulling mitigation for NB-IoT, GNSS, and BLE

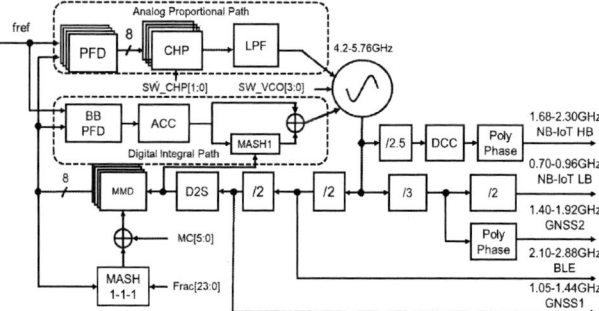

Fig. 2. Proposed hybrid PLL with a multi-mode divider chain.

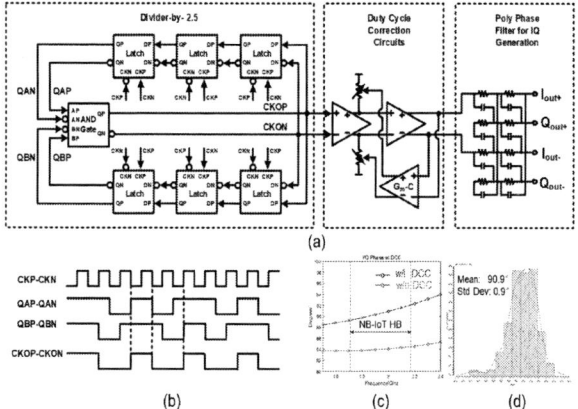

Fig. 3. (a) NB-IoT HB generation circuits, (b) timing diagram of DIV25, (c) I/Q mismatch w/i and w/o DCC, (d) I/Q mismatch of Monte Carlo

whose frequency is the sixth harmonic frequency of the PA. Since the BLE standard typically has the power class 3 with an output power of 0dBm only, a DIV2 is directly used for low complexity [5]. The PA pulling effect is not applicable to the GNSS standard, and the required frequency is generated by the divider-by-3 (DIV3) and divider-by-4 (DIV4). The PLL bandwidth needs to be as wide as possible to make the VCO robust against the PA pulling.

Fig. 2 shows a block diagram of the proposed PLL with a multi-mode divider chain. The PLL generates 0.7-2.88GHz output frequency from a 52MHz reference. To achieve linear loop dynamics and avoid a large integration capacitor, a hybrid PLL shown in Fig. 2 is designed. The analog path consisting of a PFD, a charge pump, and an RC filter works as the proportional-gain path of a type-II PLL and determines the small-signal performance of the PLL, while the digital integration path consisting of a BBPFD and an accumulator performs frequency acquisition. As the differential control

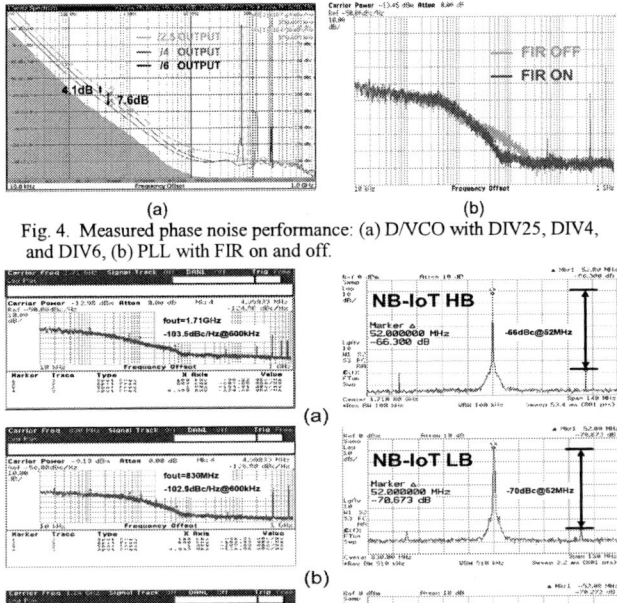

Fig. 4. Measured phase noise performance: (a) D/VCO with DIV25, DIV4, and DIV6, (b) PLL with FIR on and off.

Fig. 5. Measured phase noise and reference spur: (a) NB-IoT HB with DIV25, (b) NB-IoT LB with DIV6, (c) GNSS with DIV4.

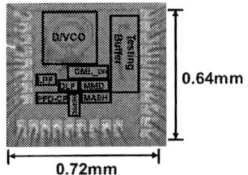

Fig. 6 Chip micrograph

Table I. Performance comparison with recent PLLs for NB-IoT.

	This Work			[3]	[7]	[8]	[9]
Technology (nm)	28			65	28	180	55
Ref Freq (MHz)	52			6	-	-	24
PLL Type	Hybrid			Analog	Digital	Analog	Digital
Area(mm²)	0.46			0.68	0.87	-	0.88
Supply(V)	0.9			0.55	-	1.7	1.2
Out Freq (GHz)	0.7-2.5			0.73-0.96	0.46-2.1	0.75-0.96	1.8
Application	NB-IoT HB / NB-IoT LB	GNSS1 / GNSS2	BLE	NB-IoT LB	NB-IoT HB / NB-IoT LB	NB-IoT LB	NB-IoT HB
PN@1MHz (dBc/Hz)	-109.8 / -107.6	-107.7 / -103.3	-101.7	-108.7	-126.3	-120	-120.5
In-band PN (dBc/Hz)	-90.2 / -96.2	-93.5 / -91.3	-87.5	-84	-101.2	-70	-94
Frac spur (dBc)	-53.7@2MHz / -61.3@2MHz	-57.8@2MHz / -70.3@6MHz	-51.8@2MHz	-66.3@600kHz			-40@23kHz
Ref Spur (dBc)	-66.3 / -70.6	-70.2 / -66.9	-66.4	-67.5@6MHz			-76.3@24kHz
Power(mW)	4.7 / 4.5	3.6 / 2.7	2.7	0.55	40	6.23-13.6	4

micrograph is shown in Fig. 6. The PLL consumes the maximum power of 4.7mW with 0.9V supply for NB-IoT HB. Fig. 4(a) shows that the phase noise at the DIV25 output is higher than that at the DIV4 and DIV6 outputs by 4.1dB and 7.6dB, respectively, showing that the phase noise contribution of the DIV25, the DCC and the polyphase filter is negligible. Due to a logic inversion mistake, the digital integration path is not fully functional, and the key performance of the PLL is measured with the analog path only, but the noise and spur contribution of the digital integral path is negligible in the simulation and also verified by other previous PLL chips. Fig. 4(b) compares the measured phase noise comparison with the 8-tap FIR filter enabled and disabled, showing that the out-of-band noise around 5MHz caused by the DSM is greatly suppressed. Fig. 5 shows the measured phase noise and spur performances. In the NB-IoT HB mode, the in-band phase noise of −90.2dBc/Hz and the reference spur of −66dBc at 52MHz offset are achieved as shown in Fig. 5(a). The in-band noise contribution to EVM is 2.7%, which is much lower than the required 17.5% under QPSK modulation. For the NB-IoT LB mode, the loop bandwidth of about 400kHz is set as shown in Fig. 5(b), and the in-band phase noise of −96.2dBc/Hz and the reference spur of −70dBc are achieved. Fig. 5(c) shows the performance of GNSS1 band. For BLE band, the performance is 6dB worse as the output frequency is doubled. Table I shows the performance summary in comparison with other NB-IoT PLLs. To the best of authors' knowledge, this work presents the first hybrid PLL to cover NB-IoT/GNSS/BLE frequency bands with a single VCO.

voltage in the analog path becomes small with the digital integral path, good matching is achieved in the charge pump, resulting in low reference spur. To achieve fine frequency resolution of the D/VCO, a first-order ΔΣ modulator (DSM) is used in the digital path. The FIR filtering method [6] is adopted to reduce the out-of-band noise from the DSM quantization noise for the design of a wideband PLL. The FIR filtering method together with the hybrid PLL structure also improves the charge pump nonlinearity. Thanks to the DIV25, the D/VCO requires only 30% tuning range from 4.2-5.76GHz.

Fig. 3(a) illustrates the block diagram of NB-IoT HB generation circuits. The DIV25 circuit is designed with six CML latches and an AND gate. Fig. 3(b) depicts the timing diagram of the DIV25. A G_m-C cell senses the output DC voltage of the duty-cycle-correction (DCC) circuit that adjusts the duty cycle with the variable resistance of a differential load. With the DCC blocks, the I/Q mismatch can be improved by 5° in the simulation as shown in Fig. 3(c). A second-order polyphase filter is designed for I/Q signal generation by considering a tradeoff between bandwidth and area. Fig. 3(d) shows the Monte Carlo simulation result of the I/Q phase matching. Over 200 runs, the mismatch variation of the polyphase filter exhibits <1° standard deviation.

III. IMPLEMENTATION

A 0.7-2.5GHz NB-IoT/GNSS/BLE hybrid PLL with a single LC D/VCO is implemented in 28nm CMOS. A chip with

REFERENCES

[1] H. Guo et al., "30.3 A SAW-Less NB-IoT RF Transceiver with Hybrid Polar and On-Chip Switching PA Supporting Power Class 3 Multi-Tone Transmission," ISSCC, Feb. 2020.

[2] D. Huang et al., "A Frequency Synthesizer with Optimally Coupled QVCO and Harmonic-Rejection SSBmixer for Multi-Standard Wireless Receiver," JSSC, May. 2011.

[3] H. R. Kooshkaki et al., "A 0.55mW Fractional-N PLL with a DC-DC Powered Class-D VCO Achieving Better than -66dBc Fractional and Reference Spurs for NB-IoT," CICC, Apr. 2020.

[4] B. Razavi., "A study of injection locking and pulling in oscillators," JSSC, Aug. 2011.

[5] J. Prummel et al., "A 10mW Bluetooth Low-Energy Transceiver with On-Chip Matching," JSSC, Dec. 2015.

[6] X. Yu et al., "A DS Fractional-N Synthesizer with Customized Noise Shaping for WCDMA/HSDPA Applications," JSSC, Aug. 2009.

[7] J. Choi, et al., "A 0.46-2.1 GHz Spurious and Oscillator-Pulling Free LO Generator for Cellular NB-IoT Transmitter with 23 dBm Integrated PAs in 28nm CMOS," A-SSCC, Nov. 2018.

[8] Z. Song et al., "A Low-Power NB-IoT Transceiver with Digital-Polar Transmitter in 180-nm CMOS," TCAS-I, Sept. 2017.

[9] N. Yan et al., "A Low Power All-Digital PLL With −40dBc In-Band Fractional Spur Suppression for NB-IoT Applications," IEEE Access, Dec. 2018.

A High-Density Large-Ratio Fuse Based Oxide Devices for One-time-programmable Memory Applications

Xuecheng Cui[1], Dong Liu[2], Jifang Cao[1], Xiao Yu[3#], Bing Chen[1,3*]

[1] School of Micro-Nano Electronics, ZHEJIANG UNIVERSITY, Hangzhou 310000, China
*E-mail: bingchen@zju.edu.cn

[2] Polytechnic Institute, Zhejiang University, Hangzhou 310015, China.

[3] Research Center for Intelligent Chips and Devices, ZHEJIANG LAB Hangzhou 311121, China
#E-mail: yuxiao@zhejianglab.com

Abstract—In this paper, the oxide fused and anti-fused behavior has been observed in a simple metal-oxide-metal device: $Pt/HfO_2/NiO_x/Ni$. The anti-fused state and fused state can be achieved by applying program voltage on the devices with or without current compliance, respectively. And the resistance window of the two states reaches about 10^9, which can effectively reduce the possibility of incorrect programming. It also showed excellent retention characteristics and a simple structure friendly for integration. It can be well used in the field of high reliability of one-time programmable memory.

Keywords—memory, one-time-programmable, oxide fuse, oxide anti-fuse, oxygen vacancy

I. INTRODUCTION

One-time-programmable (OTP) memory is a type of erasable programmable read-only-memory (EPROM), which can be written only once but can be read many times [1]. OTP memories are widely used in digital circuit modules where the programming speed is fast to reduce energy consumption and improve programming efficiency. On the other hand, the local field in the OTP device cell can be as high as 35 MV/cm [2], making device reliability an important issue for OTP memory.

In this paper, a novel, simple device structure that can be used as an OTP memory is proposed. Based on the $Pt/HfO_2/NiO_x/Ni$ structure, we achieved perfect data storage and reliability using oxide fused and anti-fused schemes [3-4]. This device has a small footprint and can be implemented with simple process. Also, it has a large ON/OFF resistance ratio, which facilitates circuit design and prevents erroneous reading [5]. Furthermore, the fuse and anti-fuse function in the circuit can be implemented using the proposed device structure, making it a possible candidate for low-cost high-density and high-reliability OTP memory.

II. EXPERIMENTS

The $Pt/HfO_2/NiO_x/Ni$-based devices were fabricated using a CMOS-compatible process, as schematically shown in Fig. 1. Si/SiO_2 was prepared as substrates, and about 100 nm Ni was deposited as the bottom electrode (BE). A 13 nm-thick nickel oxide (NiO_x) was formed by rapid thermal processing (RTP) at 365 °C in an O_2 atmosphere for 60 s. Then, a 7 nm-thick HfO_2 layer was deposited by atomic layer deposition (ALD) above the NiO_x layer. After lithography, 40-nm Pt top electrode (TE) was formed by DC magnetron sputtering and lift-off. Finally, the BE pads were formed using a dry etching process. The fabricated single device and complete array are with the cell size of 5 μm × 5 μm. All electrical measurements were performed using the Keysight B1500A semiconductor parameter analyzer and the Keithley 4200 analyzer.

III. RESULTS

Figures 2 (a) and (b) show the transmission electron microscope (TEM) and the corresponding energy dispersive X-ray spectroscopy (EDS) mapping images of the proposed $Pt/HfO_2/NiO_x/Ni$ device, respectively, which demonstrate the presence of HfO_2 and NiO_x layers, with their location and thickness in the critical section. The mechanism of device breakdown is similar to the high-k metal gate (HKMG) CMOS breakdown mechanism proposed in Ref. [3]. Initially, as shown in Fig. 3, the oxygen vacancies in NiO_x layer were randomly distributed. After applying a proper voltage with current compliance, the oxygen vacancies will generate and re-distributed in HfO_2 and NiO_x layers to form a conductive channel, which can be defined as the anti-fused state. When we apply a larger positive voltage without current compliance, the conducting channel will be fused, and the oxygen vacancies are driven to the boundary, hence the devices are switched to the fused state with high resistance.

A typical oxide anti-fused *I-V* curve for the device is shown in Fig. 4. At first, the fabricated devices were at a relatively high resistance state. Here, the device is connected with a 50 Ω series resistor. When a positive voltage is applied to the device, a conductive channel can be formed, and the device switched to the anti-fused state with low resistance. The series resistor acts as current compliance so that the current is not so high that might fuse the device. As for the fused transition shown in Fig. 5, when without the series resistor, the oxide fuse can be achieved by a positive voltage of about 5 V with a large current, which induces a thermal effect to fuse the filament in HfO_2 and NiO_x, resulting in the reduction of the current. Note that there is no current compliance, hence the current through the device is greater than that in Fig. 4. After the oxide fuse occurs, the current is below 1 nA level. Figures 4 and 5 show this OTP memory device with a significant resistance ratio of 10^9 for the oxide fused state and the oxide anti-fused state, indicating the device has a large store window to avoid errors.

Figure 6 shows the operating voltage distribution of 32 devices with two states switching, demonstrating good operation stability. The operating currents before and after fuse are shown in Fig. 7. The operating current before fuse is above 10 mA, while after the fuse, the current is mainly concentrated around 1 nA. These results suggest the store window is always maintained of 10^8 or more. In addition, the OTP device also has good stability at high temperatures. Figure 8 shows the test results at 150 °C and its linear extrapolation, which indicates that the storage window can be maintained of $\sim 10^5$ after 10 years.

In a common fuse circuit structure, only one single device proposed in this work is needed to implement the circuit programming function. The OTP base cell in a memory array is constructed by a selector transistor and a fuse-based oxide device. Its programming and reading scheme are shown in Fig. 9. We put the transistor in the on state and add a programming voltage to the bit line to enable the write function. Reading is also a similar operation. In addition to individual cell write and read methods, this OTP array can also implement bit line parallel input to achieve the entire line write and read.

IV. CONCLUSION

In summary, we proposed and demonstrated a $Pt/HfO_2/NiO_x/Ni$ structured device with fuse and anti-fuse characteristics. This device has a large storage window, good stability, and a simple process for easy high-density integration. This device with fuse and anti-fuse features is particularly suitable for OTP memory applications. It is expected to be a great candidate in the field of OTP memory in the future.

ACKNOWLEDGMENT

This work was supported by National Natural Science Foundation of China (62174146), the Major Scientific Research Project of Zhejiang Lab (2021MD0AC01), Zhejiang Province Key R & D programs (2022C01232).

REFERENCES

[1] W. C. Wang, C. C. Chuang, C. W. Chang, E. R. Hsieh, H. W. Chen and S. S. Chung, "A Novel Complementary Architecture of One-time-programmable Memory and Its Applications as Physical Unclonable Function (PUF) and One-time Password," 2020 IEEE International Electron Devices Meeting (IEDM), 2020, pp. 31.6.1-31.6.4.

[2] A. Benoist et al., "Extended TDDB power-law validation for high-voltage applications such as OTP memories in High-k CMOS 28nm FDSOI technology," 2015 IEEE International Reliability Physics Symposium, 2015, pp. GD.3.1-GD.3.5.

[3] E. R. Hsieh et al., "The demonstration of low-cost and logic process fully-compatible OTP memory on advanced HKMG CMOS with a newly found dielectric fuse breakdown," 2015 IEEE International Electron Devices Meeting (IEDM), 2015, pp. 3.4.1-3.4.4.

[4] J. Peng et al., "A Novel Embedded OTP NVM Using Standard Foundry CMOS Logic Technology," 2006 21st IEEE Non-Volatile Semiconductor Memory Workshop, 2006, pp. 24-26.

[5] H. W. Cheng et al., "A novel rewritable one-time-programming OTP (RW-OTP) realized by dielectric-fuse RRAM devices featuring ultra-high reliable retention and good endurance for embedded applications," 2018 International Symposium on VLSI Technology, Systems and Application (VLSI-TSA), 2018, pp. 1-2.

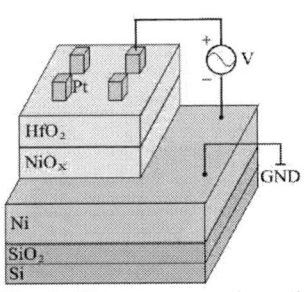

Fig. 1. The structure schematic of the Pt/HfO$_2$/NiO$_x$/Ni OTP device. The device size is 25 μm².

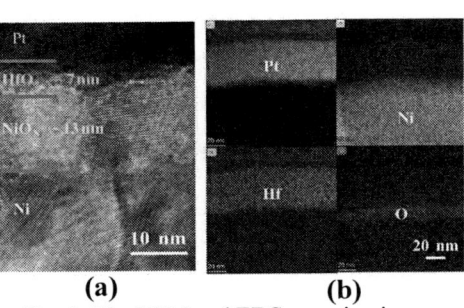

(a)   **(b)**

Fig. 2. The TEM and EDS mapping images of the Pt/HfO$_2$/NiO$_x$/Ni OTP device.

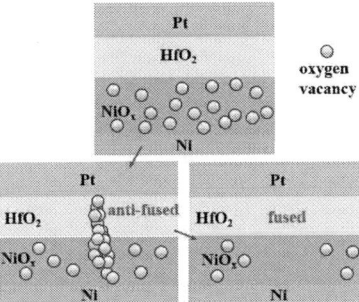

Fig. 3. Schematic diagram of the oxide anti-fused and fused mechanisms.

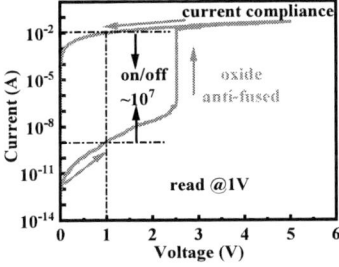

Fig. 4. The typical oxide anti-fused I-V curves of OTP device. A positive voltage is applied when the device is series connected with a resistor of 50 Ω, and the device is anti-fused.

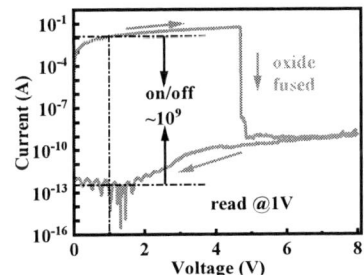

Fig. 5. The typical oxide fused I-V curve of OTP device without series resistor. A positive voltage is applied, and the device is oxide fused.

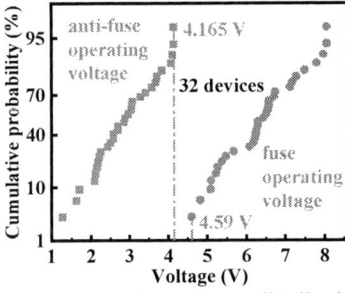

Fig. 6. The operating voltages distribution of the 32 devices with two-state switching. The operating voltages of the two stages do not overlap.

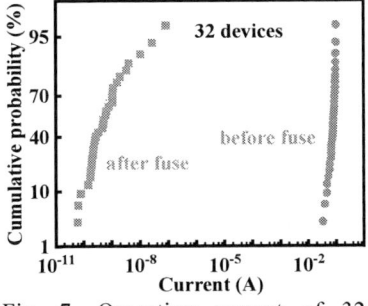

Fig. 7. Operating current of 32 devices before and after fused. The operating current before fused is above 10 mA, while after fused, the current is mainly around 1 nA.

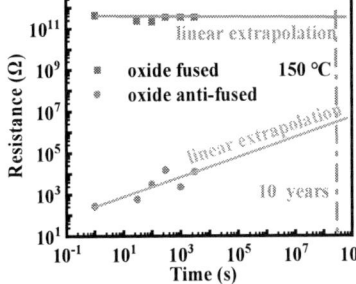

Fig. 8. Retention characteristics of OTP device at 150 °C. The storage window can maintain for 10 years.

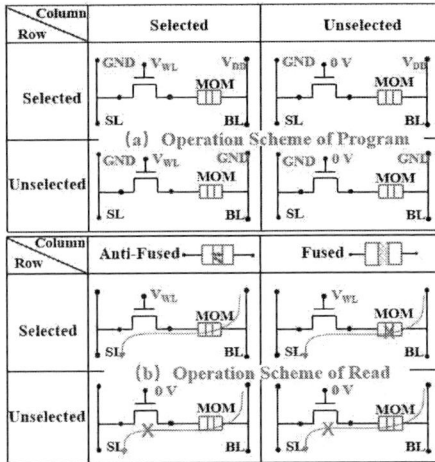

Fig. 9. (a) Program: a programming voltage will be applied to the bit line (BL). Meanwhile, the source line (SL) will be applied to 0 V. (b) Read: a small read voltage will be applied to the BL. Meanwhile, the SL will be applied to 0 V.

978-1-6654-9270-6/22 $31.00 © 2022 IEEE

2022 IEEE International Conference on
Integrated Circuits, Technologies and Applications

A 210nA Quiescent Current Bandgap Reference with 5mA Load Capability Using Shared Error Amplifier

Binwei Yang, Renwei Chen, Chenchang Zhan

School of Microelectronics, Southern University of Science and Technology, Shenzhen, China

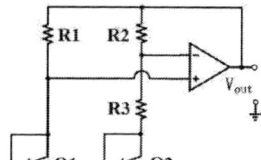

Fig. 1. Schematic of a classical BGR.

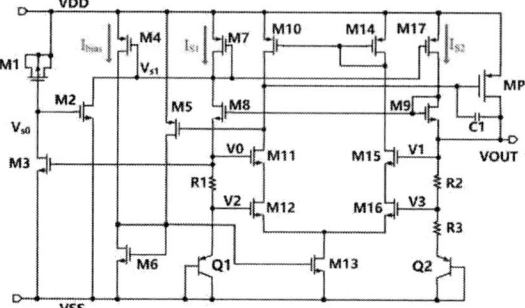

Fig. 2. Schematic of the proposed BGR with load driving capability.

Abstract—This paper presents an ultra-low quiescent current bandgap reference (BGR) with current loading capability. Based on sharing an error amplifier (EA) between the BGR core and the power transistor driving circuit, the maximum load current of the BGR is extended to 5mA without relying on an output buffer. The BGR uses a dual-clamping structure to achieve low temperature coefficient (TC) and improved DC regulation. The proposed BGR is designed in a standard 0.18-μm CMOS process. Measurement results show that the input voltage range is 1.68V-2V while the output voltage is 1.19V. The quiescent current is 210nA with a maximum load current of 5mA. In the temperature range of -40℃ to 125℃, the TC of the output voltage is 9.7ppm/℃ and 30.76ppm/℃ at the light and heavy loads, respectively. Good line and load regulations are also achieved.

Keywords—current loading capability, low quiescent current, bandgap reference, dual-clamp structure.

I. INTRODUCTION

Efficient power management circuits are needed for battery-powered devices in IoT applications, remote sensing, and implantable devices, which can extend battery life or ultimately enable autonomous operation with energy harvesting [1]. Hence, power consumption of the circuit modules cannot be ignored. Typically, quiescent current (I_Q) is required to be at a few hundred nanoamps [2]. To reduce the total power consumption, it is the mainstream design idea to reduce the power consumption of each module separately [3-5]. For example, the quiescent current of the bandgap reference (BGR) as a reference circuit is typically tens of nanoamps today [4-5]. Another feasible idea is to integrate more functions into a single circuit in order to reduce the total power consumption. In [6], a high degree of integration of low-dropout regulator (LDO) and BGR is achieved through the improvements to the op-amp architecture [6]. Unfortunately, its total power consumption reaches 5uA, which is too high.

In this paper, an ultra-low quiescent current BGR with current loading capability is proposed. A dual-clamping structure is designed and an error amplifier is shared between the BGR core and the power transistor driving circuit, which achieves low temperature coefficient (TC) and improved DC performance.

II. PROPOSED BGR WITH LOAD DRIVING CAPABILITY

Fig. 1 shows the schematic of a classical BGR. To achieve a zero-TC reference voltage, the BGR adds two voltages with opposite TCs with appropriate ratio. By proper ratio of R_2 and R_3, the complementary-to-absolute-temperature (CTAT) and proportional-to-absolute-temperature (PTAT) terms of Vout can be cancelled out to arrive at low TC. Usually, the amplifier designed in CMOS process does not support DC load current, and hence, if some current is loading Vout, the classical BGR will not work properly due to the deviated operating points.

Fig. 2 shows the schematic of the proposed BGR with

current loading capability. By sharing an error amplifier (EA) consisting of M10-M16, the classic BGR architecture is embedded into the architecture of the basic low dropout voltage regulator (LDO) with MP as the power transistor and C1 as the Miller compensation capacitor. In addition to the basic functions of a BGR, this new design implements current loading. Compared to traditional LDOs that require a separate BGR, this circuit structure is simple, and it effectively reduces area and power consumption. Furthermore, as a BGR, the sharing of EA makes it achieve lower TC. By setting the resistors R1, R2 and R3, the quiescent current is controlled at the nanoamp level. With Q1: Q2 ratio being 1: 8, the BGR output voltage VOUT is given by:

$$V_{OUT} = V_{BE2} + \frac{\Delta V_{BE}}{R3}(R2+R3) = V_{BE} + \frac{V_T \ln 8\,(R2+R3)}{R3} \quad (1)$$

The dual-clamping structure of the BGR, with M7-M9 & M17 clamping V1=V0 and M10-M16 clamping V3=V2, is the key to this architecture. The negative feedback loop formed by the EA with M12 and M16 as the input pair makes V2=V3. Resistors R1 and R2 are exactly equal, and the current mirror consisting of M7, M8, M9 and M17 can achieve V0=V1. Furthermore, with MP providing the current driving capability, V0 and V1 can have large deviation when there is loading current, and adding M11 and M15 to the EA helps reduce the error between V1 and V0, leading to improved DC regulation. Therefore, the current flowing through R1 (and Q1) and R2 (and Q2) are equal as shown in (2).

$$I_{R1} = \frac{V0-V2}{R_1} = \frac{V1-V3}{R_2} = I_{R2} \quad (2)$$

The dual-clamping structure reduces the offset between I_{R1} and I_{R2}, making the BGR achieve low TC and improved regulation. The introduction of MP transistor not only increases the gain of the negative feedback loop, but also enables the BGR to source load current. However, if there is no limit to the load current I_P, it will shunt into the internal structure of the BGR (R1 branch) and thus affects the normal operation of the circuit. The role of the dual-clamping structure is to isolate the effect of the load current by controlling the equality of the two branches' currents. Considering the ultra-low quiescent current

---

This work is supported by NSFC under grant 62174080 and SZSTI under grant JCYJ20200109141225025. (Comesponding author: Chenchang Zhan.)

978-1-6654-9270-6/22 $31.00 © 2022 IEEE

78

TABLE I. PERFORMANCE COMPARISON

	[6] 2014 ISSCC	[3] 2018 TCAS I	[7] 2021 ISCAS I	[4] 2020 TCAS II	[5] 2020 TCAS II	This work
Technology	21nm	65nm	55nm	65nm	0.35um	0.18um
Structure	BGR with Load Capacity	LDO	LDO	BGR	BGR	BGR with Load Capacity
Supply Voltage(V)	0.65-0.9	1.0-2.0	0.9-1.3	0.9	2.8-4.5	1.68-2.0
Output Voltage(V)	0.6	0.8	0.8-1.2	0.495	1.17	1.19
Quiescent current(µA)	5	0.1	1	0.084	0.018	0.21
Output Capacitor(uF)	<0.1	0.01	20	N/A	N/A	0
Maximum load current(mA)	10	10	10	N/A	N/A	5
Line Regulation(mV/V)	16	N/A	N/A	3.2	1.3	10
Load Regulation(mV/mA)	0.5	1.58	N/A	N/A	N/A	1.53
TC(ppm/°C)	30	N/A	N/A	42	65	9.7
Area(mm²)	0.015	0.0048	0.012	0.0532	0.042	0.07
PSR/100Hz(dB)	N/A	-27	N/A	-50	N/A	-42
Undershoot(mV)	10	231.4	268	N/A	N/A	150
Edge time(ns)	100	200	1000	N/A	N/A	50

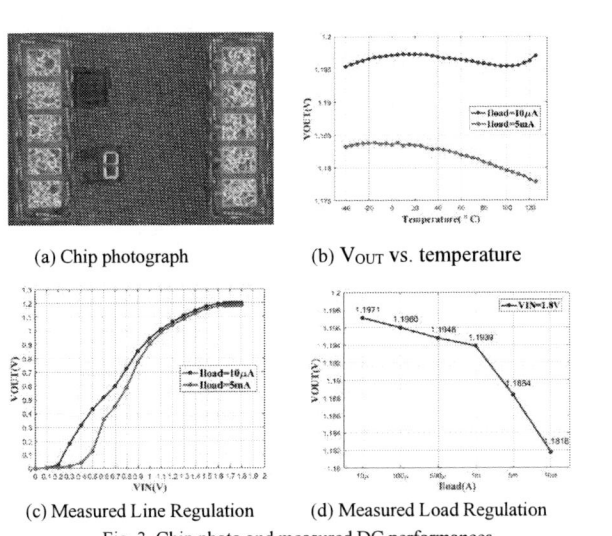

(a) Chip photograph   (b) V_{OUT} vs. temperature

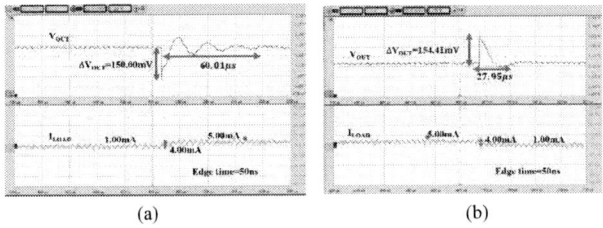

(c) Measured Line Regulation   (d) Measured Load Regulation

Fig. 3. Chip photo and measured DC performances

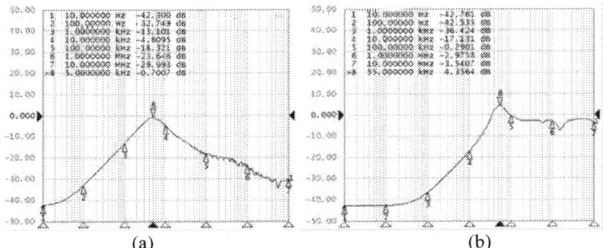

(a)   (b)

Fig. 4. Measured load transient of proposed BGR: (a) undershoot (b) overshoot

(a)   (b)

Fig. 5. Measured PSR of proposed BGR at (a) light load and (b) heavy load

requirement, transistor M5 is introduced to form an adaptive biasing circuit to enhance the transient response of the circuit. An appropriate transient response speed improvement is achieved by setting the appropriate ratio between M5 and MP.

III. MEASUREMENT RESULTS

The proposed BGR with load driving capability is fabricated in a standard 0.18-µm CMOS process. Fig. 3 (a) shows the chip photo which occupies an active area of 0.070676mm². The measured quiescent current is 208nA. Fig. 3 (b) shows the measured VOUT vs. temperature from -40℃ to 125℃. The TC is 9.37ppm/℃ and 30.76ppm/℃ with load currents of 10µA and 5mA, respectively. With a load current of 10µA and 5mA, the measured line regulation is 10mV/A and 11.25mV/A, respectively, as shown in Fig. 3 (c). The load regulation is 1.535mV/mA, as shown in Fig. 3 (d). Fig. 4 shows the measured load transient responses of proposed BGR in the case of a load current step between 1mA and 5mA with an edge of time of 50ns. The measured overshoot and undershoot voltages of the proposed design are 154.4mV and 150mV respectively. Fig. 5 shows the measured PSRR of the BGR. Under different loads, the low-frequency PSR are less than -40

dB. The performance comparison with the BGRs and LDOs of the traditional structure is shown in Table I. The proposed design achieves the benefits of both BGR and LDO with good TC and load capability. While comparing with [6] which is also a BGR with load current driving, the proposed design achieves significantly reduced quiescent current consumption.

IV. CONCLUSION

This paper presents an ultra-low quiescent current BGR with current loading capability. The dual-clamping structure allows the BGR to have load current driving as well as improve the TC and DC regulation performance. Compared to the traditional BGR and LDO, the structure of this design is simple and effectively reduces power consumption and chip area. As a result, the proposed architecture enables a higher degree of integration of circuit functions with lower power consumption.

REFERENCES

[1] M. Konijnenburg et al., "A Multi(bio)sensor Acquisition System With Integrated Processor, Power Management, 8 × 8 LED Drivers, and Simultaneously Synchronized ECG, BIO-Z, GSR, and Two PPG Readouts," in IEEE Journal of Solid-State Circuits, vol. 51, no. 11, pp. 2584-2595, Nov. 2016, doi: 10.1109/JSSC.2016.2605660.

[2] X. Ma, Y. Lu, R. P. Martins and Q. Li, "A 0.4V 430nA quiescent current NMOS digital LDO with NAND-based analog-assisted loop in 28nm CMOS," 2018 IEEE International Solid - State Circuits Conference - (ISSCC), 2018, pp. 306-308, doi: 10.1109/ISSCC.2018.8310306.

[3] Y. Huang, Y. Lu, F. Maloberti and R. P. Martins, "Nano-Ampere Low-Dropout Regulator Designs for IoT Devices," in IEEE Transactions on Circuits and Systems I: Regular Papers, vol. 65, no. 11, pp. 4017-4026, Nov. 2018, doi: 10.1109/TCSI.2018.2851226.

[4] U. Chi-Wa, W. -L. Zeng, M. -K. Law, C. -S. Lam and R. P. Martins, "A 0.5-V Supply, 36 nW Bandgap Reference With 42 ppm/°C Average Temperature Coefficient Within −40 °C to 120 °C," in IEEE Transactions on Circuits and Systems I: Regular Papers, vol. 67, no. 11, pp. 3656-3669, Nov. 2020, doi: 10.1109/TCSI.2020.3010998.

[5] S. Wang and P. K. T. Mok, "An 18-nA Ultra-Low-Current Resistor-Less Bandgap Reference for 2.8 V–4.5 V High Voltage Supply Li-Ion-Battery-Based LSIs," in IEEE Transactions on Circuits and Systems II: Express Briefs, vol. 67, no. 11, pp. 2382-2386, Nov. 2020, doi: 10.1109/TCSII.2020.2965539.

[6] W. Chen, Y. Su, Y. Lee, C. Wey and K. Chen, "17.10 0.65V-input-voltage 0.6V-output-voltage 30ppm/°C low-dropout regulator with embedded voltage reference for low-power biomedical systems," 2014 IEEE International Solid-State Circuits Conference Digest of Technical Papers (ISSCC), 2014, pp. 304-305, doi: 10.1109/ISSCC.2014.6757445.

[7] W. Chen, M. Chen, Y. Hao, L. Qi and J. Zhao, "A 1-µA-Quiescent-Current Capacitor-Less LDO Regulator with Adaptive Embedded Slew-Rate Enhancement Circuit," 2021 IEEE International Symposium on Circuits and Systems (ISCAS), 2021, pp. 1-5, doi: 10.1109/ISCAS51556.2021.9401457.

Implementation of Polynomial Fitted Poly-Harmonic Distortion Model with Frequency Defined Device

Xiaoqiang Tang, Jialin Cai*

Key Laboratory of RF Circuit and System, Ministry of Education, Hangzhou Dianzi University, China.

Abstract—In this paper, a polynomial fitted poly-harmonic distortion (PHD) model is proposed, and it is implemented with frequency defined device (FDD). Polynomial fitting technique provides an effective method to including PHD model with different input power states through single set of model parameter. It can greatly reduce the model extraction complexity, and compact the model file size. The basic theory of PHD model, polynomial fitting method, and the FDD technique is provided in this work. A 10 W Gallium Nitride (GaN) packaged transistor is used in the test example. The results show that the proposed model has high accuracy for both fundamental and second harmonic behavioral predictions.

Keywords—Frequency defined device, GaN HEMTs, PHD model, Polynomial fitting.

I. INTRODUCTION

With the development of modern wireless communication systems, the topology of the power amplifiers (PAs) become more complicated, which result in a requirement of a more accurate transistor model for circuit designing.

Many different approaches have been developed for transistor modeling, ranging from physics-based model, equivalent circuit model, to behavioral model [1]-[3]. In recent years, with the rapid development of the third-generation semiconductor material [4], behavioral model has become a hot research topic. It is a model that only take into consider the input-output relationship of the device under test (DUT). The main advantage of the behavioral model is that it is independent of the device process and technology. Furthermore, it has extraordinary prediction accuracy around where it has been extracted. Thus, this work mainly focuses on behavioral modeling technology.

In [5], Verspecht and Root introduced the Poly-Harmonic Distortion (PHD) model. It starts with the concept of a describing function. With the adoption of the harmonic superposition principle [5], the function can be expanded as a multi-dimensional Taylor series around the large-signal operating point (LSOP), truncated to first order [5]. However, the PHD model is input power dependent, so it is important to extract as many sets of parameters at different input powers as possible, making the extraction process complicated, and the final generated file inconvenient to utilize.

In order to alleviate this issue, in this paper, we use polynomial fitting method to fit the directed extracted PHD model [6], [7], thus, the extraction of the PHD model parameter sets can be greatly reduced, minimizing the model file size at the same time. The fitted PHD model is implemented with the frequency defined device (FDD) technique in the commercial simulation tool, advanced design system (ADS), and load-pull verification was taken as well.

The rest paper is organized as follows. The basic principle of the PHD model, polynomial fitting method, and the FDD topology are provided in Section II, while in Section III simulation validation results are given. Conclusions are presented in Section IV.

This work was supported by the National Natural Science Foundation of China (NSFC) under Grants 61971170, and Qianjiang Talent Project Type-D of Zhejiang under Grant QJD2002020.

II. BASIC THEORY OF PROPOSED POLYNOMIAL FITTED PHD MODEL

In this work, the polynomial method is used to fit the directed extracted PHD model, thus, the PHD model that haven't been extracted can be calculated and used for simulation, which can simplify the model extraction. At the same time, the fitted model is implemented by the FDD component, allowing it to be applied for circuit simulation [8], [9].

A. Basic Theory of Poly-Harmonic Distortion Model

The transistor device is usually regarded as a two-port network, and its scattered waves on port p and harmonic m can be described as (1).

$$B_{pm} = F_{pm}(A_{11}, \dots, A_{1l}, A_{21}, \dots, A_{2l}). \tag{1}$$

where F_{pm} is the describing function of multiple complex-valued inputs, A_{ql} is the incident wave on port q and harmonic l.

The large-signal steady state contains all large-signal excitations and the corresponding responses, and is called the LSOP. It includes all excitations, and when small-signal perturbations are applied near the large-signal operating point, the scattered waves of each port can be described by (2), which is the standard expression of the PHD model [5].

$$B_{pm} \cong X_{pm}^F(LSOPS)P^k + \sum_{ql} X_{pm,ql}^S(LSOPS)P^{k-l}A_{ql}$$
$$+ \sum_{ql} X_{pm,ql}^T(LSOPS)P^{k+l}A_{ql}^* \tag{2}$$

B. Basic Theory of Polynomial Fitting Technique

The mathematical basis of polynomial fitting method is the least squares curve fitting, which can be expressed as shown in (3) [7].

$$y(x,w) = \sum_{j=0}^{n} w_j x^j. \tag{3}$$

where x^j is the jth power of x, w_j is the coefficient of x^j. For a sample x_n, let its output be t_n, then its squared error summation is shown in (4).

$$E(w) = \sum_{n=0}^{N} \{y(x_n,w) - t_n\}^2. \tag{4}$$

Thus, the coefficient vector, w, can be solved by its partial derivative.

C. Basic Principle of FDD

In this work, FDD is used for model implementation. FDD can create equation-based, user-defined, nonlinear components. It is a multi-port device that describes current and voltage spectral values in terms of algebraic relationships of other voltage and current spectral values.

Taking the two-port FDD as an example, the FDD component symbol is shown in Fig. 2, and the required model can be created by simply entering the correct algebraic relationship between the voltage and current spectral values at each port. In this work, the extracted polynomial fitted PHD model is built in FDD and simulation test is described in the following section.

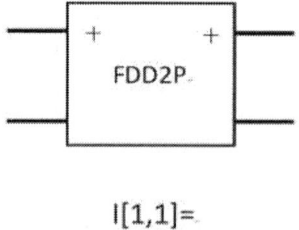

Fig. 1. 2-port FDD component.

III. MODEL IMPLEMENTATION AND VALIDATION

A. Model Implementation

A 10 W GaN package transistor is chosen as the device under test (DUT) to illustrate the validation example. Firstly, the PHD model that including up to the third harmonic is extracted, with the gate bias of -2.7 V, drain bias of 28 V, and fundamental operating frequency of 3 GHz. The input power is ranging from 20 dBm to 30 dBm with a step size of 2 dBm. The output load impedance is fixed at 50 ohm. Therefore, a total of six different input power points are used for model extraction.

With the extracted PHD model above, the polynomial fitting technique is used to interpolate the PHD model parameters, thus, the polynomial fitted PHD model can be obtained. After that, the obtained model can be used to calculate the PHD model when the input powers are different from the direct extracted values, such as the inputs ranging from 20 dBm to 30 dBm, with a step size of 1 dBm. Taking the fundamental output PHD model parameters x_{21}^{FB}, $x_{21,21}^{S}$, $x_{21,21}^{T}$ as an example, the fitted results are shown in Fig. 2.

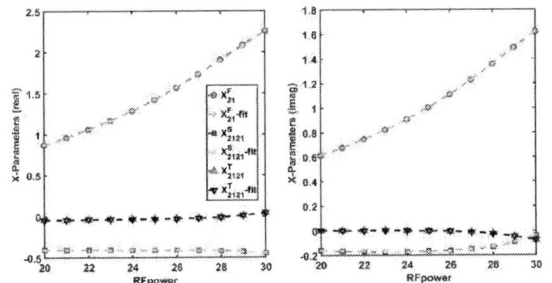

Fig. 2. Real part and imaginary part of x_{21}^{FB}, $x_{21,21}^{S}$ and $x_{21,21}^{T}$, from both directed extracted PHD model (X) and fitted model (X-fit).

As can be seen from Fig. 3, the polynomial fitting for both x_{21}^{FB}, $x_{21,21}^{S}$ and $x_{21,21}^{T}$ are excellent.

Then, the FDD is employed to implement the polynomial fitted PHD model, the detail model topology is shown in Fig. 3. Through the data access component (DAC), the new PHD model file data is entered and can correctly represent the algebraic relationship between the port voltage and current spectrum values of the FDD.

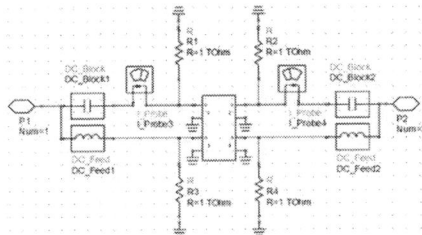

Fig. 3. The topology of the implemented model with FDD.

B. Model Validation

The implemented polynomial fitted PHD model is validated in this part. The fitted model is used to compared with the direct extracted PHD model, when the input powers are not used for fitted model extraction, at 23 dBm and 25 dBm. Both the fundamental and second harmonic load-pull simulation results are given in Fig. 4 and 5, respectively.

As can be seen in the figures, both the fundamental and second harmonic scattered waves, B_{21} and B_{22}, from the proposed model fit well with the directed extracted PHD model, when the input power is 23 dBm and 25 dBm, indicating that the

proposed modeling technique is effective. The proposed model provides high accuracy prediction, but need much less model parameter sets.

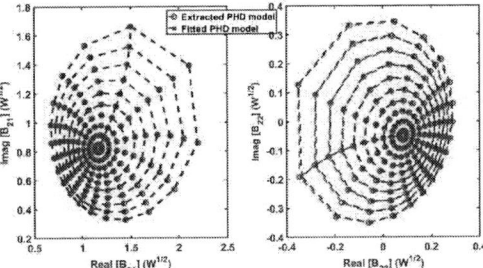

Fig. 4. Fundamental and second harmonics load-pull validation results when the input power is 23 dBm.

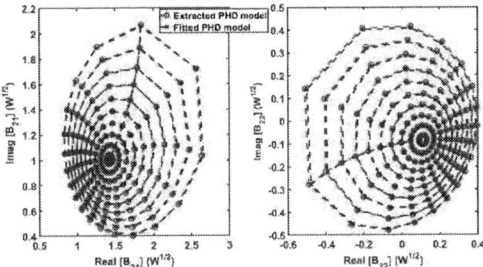

Fig. 5. Fundamental and second harmonics load-pull validation results when the input power is 25 dBm.

IV. CONCLUSION

In this paper, a new method for generating FDD-based PHD model using polynomial fitting technique is proposed. The proposed method can greatly simplify the extraction process of traditional PHD model, and minimize the model file size at the same time. Simulation tests show that this new method can effectively predict device behavior, not only at the fundamental but also at the second harmonic frequency.

REFERENCES

[1] O. Jardel et al., "An Electrothermal Model for AlGaN/GaN Power HEMTs Including Trapping Effects to Improve Large-Signal Simulation Results on High VSWR," in IEEE Transactions on Microwave Theory and Techniques, vol. 55, no. 12, pp. 2660-2669, Dec. 2007.

[2] A. Prasad, M. Thorsell, H. Zirath and C. Fager, "Accurate Modeling of GaN HEMT RF Behavior Using an Effective Trapping Potential," in IEEE Transactions on Microwave Theory and Techniques, vol. 66, no. 2, pp. 845-857, Feb. 2018.

[3] J. Cai, J. Liu, B. Pan, X. Du and J. King, "A Modified Poly-Harmonic Distortion Model Based on the Canonical Piecewise Linear Functions for GaN HEMTs," in IEEE Access, vol. 8, pp. 181420-181431, Oct. 2020.

[4] G. Crupi, M. D. Schreurs and A. Caddemi, "On the small signal modeling of advanced microwave FETs: A comparative study", Int. J. RF Microw. Comput.-Aided Eng., vol. 18, no. 5, pp. 25-417, Sep. 2010.

[5] Verspecht and D. E. Root, "Poly-harmonic distortion modeling," in IEEE Microwave Magazine, vol. 7, no. 3, pp. 44-57, June 2006.

[6] L. Fang and D. C. Gossard, "Multidimensional curve fitting to un-organized data points by nonlinear minimization," Computer-Aided Design, vol. 27, pp. 48-58, 1995.

[7] V. Pricop, E. Helerea and G. Scutaru, "Fitting magnetic hysteresis curves by using polynomials," 2014 International Symposium on Fundamentals of Electrical Engineering (ISFEE), Bucharest, Ro-mania, 2014, pp. 1-6.

[8] S. Woodington, R. Saini, D. Williams, J. Lees, J. Benedikt and P. J. Tasker, "Behavioral model analysis of active harmonic load-pull measurements," 2010 IEEE MTT-S International Microwave Symposium, Anaheim, CA, USA, 2010, pp. 1688-1691.

[9] S. Woodington et al., "A novel measurement based method enabling rapid extraction of a RF Waveform Look-Up table based behavioral model," 2008 IEEE MTT-S International Microwave Symposium Digest, Atlanta, GA, USA, 2008, pp. 1453-1456.

Dynamic Sensor Arrays Based on Solution-Processed Metal Oxide Semiconductor Thin-Film Transistors

Bowen Zhu[1,2]

[1] School of Engineering, Westlake University, Hangzhou, China
[2] Institute of Advanced Technology, Westlake Institute for Advanced Study, Hangzhou, China

Abstract—Flexible active-matrix sensor arrays provide large area and high spatial resolution for emerging sensing applications in electronic skin, health monitoring, and human-machine interfaces. However, it is still a challenge to achieve flexible active-matrix sensor arrays with low cost, low crosstalk, high sensitivity, and high uniformity characteristics. Here, we demonstrate a low-cost, high-resolution flexible sensor array by integrating solution-processed indium oxide (In_2O_3) transistor array with stimuli-sensitive (force, light, etc.) layers. This strategy provides an effective and universal solution to achieve large-area, active-matrix sensor arrays for future soft electronics applications.

Keywords—Oxide Semiconductors, Thin-Film Transistors, Active-Matrix, Flexible Electronics

I. INTRODUCTION

Flexible and wearable sensors play important roles in applications of electronic skin, healthcare, and human-machine interfaces. Thin-film transistors (TFTs) are the core electronic components for constructing large-area, active-matrix flexible sensor arrays with high spatial resolution. Metal oxide semiconductors (MOSs) based TFTs hold great promises for emerging applications in active-matrix flexible sensor arrays by virtue of their high optical transparency, large-area uniformity, sound electrical performance, and capability of being fabricated via low-temperature solution-based processes. However, current flexible sensor arrays based on solution-processed MOS TFTs often exhibited inferior electrical performance and mechanical flexibility, resulting in low spatial-resolution in signals mapping and severe mechanical mismatch with soft, dynamic bio-interfaces.

II. MAIN

A. System design

To address these challenges, we developed low-cost, solution-processed MOS TFTs to serve as the driving back-panel for dynamic, active-matrix sensor arrays, where each sensor pixel is monolithically integrated with one or more TFT devices by connecting the sensors' interdigitated electrodes with the TFT's source/drain electrodes (Fig. 1). In this way, each sensor can be arbitrarily accessed and controlled by TFTs, providing fast-switching speed and high density.

Furthermore, the TFTs based dynamic sensor arrays could be readily integrated with printed circuit board (PCB) and signal processors, providing real-time, user-readable mappings of input signals' distributions (e.g., force, pressure, light, etc.). With the TFT back-panel, we could reshape the flexible sensors by transforming the passive, static, and dispersed sensors into active, dynamic, and highly integrated sensor arrays.

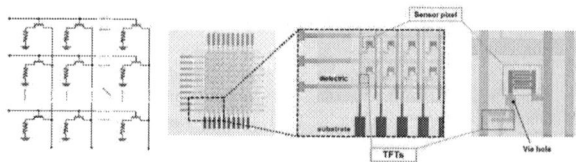

Fig. 1 Schematic illustration of active-matrix flexible sensor arrays composed of TFT backplane and sensors. An individual sensor pixel is a two-terminal device composed of interdigitated electrodes and resistive sensitive layers (e.g. force, pressure, light). In this way, each sensor pixel could be selected and operated arbitrarily, providing high spatial resolution compatible with standard microfabrication with suppressed signal crosstalk.

B. TFT performance

We developed solution-processable metal oxide semiconductors (MOSs), e.g., In_2O_3, InZnO (IZO), and InGaZnO (IGZO), to serve as the active channels for constructing large-area TFTs. MOSs have been highly recognized as promising active channel materials for constructing high-performance TFTs in applications of active-matrix electronics such as displays and sensor arrays. The electron transport in MOSs is insensitive to structure distortion, and they demonstrate high mobility with both amorphous and crystalline structures, because the delocalized s orbitals with heavy metal cations (e.g., In^{3+}) could form a largely dispersed conduction band with a small effective mass. In this way, MOSs could exhibit high electrical performance with ultrathin thickness (<10 nm) at low-temperature, solution-based processing, which are advantageous for flexible sensing array applications.

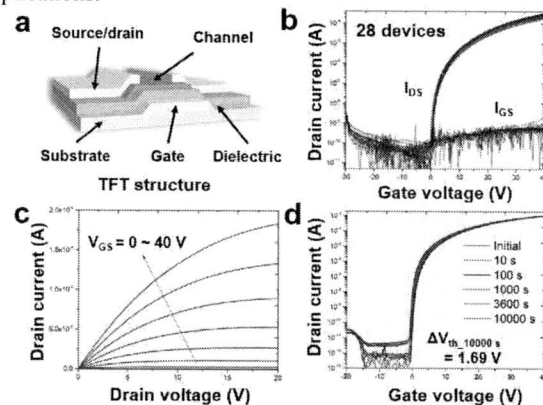

Fig. 2 Solution processed In_2O_3 TFTs. (a) A schematic showing the structure of a three-terminal TFT device. (b) The transfer curves of 28 TFTs distributed on an area of 4×4 cm². (c) Output characteristics of the In_2O_3 TFT. (d) Electrical stability of the In_2O_3 TFT.

Fig. 2 shows the solution-processed In_2O_3 TFTs possess high electrical performance, demonstrating large current on/off ratio (>10⁷), near-zero turn-on voltage (V_{on}), high device-to-device uniformity (Fig. 2b), and high output current (Fig. 2c). In addition, the electrical stability under voltage bias could be improved by surface passivation to isolate moisture and oxygen in the environment, showing a threshold voltage shift (V_{th}) of only 1.69 V under the test condition of V_{GS} = 20 V, and V_{DS} = 0.1 V for 10000 s (Fig. 2d), illustrating the feasibility for long-term applications.

C. Sensor array integration

As an example, we demonstrated a phototransistor array based on both solution-processed n-type MOSs and polymer semiconductors of In_2O_3 and poly{5,5'-bis[3,5-

978-1-6654-9270-6/22 $31.00 © 2022 IEEE

bis(thienyl)phenyl]-2,2'-bithiophene-3-ethylesterthiophene]} (PTPBT). The In_2O_3 and PTPBT provide fast electron transport and high near-infrared (NIR) light responses, respectively. As a result, the phototransistors exhibited high saturation mobility of 7.1 cm^2 V^{-1} s^{-1}, large current on/off ratio of $>10^7$, and low device-to-device variations. As shown in Fig. 3, by virtue of the large-area solution-based processibility with high uniformity, we successfully constructed a 10×10 phototransistor array with an active area of 2×2 cm^2, which demonstrated effective photomapping functions. Importantly, the array integration is compatible with standard semiconductor fabrication technique, allowing for future monolithic integration of large-area, high-density phototransistor arrays for image sensor applications.

Also, flexible In_2O_3/PTPBT-ET phototransistors could successfully achieved on polyimide substrates, demonstrating outstanding mechanical flexibility up to 1000 bending/releasing cycles at a bending radius of 5 mm. With large-area uniformity, low-cost fabrication, and superior mechanical flexibility, the solution-processed MOS TFTs provide an intriguing alternative to vacuum-based capital-intensive counterparts for flexible electronics applications.

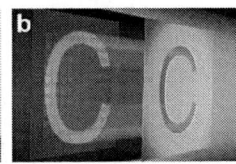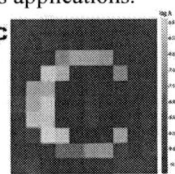

Fig. 3 Phototransistor array based on solution processed MOS TFTs. (a) A photo of a 10×10 phototransistor array with an active area of 2×2 cm^2. Scale bar: 1 cm. (b) Schematic illustrating applying light on the array via a shadow mask with a shape of character "C". (c) The photocurrent mapping of the phototransistor array under light illumination via the mask.

III. CONCLUSIONS

We utilized low-cost, solution-processed MOSs to construct high-performance flexible, active-matrix physical sensor arrays. This opens new opportunities for employing MOSs TFTs in applications of wearable electronic skin, health monitoring, and human-machine interfaces.

ACKNOWLEDGMENT

B. Zhu acknowledges the financial supports from the National Natural Science Foundation of China (Grant No. 62174138), the Westlake Multidisciplinary Research Initiative Center (MRIC) Seed Fund (Grant No. MRIC20200101), and the Leading Innovative and Entrepreneur Team Introduction Program of Zhejiang (Grant No. 2020R01005).

REFERENCES

[1] D. Li, J. Du, Y. Tang, K. Liang, Y. Wang, H. Ren, R. Wang, L. Meng, B. Zhu, Y. Li, "Flexible and air-stable near-infrared sensors based on solution-processed inorganic–organic hybrid phototransistors." *Adv. Funct. Mater.*, vol. 31, pp. 2105887, Nov 2021.

[2] K. Liang, D. Li, H. Ren, M. Zhao, H. Wang, M. Ding, G. Xu, X. Zhao, S. Long, S. Zhu, P. Sheng, W. Li, X. Lin, and B. Zhu, "Fully printed high-performance n-type metal oxide thin-film transistors utilizing coffee-ring effect," *Nano-Micro Lett.*, vol. 13, no. 1, Aug 2021.

[3] F. Li, R. Wang, C. Song, M. Zhao, H. Ren, S. Wang, K. Liang, D. Li, X. Ma, B. Zhu, H. Wang, and Y. Hao, "A skin-inspired artificial mechanoreceptor for tactile enhancement and integration." *ACS Nano*, vol. 15, pp. 16422-16431, Oct 2021.

[4] K. Liang, H. Ren, Y. Wang, D. Li, Y. Tang, C. Song, Y. Chen, F. Li, H. Wang, and B. Zhu, "Tunable plasticity in printed optoelectronic synaptic transistors by contact engineering." *IEEE Electron. Device Lett.*, vol. 43, Apr 2022, doi: 10.1109/LED.2022.3166507.

[5] K. Liang, R. Wang, B. Huo, H. Ren, D. Li, Y. Wang, Y. Tang, Y. Chen, C. Song, F. Li, B. Ji, H. Wang, and B. Zhu, "Fully printed optoelectronic synaptic transistors based on quantum dot–metal oxide semiconductor heterojunctions," *ACS Nano*, vol. 16, Apr 2022, doi: 10.1021/acsnano.2c00439.

A 22.8 GHz to 32.8 GHz Compact Power Amplifier with a 15 dBm Output P$_{1dB}$ and 36.5% Peak PAE in 65-nm CMOS

Huabing Liao, Haikun Jia, Xiangrong Huang, Bao Shi, Wei Deng, Baoyong Chi, Zhihua Wang*

School of Integrated Circuits, BNist, Tsinghua University
*Research Institute of Tsinghua University in Shenzhen

Email: {jiahaikun, zhihua@tsinghua.edu.cn}

Abstract—A CMOS broadband millimeter-wave power amplifier (PA) based on a Sandwiched Transformer (ST) output matching network is presented in this paper. The ST output matching network with a three-layer structure providing a larger coupling coefficient (k) than the traditional two-layer stack structure is proposed for PA's output matching network. The layout of the transistors is optimized to improve the PA's performance. Fabricated in 65-nm CMOS process, the PA has achieved 15 dBm OP$_{1dB}$ and 36.5% peak power added efficiency (PAE). The 3-dB bandwidth of the PA is from 22.8 GHz to 32.8 GHz.

Keywords—*power amplifier (PA), broadband matching network, transistors layout optimization, sandwiched transformer matching network*

I. INTRODUCTION

The 28 GHz band for 5G millimeter-wave wireless communication requires excellent power and spectrum efficiency for high data-rate throughput [1]. It is still challenging for the power amplifier to meet the output power and speed requirement. In this paper, a 22.8 GHz to 32.8 GHz PA is presented. A two-stage power amplifier with layout optimization techniques, the sandwiched transformer, and the broadband transformer matching networks achieves 15 dBm power at the output 1-dB compression point (OP$_{1dB}$) and 36.5% peak power added efficiency (PAE). The paper is organized as follows. Section II introduces the circuit design details of the sandwiched transformer and the layout optimization technique. Section III presents the measured results of the proposed PA. Finally, conclusions are given in section IV.

II. PROPOSED CIRCUITS

A. Architecture Overview

Fig. 1 illustrates the architecture of the proposed PA. The PA is designed in 65-nm CMOS process, which includes RF transistor models, and low-threshold-voltage (lvt) transistor models. For low power consumption and some other better performance, lvt transistors are used in our design. In addition, there are three thick metal layers in this process, M8, M9, and AP. The thickness of each layer is 0.9 μm, 3.4 μm and 1.45 μm respectively. This PA is based on two amplification stages, the transistor sizes of the first stage and the second stage are 64μm/60nm and 128μm/60nm respectively. Neutralization capacitors are used in each stage, and the capacitances are 23 fF for the first stage and 46 fF for the second stage.

B. Sandwiched Transformer Design

Larger coupling confident (k) and larger quality factor (Q) are usually needed when the transformer matching network is used in the output stage matching of the power amplifier. Increasing the coupling coefficient of the transformer by layout techniques can efficiently decrease the insert loss.

Fig. 2 shows the 3D view of the sandwiched transformer we proposed in the output stage. The primary inductor is based on M9 whose center tap connects to the power supply. The secondary inductor is based on layers of M8 and AP, which ties together at the end and connects to the G-S-G pad. By placing M9 between the layer of AP and M8, the coupling coefficient of the primary inductor and secondary inductor is significantly improved. For better matching performance, a MOM capacitor is connected in parallel with the secondary inductor.

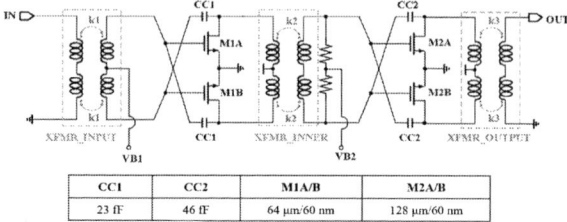

CC1	CC2	M1A/B	M2A/B
23 fF	46 fF	64 μm/60 nm	128 μm/60 nm

Fig. 1. Schematic of the proposed power amplifier.

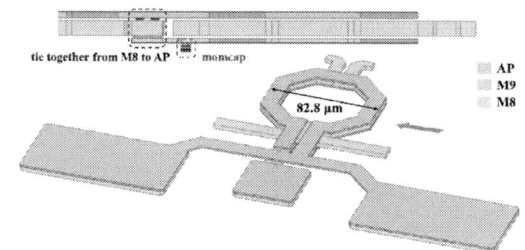

Fig. 2. The 3-D view of the output matching network.

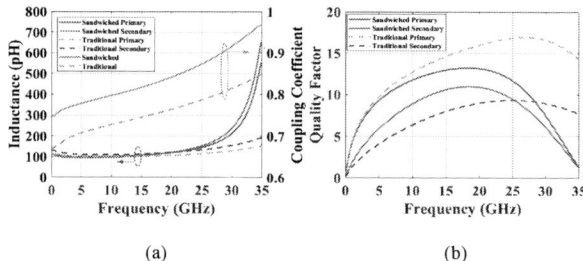

(a)   (b)

Fig. 3. comparison between the sandwiched transformer and the traditional transformer, (a) inductance and k and (b) Q.

Fig. 3 shows the comparison results of the sandwiched transformer and the traditional transformer. The only difference is that no layer M8 is under the primary inductor of the traditional transformer. From Fig. 3 (a), under almost the same inductance of the primary inductor and the secondary inductor, the coupling coefficient of the sandwiched transformer can achieve 0.9 at 28 GHz, around 0.1 larger than the traditional transformer. In Fig. 3 (b), the Q of the secondary inductor of the sandwiched transformer is larger than that of the traditional, which makes it easier to achieve high output power.

C. Transistors Layout Optimization

To reduce parasitic impedance, especially the parasitic resistance of the source, dummy transistors are used in the layout of the transistors. In addition, a larger gate pitch can also optimize the impedance to increase the quality factor (Q) of the matching network.

Fig. 4 (a) illustrates the layout of a transistor whose size is 64μm/60nm. The gate pitch changes increasingly from default 0.24 μm to 0.36 μm. Both the first stage and the second stage use the same structure as shown in Fig. 4 (a).

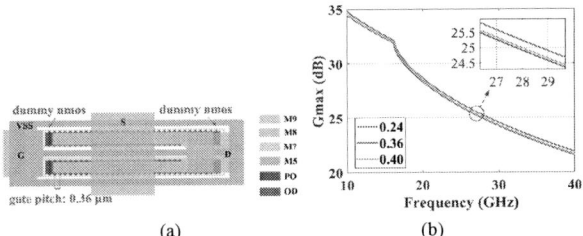

(a)

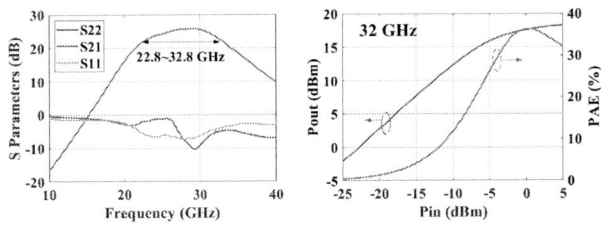

(a)   (b)

Fig. 4. (a) The layout of the 64μm/60nm transistor and (b) simulated Gmax results of different gate pitches.

Fig. 7. The measured (a) S parameters and (b) the output power and PAE at 32 GHz of the PA.

IV. CONCLUSION

The simulation results of different gate pitches are shown in Fig. 4 (b). As gate pitch increases from 0.24 μm to 0.40 μm, the maximum power gain (Gmax) increases because of the gradually decreasing source resistance. With the increasing gate pitch, the parasitic capacitance increases, so an optimal gate pitch would be picked. In this design, the gate pitch is 0.36 μm.

A 22.8 GHz to 32.8 GHz power amplifier based on 65-nm CMOS process is designed in this paper. The PA with broadband transformer matching network and sandwiched transformer matching network can achieve good performance of gain flatness and output power. In addition, the optimization of the transistors' gate pitch can make the Gmax improved. The measured gain 3-dB bandwidth is from 22.8 GHz to 32.8 GHz, and the OP$_{1dB}$ is 15 dBm at 32 GHz.

D. Broadband Transformer Matching Network

Transformer matching networks can effectively increase bandwidth. And existing methods make it easier to design a transformer based on the impedance of each side [2]. Fig. 5 (a) shows the 3D view of the input matching network. Fig. 5 (b) shows the 3D view of the inner matching network. Stack layers of M8 and M9 are used to increase the Q value of the secondary inductor, and for the primary inductor, AP layer is used in the input matching network, and M9 is used in the inner matching network.

TABLE I. PERFORMACE COMPARISON

	PERFORMANCE COMPARISON			
	A-SSCC 2016[2]	ISSCC 2016[3]	ISSCC 2018[4]	This Work
Technology	65 nm CMOS	28 nm CMOS	65 nm CMOS	65 nm CMOS
Supply (V)	1.0	1.15	1.1	1.3
OP$_{1dB}$ (dBm)	12.9	14.3	14	15.0
PAE Peak (%)	32.9	36.6	41	36.5
Gain (dB)	20.8	15.7	15.8	25.7
Core Area (mm²)	0.11	0.16	0.24	0.09

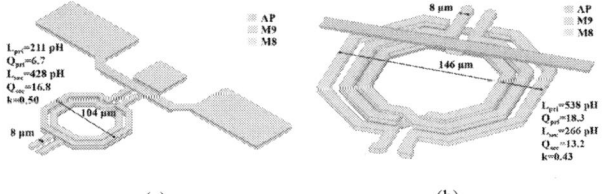

(a)   (b)

Fig. 5. The 3-D view of (a) the input matching network and (b) the inner matching network.

III. MEASURE RESULTS

The proposed PA is fabricated in 65-nm CMOS technology and occupies 0.54 mm ×0.17 mm without pads, as shown in Fig. 6. The PA consumes 115 mA current from a 1.3 V power supply.

ACKNOWLEDGMENT

This work was supported by the National Natural Science Foundation of China (No. 62074090), the Shenzhen Science and Technology Program (SGLH20180622095014688, JSGG20191129141019090), Guangdong Basic and Applied Basic Research Foundation (No. 2019A1515110431), and the Beijing Innovation Center for Future Chips (ICFC).

REFERENCES

[1] Z. Ma, K. Ma, K. Wang and F. Meng, "A 28GHz Compact 3-Way Transformer-Based Parallel-Series Doherty Power Amplifier With 20.4%/14.2% PAE at 6-/12-dB Power Back-off and 25.5dBm PSAT in 55nm Bulk CMOS," 2022 IEEE International Solid- State Circuits Conference (ISSCC), 2022, pp. 320-322.

[2] H. Jia, C. C. Prawoto, B. Chi, Z. Wang and C. P. Yue, "A 32.9% PAE, 15.3 dBm, 21.6–41.6 GHz power amplifier in 65nm CMOS using coupled resonators," 2016 IEEE Asian Solid-State Circuits Conference (A-SSCC), 2016, pp. 345-348.

[3] S. Shakib, H. -C. Park, J. Dunworth, V. Aparin and K. Entesari, "20.6 A 28GHz efficient linear power amplifier for 5G phased arrays in 28nm bulk CMOS," 2016 IEEE International Solid-State Circuits Conference (ISSCC), 2016, pp. 352-353.

[4] S. N. Ali, P. Agarwal, J. Baylon, S. Gopal, L. Renaud and D. Heo, "A 28GHz 41%-PAE linear CMOS power amplifier using a transformer-based AM-PM distortion-correction technique for 5G phased arrays," 2018 IEEE International Solid - State Circuits Conference - (ISSCC), 2018, pp. 406-408.

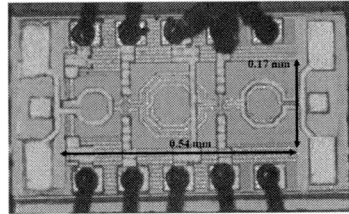

Fig. 6. Die photo of proposed PA.

Fig. 7 (a) shows the measured S parameters of the proposed PA. Using the proposed broadband matching technique, the measured gain (S21) of the proposed PA is larger than 22.4 dB in the frequency range from 22.8 GHz to 32.8 GHz, and the measured gain at 29.1 GHz is 25.7 dB. Fig. 7 (b) shows the measured output power and power added efficiency (PAE) of the PA at 32 GHz. The output power at 1-dB compression (OP1dB) is 15 dBm when the input frequency is 32 GHz. And the output power is larger than 10 dBm in the frequency range from 22.8 GHz to 32.8 GHz.

A Fully Synthesizable Injection Locked PLL with Dual-DCO Frequency Tracking in 55nm CMOS

Xuanchi Yu[1], Yan Chen[1], Gaofeng Jin[1], Fei Feng[1], Xun Luo[2], Xiang Gao[1,3]

[1] School of Micro-Nano Electronics, Zhejiang University, Hangzhou, China
[2] University of Electronic Science and Technology of China, Chengdu, China
[3] Peng Cheng Laboratory, Shenzhen, China

Abstract—This paper presents a fully synthesizable injection locked phase-locked loop (PLL), with a dual-DCO frequency tracking. The design has been fabricated in 55-nm CMOS and the layout is realized completely by digital flows. The proposed PLL covers a 0.2-to-1.2-GHz tuning range, achieving an absolute rms jitter (integrated from 100kHz to 100MHz) of 3.2ps at 4.3-mW power consumption, with a corresponding jitter-power figure of merit (FoM) of -224dB and the occupied core area is only 0.0225 mm².

Keywords—all-digital PLL, fully synthesizable PLL, injection locking, dual-DCO.

I. INTRODUCTION

As the feature size of integrated circuits continues to decrease, chip integration is increasing and the functionality of modern SoCs (System on Chip) is becoming more complex. The number of PLLs needed to meet various clocking requirements in large SoCs could be more than 10. Fully synthesizable PLLs [1-4], designed using pure digital flows, have been proposed to reduce the design time and allow easier porting and integration. An all-digital PLL based on time-to-digital converter (TDC) shown in Fig. 1 is not easy to fully synthesize, as the TDC is vulnerable to layout uncertainty from automatic placement and route(P&R), which introduces nonlinearity and may lead to poor in-band noise. In this design, we present a fully synthesizable PLL, utilizing the injection locking technique and a FSM (Finite State Machine) for DCO frequency tracking/calibration.

II. THE PROPOSED PLL ARCHITECTURE

A. Overall-Architecture

Fig.2 depicts the architecture of the proposed fully synthesizable PLL. The PLL consists of an Injection Pulse Generator (IPG), a DCO, and a Frequency-Locked-Loop (FLL) with an auxiliary DCO and two counters for frequency tracking.

Compared with the TDC-based digital PLL, the injection locked PLL does not need TDC and digital loop filter. However, the initial DCO frequency can't be too far away from the target value, otherwise the injection spur will be large or it may even fail to lock. Initial frequency calibration is thus a must. All the blocks in Fig.2 are either designed with digital standard cells or described in Verilog, such that the entire PLL is fully synthesizable in the end.

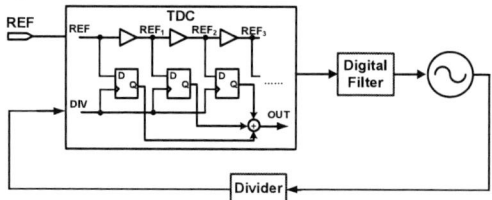

Fig. 1. Topology of conventional TDC-based all-digital PLL.

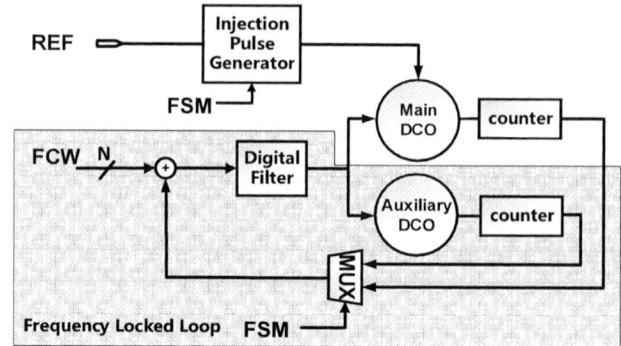

Fig. 2. Blocks diagram of the proposed fully synthesizable PLL.

B. Injection Locking

Fig.3(a) presents the design of the injection locking path and the DCO. The injection window control V_x and injection edge V_y are generated from the reference clock via IPG. When V_x is activated (logic low), the ring oscillator "stops" oscillating, the loop is broke and a clean falling edge originated from the reference clock is fed into the oscillator, replacing the original noisy one and resetting the jitter accumulated when the ring oscillator is free running.

In conventional PLLs, the loop bandwidth is typically less than one tenth of f_{REF} because of the loop stability constraint. The injection locked PLL could achieve several times higher effective bandwidth to suppress the oscillator noise. Fig. 3(b). shows the oscillator jitter accumulation mechanism and the clean up by the injection locking.

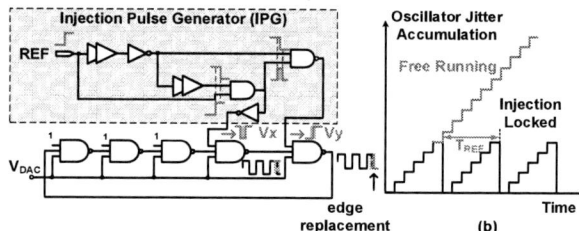

Fig. 3. (a). The injection locked PLL design, (b). Oscillator jitter accumulation mechanism and clean up by injection locking.

C. DCO

Targeting being fully synthesizable, the DCO needs to be designed with digital flows as well. Fig.4(a) shows the binary switch current-output DAC based synthesizable DCO , similar to [1]. In order to alleviate the trade-off between tuning range and resolution, an always on branch is added in parallel with the DAC to maintain a certain number of PMOS always on to keep the oscillation even when the code is reaching the minimum (all logic 0). Fig.4(b) shows the details of the proposed synthesizable DCO design.

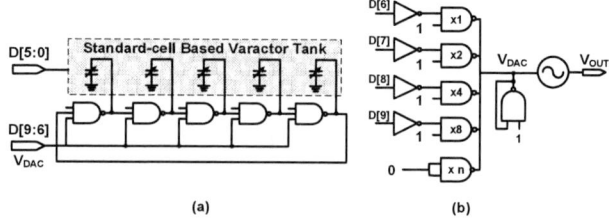

Fig. 4. Fully synthesizable DCO design.

D. Dual-DCO and FSM

In order to achieve lower spur and robust injection locking, the DCO frequency should be adjusted to be close to the target frequency before injection. In this design, counters and comparators are used to quantify the oscillator frequency. In addition, to handle the DCO frequency variation with PVT corners after locking, an Auxiliary-DCO, identically designed as the main DCO but free-running, is added outside the loop to capture this frequency variation. To regulate the loop's operation condition, a FSM is employed and the implementation is shown in Fig. 5. In state1, the proposed PLL calibrates the mismatch between the two DCOs by comparing their counter values. In state2, the mechanism forces main DCO to oscillate in the vicinity of target frequency by using the counter results and tuning the DCO code. In state3 the injection pulse generator is enabled and the generated injection pulse is connected to the main DCO, the injection locked loop is closed. At the same time, the FLL keeps running in background, monitoring the difference between free-running DCO frequency and the target frequency, and adjusting the DCO code when necessary. In this way, the DCO frequency variation is compensated.

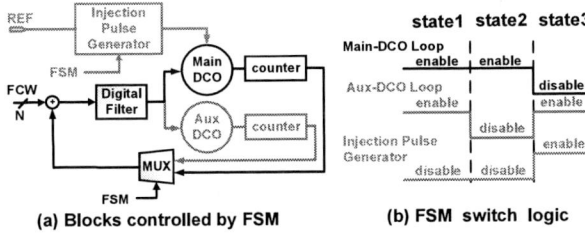

Fig. 5. The implementation of FSM: (a). Blocks controlled by FSM (b). FSM switch logic.

III. MEASUREMENTS

The proposed fully synthesizable PLL was fabricated in TSMC 55nm CMOS. Fig. 6 shows a micrograph of the PLL which occupies only 0.0225 mm² layout area. The measured frequency tuning range of the PLL is 0.2 to 1.2 GHz. At 1.0 GHz output using a 100MHz reference clock, the absolute rms jitter (integrated from 100kHz to 100MHz) is 3.2 ps and the power consumption with a 1.1V supply is 4.3 mW, with a corresponding jitter-power figure of merit (FoM) of -224dB.

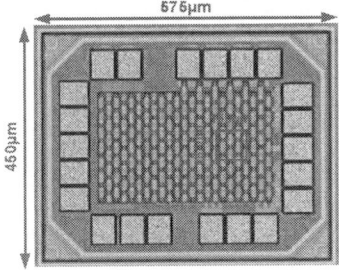

Fig. 6. The layout topology of the proposed PLL.

As shown in Fig. 7, comparing with the injection locking off case, the presented PLL provides 20dB/dec suppression for the DCO phase noise, in line with the injection locking theory. TABLE I shows the performance comparation of this work with literature.

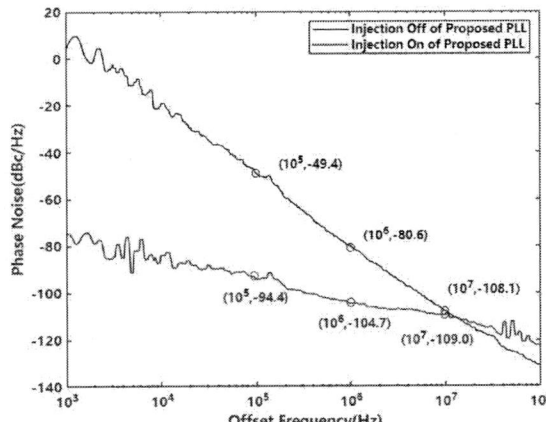

Fig. 7. Phase noise of the proposed fully synthesizable PLL.

TABLE I. PERFORMANCE COMPARISON

Performance	Works			
	This work	*JSSC2021 [2]*	*ISSCC2017 [3]*	*ISSCC2016 [4]*
Output Freq. (Range) (GHz)	1.0 (0.2-1.2)	3.6 (1.2-3.8)	0.998 (0.25-1.0)	1.4175 (0.2-1.45)
REF (MHz)	100	80	250	87.5
Power (mW)	4.3	3	15.2	8
Area (mm²)	0.0225	0.00525	0.0047	0.054
Jitter (ps)	3.2	2.55	3.3	11.7
FoM* (dB)	-224	-227.2	-218	-222
CMOS Tech.	55nm	22nm	28nm	65nm
Topology	Injection Locking	MDLL	RACC	Injection Locking
Fully Synthesizable?	YES	NO	YES	NO

FoM = 10log((Jitter/1s)²(Power/1mW))

IV. CONCLUSION

In conclusion, a fully synthesizable injection locked PLL, with a dual-DCO frequency tracking calibration, is presented in this paper, and the layout is implemented totally by digital automation tools. The measurement results show that the fully synthesizable PLL is feasible and achieves good jitter and power, with shorter design cycle, lower cost and better portability.

REFERENCES

[1] W. Deng et al., "A Fully Synthesizable All-Digital PLL With Interpolative Phase Coupled Oscillator, Current-Output DAC, and Fine-Resolution Digital Varactor Using Gated Edge Injection Technique," in IEEE Journal of Solid-State Circuits (JSSC), vol. 50, no. 1, pp. 68-80, Jan. 2015.

[2] S. Kundu, L. Chai, K. Chandrashekar, S. Pellerano and B. R. Carlton, "A Self-Calibrated 2-bit Time-Period Comparator-Based Synthesized Fractional-N MDLL in 22-nm FinFET CMOS," in IEEE Journal of Solid-State Circuits (JSSC), vol. 56, no. 1, pp. 43-54, Jan. 2021.

[3] H. Cho et al., "A 0.0047mm2 highly synthesizable TDC- and DCO-less fractional-N PLL with a seamless lock range of fREF to 1GHz," 2017 IEEE International Solid-State Circuits Conference (ISSCC), 2017.

[4] S. Kundu, B. Kim and C. H. Kim, "A 0.2-to-1.45GHz subsampling fractional-N all-digital PD-based spur cancellation and in-situ timing mismatch detection," 2016 IEEE International Solid-State Circuits Conference (ISSCC), 2016.

DFT Architecture for Click-Based Bundled-Data Asynchronous Circuits

Ruimin Zhu[1], Zeyang Xu[1], Yuhao Huang[2], Shanlin Xiao[1,3*], Zhiyi Yu[1,3*]

[1]School of Microelectronics Science and Technology, Sun Yat-sen University
[2]School of Electronics and Information Technology, Sun Yat-sen University
[3]Guangdong Provincial Key Laboratory of Optoelectronic Information Processing Chips and Systems

Abstract—Event-driven asynchronous circuits are gaining attention because of their low power consumption and robustness. Among asynchronous circuits, the Bundled data (BD) circuit used by Loihi has attracted attention because it can obtain a similar area as a synchronous circuit. Click circuit is a mainstream BD circuit, but due to the lack of DFT (Design For Test) architecture, the Click-based asynchronous circuit cannot be widely used. This paper proposes a DFT architecture suitable for BD circuits, which can be accomplished using traditional EDA tools rather than developing new ones. This paper verifies the proposed DFT architecture on a five-stage pipeline processor based on the RISC-V instruction set. The result is 99.62% coverage for stuck-at faults, 1.8398% area overhead, and 6.2259% power overhead.

Keywords—Asynchronous circuit, Bundled-data circuit, Click, DFT (Design For Test), Test coverage, Traditional EDA

I. INTRODUCTION

There is growing interest in asynchronous circuits because they can eliminate wasted power consumption caused by clock networks. Among asynchronous circuits, the Bundled data (BD) circuit used by Loihi has attracted attention because it can obtain a similar area as a synchronous circuit. Fig. 1 shows the difference between a synchronous circuit and an asynchronous circuit: the asynchronous handshake replaces the synchronous clock.

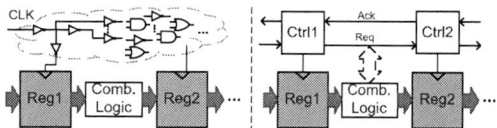

Fig. 1: (a)Synchronous   (b) Asynchronous

In BD circuits, the Click circuits [1][2] can be synthesized and implemented using traditional EDA (Electronic Design Automation) tools [3]. Click circuit is the current mainstream BD circuit. However, the lack of EDA tools and ATE (Automatic Test Equipment) designed for asynchronous circuits is a barrier to the widespread adoption of Click circuits. We found that [1] suggested a DFT architecture for the Click circuit but did not give its feasibility and results.

The contributions of this paper are:

1) We propose a DFT architecture for the Click circuit. This architecture enables independent testing of data paths without affecting Click pulse generation. This architecture can be done using traditional EDA tools rather than developing new ones.
2) We verified this architecture on an asynchronous RISC-V processor with 1.8398% area overhead, 6.2259% power consumption, and 99.62% coverage for stuck-at faults.

II. PROPOSED CLICK DFT ARCHITECTURE

A. Overview DFT Architecture

In this section, we detail the DFT architecture of the Click circuit with CBB (Clock Bypass Block) in Fig. 2. The datapath clock is the pulse generated by the handshake between asynchronous signals, called Click before CBB and Fire after CBB. The lower right corner of the Fig. 2 is the timing diagram of the asynchronous circuit handshake.

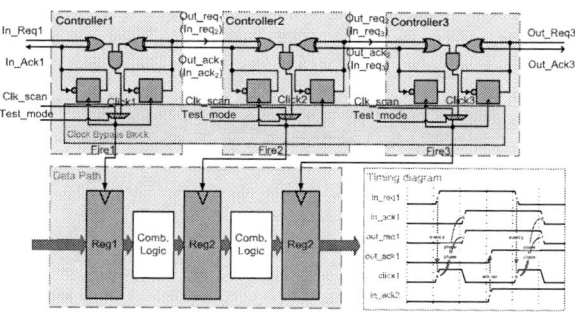

Fig. 2: Overview DFT Architecture and Timing Diagram

B. ATPG (Auto Test Pattern Generated) Flow

ATPG that only recognizes combinational logic needs to be based on the scan chain(also called shift register chain: SI-SDFF1(si)-SDFF1(so)-SDFF2(si)-SDFF2). When SE=1, the scan chain converts sequential logic to combinational logic. DFT-Compiler replaces DFF with SDFF and stitches SDFF.

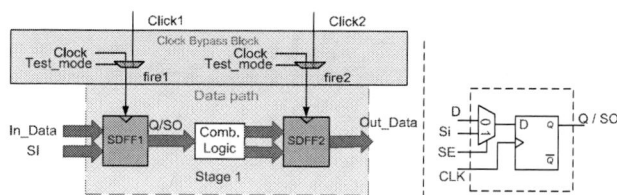

Fig. 3:   (a) Data Path with DFT   (b) SDFF Structure

1) *SE=1, Scan Chain, Shift In*: When the clock arrives, the SI end of SDFF1 will shift in the preset value. After the function block calculates the value, it is transferred to the D terminal of the SDFF2 register to wait for capture.
2) *SE=0, Function Path, Capture*: When the clock arrives, the SDFF2 register captures the value that is ready to be captured from the D end.
3) *SE=1, Scan Chain, Shift Out*: When the clock arrives, the SDFF2 register passes the value captured in step 2 through the scan chain to the next level register. The terminal of the shift register chain shifts out value. (Also called next pattern's shift in stage: Chain's front end shifts in)

This work was supported in part by the Key-Area Research and Development Program of Guangdong Province under Grant 2021B1101270005 and 2021B0101410004; in part by the National Key Research and Development Program of China under Grant 2017YFA0206200 and 2018YFB2202601; in part by the National Natural Science Foundation of China (NSFC) under Grant 61834005 and Grant 61902443; in part by the Guangdong Basic and Applied Basic Research Foundation under Grant 2022A1515011708; in part by the Zhuhai Industry-Academic collaboration program ZH22017001200097PWC; in part by the Huawei Technologies Co., Ltd.

*Corresponding authors

Currently, neither traditional EDA tools nor ATE machines support the constraints of asynchronous signals of different frequencies and phases to control the ATPG flow. Only by adding CBB in Fig. 3 to the asynchronous circuit can traditional EDA tools complete the ATPG process.

C. Our design flow and architectural features

CBB is added in the Async-RTL-Model in the design flow.

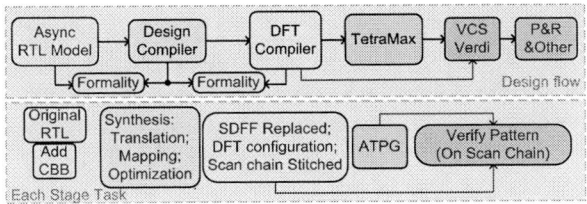

Fig. 4: Design Flow with Asynchronous DFT

Our proposed design flow and architecture have the following characteristics:

1) Asynchronous flow matches synchronous flow and uses the synchronous EDA tool to complete this flow.
2) Signals are not switched during ATPG. Test_mode is always 1 to make the CBB always strobe the scan clock.
3) Use multiplexer [4] to let test_mode select external clock and Click pulse.The output of the multiplexer is called the Fire signal to ensure that the data path works. The CBB will not affect the generation and testing of the Click pulse.

III. EXPERIMENTAL RESULTS

A. Experimental Conditions

The experimental RISC-V processor and experimental conditions are shown in Fig. 5 and Table I.

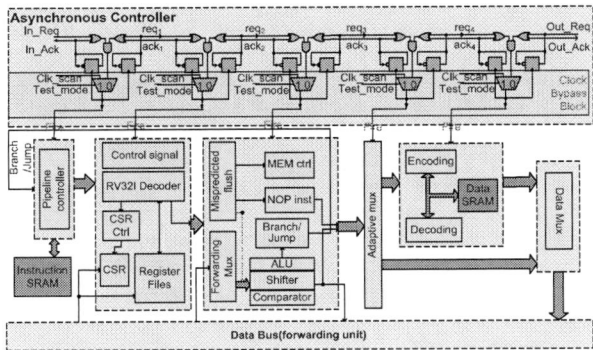

Fig. 5: RISC-V Processor Architecture with DFT

B. Area and Power Comparison

Table II shows the area and power consumption comparisons for the three synthesis stages for the three circuit styles. An explanation can be seen in the notes below the TableII.

From the table II it can be seen that **area (c) and area overhead (g)** of *Our CBB* are less than *Modified-[1]*. Also, *Our CBB* has a **smaller power overhead (g)** than *Modified-[1]*. In other words, if at the *a* stage, *Modified-[1]* has the same initial power as *Our CBB*, then at the *c* stage, the final power of *Our CBB* will less than *Modified-[1]*. This result means that *Our CBB* can reduce the **power consumption of a-b-c flow**.

TABLE I. Experimental Conditions

	Para.		**Para.**
Instruction Set	RISC-V	Pipeline	5
Technology	TSMC 65nm	Bit width	32
Scan chain EDA	DC (DFT) - Compiler	ATPG EDA	Tessent
Formal EDA	formality	Sim EDA	VCS
Serial-Sim time	1699330200 ps	Register count	3423
Parallel-Sim time	250002000 ps	Chain count	16

TABLE II. Area & Power Comparison

Architecture	***Typical Compile*** *a*	***SDFF Replaced*** *b*	***Chain Stitched*** *c*	***a-b-c flow Overhead*** *g*
Area (um²)				
Original *d*	629246.39	NA	NA	NA
Modified-[1] *e*	629301.59	640853.28	640892.16	1.8418%
Our CBB *f*	629299.19	640838.40	640877.28	**1.8398%**
Power (mw)				
Original *d*	21.45	NA	NA	NA
Modified-[1] *e*	19.59	19.53	21.21	8.2695%
Our CBB *f*	20.72	20.67	22.01	**6.2259%**

a Purple in Fig.4, compile without any DFT settings.
b Orange in Fig. 4, replace DFF with SDFF when compiling.
c Orange in Fig.4, insert scan chain (stitch SDFFs like shift register chains.)
d Original Click without any modification.
e Add Modified-[1] architecture to Click circuit (Modification: Change the Se signal above the AND gate to TM=1, otherwise it will fail to compile.)
f Add our DFT architecture to Click circuit (MUX-based CBB)
g ((c-a)/a) *100%, which represents the *a-b-c flow* overhead.

C. Test Coverage Comparison

Test coverage (1) can measure DFT quality. Table III shows that our circuit can achieve the same coverage with a lower pattern count for Stuck-at faults.

$$Test\ Coverage = \frac{DT + (PT * PT\_credit)}{All - UD - (AU * AU\_credit)} *100 \quad (1)$$

TABLE III. Test Coverage Comparison

Stuck-at faults	**ATPG (Tessent)**	
	Test Coverage *a*	***Pattern Count***
Modified-[1]	99.62%	482
Our CBB	99.62%	474

IV. CONCLUSION

Our DFT architecture can independently test stuck-at faults of the data paths in Fig. 2 without affecting Click pulse generation. The architecture supports the use of traditional EDA tools to test asynchronous circuits with low hardware and power overhead, good coverage, and a low pattern count. We apply this architecture to an asynchronous processor based on the RISC-V instruction set in Fig. 5. The result is 99.62% test coverage using 474 patterns, with an area overhead of 1.8398% and a power overhead of 6.2259%.

REFERENCES

[1] A. Peeters, F. Te Beest, M. De Wit, and W. Mallon, "Click elements: An implementation style for data-driven compilation," in *2010 IEEE Symposium on Asynchronous Circuits and Systems*. IEEE, 2010, pp. 3–14.
[2] A. Mardari, Z. Jelčicová, and J. Sparsø, "Design and fpga-implementation of asynchronous circuits using two-phase handshaking," in *2019 25th IEEE International Symposium on Asynchronous Circuits and Systems (ASYNC)*. IEEE, 2019, pp. 9–18.
[3] Z. Li, Y. Huang, L. Tian, R. Zhu, S. Xiao, and Z. Yu, "A low-power asynchronous risc-v processor with propagated timing constraints method," *IEEE Transactions on Circuits and Systems II: Express Briefs*, vol. 68, no. 9, pp. 3153–3157, 2021.
[4] F. J. te Beest, *Full scan testing of handshake circuits*. Twente University Press, 2003.

Towards Near LLC Speed STT-MRAM Sensing Using Reconfigurable Clock Trimming

Xiaoyun Tian, Zhongjian Bian, Hao Cai

National ASIC System Engineering Center, Southeast University, Nanjing 210096, China

Abstract—Spin-transfer-torque magnetic random access memory (STT-MRAM) shows great potential to replace mainstream working memories thanks to its high energy efficiency and endurance. As RAM-like applications require higher speed, it is preferred to use a robust current-type sense amplifier (SA) with complex operating timing, which limits their working speed. The timing generated by the inverter chain is greatly affected by the process, voltage, and temperature (PVT) variations. In this work, a clock trimming sensing scheme is proposed to increase sensing speed and solve PVT variation in current-type SA. Since the timing is generated through voltage difference sampling between differential inputs, this scheme can achieve stable and fast sensing over a wide temperature range. According to the simulation results, the proposed scheme can sense data within 8-ns (near LLC working speed) and save up to 45.6% of energy consumption compared to the traditional SAs.

Keywords—STT-MRAM, Clock Trimming, Wide Temperature, Sense Amplifier, Near LLC speed

I. INTRODUCTION

Spin-transfer-torque MRAM (STT-MRAM) is a proper solution for next-generation memory due to its advantages of non-volatility, high storage density, fast access speed, and strong radiation immunity [1]. MRAM can operate with near last level cache (LLC) speed and wide operation temperature (-55 °C to +125 °C). Sense amplifier (SA) and peripheral circuits are required to take process, voltage, and temperature (PVT) variations into design phase [2]. Current-type SA (CSA) is more robust compared with voltage-type SA (VSA) [3-4]. The mainstream CSA is operated in multiple phases, which puts forward higher requirements for the generation of timing so that sensing speed is limited. Existing CSA mainly uses the inverter structure or multi-cycle operation to generate control signals (as shown in Fig.1).

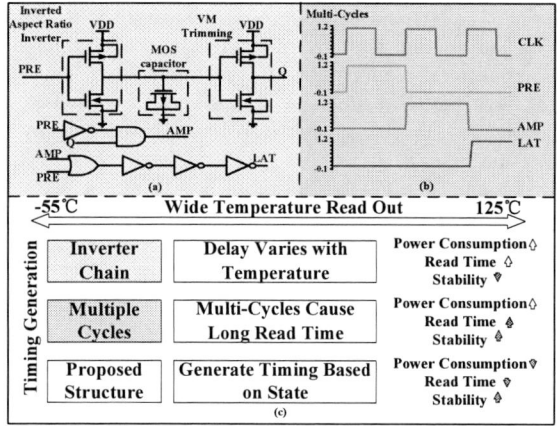

Fig. 1. (a) Timing generation structure based on inverter chain. (b) Control signals generated by multicycle structures. (c) The introduction of the structure proposed in this paper

The delay generated by the inverter chain varies with

This work is supported by the National Natural Science Foundation of China under Grant 61904028.

temperature, resulting in fluctuations in sensing speed and high power consumption, which impacts the timing issue. Although the multi-cycle operation ensures stable timing, the fixed duration of each phase leads to significant waste in the total sensing time.

In this paper, a clock trimming sensing scheme (CTSS) is proposed to sample the voltage difference between differential inputs in time to generate control signals, thus it can increase working speed and reduce power consumption. The sensing speed increase brings it closer to LLC-like memory. In addition, multiple timing generation modes are reconfigured by using the transmission gate logic.

II. THE PROPOSED CLOCK TRIMMING SENSING SCHEME

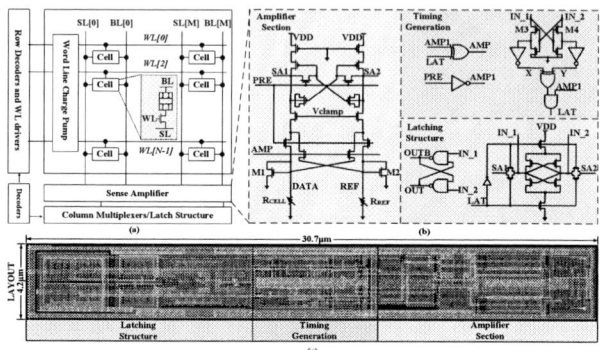

Fig. 2. (a) MRAM array structure. (b) Circuit configuration of the clock trimming structure proposed in this work. (c) Layout design based on 40nm CMOS process.

Offset-canceling current-sampling current sense amplifier (OCCS-CSA) is an excellent resistive memory sensing scheme at present [5]. The CTSS in this paper optimizes OCCS-CSA in the amplifier section, which can generate timing inside SA to improve sensing speed and stability. It contains three main parts: amplifier section, timing generation structure, and latching structure (see Fig.2). Layout design is shown in Fig.2, the increased CTSS can save more than 76.3% of the area compared with the inverter chain. The workflow consists of three phases: pre-charge phase (PRE), amplification phase (AMP), and latch data phase (LAT). And, its operation process is shown in Fig. 3.

A. Pre-charge phase (PRE)

In this phase, the system clock is used as pre-charge signal with high pulse width and SA charges the bit lines of data and reference. At the PRE phase, SA1 and SA2 are charged to stable levels. Signal AMP1 is the inversion result of the signal PRE, the XOR result of signal AMP1 and signal LAT are used to obtain signal AMP. At the end of this phase, signal AMP turns high, and the AMP phase begins.

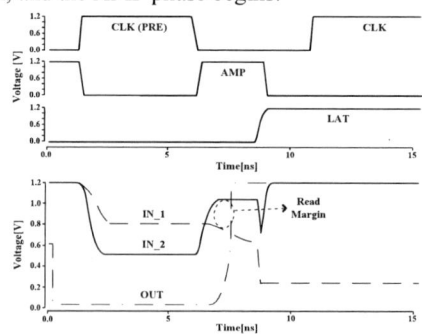

Fig. 3. Operation process of the proposed clock trimming sensing.

978-1-6654-9270-6/22 $31.00 © 2022 IEEE

B. Amplification Phase (AMP)

In the AMP phase, the voltage difference between SA1 and SA2 is amplified. The timing generation structure can sample the voltage difference between IN_1 and IN_2. At this time, both M3 and M4 are turned off, and node X and node Y are low. When the voltage difference between IN_1 and IN_2 is amplified greater than the threshold voltage of M3 (M4), M3 (M4) turns on, so that node X or node Y is charged to a high level (greater than the switching threshold Vm of the inverter). Then the LAT signal turns to high, the AMP signal turns to low, and LAT phase is activated.

C. Latch Data Phase (LAT)

The design of the latch structure is used to store data more stably. OUT and OUTB are the final output results. Since they are generated by IN_1 and IN_2, the final result can be accurately output before LAT goes high. This means the actual sensing time is shorter, which leads to a higher timing margin.

There may be a lot of charge accumulated on the BL due to writing operation, so NMOS M1 and M2 are added to the structure to release the charge quickly and save the pre-charge time. This addition does not affect the sensing process, because the voltage range on BLs is also small under normal operation.

The proposed CTSS brings three main advantages: AMP and LAT signals are autonomously generated by sensing so that signal generation has considerable tolerance to temperature and process variations; Area advantage is achieved by not using the inverter chain; The structure is optimized to prevent bit line charge accumulation caused by writing.

III. EXPERIMENTAL EVALUATION

In this section, the performance of the proposed CTSS under different temperature and process variations is evaluated, as well as existing sensing structures. All data are measured at a sensing yield above 99.5%.

A. Performance of the proposed CTSS

Differ from the PRE phase, process corner and operating temperature will affect the stability of sensing. Influences of these factors on the proposed structure are provided in Fig.4 (a). The data in the figure comes from an operating temperature of 25°C, and the fluctuation range of the sensing time with the process corner is marked around the center point. In particular, the increase in the PRE duration is helpful for stable bit line charging, resulting in a longer readout time and an increase in yield. As operating voltage increases, the sensing time will increase slightly, but it will reduce sensing time fluctuation at high temperatures.

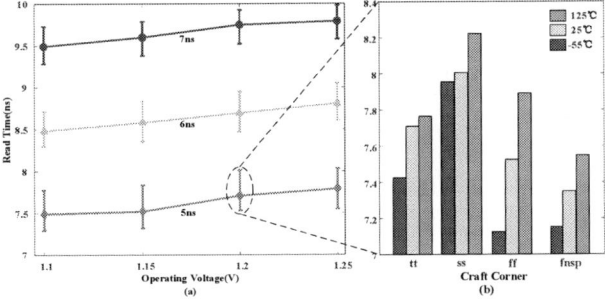

Fig. 4. (a) Influence of different PRE time and process corner on sensing time at 25°C; (b) Influences of different temperature and process corner on sensing time at 1.2V operating voltage and 5ns pre-charge

Fig.4 (b) shows the influences of different temperatures and process corners on sensing time at 1.2V operating voltage and 5ns pre-charge. It can be seen that the readout time fluctuates the least with temperature in the ss corner, and the readout is the fastest in the fnsp corner. Under the most commonly used 1.2V, 5ns pre-charge condition, the data can be read out within 8ns at most. The highest sensing time occurs at 7ns pre-charge, 1.25V operating voltage, which takes 10ns.

B. Comparision between different structures

The sensing time and power consumption of different structures under 5ns pre-charge, 1.2V operating voltage and tt process corner are shown in Fig.5. As shown in Fig.5 (a), the sensing time of the inverter chain structure fluctuates from 9.565ns to 12.379ns as the temperature changes. The read time of the multi-cycle structure controlled by the system clock is always around 15ns. And the proposed CTSS can stably fast generate timing and read out data within 8ns. Due to the reduced access time, the proposed structure has a significant power dissipation advantage over existing structures, which can be reduced by up to 45.6% at 125°C.

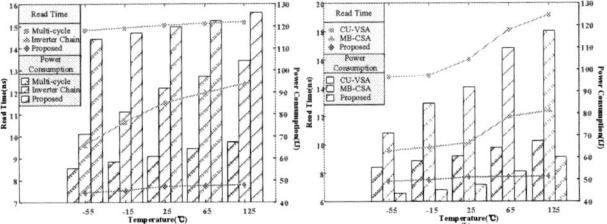

Fig. 5. Sensing time and power consumption of different (a) timing generation structures and (b) SAs under a wide range of temperatures.

Fig.5 (b) shows the comparison between MB-CSA, CU-VSA, and proposed CTSS-OCCS-CSA in this work [6-7]. CSA directly detects the memory cell current, which means it consumes more power and operates faster than VSA. It can be seen that both MB-CSA and CU-VSA have large read time fluctuations at high temperature, CU-VSA even reaches 4.37ns. At a high temperature of 125 °C, the structure proposed in this paper improves the reading speed by 37.2% compared with MB-CSA, and saves 42.2% of power consumption.

IV. CONCLUSION

This work proposes a clock trimming scheme that generates timing signals by sampling the bit-line voltage difference. Combined with OCCS-CSA, it can achieve stable data sensing within 8-ns (overall yields above 99.5% and approach the LLC working speed) over a wide temperature range (-55°C to 125°C) It reduces read power consumption up to 45.6%. Compared with the mainstream MB-CSA and CU-VSA, the proposed scheme achieves a significant improvement of reading speed and stability in a wide range of temperatures. Besides, the customized CTSS timing generator structure shows advantages of layout area over the inverter chain.

REFERENCES

[1] Y. Zhou et al., IEEE TCASI, vol. 67, no. 5, pp. 1602-1614 (2020).

[2] Y. -D. Chih et al., ISSCC, pp. 222-224 (2020).

[3] M. Chang et al., JSSC, vol. 48, no. 3, pp. 864-877 (2013).

[4] H. Cai et al., IEEE Trans on Magnetics, vol. 57, no. 2, pp. 1-5 (2021).

[5] T. Na, et al., JSSC vol. 52, no. 2, pp. 496-504 (2017).

[6] T. -C. Chang et al., ISSCC, pp. 224-226 (2020).

[7] Q. Dong et al., ISSCC, pp. 480-482 (2018).

Characterization and Modeling of Trapping Effects in GaAs Enhanced HEMT under High Input Dynamic Range

Lei Huang[1], Huanpeng Wang[2], Qingzhi Wu[1], Shuman Mao[2], Yuehang Xu[1,2]

[1]School of Electronic Science and Engineering, University of Electronic Science and Technology of China, Chengdu 611731, P.R. China
[2]Yangtze Delta Region Institute (Huzhou), University of Electronic Science and Technology of China, Huzhou 313001, P. R. China.

Abstract—Trapping effects (TE) have significant influence on device performances, including Pulse-IV, scattering parameters and linearity. Due to its slight influence on GaAs high electron mobility transistors (HEMTs), the TE are always neglected in compact models like EE-HEMT. In this paper, we present a physical-based quasi-physical zone division (QPZD) large-signal model and the TE is characterized by using simplified Shockley-Read-Hall (SRH) model, which can characterize the dynamic process of electron capture and emission. The results show that a more accurate model is obtained with TE taken into consideration, which can characterize the Pulse-IV and radio frequency (RF) performance with less errors, especially the linearity of GaAs HEMTs under two-tone excitation with high input dynamic range.

Keywords—GaAs HEMT , Trapping effects, SRH, IMD3

I. INTRODUCTION

Owing to the rapid development of fifth-generation (5G) communication technology, the demand for high linearity devices and circuits usher in a huge growth. As the bridge linking between device and circuit, device model which can characterize the linearity accurately is crucial for high linearity applications. Due to the excellent properties of high electron mobility and low noise characteristics, GaAs high electron mobility transistors (HEMTs) are widely used in civil microwave and millimeter-wave circuits and systems such as radars, mobile communication base station [1].

Compact models including Angelov and Keysight EE-HEMT model are purely empirical and not physical. The most importantly, some significant effects like frequency dispersion or trapping effects (TE) are not considered. To solve these problems, we introduce a quasi-physical zone division (QPZD) model and a simplified Shockley-Read-Hall (SRH) model to characterize GaAs HEMTs in this paper. QPZD model, which is originally proposed for GaN HEMTs, is constructed on the principle of dividing the channel of devices into several zones and just contains several physical equations. It has been applied in different applications [2]-[3]. Furthermore, to model TE or frequency dispersion in GaAs HEMTs, the SRH model is modified and less fitting parameters are needed.

II. MODEL DESCRIPTION

A. Basics of trapping effects in Enhanced HEMT

The device we used is a double δ-doping GaAs enhanced HEMT with threshold voltage of 0.3 V. The cross section of this device is shown in Fig.1.

TE can cause remarkable frequency dispersion. The main mechanisms of these are surface- and buffer-related TE [4]. Although surface-related traps can be improved greatly using surface passivation, the buffer-related traps cannot be perfectly removed.

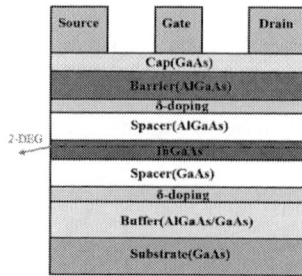

Fig. 1.   Cross section of GaAs enhanced HEMT

B. SRH Trapping Model

To investigate the trapping effects on the radio frequency (RF) performance, the SRH model is presented in this paper. Compared with traditional method, SRH model, which is capable of characterizing the electron capture and emission in the trap center, the dynamic TE can be considered. The SRH model can be expressed using physical equations in Ref [5].

As shown in Fig.2(a), When the quiescent drain-source voltage remains same, the measured Pulse-IV data basically unchanged at two different gate-source voltage, which indicates gate-related traps in GaAs HEMTs are negligible. However, the measured Pulse-IV data change a lot at two different drain-source voltage as shown in Fig.2(b). It means buffer-related trap occur in GaAs HEMTs. Therefore, a simplified SRH model can be used. Compared with Ref [5], the effective trap potential v_I can be written as:

$$v_I = B_{trap} V_{ds} \qquad (1)$$

Where B_{trap} is a fitting parameter.

By using the TE parameter extraction method, an accurate TE model has been gotten. As shown in Fig.2(b), the measured and simulated Pulse-IV data have a good agreement.

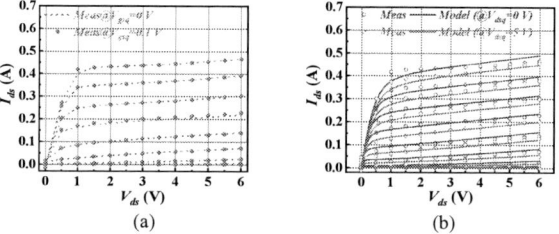

Fig. 2.   Measured and simulated Pulse-IV. (a) two different gate-source voltages biases (V_{dsq}=0V); (b) two different drain-source voltages biases (V_{gsq}=0.1V).

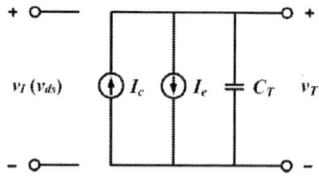

Fig. 3.   The sub-network for describing SRH model

To embed the SRH model in the SPICE-like simulation software, a sub-network is proposed for describing the dynamic process in SRH model as shown in Fig.3. The calculated effective trap potential v_T is then used to calculate effective gate-source voltage which can be expressed as:

978-1-6654-9270-6/22 $31.00 © 2022 IEEE

$$V_{gseff} = V_{gs} + V_T - V_{T,S} \qquad (2)$$

Where $V_{T,S}$ is steady effective trap potential when trapping current i_T is zero.

III. MODEL VALIDATION

For validating the TE model, the SRH model is implemented in SPICE-like QPZD model [1]. The large-signal model (LSM) is embedded into commercial EDA software and validated using an in-house AlGaAs/GaAs enhanced HEMT with a large gate periphery of 1.25 mm.

For RF performance verification, S_{21} and S_{22} are considered to be significantly influenced by the TE. The inconsistencies between the measured- and modeled-data are due to the output resistance (R_{ds}), which mostly changes the starting point of S_{21} and S_{22} and is related to the low-frequency dispersions or TE.

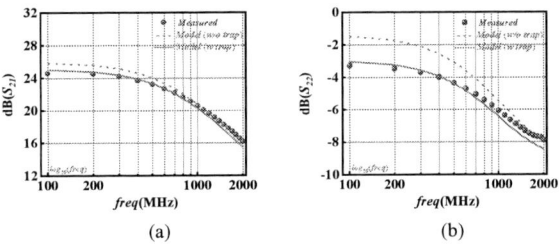

Fig. 4. Comparsion of measured and simulated S-parameters at biases V_{gs}=0.4V and V_{ds}=5V. (a) S_{21}; (b) S_{22}.

Fig.4 shows the measured and modeled S_{21} and S_{22} parameters biased at V_{gs} =0.4V and V_{ds} =5V. Since the TE are taken into consideration, better fitting results are obtained which also indicate the TE are well modeled. At the start frequency point (100MHz), the errors between measured and simulated are reduced by nearly 1dB and 2dB, respectively.

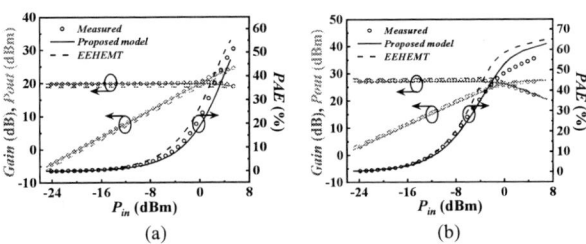

Fig. 5. Comparsion of measured and simulated results of the fundamental output performance with biases at V_{ds}=5 V, V_{gs}=0.556 V, f_0 = 1.8 GHz. (a) 50 Ohm matching; (b) maximum Pout matching.

In addition, the RF large-signal performance under one-tone excitation is validated. Compared with EE-HEMT model, better fitting results are got as shown in Fig.5. When the output power (Pout) reaches the maximum, the errors of EEHEMT and proposed model's Pout and power added efficiency (PAE) are 0.6dBm, 0.09dBm; 3.63%, 1.83% (50 Ohm matching) and 0.57dBm, 0.2dBm; 8%, 4.7% (maximum Pout matching), respectively. When the input power is −5dBm, the errors are 0.5dBm, 0.45dBm; 3.14%, 0.14% (50 Ohm matching) and 1dBm, 0.4dBm; 5%, 1% (maximum Pout matching), respectively. However, when the input power is -25 ~ -10 dBm, both of them are accurate.

To highlight the importance of TE on the linearity of GaAs HEMTs. The RF performance under two-tone excitation is investigated. Fig.6 and Table I show the comparison of third-order intermodulation distortion (IMD3) between measured

and simulated data. It can be seen that more accurate fitting results with less errors (less than 4dBc) are obtain compared with EEHEMT model, especially at the low input power (-20 ~ -5 dBm).

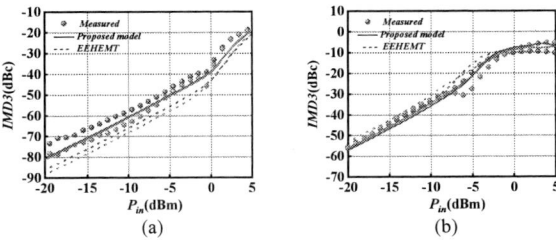

Fig. 6. Comparsion of measured and simulated results of the IMD3. (a)V_{ds}=5 V, V_{gs}=0.556 V, f_0 = 1.8 GHz, $f_{spacing}$=10MHz; (b) V_{ds}=5 V, V_{gs}=0.608 V, f_0 = 1.8 GHz, $f_{spacing}$=10MHz.

TABLE I   Comparison of IMD3_low and IMD3_high

Condition	Bias a (@Pin=-5dBm/-10dBm/-15dBm)		Bias b (@Pin=-5dBm/-10dBm/-15dBm)	
Errors of IMD3 (dBc)	IMD3_low	IMD3_high	IMD3_low	IMD3_high
Proposed model	**3.7/1/0**	**3.8/3.5/3**	**3.2/2/2.5**	**1.9/2/2**
EEHEMT	10.7/6/6	10.5/9/10	9/2/2	5.6/4/2

Bias a: V_{gs}=0.556 V, V_{ds}=5 V. Bias b: V_{gs}=0.608V, V_{ds}=5 V

IV. CONCLUSION

In this paper, we introduce a physical-based QPZD large signal model used for GaAs HEMTs. Most importantly, we consider the TE of GaAs HEMTs which are often neglected in Angelov and EEHEMT model. To model the TE, a simplified physical-based SRH model is presented. RF results indicate that TE play a key role in an accurate model. Especially, the linearity analysis is conducted in this paper, the results also show that a better IMD3 fits are obtained under high input dynamic range.

ACKNOWLEDGMENT

This work was supported part by the National Natural Science Foundation of China (Grant No. 61922021 and 62131014) and part by Sichuan Province Engineering Research Center for Broadband Microwave Circuit High Density Integration.

REFERENCES

[1] L. M. Burns, "Applications for GaAs and silicon integrated circuits in next generation wireless communication systems," IEEE Journal of Solid-State Circuits, vol. 30, no. 10, pp. 1088-1095, Oct. 1995.

[2] Z. Wen et al., "A Quasi-Physical Compact Large-Signal Model for AlGaN/GaN HEMTs," IEEE Trans. Microw. Theory Tech, vol. 65, no. 12, pp. 5113-5122, Dec. 2017.

[3] S. Mao, W. Zhang, Y. Yao, X. Yu, H. Tao, F. Guo et al., "A Yield-Improvement Method for Millimeter-Wave GaN MMIC Power Amplifier Design Based on Load—Pull Analysis," IEEE Trans. Microw. Theory Tech, vol. 69, no. 8, pp. 3883-3895, Aug. 2021.

[4] O. Pajona, C. Aupetit-Berthelemot and J. M. Dumas, "Modelling of the trap related parasitic effects in metamorphic HEMT on GaAs substrate," The 11th IEEE International Symposium on Electron Devices for Microwave and Optoelectronic Applications, 2003. EDMO 2003., 2003, pp. 151-156.

[5] Y. Xu, L. Huang, X. Yu, Y. Duan and S. Mao, "Compact Physical Modeling of Trapping Effects for Microwave GaN HEMT," 2021 IEEE International Workshop on Electromagnetics: Applications and Student Innovation Competition (iWEM), 2021, pp. 1-3.

A Broadband 20W GaN High Power Amplifier for Ku-band satellite communication

Yujie Liu[1], Zhilong Xiao[1], Shiquan Zhu[1], Huanpeng Wang[2], Shuman Mao[2], Qingzhi Wu[1], Ruimin Xu[1], BO Yan[1], Yuehang Xu[1,2]
[1]School of Electronic Science and Engineering, University of Electronic Science and Technology of China, Chengdu 611731, P.R. China
[2]Yangtze Delta Region Institute (Huzhou), University of Electronic Science and Technology of China, Huzhou 313001, P. R. China.

Abstract—A Ku-band high power amplifier (HPA) is designed based on the 0.15μm GaN HEMT process. To improve the power added efficiency and gain, a high gate-width drive ratio of 1:6:38.4 is selected for a three-stage topology. Multi-order Chebyshev impedance transformers are used for realizing this high impedance transformation ratio match networks. Meanwhile, a compact 8-way power combining network with low insertion loss is adopted to improve the output power and power added efficiency. The measured results under continuous wave（CW）show that the small signal gain exceeds 30 dB over 13-17 GHz, and the input return loss (IRL) is better than -11dB. The output power is between 42-44 dBm and the power-added efficiency (PAE) is more than 30%. The chip size is 2.6 mm×4.3mm.

Keywords—GaN HEMT, Ku-band, power amplifier

I. INTRODUCTION

In recent years, with the continuous development of satellite communication technology, the requirements for the Ku-band system are getting higher and higher, requiring the transmitter to have the characteristics of miniaturization, high output power, and high linearity. As GaN HEMT has a higher power density than GaAs HEMT, it is more suitable for high-power applications. At the same time, the excellent thermal conductivity brought by SiC substrate makes the efficiency also higher than GaAs at high output power. In 2014, Koh Kanaya et al. reported a Ku-band power amplifier with a built-in linearizer. The saturation power, gain, and power-added-efficiency (PAE) are about 43 dBm, 20 dB, and 16 %[1]. In 2021, Rocco Giofrè et al. reported a power amplifier working at 17.3-20.2 GHz, with a saturated output power of 39.5 dBm and a PAE of more than 28%[2]. To get better efficiency and high output power at broadband, this paper designs a Ku-band (13-17GHz) Monolithic Microwave Integrated Circuit(MMIC) high power amplifier(HPA) based on the 0.15μm GaN HEMT process.

II. CIRCUIT DESIGN

A. Topology design

The GaN HEMT is manufactured on a 100 μm SiC substrate. The typical f_T is 34.5GHz under 28V operation and it has a typical power density of 4 W/mm at 29GHz.

A three-stage reactance matching topology is selected as shown in Fig.1. The driving ratio is 1:6:38.4 and the total gate width of each stage is 0.2mm, 1.2mm, and 7.68mm, respectively. The static operating point of the transistor is selected in deep class AB with V_{GS}=-1.7V and V_{DS}=28V.

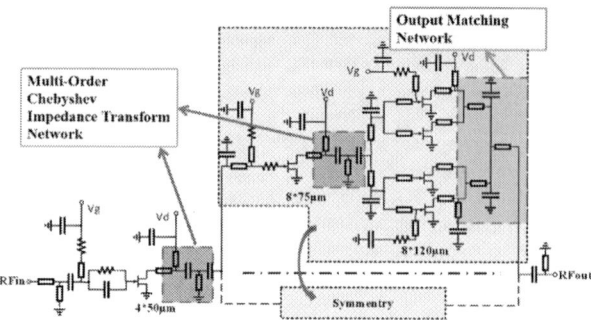

Fig. 1. The topology of the proposed GaN MMIC HPA

B. Matching Circuit Design

Since the gate-width ratio of the power amplifier is 1:6:38.4. There is a large impedance transformation ratio, making the design of the matching network more challenging. The inter-stage matching of the power amplifier adopts a multi-order Chebyshev impedance transformation network to complete the impedance matching and ensure the flatness in the band. The Insertion Loss (IL) of the Interstage Matching Network (ISMN) is shown in Figure 2. It can be seen that the insertion loss of ISMN has the typical equiripple response of the Chebyshev impedance transformation network.

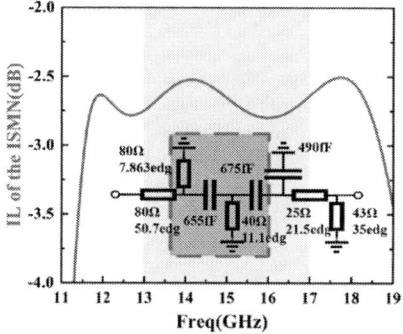

Fig. 2. Insertion Loss of Chebyshev Impedance Transform Network for Interstage matching

The output matching network(OMN) is designed to have low insertion loss, which makes the output power and PAE of the power amplifier higher. Figure 3 shows the insertion loss of OMN. OMN has an insertion loss of 0.8dB.

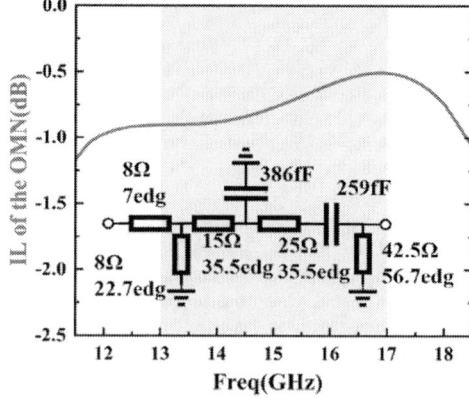

Fig. 3. Insertion Loss of Output Matching Network

III. MEASUREMENT

The photographs of the MMIC power amplifier and the test fixture are shown in Fig.4. The chip size is 2.6×4.3 mm^2. Under the biases of V_{GS}=-1.7V and V_{DS}=28V, S-parameter and power tests were carried out in continuous mode. The small-signal measurement system consists of a vector network analyzer and an attenuator. The output power measurement system is shown in Fig.5.

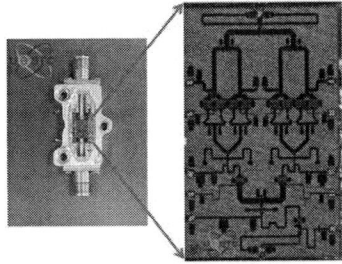

Fig. 4. Photographs of the MMIC power amplifier and the test-fixture

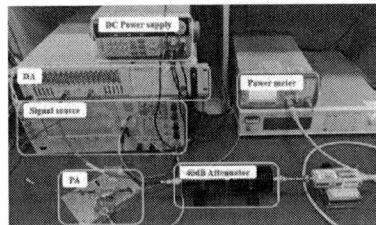

Fig. 5. Measurement setup of power amplifier

Fig.6(a) shows the comparison results of simulated and measured S-parameter. In the 13-17 GHz frequency band, the HPA achieves a gain better than 30dB and an input return loss (IRL) better than -11dB. The comparison between the measured and simulated results of output power and power-added efficiency (PAE) is shown in Fig.6(b). When input power is 22dBm, the output power of this HPA in the 13-17 GHz band is between 42-44 dBm, and the PAE is more than 30%. The power amplifier was also measured with large signal power and S- parameters at -40°C, 25°C, and 85°C. The comparison results are shown in Fig.7.

It can be seen that the large and small signal parameters of the power amplifier are very stable at high and low temperatures. And the amplifier shows good performance under different ambient temperatures.

We can see from the comparisons of state-of-the-art Ku-band HPA in Table I. The results of the work have the highest output power per unit area while possessing broadband characteristics under good PAE.

TABLE I   Comparison of state-of-the-art Ku-band GaN HPA at CW

Ref	Freq (GHz)	Relative BW	P_{out} (dBm)	PAE (%)	Area (mm^2)	Typical Power density (W/mm^2)
[1]	13.7-14.5	5.7%	42-43	>16	4.6×4	0.97
[2]	17.3-20.2	15.5%	39.5-40.4	>28	5×4.5	0.44
[3]	13.4-16.5	20.7%	43.7-44.6	>30	--	--
[4]	10.5-15.5	38.5%	38.5-39.8	>35	2.7×2.4	1.27
[5]	13-15.5	17.5%	45.6-46.2	>32	5×6.65	1.17
[6]	13.4-15.5	14.5%	46.9-47.3	>31	--	--
This work	13-17	26.7%	42-44	>30	2.6×4.3	1.79

IV. CONCLUSION

A broadband Ku-band GaN MMIC HPA with 20W output power and good PAE performance is reported. Multi-order Chebyshev impedance transformation network is used for high impedance transformation ratio matching to ensure bandwidth and small signal gain flatness. The measured results show that this method can have a more than 25% relative bandwidth while having good performance. These results show that the proposed HPA is suitable for satellite communication.

ACKNOWLEDGMENT

This work was supported partly by the National Natural Science Foundation of China (Grant No. 61922021 and 62131014) and part by Sichuan Province Engineering Research Center for Broadband Microwave Circuit High-Density Integration.

REFERENCES

[1] K. Kanaya et al., "A Ku-band 20 W GaN-MMIC amplifier with built-in linearizer," 2014 IEEE MTT-S International Microwave Symposium (IMS2014), 2014, pp. 1-4.

[2] R. Giofrè, P. Colantonio, F. Costanzo, F. Vitobello, M. Lopez and L. Cabria, "A 17.3–20.2-GHz GaN-Si MMIC Balanced HPA for Very High Throughput Satellites," IEEE Microwave and Wireless Components Letters, vol. 31, no. 3, pp. 296-299, March 2021.

[3] Qorvo, part number: TGA2219-CP. https://cn.qorvo.com.

[4] J. Zhang, L. Nie, Y. Chen, J. Ren and S. Ma, "A 6.5-mm2 10.5-to-15.5-GHz Differential GaN PA with Coupled-Line-Based Matching Networks Achieving 10-W Peak Psat and 42% PAE," IEEE Transactions on Circuits and Systems II: Express Briefs, 2022.

[5] Triquint, part number: TGA2239. www.triquint.com.

[6] Qorvo part number: TGA2239-CP. https://cn.qorvo.com.

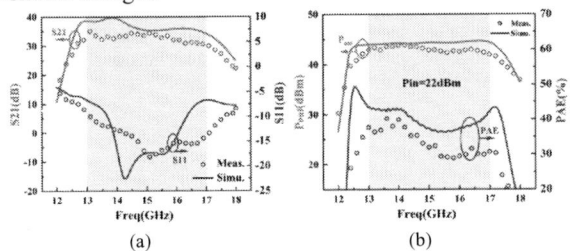

Fig. 6. Comparison of simulated and measured results. (a) S parameter (b) PAE and P_{out}

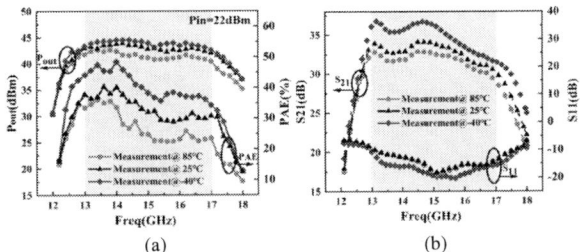

Fig. 7. Results at -40 °C, 25 °C, and 85 °C (a)PAE and P_{out} (b) S-parameters

Hardware Based RISC-V Instruction Set Randomization

Sheng Zuo, Junjie Zhuang, Yao Liu*, Mingyu Wang, Zhiyi Yu

School of Microelectronics of Science and Technology, Sun Yat-Sen University
* liuyao25@mail.sysu.edu.cn

Abstract—Instruction set randomization has been proposed for many years as a strategy against code injection. However, most of the methods are based entirely on software, which is vulnerable to possible threats like key leakage or bypassing attack. The translation of instructions also brings the loss of performance. Some designs randomize the instruction set based on hardware, but using weak approaches which can be easily bypassed. In this paper, we propose a hybrid instruction set randomization with both compiler support and hardware extension on a RISC-V processor. We adopt AES-128 to randomize RISC-V instruction set with little performance loss. The design has been implemented on Xilinx AV7K325 FPGA board, the results shows that RISC-V instruction set is randomized with no changes in clock frequency, 1377 LUTs increase in resources and 0.38% performance overhead.

Keywords—runtime safety, hardware security, instruction set randomization, RISC-V

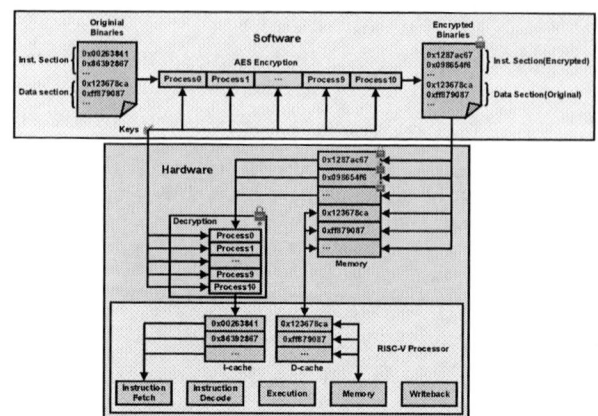

Fig. 1. Overview of RISC-V instruction set randomization

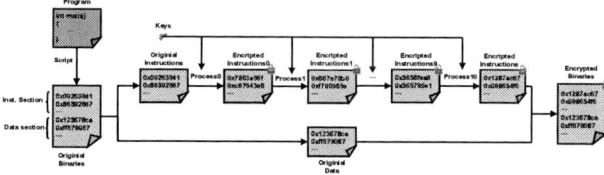

Fig.2. The proposed software encryption flow for ISR.

I. INTRODUCTION

Instruction set randomization (ISR) was proposed as a general method against code injection [1][2]. ISR creates a unique environment for the running process, and conceals the operations in the processor from attackers. Therefore, ISR is able to prevent attacks via external programs. For example, the opcode 0x23 represents the store operation in the RISC-V architecture. After randomization, the relationship between the opcode and the instruction will be changed. The opcode 0x40 may indicate store operation on one randomized instruction while may be invalid on another one. Therefore, new instructions are created without affecting the architecture of the processor. In this way, different instructions in the same program cannot be attacked by attackers who take advantage of the same vulnerability.

In general, software implementation for ISR leads to significant performance loss, since the decryption programs need to be executed before the regular programs. In addition, decryption programs are also vulnerable because of lacking self-protection. Most of previous hardware solutions, *e.g.* ASIST [3], utilize approaches like transposition or exclusive-or (XOR) to save the hardware resources, yet with a compromised security level.

In this paper, a hardware-based ISR method is proposed to solve the above problems. We utilize advanced encryption standard (AES), with a better security assurance than XOR or bit transposition for ISR. Unlike the above schemes, we encrypt the binary file instead of program. It ensures that no extra decryption programs that lacks of self-protection are needed before executing the original programs, which results in less performance loss and better security. We choose RISC-V instruction set as the experimental object since it is open source and free to use. The overview of our proposed design is shown in Fig. 1. The implementation on the Xilinx AV7K325 Field-Programmable Gate Array (FPGA) shows that the architecture

This work was supported in part by the Key-Area Research and Development Program of Guangdong Province 2021B0101410004, in part by the Guangdong Basic and Applied Basic Research Foundation 2022A1515011708

we proposed bears an improved security level with a small overhead on performance and hardware resources.

II. PROPOSED DESIGN

A. Software

The generation flow of the encrypted binaries is shown in Fig. 2. In order to create randomized RISC-V instructions, we need to separate the instruction section from the data section after generating the original binaries. Then we encrypt the RISC-V instructions. It has been proved that encryption schemes like bit transposition and XOR can be easily bypassed [4], so we randomize the binaries based on AES, with randomly generated 128-bit keys. Meanwhile, because the instructions are symmetrically encrypted, the keys generated by software can also be applied to hardware, which reduces the complexity of the hardware design and resource overhead. After randomization, we combine the encrypted instructions with the original data to form new encrypted binaries.

B. Hardware

Most of the schemes on ISR based on software bear potential problems. First of all, the encryption and decryption must be executed before the actual programs to randomize the instructions leading to extra performance loss. Second, the solutions may be risky, because they are not protected by any other programs. Attackers can easily turn off the protection program, and the target programs will be exposed. Last but not least, memory is always the focus of the modern attacks while memory-read ability is even an indispensable part of it [5].

In order to get rid of the inherent weaknesses of software ISR solutions, we propose a hardware ISR architecture and implement it on a RISC-V processor instead, as shown in Fig. 3. The encrypted binaries are stored in memory based on the corresponding address. Therefore, the RISC-V instructions in memory are randomized to prevent attackers from obtaining the original instructions through memory-read. When the RISC-V core performs a fetch operation, it will first look for instructions in I-cache. If an I-cache miss occurs, it will find the instructions

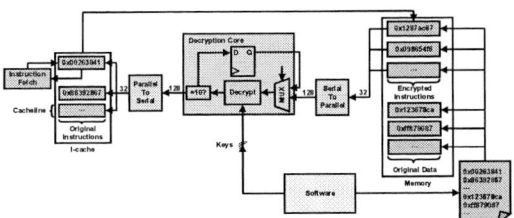

Fig.3. The proposed hardware decryption flow for ISR

in memory at next level. Since the instructions in the memory are encrypted, we need a decryption circuit between I-cache and memory which includes the serial-parallel conversion circuit, decryption core and parallel-serial conversion circuit. Four words are taken from the memory after an I-cache miss, since AES-128 deals 4 32-bit RISC-V instructions at a time. After going through the serial-parallel conversion, the instructions enter the decryption core. After decryption, the plaintext will be converted to the original instructions and sent to I-cache. The original instructions will be fetched by the core, as long as they are not replaced in the I-cache. The keys can be used on hardware as well, since the instructions are symmetrically encrypted. Therefore, the keys are embedded in the core instead of being generated by other circuits, which reduces the complexity and the resource utilization. Since the decryption of AES has eleven rounds while the process of each round is similar, we optimize the architecture of the decryption core by reusing the decrypt module and inserting register between each round to reduce the resource utilization and shorten its timing path without affecting the frequency.

Decryption based on hardware ensures that all the RISC-V instructions in memory are protected without protection programs that rely on software decryption. Also, compared to software decryption, our approach results in less performance overhead since it only requires a few cycles of delay during an I-cache miss instead of a timing-costly encryption program. This method can be implemented in processors based on any instruction set as long as the I-cache is included.

III. RESULTS

To validate the design and evaluate the performance, we simulate it based on Vivado 2018.3 and implement it based on Xilinx AV7K325 FPGA board.

A. Comparison With Original Design

Table I shows the frequency and the power of the proposed design and original design. The results indicate that the frequency of the RISC-V remains the same, while the power increases about 26%.

We run the benchmark Dhrystone to compare the performance of our proposed design with the original design. Results are presented in Table I. The original design can reach 0.263 DMIPS/MHz, while the proposed design is 0.262 DMIPS/MHz as shown in Fig.4. The performance of our design only decreases 0.38% compared to the original design. It proves that our RISC-V ISR has little impact on the performance of processor.

B. Comparison With Other Designs

Table II shows the comparison of the performance between the proposed design and other ISR designs. The proposed design is better than [3] in security, since it is based on XOR. The performance overhead of our design is low due to the hardware-based design. The proposed design also performs well in resource utilization. We use 1377 more LUTs than the original design, which is far less than the hardware overhead in Ployglot [4].

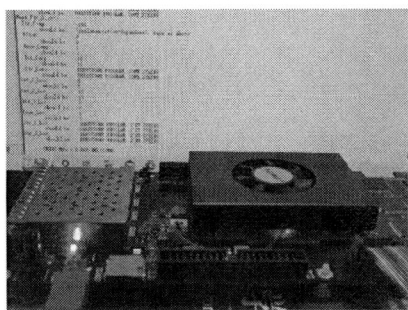

Fig.4. FPGA demo of our proposed design

TABLE I. COMPARISON TO ORIGINAL DESIGN

	Original design	Proposed design	Comparison
Frequency (MHz)	90	90	0
Power (W)	0.315	0.397	26%
Performance (DMIPS/MHz)	0.263	0.262	-0.38%

TABLE II. COMPARISON TO OTHER DESIGNS

	PNS[6]	ASIST[3]	Ployglot[4]	Our design
Real Hardware	Yes	Yes	Yes	Yes
Encryption	QARMA Block Cipher Family	XOR/ Transposition	AES, ECC	AES
Performance Overhead	6%	<3%	<5%	0.38%
Increased LUTs	/	104/130/1063	13,986 to 49,724	1377

IV. CONCLUSION

In this paper, we propose a hybrid ISR architecture based on AES-128 on a RISC-V processor with compiler support and hardware extension. We encrypt the RISC-V instructions by software and implement the decryption on circuit. We develop the architectural optimizations on the decryption circuit to minimize its impact on resources, clock frequency and other performance of the RISC-V processor while improving the security. We simulate our design and implement it on Xilinx AV7K325 FPGA board. The result shows that our design have the same frequency and little loss in performance compared to the original design. The overhead on hardware resource is 1377 LUTs, which is far less than other ISR works.

REFERENCES

[1] G. S. Kc, A. D. Keromytis, and V. Prevelakis, "Countering Code Injection Attacks With Instruction-Set Randomization," in *Proc. of CCS*, pp. 272–280, 2003.

[2] E. G. Barrantes, D. H. Ackley, T. S. Palmer, D. Stefanovic, and D. D. Zovi, "Randomized Instruction Set Emulation to Disrupt Binary Code Injection Attacks," in *Proc. of CCS*, pp. 281–289, 2003.

[3] A. Papadogiannakis, L. Loutsis, V. Papaefstathiou, and S. Ioannidis, "ASIST: Architectural Support for Instruction Set Randomization," in *Proc. of ACM CCS*, pp. 981–992, 2013.2

[4] K. Sinha, V. P. Kemerlis and S. Sethumadhavan, "Reviving instruction set randomization," 2017 *IEEE International Symposium on Hardware Oriented Security and Trust (HOST)*, 2017, pp. 21-28.

[5] K. Z. Snow, F. Monrose, L. Davi, A. Dmitrienko, C. Liebchen and A. Sadeghi, "Just-In-Time Code Reuse: On the Effectiveness of Fine-Grained Address Space Layout Randomization," 2013 *IEEE Symposium on Security and Privacy*, 2013, pp. 574-588.

[6] M. T. I. Ziad, M. A. Arroyo, E. Manzhosov, V. P. Kemerlis, and S. Sethumadhavan, "Using Name Confusion to Enhance Security," *arXiv:1911.02038 [cs]*, Aug. 2020

A 28nm, 4.69TOPS/W Training, 2.34µJ/Image Inference, on-chip Training Accelerator with Inference-compatible Back Propagation

Haitao Ge, Weiwei Shan*, Yicheng Lu, Jun Yang

Southeast University, Nanjing, China

Abstract—Previous on-chip training accelerators improved training efficiency but seldomly considered inference efficiency. We propose to convert back propagation to be compatible with inference, use interleaved memory allocation to reduce external memory access and zero-skipping loss propagation. Working at 40MHz, 0.48V core voltage, our 28nm one-core OCT chip has peak training efficiency of 4.69TOPS/W and the best inference energy of 2.34 µJ/inf/ image, 9.1× better than SoTA work.

Keywords—On-chip Training, Neural Network, Energy Efficiency

I. INTRODUCTION

Energy-efficient on-chip training (OCT) accelerators not only protect users' privacy, but also overcome the accuracy degradation due to non-ideal or unfamiliar private data (Fig. 1a). However, NN training is much more complex than inference due to an additional back propagation (BP) process including loss propagation (LP) and weight gradient (WG), shown in Fig. 1b. Thus, it is computation extensive and usually needs massive external memory access (EMA).

Among recent training processors, most of them were proposed to achieve high energy efficiency for training [1-5]. GANPU achieved a training efficiency of 1.81TFLOPS with no sparsity and 75.68TFLOPS/W (FP16) with 90% sparsity [4]. Some work used separate computation arrays or even additional cores to process BP computations [1-4], resulting in poor data reuse between inference and training and thus large inference energy. Previous work seldom considered inference efficiency that they usually used tens to thousands µJ/Inf to inference a single image [2-4], although it's more often used in OCT chip.

By unifying the inference and BP processes of training, we propose a one-core OCT chip that has high energy efficiency for both training and inference. Designed and fabricated in 28nm CMOS, it is efficient in both training and inference. Main contributions are: 1) An inference-pattern-compatible back propagation architecture, unifying convolutions in both BP and inference. 2) An interleaved on-chip memory allocation, mixing back propagation's LP and WG to reuse more data and reduce external memory access. 3) A zero-skipping strategy with low overhead mask memory, reusing data in WG to obtain these zero-skip index data and skip the related convolutions in LP.

II. ENERGY EFFICIENT OCT DESIGN

A. Inference-pattern-compatible Back Propagation

BP is mainly composed of two steps of loss propagation (LP) and weight gradient (WG), where WG's calculation pattern is quite different from that of the inference mainly in two aspects (Fig. 2). 1) Its convolution is dense while with different operation order, in that each input map of WG convolutes with all the weight maps and accumulate along batch channel, while in the inference each input map only convolutes with the weights in the same channel and accumulate along ifmap channel. 2) The convolution sizes are quite different that WG has a large weight kernel (i.e., 98×98) while inference has a small weight kernel (i.e., 3×3).

We propose an inference-pattern-compatible BP calculation pattern to cope with inference and BP in only one core, which solves the above problems in two ways. First, we regroup the convolution order as Fig. 2a to make the WG convolutions be

Supported by the National Natural Science Foundation of China (62122021 and 62074035) & Fundamental Research Funds for the Central Universities.

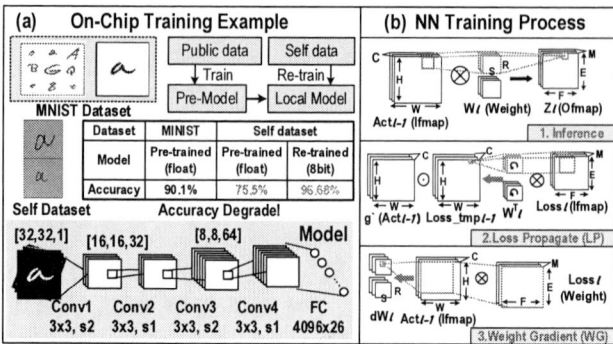

Fig. 1. Example of on-chip training (a) and its training process, including inference, LP, and WG (b).

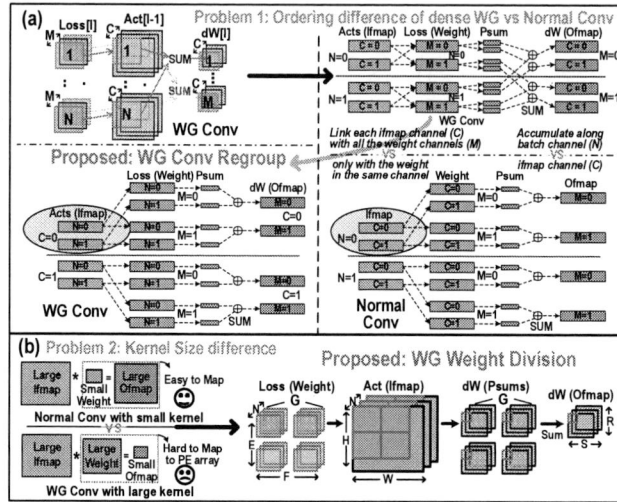

Fig. 2. Problems of WG convolutions and our proposed solutions.

the same as the inference part while fulfilling its original function. That is, we move the input channel loop (C) to the most outside layer and the batch channel loop (N) to the most inside layer. Second, to solve the large window convolution problem, we divide the large weight kernel to G×G blocks, and then accumulate all the G×G output feature maps (Psums) to get a full Ofmap (Fig. 2b). Our optimized WG algorithm unifies the convolutions in both inference and BP, avoiding the problem of designing a new core specifically for BP.

B. Interleaved On-chip Memory Allocation

External memory access (EMA) is power-hungry yet unavoidable due to the limited on-chip SRAM. We propose an interleaved memory allocation (IMA) between WG and LP to achieve deep data reuse can be fulfilled to reduce EMA. As in Fig.3, WG convolutes $Loss_l$ (as weight) with Act_{l-1}, as $dW_l = Loss_l \otimes Act_{l-1}$, while LP convolutes W_l with $Loss_l$ then do multiply with gradient $g'(Z_{l-1})$, as $Loss_{l-1} = (W_l \otimes Loss_l) \odot g'(Z_{l-1})$. Besides sharing the same $Loss_l$, WG and LP also have close data between Act_{l-1} and $g'(Z_{l-1})$ since $Act_{l-1} = Relu(Z_{l-1})$.

Our IMA does part0 of WG first, then does part 0 of LP, which reuses two parts of the data ($Loss_l$ and Act_{l-1}) to reduce EMA access. Using the Alexnet and Resnet18 for testing, our IMA method saves 33.6% and 29.8% of EMA respectively.

C. Zero-skipping Strategy in LP

We propose a zero-skipping strategy to cope with the sparsity in LP. As Fig. 4a, the propagation gradients of the LP phase are from the Relu activation data in the inference phase, meaning that the 0 positions in the gradient vector are known in advance and can be utilized to skip the corresponding convolutions in LP.

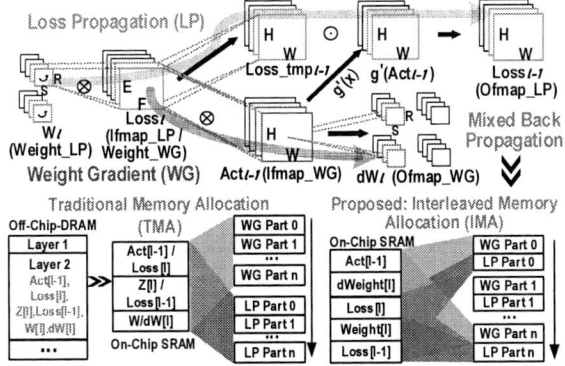

Fig. 3. Interleaved memory allocation (IMA) via interleaved convolutions in both WG and LP for deep data reusing to reduce EMA.

We generate a mask of zero-skip index according to the equation in Fig. 4b to skip the related convolutions at 0s. Furthermore, we elaborately place WG before LP and use an index scheduling to obtain these zero-skip index data at the same time when loading data to WG (Fig. 4c), which avoids the EMA of the mask at almost no cost. Combined with the block division for large-kernel convolution (Fig. 2), we greatly reduce the mask memory size to only 4kB instead of 72kB [2].

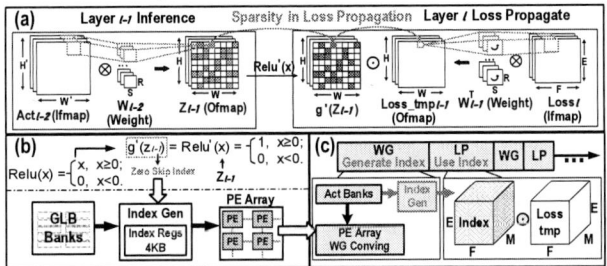

Fig. 4. Zero-skipping on part of the convolutions in LP based on inference data generated zero-skip index and proper data scheduling.

D. Hardware Architecture of the OCT

The OCT hardware architecture (Fig. 5) is composed of a 12×8 PE array based NoC with routers, 64kB global buffer (GLB) banks, configuration registers, a loss generate engine, a weight update engine, and interfaces with external DRAM. we also implement a genetic algorithm (GA) based design space exploration (DSE) software to find an optimum NN architecture to minimize the total energy.

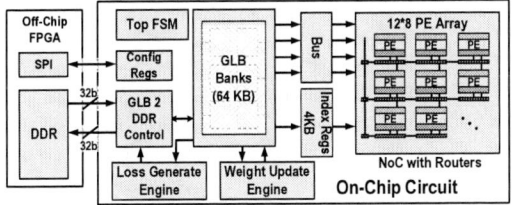

Fig. 5. OCT hardware architecture.

III. CHIP MEASUREMENTS

Fabricated in 28nm CMOS with a die area of 2mm², its die photo and testing platform are shown in Fig. 6. It can train CNN/DNN, such as AlexNet, ResNet, VGG, with using int8 data in inference and int16 in training.

First, we train a light-weight CNN used for handwritten letter classification on MNIST dataset as an example (Fig. 1a), similar to the customized HachNet [2]. Our chip improves the original private-letter recognition accuracy of 75.5% to 96.68% after retraining with two more private handwritten letter images for training and 100 figures for testing for each class. And its

Fig. 6. Chip die and test platform, measured efficiency and frequency.

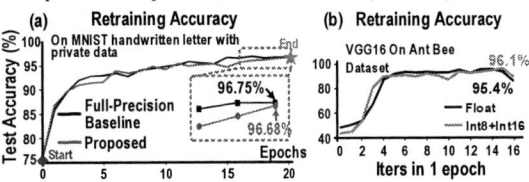

Fig. 7. Measured retraining accuracy with different data format.

hardware retraining accuracy has little loss as compared to the full-precision accuracy of 96.75% (Fig. 7a).

Second, this chip also supports ImageNet-level transfer learning. We deploy a pre-trained VGG16 to the chip, with the ant bee dataset having the same image size as ImageNet used for on-chip retraining [6]. It achieves 96.1% accuracy, as Fig. 7b. Here its inference and training consume 1.325mW and 1.546mW respectively, under 40MHz, 0.48V core / 0.7V SRAM, which is significantly lower than 43.1mW in [3].

TABLE I. Comparisons with the other work.

	VLSI'20 [1]	JSSC'20 [2]	ISSCC'19 [3]	ISSCC'20 [4]	ISSCC'21 [5]	This Work
CMOS Process	14 nm	65 nm	65 nm	40 nm	7 nm	28 nm
Training	C/DNN	C/DNN	C/R/DNN	DNN(GAN)	C/R/DNN	C/DNN
Number of Cores	2	3	Multiple	Multiple+ RISC	1	1
Die Area (mm²)	9.84*	10.24*	16	32.4	6.25*	2
Data Format	FP16/32	Fixed8	FP8/16	FP8/16	FP8	Training: fixed16 Inference: fixed8
Core Voltage (V)	0.54-0.62	0.65-1.0	0.78-1.1	0.7-1.1	0.75-1.1	0.43-0.9
On-Chip Memory (KB)	2048	364	372	676	293	64
Frequency (MHz)	1000-1500	5-160	50-200	25-200	20-180	20-200
Peak Performance (GOPS)	3000	134	300	540	567	38.4
Training Power (mW)	NA	9.55@0.65V	43.1@0.78V, 50MHz	58@0.7V 647@1.1V	13.1@0.75V 230@1.1V	0.836@0.43V 18.0@0.9V
Training Energy Efficiency (TOPS/W)	1.4@0.54V	1.03@0.78V, 50MHz	1.74@FP16, 3.48@FP8, 0% sparsity	1.81@0.7V, 0% sparsity	2.47@1.1V 4.81@0.75V	2.13@0.9V, 200MHz 4.69@0.48V, 40MHz
Inference Energy (µJ/inf/image)	NA	21.4@HachNet 6.9*10³@AlexNet	3.2*10³ @VGG-16	56.28*10³@max frequency	NA	2.34@8-bit customized CNN

* Core area, others are Die area

This chip operates at 0.9V, 200MHz with 38.4GOPS peak performance, or down to a near-threshold of 0.43V, 20MHz. Measuring the light-weight customized CNN in the first case, its peak efficiency for training is 4.69TOPS/W at 40MHz, 0.48V core / 0.7V SRAM (Fig. 6), equivalent to 7.68µJ/epoch/image. As for inference, it consumes only 2.34 µJ/inf/image at the same conditions. As compared to state-of-the-art work (Table I), our chip achieves not only comparable training efficiency but also the lowest inference energy of 2.34 µJ/inf/image, over 1000× better than [3-4] using large-scale NNs and 9.1× better than [2] using a similar small-scale NN.

REFERENCES

[1] Oh, Jinwook, et al. "A 3.0 TFLOPS 0.62V scalable processor core for high compute utilization ai training and inference." 2020 IEEE Symposium on VLSI Circuits, Honolulu, HI, USA, 2020.

[2] Lee, Jinsu, et al. "LNPU: A 25.3 TFLOPS/W sparse deep-neural-network learning processor with fine-grained mixed precision of FP8-FP16." 2019 ISSCC, San Francisco, CA, USA, 2019.

[3] Choi, Seungkyu, et al. "An energy-efficient deep convolutional neural network training accelerator for in situ personalization on smart devices." IEEE Journal of Solid-State Circuits 55.10 (2020): 2691-2702.

[4] Kang, Sanghoon, et al. "GANPU: A 135TFLOPS/W multi-DNN training processor for GANs with speculative dual-sparsity exploitation." 2020 ISSCC, San Francisco, CA, USA, 2020.

[5] Agrawal, Ankur, et al. "A 7nm 4-core AI chip with 25.6TFLOPS hybrid FP8 training, 102.4TOPS INT4 inference and workload-aware throttling." 2021 ISSCC, San Francisco, CA, USA, 2021.

[6] Pytorch. "Transfer learning for computer vision tutotial". https://pytorch.org/tutorials/beginner/transfer_learning_tutorial.html

A 20W Ka-Band Dual-Port Power Amplifier for Communication Satellites

Chi Chen, Kuan Hu, Weilin Luo, Kang Yin, RuiYuan Kang, Ying Zhao, Fei Yang

China Academy of Space Technology (Xi'an)

Abstract—this paper presents a ka-band dual-port power amplifier, developed for low-orbit communication satellite. The power amplifier was designed based on 0.15 um gate length GaN MMIC power amplifiers. The RF output port is optional and controlled by external command. The maximum saturated power 25 W with a PAE of 33% has been achieved. The environmental tests for power amplifier have been carried out. The measured result and thermal vacuum test result have been shown in this paper. The power amplifier has been working well on-orbit for two years.

Keywords-dual-port power amplifier, ka-band, GaN MMIC, thermal vacuum test

I. INTRODUCTION

In satellite payload, GaN device has been used in lower frequency, such as L-band, S-band , C-band and X-band power amplifiers for several years [1]-[3]. However, in high frequency, travelling wave tube amplifier(TWTA) is primary option because of high performance and reliability [4]. Recent years, with the advent of highly reliable and performing gallium nitride FET process, the solid state power amplifier (SSPA) has been developed in high frequency [5]. In lower-orbit satellites, downlink communication is operated in Ka band and request of smaller, lighter and high integration transponder is raised. Therefore, the GaN SSPA is adopted to replace the TWTA for downlink communication .

This paper presents the design and experimental characterization of a SSPA operation in the Ka band (17.7~20 GHz). The power stage of SSPA has been implemented by combining, in WR51 waveguide structure, two high power MMIC power amplifiers designed on a 0.15 um gate length GaN HEMT on SiC. In band, the SSPA is featured by more than 25 W output power with the power added efficiency 33%. The SSPA can be set a dual port mode. The switch and hybrid bridge is used for the option of output ports. The SSPA was finished by the initial test, vibration test, thermal test and thermal vacuum experiment. And the thermal vacuum test is shown in this paper. The SSPA is used in the downlink communication with the gateway. The satellite was launched in 2020 year. The SSPA work well on orbit.

II. SSPA DESIGN

The SSPA works as a part of high integrated Ka-band microwave equipment which includes the transmitter, receiver and electronic power conditioner (EPC).

The receiver is convert Ka band frequency to IF band frequency. The EPC is supply four voltages to transmitter and receiver which can achieve 90% convert efficiency through the low loss design. And it is turned on and off by receiving ON/OFF telecommand. More functions are provided, such as the sequence control, telemetries (RF output power, operation temperature, work current, and ON/OFF status), under-voltage protection and over-current protection.

The dual-port SSPA is shown in Fig. 1. It is composed by three subunits: the low power section (LPS), high power section (HPS) and control section. The SSPA output ports can be selected by external switch command. When one is output port, another is isolation port, and vice versa.

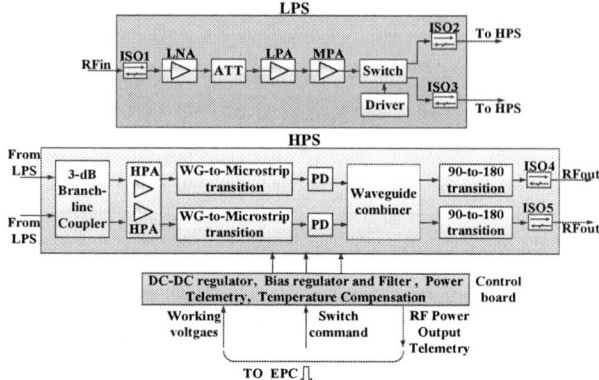

Fig.1. Block diagram of SSPA

A. LPS And HPS Design

The LPS is composed of isolators, low noise amplifier (LNA), low power amplifier (LPA), medium power amplifier (MPA), the switch and switch driver which are chips. Main function of LPS is amplifying input power to required output level for HPS. The LPS can provide gain of 30dB and the gain range of 10dB. Gain regulator is also used for temperature compensation. LPS can endure the overdriven of 10dB. The switch function is realized by a external command.

The HPS includes the branch-line coupler, two high power amplifier chips (HPA), waveguide to microstrip transitions, waveguide combiner, power detector (PD), 90 ° to 180 ° transitions and high power waveguide isolators.

In Fig. 2, the output port is 90° out of phase in respect to the voltage component of the RF signal. The output port can be chosen by the switch chip, input branch-line couple and output combiner. The waveguide to microstrip transition coupled the TE10 waveguide mode the quasi-TEM microstrip mode. The power combiner is used the H port Magic-T combiner.

GaN MMIC power amplifier was based on 0.15 um gate length process on CETC 55. The chip size is 3.05×2.8×0.05 mm. The minimum output power of MMIC is 14W with an associated efficiency more than 35%.

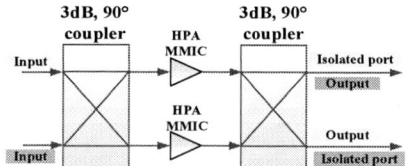

Fig.2. Port relationship of 90° hybrid coupler

B. Control Board Design

The control board is the interface between EPC and RF line-up. It consists of several functions:

Receiving switch command and providing standard interface circuits;

Telemetry circuit for RF output power telemetry;

Temperature compensation circuit (from -25 ℃ to +60 ℃);

Positive linear regulator for low level voltage.;

Negative voltage bias regulator and voltage filter circuits;

C. *Thermal Simulation*

The most heat is dissipated in the final stage MMIC power amplifier. The two MMICs are mounted with AuSn eutectic direct to the tungsten copper heat sink. The maximum work temperature is approximate 225 ℃. And then the heat sink is attached to the chassis of SSPA. The thermal analysis is shown in Fig. 3. The thermal resistance of MMIC is not more than 3 ℃/W. The temperature of backside of MMIC is 96.6 ℃, When SSPA output signal at operating point, the channel temperature is 155.3 ℃ which is lower than the derating temperature of 160 degree.

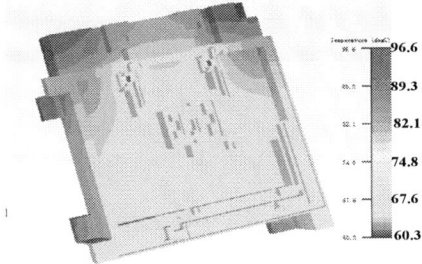

Fig.3. Thermal analysis for SSPA

III. PERFORMANCE

The picture of the SSPA prototype is shown in Fig. 4. The size and weight of the SSPA are 250.8×112.2×26.3 mm and 0.9 kg respectively, including waveguide hybrid and Isolator (not including wg-to-coaxial). The structure of SSPA is aluminum alloy and the external surface of the SSPA will be painter black to achieve thermal emissivity.

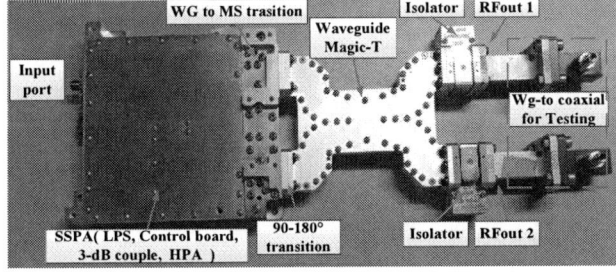

Fig.4. Picture of SSPA prototype.

Fig. 5 shows a plot of RF output power and PAE against frequency at 25 ℃.The operation frequency is from 17.7 to 20.5 GHz. The input power is -2 dBm. The data shows that the variation of output power is almost 1.5 dB and the PAE is from 29% to 33% in band. The data decrease in lower frequency is because of band limit of the input coupler.

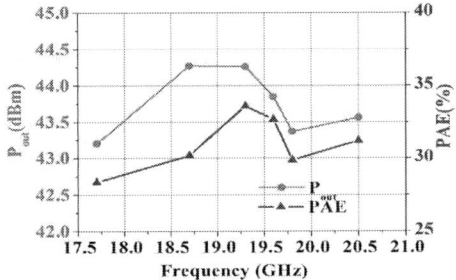

Fig .5. Measured characteristics of Pout and PAE vs Frequency

Fig. 6 shows the thermal vacuum test of SSPA. The work temperature is from -25 ℃ to 60 ℃. The thermal Cycles are half and three times and the holding time is needed. In order for the linearity, the real output power of SSPA is 42.7 dBm at 25 ℃, 2 dB back compared the maximum output power . During the test, the output power is 41.1 dBm and 44.1 dBm at 60 ℃ and -25 ℃, respectively.

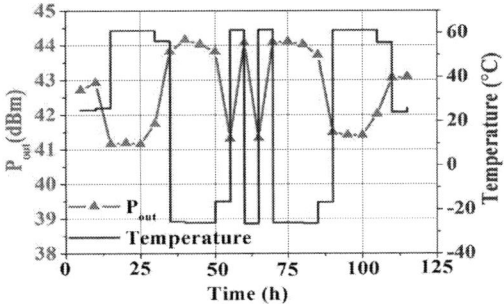

Fig.6. Performance of SSPA in thermal vacuum test

IV. CONCLUSION

The design and experimental results of a ka-band dual-port power amplifier suitable for low-orbit satellites have been presented. The power amplifier achieves a saturated output power in excess of 20 W with a PAE of 33%, power gain better than 45 dB in band, respectively. Combining structures have been realized in WR-51 waveguide. The size and weight of the SSPA are 250.8×112.2×26.3 mm and 0.9 kg which need the high integrated transmitter demand. The space environmental tests were carried out. The thermal vacuum tests is presented. The satellite was launched in 2020. The dual port power amplifier works well till now. Compared with other work [6], the ka-band dual-port GaN SSPA on orbit is reported firstly.

ACKNOWLEDGMENT

The work is carried out and finished in several months. The author thanks the different partners for their help and advices about this work.

REFERENCES

[1] Akihiro KIYOHARA, Yutaka KAZEKAMI, et al "On superior Tracking Performance of C-band Solid State Power Amplifier for Inmarsat-4," *21st International Communications Satellite Systems Conference and Exhibit 2003.* pp. 1–9.

[2] M. Kido, S. Kawasaki, A. Shibuya, K. Yamada, et al "100w C-Band GaN Solid State Power Amplifier with 50% PAE for Satellite Use". *Asia-Pacific Microwave Conference (APMC)*, Dec 2016, pp. 1–4.

[3] A. Katz, J. MacDonald, R. Dorval, B. Eggleston, et al "High-efficiency high-power linearized L-band SSPA for navigational satellites", 2017 *IEEE MTT-S International Microwave Symposium (IMS)*, Honolulu, HI, 2017, pp. 1834-1837

[4] R. Giofre, F. Costanzo, A. Massari, et al "A 20W GaN-in-Si Solid State Power Amplifier for Q-Band Space Communication Systems". *IEEE 2020 IEEE/MTT-S International Microwave Symposium.* pp.413–415 2020.

[5] H. Fenech, S. Amos, A. Tomatis, and V. Soumpholphakdy, et al "High throughput satellite systems: An analytical approach" *IEEE Transactions on Aerospace and Electronic Systems*, vol. 51. no.1 pp.192–202. January 2015.

[6] R. Giofre, P. Colantonio, F. Di Paolo1, et al "Power Combining Techniques for Space-Borne GaN SSPA in Ka-Band". *2020 International Workshop on Integrated Nonlinear Microwave and Millimeter-wave Circuits (INMMiC)*, 2020. pp.1–3

A Voltage Error Quantizer For Digital Low Dropout Regulators With Fast Transient Response and Low Steady-State Error

Kaize Zhou[1], Dejian Li[2], Chongfei Shen[2], Yuxuan Du[1], Zhuo Chen[1], Weiwei Shan[1*]

[1] Southeast University, Nanjing, China
[2] Beijing smartchip microelectronics technology company limited, Beijing, China

Abstract—This paper proposes a voltage error quantizer for digital low dropout regulators (DLDOs) with fast transient response and low steady-state error. Compared with traditional DLDOs quantizing the reference voltage and output voltage separately, the proposed voltage error quantizer quantifies the voltage difference directly with high quantization speed and accuracy. Implemented in 28nm CMOS process, the proposed quantizer with on-chip self-calibration identifies the voltage difference as small as 4mV and has stable output codes at sampling frequencies up to 500MHz, which satisfies the fast transient response and low steady-state error demands of DLDOs.

Keywords—Voltage error quantizer, on-chip self-calibration, digital low dropout regulator, fast transient response

I. INTRODUCTION

With the advancement of process and technology, the demand for power supply voltage management chips is getting higher and higher, especially in multi-core systems and multi-voltage domain systems. Low dropout regulators (LDOs) have received extensive attention due to their low dropout and low noise characteristics [1]. In some ultra-low power consumption applications, it is difficult to design analog LDOs (A-LDOs) because of the reduction in the power supply voltage. Therefore, DLDOs are widely employed in sub-threshold operating conditions due to their superior performance at low voltages.

Generally, DLDO consists of three components: voltage difference quantizer, digital control unit and power array. The traditional voltage comparator [2], which is equivalent to a 1-bit voltage difference quantizer, needs multiple cycles to adjust the power arrays. To improve the transient response speed, the sampling frequency is required to be increased correspondingly, leading to higher power consumption. To break the tradeoff between the transient response speed and power consumption, multi-bit voltage quantizers [3], [4] and ADCs [5] are also used in DLDOs. Among them, time domain based voltage quantizers show advantages in speed and accuracy, Fig. 1(a) and (b) shows two typical structures of them. In these schemes, two separate voltage quantizers are necessary due to the requirement of quantizing the reference voltage and the output voltage leading to a waste of power consumption and chip area.

A voltage error quantizer that directly quantifies the voltage difference is presented in this paper. The voltage error quantizer improves the transient response speed and output accuracy of DLDOs. The remainder of this paper is organized as follows. Section II introduces the structure and operating principle of the proposed quantizer. Section III provides the simulation results of the quantizer. Finally, Section IV concludes this paper.

II. PROPOSED VOLTAGE ERROR QUANTIZER

The structure of the proposed voltage error quantizer with on-chip self-calibration is shown in Fig. 1 (c). The quantizer has two working loops, a calibration loop and an output loop. When

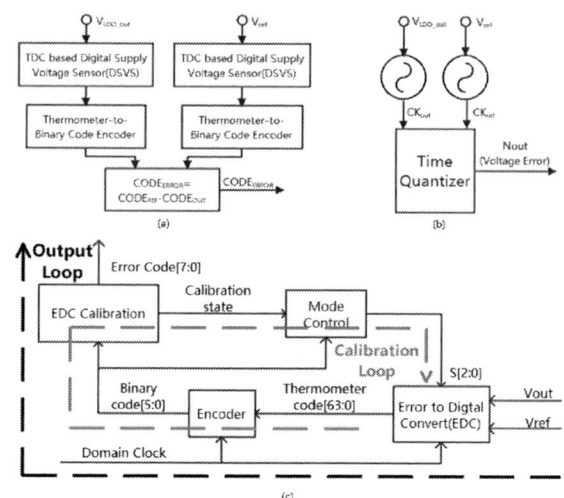

Fig. 1. Different voltage error quantizer structures (a) DSVS based quantizer (b) Beat-Frequency Quantizer (c) Proposed voltage error quantizer.

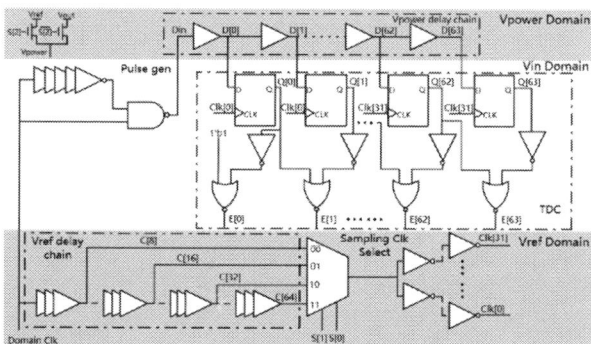

Fig. 2. The implementation of error-to-digital-convert circuit.

entering the calibration loop after power-on, the voltage error quantizer is controlled to measure the output codes in different states, which is used to mitigate the influence of PVT and calibrate the output loop. In the output loop, the error-to-digital-convert (EDC) circuit will quantify the difference between the reference voltage and the output voltage in one circuit, and the result will be reflected as a 64-bit thermometer code E[63:0]. Then, the thermometer code will be converted into a 6-bit binary code through the encoding unit. Through EDC calibration, the final voltage error code is reflected to a range of (-128-127), a total of 256 code values, which reduces the power consumption and circuit area with a high resolution of 8-bit.

The implementation of the proposed EDC circuit is shown in Fig. 2. Since the function of the EDC circuit is to convert the difference of two voltages to a digital code, the EDC circuit is divided into three voltage domains: the voltage domain to be measured (Vpower), the input voltage domain (Vdd) and the reference voltage domain (Vref).

In the reference voltage domain, the 64-stage buffer chain powered by the Vref generates a fixed delay inversely proportional to the reference voltage, lagging behind the domain clock. S[1:0] produced by the mode control module selects the number of delay stages. To ensure the quality of the generated clock, the delayed clock is passed through a 5-level clock tree to generate 32 clock signals with the same phase as the sampling clock. The sampling clock is delayed by a total of t_s to the domain clock as shown in Fig. 3.

978-1-6654-9270-6/22 $31.00 © 2022 IEEE

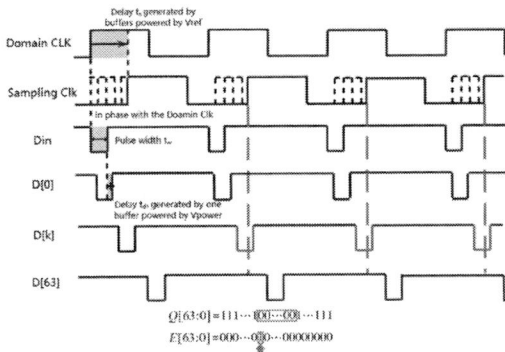

Fig. 3. Timing diagram of EDC.

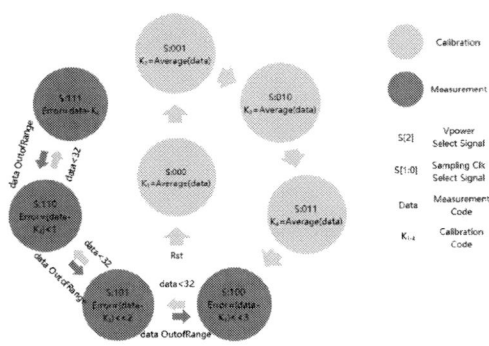

Fig. 4. Operation process of the mode control module.

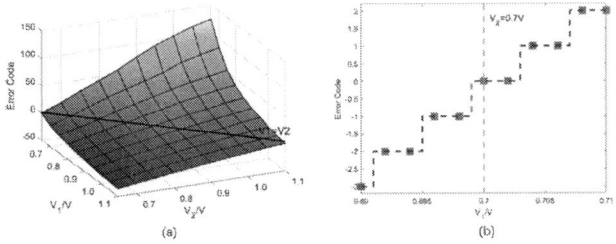

Fig. 5. The outputs of the proposed design (a) Overall outputs at various voltage differences (b) Outputs of voltage error quantizer when V_2=0.7V.

TABLE I.   PERFORMANCE COMPARISON

	[3]	[4]	[5]	This work
Type	DSVS	VCO	ADC	EDC
Process	10nm	65nm	65nm	28nm
Supply	0.7-1.05V	0.6-1.2V	0.65-1.15V	0.7-1.1V
Range	0.65-0.95V	0.4-1.1V	0.6-1.1V	0.65-1.05V
Resolution	6bit	6bit	7bit	8bit
$fsamp$	1GHz	Adaptive	100MHz	500MHz
Area(μm^2)	402	4000	16000	1400

In the input voltage domain, a pulse signal Din with a fixed width of t_w is generated by an inverter chain. As the input signal of the delay chain in the Vpower domain, Din is in the same phase as the domain clock. Then, the input signal Din propagates with a delay of td1 in a single buffer powered by the Vpower.

In the Vpower domain, control signal S[2] chooses the working status and selects the voltage to be measured. When S[2] is '0', it works in the calibration loop, at this time, the Vpower is equal to the reference voltage, which is used to calibrate the working deviation of the circuit. When S[2] is '1', it works in the output loop, the Vpower is the DLDO's output voltage at this time. To quantify the difference between the Vref and Vpower, a time-to-digital-convert (TDC) circuit is used to reflect the delay error generated by the buffers powered by Vref and Vpower. For the convenience of encoding, inverters and NORs are additionally used to convert Q[63:0] to one-hot codes E[63:0] as the output of EDC.

The operation process of the mode control module is shown in Fig. 4. S[2] controls the working loop and K_{1-4} are the average codes of the EDC during the calibration stage. After calibration, the mode control module will adjust the control signal S[1:0] to ensure the output code of EDC remains more than half of the maximum range, which makes the accuracy of the quantizer as high as possible. Finally, by subtracting the corresponding calibration code and shifting, the quantizer can achieve a high accuracy output of 8 bits.

III. SIMULATION RESULTS

The proposed voltage error quantizer has been implemented in 28nm CMOS technology. In the simulation experiment, the sampling frequency is set to 500MHz, the supply voltage is

1.1V, and the reference voltage and quantization voltage range are both 0.65-1.05V. Fig. 5 (a) shows the overall outputs of the quantizer at different voltage differences. As shown in Fig. 5(b), the highest resolution reaches 4mV when the Vref is 0.7V.

IV. CONCLUSION

In this paper, a voltage error quantizer for DLDOs with fast transient response and low steady-state error is proposed. Through the EDC circuit and on-chip calibration, the voltage difference can be quantified directly. Table I compares the performances of the proposed voltage error quantizer with the state-of-the-art works. The simulated results prove the proposed quantizer simultaneously adopts to high sample frequency and achieves higher accuracy.

ACKNOWLEDGMENT

Funded by the state grid corporation of China in 2021 (5700-202141255A-0-0-00).

REFERENCES

[1] S. B. Nasir, S. Gangopadhyay and A. Raychowdhury, "All-Digital Low-Dropout Regulator With Adaptive Control and Reduced Dynamic Stability for Digital Load Circuits," in *IEEE Transactions on Power Electronics*, vol. 31, no. 12, pp. 8293-8302, Dec. 2016.

[2] Yasuyuki Okuma *et al.*, "0.5-V input digital LDO with 98.7% current efficiency and 2.7-µA quiescent current in 65nm CMOS," *IEEE Custom Integrated Circuits Conference 2010*, 2010, pp. 1-4.

[3] S. Bang *et al.*, "25.1 A Fully Synthesizable Distributed and Scalable All-Digital LDO in 10nm CMOS," *2020 IEEE International Solid- State Circuits Conference - (ISSCC)*, 2020, pp. 380-382.

[4] S. Kundu, M. Liu, S. -J. Wen, R. Wong and C. H. Kim, "A Fully Integrated Digital LDO With Built-In Adaptive Sampling and Active Voltage Positioning Using a Beat-Frequency Quantizer," in *IEEE Journal of Solid-State Circuits*, vol. 54, no. 1, pp. 109-120, Jan. 2019.

[5] X. Sun, A. Boora, W. Zhang, V. R. Pamula and V. Sathe, "14.5 A 0.6-to-1.1V Computationally Regulated Digital LDO with 2.79-Cycle Mean Settling Time and Autonomous Runtime Gain Tracking in 65nm CMOS," *2019 IEEE International Solid- State Circuits Conference - (ISSCC)*, 2019, pp. 230-232

A 200-Gb/s PAM-4 Feedforward Linear Equalizer with Multiple-Peaking and Fixed Maximum Peaking Frequencies in 130nm SiGe BiCMOS

Zhengzhe Jia[1], Taiyang Fan[1], Dongfan Xu[1], Dongshen Zhan[1], Linxuan Hu[1], Zhengyang Zhang[1], Yanchao Wang[1], Chun-Zhang Chen[2], Xuhui Liu[2], Hanming Wu[2], Quan Pan[1,2]

[1]School of Microelectronics, Southern University of Science and Technology, Shenzhen, China

[2]Peng Cheng Laboratory, Shenzhen, China

Corresponding email: panq@sustech.edu.cn

Abstract—This paper presents a peak-tunable PAM-4 linear equalizer for 200-Gb/s communication in 130-nm SiGe BiCMOS process. It consists of two stage continuous-time linear equalizers (CTLE) and an output driver. The feedforward and degeneration techniques are employed to realize multiple peaks at high, middle and low frequency points, respectively. Moreover, the equalizer can reach a maximum 20.6-dB compensation with a fixed peaking frequency at 51 GHz. The simulation results show that with an 18-dB lossy channel, the proposed equalizer provides a 53-GHz bandwidth and 200-Gb/s PAM-4 eye. It consumes 159-mW power under 2.8/3.3-V supply and achieves 0.80-pJ/bit power efficiency.

Keywords—CTLE, feedforward technique, linear equalizer, 200Gb/s PAM-4 signal.

I. INTRODUCTION

With the rapid development of AI supercomputers, cloud computing, and core network switch, higher demands are put forward in order to meet high-speed data interconnects and exchange requirements for extended bandwidth and throughput in future data centers. Due to the skin effect, signal channels introduce considerable high-frequency loss during high-speed transmissions, necessitating greater demands on equalization circuits. Besides, the PAM-4 modulation, which improves the spectrum efficiency compared to the conventional NRZ modulation, has been adopted extensively in 100+ Gb/s communications [1].

Conventional continuous-time linear equalizers (CTLEs) in SiGe BiCMOS process mainly focus on channel compensation at high frequencies, but ignore the low and middle frequency parts [2], [3]. This work presents a 200-Gb/s PAM-4 linear equalizer with gain peaks at 0.3 GHz, 5 GHz and 51 GHz, respectively. The feedforward technique is applied to better compensate for the channel loss. Moreover, digital control technique is adopted to guarantee the compensation accuracy of the circuit under process, voltage, and temperature (PVT) variations, and to provide sufficient compensation accuracy under high transmission data rates.

This paper is organized as follows. Section II presents the architecture and the circuit analysis of the linear equalizer. Section III summarizes the simulation results. Finally, the conclusion is draw in section IV.

II. ARCHITECTURE AND CIRCUITS IMPLEMENTATION

Figure 1 shows the block diagram of the proposed linear equalizer consists of four stages. The input buffer takes the role of input matching. The signal is then equalized by two stages of CTLE, including a HF-CTLE and a High Frequency/ Middle Frequency/ Low Frequency-CTLE (HF/MF/LF-CTLE). Finally, the output driver utilizes the bias current to increase the overall linearity.

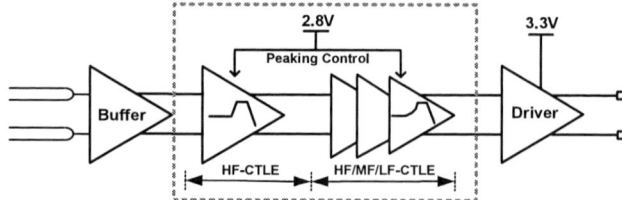

Fig. 1. Linear equalizer architecture.

A. HF-CTLE

The first stage of the CTLE employs a degenerated RC component used in [4], a DC-offset cancellation (DCOC) part and an emitter follower pair. The HF peaking can be adjusted by tuning the source degenerated resistors and capacitors with external digital controls.

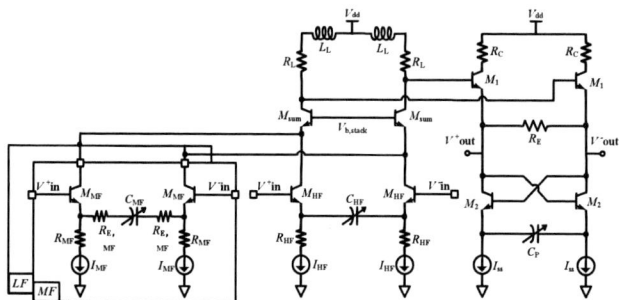

Fig. 2. Schematic of the second stage of CTLE circuits.

B. HF/MF/LF-CTLE

The schematic of the second-stage CTLE is shown in Fig. 2. For the HF equalization part, there is one zero of $\omega_z = 2/R_{HF}C_{HF}$, and two poles of $\omega_{p1} = (2 + g_{m,HF}R_{HF})/R_{HF}C_{HF}$ and $\omega_{p2} = 1/R_L C_L$. By changing the value of the capacitor network C_{HF}, the zero and poles can be put at appropriate positions to provide HF gain boosting. For the MF and LF parts, there are two resistors $R_{E,MF/LF}$ connected in series on both sides of the capacitor network in order to reduce the capacitance requirements and obtain lower frequency zeros and poles.

The feedforward structure is also introduced in this work. The circuits provide the equalizations for HF/MF/LF by the degeneration of the three pairs of input transistors, followed by current feedforward summation at source of the cascode transistors M_{sum}, and then output the voltage across R_L resistor to the next emitter follower.

The transfer function of the emitter follower is expressed as follows:

$$\frac{V_{OUT}}{V_{IN}} \approx \frac{g_{m1}}{g_{m1} + \frac{R_E}{2}} \frac{(1 + \frac{s}{\omega_{z,t}})}{(1 + \frac{s}{\omega_{p,t}})} \quad (1)$$

where $\omega_{z,t} = g_{m2}/2C_p$, $\omega_{p,t} = g_{m2} \times (g_{m1} + 2R_E^{-1})/[2C_p \times (g_{m1} - g_{m2} + 2R_E^{-1})]$.

The output node uses negative capacitance C_p to neutralize the load capacitance, and the C_p can be changed by the digital control signal of 5 bits to adjust the gain at a fixed maximum peaking frequency. It can be used to calibrate the deviation of peaking gain caused by the PVT variations in high-speed wireline communications.

978-1-6654-9270-6/22 $31.00 © 2022 IEEE

III. SIMULATION RESULTS

The proposed linear equalizer is implemented in 130-nm SiGe BiCMOS technology and consumes 159 mW under 2.8-V supply voltage. Figure 3 shows the AC response of the circuit with different frequency peaking control signals. It verifies the ability of the multiple peaks and gain variation at the fixed maximum peaking frequency.

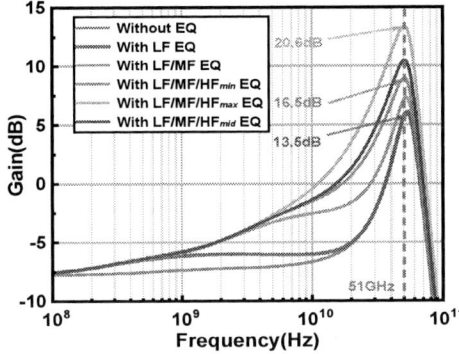

Fig. 3. Simulated frequency response of the linear equalizer.

With the impedance matching of 50 Ω, Figure 4 shows the S-parameters in different situations. The channel has 18 dB loss at 51 GHz. After the signal passing through the proposed linear equalizer, the -3dB bandwidth is extended to 53 GHz. Figure 5 shows the 200-Gb/s PAM-4 eye diagram of the proposed linear equalizer. The eye height and eye width reach 43 mV and 0.55 UI, respectively. The equalizer circuit can successfully meet the requirements for 200-Gb/s PAM-4 signal transmission.

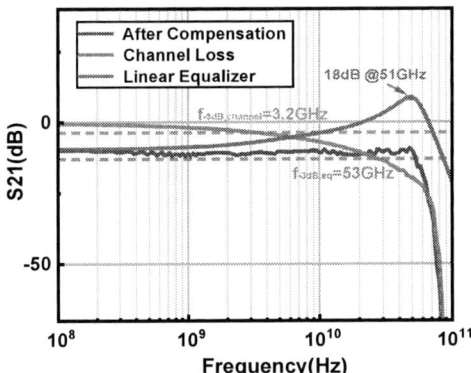

Fig. 4. Simulated results of S-parameter.

Table I summarizes the simulation performance of equalizers compared with previous designs in BiCMOS processes. This design achieves a higher data rate and a better power efficiency compared to other work. Moreover, this work has resulted in a wider range of equalization ability in the low, middle and high frequency.

IV. CONCLUSION

A low/middle/high frequency linear equalizer with feedforward techniques is designed and simulated in 130-nm BiCMOS technology. It realizes the equalizations of the multiple peaking and the fixed maximum peaking frequency. The simulation results indicates that it can successfully support a 200-Gb/s PAM-4 signal transmission.

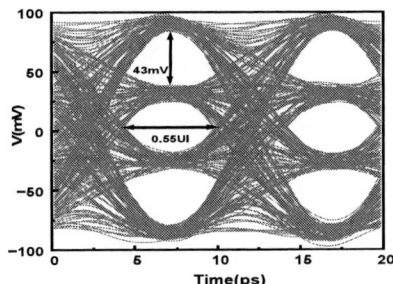

Fig. 5. Simulated eye diagram of 200-Gb/s PAM-4 signal.

TABLE I.   PERFORMANCE COMPARISON WITH RECENT WORKS

	[4]	[5]	[6]	This work*
Technology	180-nm SiGe	130-nm SiGe	180-nm SiGe	130-nm SiGe
Data rate (Gb/s)	56	64	56	200
Modulation	PAM-4	PAM-4	PAM-4	PAM-4
Equalization (dB@GHz)	@0.5 @4 @18	29@50	4@0.5 13@4 27@18	1.5@0.3 5@5 20.6@50
Maximum peaking frequency (GHz)	18GHz	50GHz	18GHz	51GHz
Fixed Maximum peaking frequency	No	No	No	Yes
Supply(V)	3.3	-2.6	/	2.8/3/3
Power(mW)	146	130	149	159
Power Efficiency (pJ/bit)	2.60	2.03	2.66	0.80

*Simulated results

ACKNOWLEDGMENT

This work was supported in part by NSQKJJ under Grant K21799121, in part by STPSZ under Grants JCYJ20190809142017428 and JCYJ20200109141225025, and in part by NSFC under Grant 62074074, and in part by NSFGP under Grant 2021A1515011266.

REFERENCES

[1] Z. Li, M. Tang, T. Fan and Q. Pan, "A 56-Gb/s PAM4 Receiver Analog Front-End With Fixed Peaking Frequency and Bandwidth in 40-nm CMOS," *IEEE Transactions on Circuits and Systems II: Express Briefs*, vol. 68, no. 9, pp. 3058–3062, Sept. 2021.

[2] A. Balteanu and S. P. Voinigescu, "A cable equalizer with 31 dB of adjustable peaking at 52 GHz," in *2009 IEEE Bipolar/BiCMOS Circuits and Technology Meeting*, pp. 154–157, 2019.

[3] I. Sarkas and S. P. Voinigescu, "A 1.8 V SiGe BiCMOS Cable Equalizer with 40-dB Peaking Control up to 60 GHz," in *2012 IEEE Compound Semiconductor Integrated Circuit Symposium (CSICS)*, pp. 1–4, 2012.

[4] K. Maeda, T. Norimatsu, K. Kogo, N. Kohmu, K. Nishimura and I. Fukasaku, "An Active Copper-Cable Supporting 56-Gbit/s PAM4 and 28-Gbit/s NRZ with Continuous Time Linear Equalizer IC for to-Meters Reach Interconnection," in *2018 IEEE Symposium on VLSI Circuits*, pp. 49–50, 2018.

[5] G S. Giannakopoulos, Z. S. He and H. Zirath, "Tunable Equalizer for 64 GBPS Data Communication Systems in 130NM SiGe," in *2018 Asia-Pacific Microwave Conference (APMC)*, pp. 627–629, 2018.

[6] K. Maeda, S. Yamamoto, N. Kohmu, K. Nishimura and I. Fukasaku, "An Active-Copper-Cable with Continuous-Time-Linear-Equalizer IC for 30-AWG 7-meters Reach Interconnect of 400-Gbit/s QSFP-DD," in *2019 IEEE Asia Pacific Conference on Circuits and Systems (APCCAS)*, pp. 217-220, 2019.

2022 IEEE International Conference on
Integrated Circuits, Technologies and Applications

An Analog-Assisted Digital LDO with 0.37mV Output Ripple and 5500x Load Current Range in 180nm CMOS

Luhua Lin, Bowen Wang, Woogeun Rhee, Zhihua Wang

School of Integrated Circuits, Tsinghua University, Beijing, China

Abstract—This paper presents an analog-assisted digital low dropout regulator (LDO) by adopting a delta sigma modulator (DSM) and a finite impulse response (FIR) filter for reduced output ripple. By employing a dual-mode gain-controlled voltage detector (GCVD) and a gear-shift algorithm, reduced recovery time is achieved. An exponential-ratio array (ERA) is designed to expand the load current range. A charge pump (CP) LDO as an analog-assisted loop enhances transient performance. The proposed digital LDO is implemented in 180nm CMOS. For an output voltage of 0.9V, a maximum load current of 100mA and 5500× load current range are achieved with an input voltage of 1V. The undershooting voltage is 78mV when the load current changes from 10mA to 100mA, and the output ripple is 0.37mV.

Keywords—low dropout regulator (LDO), digital LDO, gear-shift algorithm, delta sigma modulator (DSM), finite impulse response (FIR) filter, charge pump (CP) LDO

I. INTRODUCTION

The low dropout regulator (LDO) plays an important role as the core of supply voltage management circuits in the system-on-chip (SoC) design. With advanced CMOS technology, digital LDOs have attracted widespread attention by offering low-voltage operation and process scalability compared to analog LDOs. However, the conventional digital LDO based on a bidirectional shift register [1] faces the trade-off among transient response speed, output current resolution, and power consumption. In this paper, we propose a dual-loop digital LDO with a dead-zone charge pump (CP) LDO [2] as an analog-assisted loop to speed up transient response. In addition, we adopt a delta sigma modulator (DSM) and an FIR filter in the main digital loop to reduce the output ripple.

II. ARCHITECTURE

Fig. 1 shows the system architecture of the proposed digital LDO, consisting of a main digital loop in parallel with an analog-assisted loop. The main digital loop is composed of a gain-controlled voltage detector (GCVD), a gear-shift algorithm, an accumulator, a DSM, an FIR filter, an exponential-ratio array (ERA) power transistors and a power transistors array for FIR filter. The analog-assisted loop is based on a dead-zone CP LDO.

Firstly, in order to reduce the circuit recovery time and output ripple, the GCVD with dual-mode operation of the transient mode and the steady mode is designed in the digital LDO. The GCVD consists of three synchronous comparators, a decoder and a multiplier, where the reference voltage for each comparator is generated by a resistor ladder with a high reference voltage V_{REFH} and a low reference voltage V_{REFL}. The decoder increases the loop gain by putting a boosted weight on higher and lower quantization bits. Furthermore, the multiplier provides a variable gain of 2^m, where m is the gear of the gear-shift algorithm. And the gear m depends on the accumulator output. When a voltage undershooting or overshooting appears, the gear-shift algorithm is triggered by the CP LDO. If the difference of the accumulator output for every N cycles is smaller than a threshold, m is decremented by 1. After several cycles, m is decreased to 0, which means the digital LDO is settled. The gear-shift algorithm with multiple automatic gears enables the digital LDO to achieve fast recovery in the transient mode and the minimum output ripple in the steady mode.

Secondly, the DSM and the FIR filter are designed in the main digital loop. The DSM enables the digital LDO to achieve high resolution with oversampling. Also, its noise shaping property pushes quantization noise to high frequencies. The FIR filter consisting of a series of delay units suppresses the quantization noise in high frequencies, resulting in the reduced output ripple of the digital LDO.

Thirdly, the ERA is designed for PMOS power transistors instead of using same-sized transistors. When there is a heavy load, the current resolution of each power transistor is relaxed under the same output ripple. In the ERA, the transistor size is increased exponentially with the base of 1.02, which

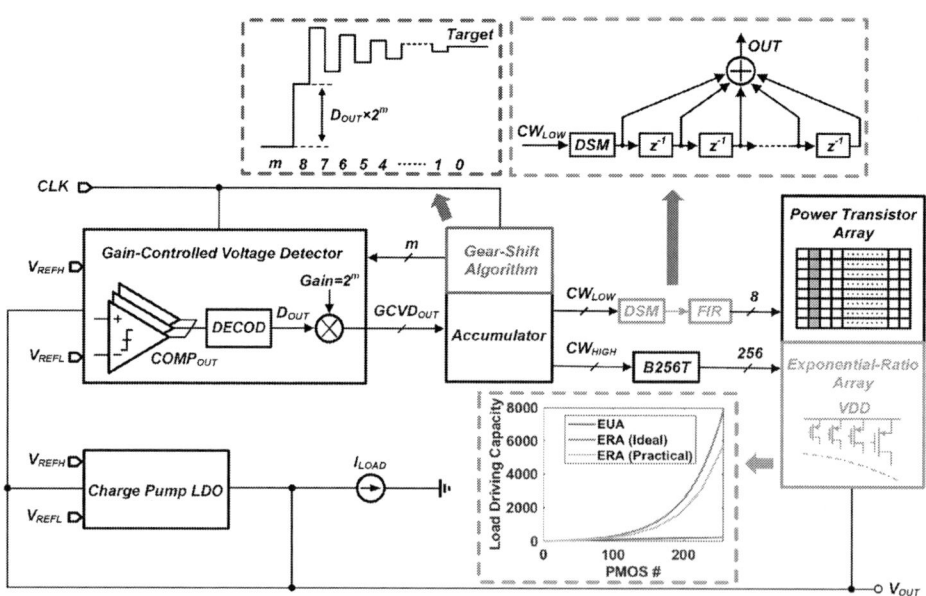

Fig. 1. System architecture of proposed digital LDO.

978-1-6654-9270-6/22 $31.00 © 2022 IEEE

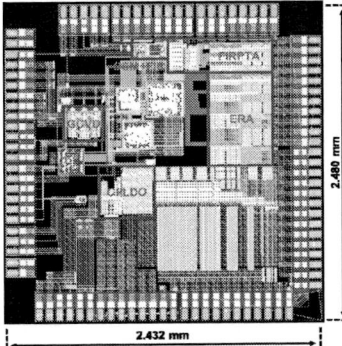

Fig. 2. Layout of proposed digital LDO.

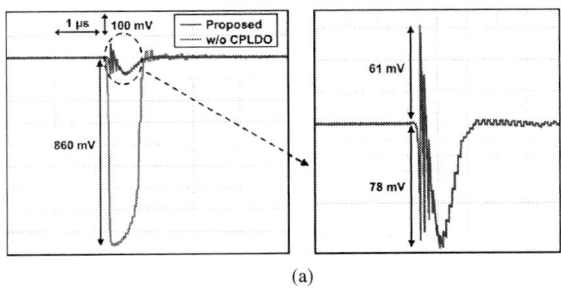

(a)

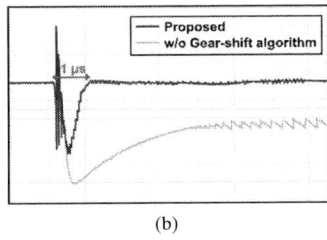

(b)

Fig. 3. Simulated load transient response: (a) Undershooting voltage, and (b) Recovery time.

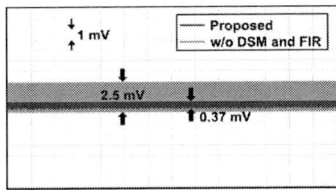

Fig. 4. Simulated steady-state output voltage.

substantially expands the load current range of the digital LDO [3].

Lastly, the dead-zone CP LDO with inverter-based asynchronous comparators is applied as an analog-assisted loop. A voltage undershooting is observed at the output when the load current has a jump. If the undershooting exceeds V_{REFL} of the analog-assisted loop, the comparator is triggered, and the CP LDO is turned on. The CP LDO improves the transient response speed, thus reducing the undershooting voltage.

III. SIMULATION RESULTS

The proposed digital LDO was implemented in 180nm CMOS. The chip area including pads and load circuits for testing is 2.480mm × 2.432mm with an active area of 0.911 mm². The chip layout is shown in Fig. 2.

TABLE I.   PERFORMANCE COMPARISON

	This Work	[5]	[4]	[2]
Process	180nm	65nm	180nm	65nm
V_{IN} (V)	1-1.8	0.5-1	0.7-1.8	0.6-1.2
V_{OUT}(V)	0.9-1.7	0.4-0.95	0.6-1.7	0.5-1.1
Load Range	0.016mA-100mA (5500×)	0.03mA-15mA (512×)	N.A.	0.02mA-100mA (5000×)
F_s (MHz)	20	10	10	5
C_{TOTAL} (pF)	11.6	100	30	26
Recovery Time (µs)	1	3	6	2
ΔV_{OUT}(mV)	78	105	106	95
ΔI_{LOAD} (mA)	90	10	90	90
I_Q (µA)	248	3.2	6.3	28.5
FOM* (fs)	27.7	230	2.0	8.69
V_{RIPPLE} (mV)	0.37	3	10.5	5

$$* \ FOM = \frac{C_{TOTAL} \cdot \Delta V_{OUT}}{\Delta I_{LOAD}} \times \frac{I_Q}{\Delta I_{LOAD}}$$

The proposed digital LDO works at 1V supply with an output voltage of 0.9V. The simulated load transient response with 90mA current change is shown in Fig. 3. Thanks to the CP LDO, the proposed digital LDO achieves the undershooting voltage of 78mV, compared with 860mV for the digital LDO without CP LDO. Moreover, the proposed digital LDO with the gear-shift algorithm achieves the recovery time of 1µs, which is much reduced compared to the digital LDO without gear-shift algorithm. Fig. 4 shows the zoom-in steady-state output voltage. The proposed digital LDO achieves 0.37mV output ripple, compared with 2.5mV for the digital LDO without DSM and FIR filter. Table I summarizes the simulated performance of the proposed digital LDO in comparison with the state-of-the-art works. Even though the I_Q is 248µA, most of the current is consumed by the high frequency DSM and FIR filter to achieve a sub-1mV output ripple.

IV. CONCLUSION

A digital LDO with an analog-assisted loop has been proposed in this paper. The ERA achieves a 5500× load current range with 100mA maximum current. By applying the gear-shift algorithm, the recovery time is reduced to 1µs. Thanks to the DSM and FIR filter, the output ripple is only 0.37mV. Additionally, a FOM of 27.7fs is achieved.

ACKNOWLEDGMENT

This work was supported by National Key Research and Development Program of China under Contract #2020YFB2205602.

REFERENCES

[1] Y. Okuma et al., "0.5-V Input Digital LDO with 98.7% Current Efficiency and 2.7-µA Quiescent Current in 65nm CMOS," in Proc. IEEE CICC, 2010, pp. 1-4.

[2] B. Wang et al., "A Sub-10fs FOM, 5000x Load Driving Capacity and 5mV Output Ripple Digital LDO with Dual-Mode Nonlinear Voltage Detector and Dead-Zone Charge Pump Loop," in Proc. IEEE RFIC, 2020, pp. 315-318.

[3] Y. Zhang et al., "A Capacitor-Less Ripple-Less Hybrid LDO with Exponential Ratio Array and 4000x Load Current Range," IEEE TCAS-II, vol. 66, no. 1, pp. 36-40, Jan. 2019.

[4] J. Tang et al., "A 0.7V Fully-on-Chip Pseudo-Digital LDO Regulator with 6.3µA Quiescent Current and 100mV Dropout Voltage in 0.18-µm CMOS," in Proc. IEEE ESSCIRC, 2018, pp. 206-209.

[5] M. Huang et al., "An Output-Capacitor-Free Analog-Assisted Digital Low-Dropout Regulator with Tri-Loop Control," in ISSCC Dig. Tech. Papers, Feb, 2017, pp. 342-343.

A 2.06μW/MHz, 5.05-MHz, -40-125°C, 22ppm/°C Relaxation Oscillatior with Single Comparator Control

Yifan Yao, Chuqi Chen, Chenchang Zhan

School of Microelectronics, Southern University of Science and Technology, Shenzhen, China

Abstract—In this paper, a 5.05MHz, 2.06μW/MHz, 22ppm/°C relaxation oscillator working as temperature varies from -40°C to 125°C and supply voltage varies from 1.1V to 2V is proposed and designed in a 180nm CMOS process. By using a single comparator control method and a reduced charging/discharging capacitance, the energy-per-cycle figure of merit reaches 2.06μW/MHz. With on-chip voltage and current reference and optimized comparator delay times, the output frequency has a temperature coefficient of only 22ppm/°C. The frequency changes for only 16% when supply voltage varies from 1.1V to 2V.

Keywords—relaxation oscillator, energy-per-cycle, temperature co-efficient, delay optimization

I. INTRODUCTION

RC oscillator has been favored because of its high integration, low power consumption and low sensitivity to temperature, compared to crystal oscillator and LC oscillator [1-3]. Conventional relaxation oscillator frequency is affected by the propagation delay and the offset voltage of comparators, thus it depends on temperature, supply voltage and aging effect. To overcome such problems, chopping and comparator offset cancellation techniques are developed especially in the megahertz frequency range [1-2]. In [3], a low dropout regulator (LDO) and a voltage and current reference (VCR) are added with modified control circuits, cutting the whole power consumption to 5.72μW. But the energy-per-cycle figure of merit (FoM) of this circuit is still large, which is 5.72μW/MHz, partly due to the need of two power hungry comparators.

To reach higher output frequency and decrease energy-per cycle, this work improves the design from [3] by reducing the number of comparators to one with optimized delay times, and also minimizing the charging/discharging capacitance. The power consumption of this work is twice of the work in [3], while the oscillation frequency is 5x of [3], which means the energy efficiency has around 2.5x improvement. The whole circuit reuses the LDO and VCR from [3], which provide stable reference voltage, current and supply voltage for the oscillator, enhancing the stability of the output signal.

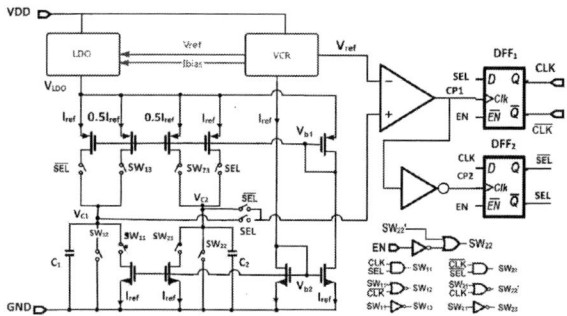

Fig. 1: Schematic of the proposed relaxation oscillator

This work is supported by NSFC under grant 62174080 and SZSTI under grant JCYJ20200109141225025. (Corresponding author: Chenchang Zhan.)

II. PROPOSED RELAXATION OSCILLATOR

A. Proposed Relaxation Oscillator Overview

Fig.1 shows the overall schematic of the proposed relaxation oscillator. In this work, the VCR provides 300mV reference voltage for the LDO and the comparator in the oscillator, and biasing current for the LDO and for charging/discharging the two identical capacitors C_1 and C_2. The LDO generates stable 1V output voltage as the supply voltage for the oscillator core.

The two capacitors, which are both 315fF, are controlled by the digital control signals to charge and discharge alternatively, creating sloping oscillating signals V_{C1} and V_{C2}. Then, the comparator and the inverter convert them into square waves CP1 and CP2, while the two DFFs further transform them into 50% duty cycle clock signal CLK and control signal SEL. The oscillator circuit itself includes biasing circuit to generate V_{b1} and V_{b2} for the charging/discharging current of C_1 and C_2.

B. Single Comparator Control

Compared to the previous design in [3], the main improvement of this work is to cut the number of comparators from two to one, and reduce the charging capacitance, both of which help significantly reduce energy-per-cycle. Firstly, the biasing current of the comparator is optimized such that the difference between the delay of the rising edge of the comparator and the delay of the falling edge is largely reduced and is approximately zero. Meanwhile, the tiny difference between the two delays varies for only 1.3ns from -40°C to 125°C. The difference is so tiny and varies so little with temperature that the output frequency of the oscillator is nearly unchanged from -40°C to 125°C. Fig.2 illustrates the comparator delay definition and simulated delays vs. temperature. Thus, the structure of two comparators can be replaced by the structure of one comparator and one inverter [2]. This contributes to saving energy.

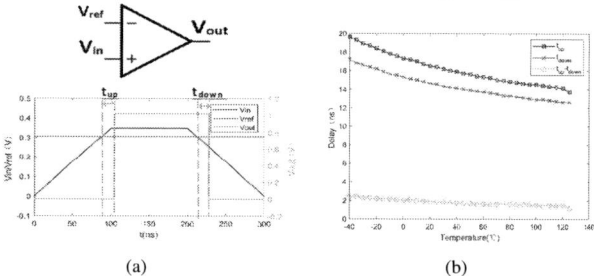

| (a) | (b) |

Fig. 2: Comparator (a) delay definitions and (b) simulated delay time vs. temperature.

Secondly, the capacitance of two capacitors is reduced from 584fF in [3] to 315fF. For the few-MHz oscillation frequency, the comparator delay should be small, and the capacitors are charged/discharged quickly. Therefore, for the same biasing currents, the capacitance can be reduced. As explained in the next section of working principle, the delay time difference of the comparator rising and falling edges is reduced and ignorable, so the oscillation period is roughly given by,

$$T_{period} = \frac{2 * C * V_{ref}}{I_{ref}} + (t_{up} - t_{down}) \approx \frac{2 * C * V_{ref}}{I_{ref}} \quad (1)$$

where C is the capacitance of C_1 and C_2, V_{ref} and I_{ref} are the reference voltage and reference current, respectively. Energy-per-cycle if considering the charging of capacitors only is given by,

$$E_{freq} = \frac{P}{f} = V * I * T_{period} = 2 * C * V_{ref}^2 \quad (2)$$

With the capacitance reduced, the energy-per-cycle is 0.5x reduced compared to the structure in [3].

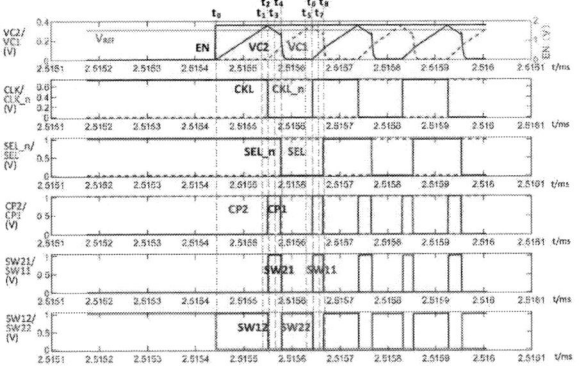

Fig. 3: Timing diagram of the proposed relaxation oscillator.

C. Working Principle of Oscillator

Fig. 3 shows the timing diagram of the proposed relaxation oscillator, using the transient simulation results of the control signals. Accurate charging and discharging of capacitors with controlling signals create the precise oscillation signals. Specifically, capacitors are controlled to charge and discharge by switches, and the states of switches are determined by CLK, EN and SEL signals. CLK is the final output signal and decides the capacitors' alternate charging period. SEL signal determines whether C_1 or C_2 is connected to the comparator input. EN overturns CLK and $\overline{SEL}$ to be high.

When EN turns high, oscillator starts up in one cycle. At t_0, EN leads output Q of the two DFFs, i.e., CLK and $\overline{SEL}$, to be high. Then, $\overline{SEL}$, SW_{13}, SW_{23} and SW_{12} close and Iref starts to charge C_1. However, only V_{C2} is connected to the comparator as $\overline{SEL}$ is high. V_{C2} keeps on increasing and becomes larger than Vref after t_1. After the delay time from t_1 to t_2, V_{C2} comes to the peak at t_2. At the positive edge of CP1, DFF1 changes CLK into low level. Due to this transition, the digital control circuits close SW_{21} and open SW_{12} and SW_{23}. Now, C_1 is charged by 1.5 times of Iref and C_2 is discharged by Iref. After the second delay time, i.e., t_3 to t_4, V_{C2} is smaller than Vref by the same amount as Vref to peak. Similar to DFF1 at t_2, at the positive edge of CP2 at t_4, DFF2 changes the state of SEL to low level. After the transition, SW_{21} and $\overline{SEL}$ open, and SW_{23}, SEL and SW_{22} close. Now, V_{C1} is linked to the comparator, C_2 continues to discharge and C_1 starts to be charged by Iref until t_6. This periodic operation keeps on and forms the oscillating signal.

Due to the optimized comparator, the two delay times, t_1 to t_2 and t_3 to t_4, are almost the same. They are also equal to t_2 to t_3, as C_2 is charged and discharged both by Iref and the slopes should equal to each other. Eq.(1) can be readily derived. By controlling the temperature coefficient of two terms in Eq.(1), the oscillation frequency can achieve low TC.

III. SIMULATION RESULT

The proposed relaxation oscillator is designed in a standard 180nm CMOS technology with 5.05MHz oscillation frequency. Fig. 4 shows the simulation results. With the modified circuit structure and optimized comparator and capacitor, the energy-per-cycle is reduced to 2.06μW/MHz, which is 36% of the two-comparator structure in [3]. The VCR and LDO consume 1.24uA and 1.66uA respectively from a 1.1V supply while

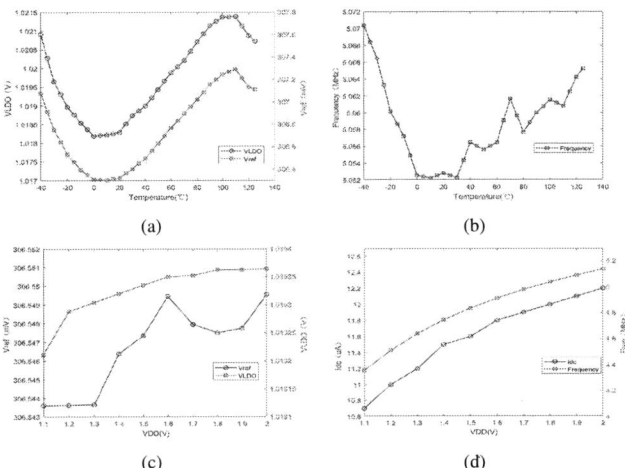

Fig. 4: Simulation results of (a) V_{LDO} and Vref vs. temperature; (b) oscillation frequency vs. temperature; (c) V_{LDO} and Vref vs. supply voltage; and (d) Idc and oscillation frequency vs. supply voltage.

TABLE I. PERFORMANCE COMPARISON

	This work*	[1]	[2]	[3]	[4]
Technology(nm)	180	180	350	180	65
Frequency(MHz)	5.05	12.77	1	1.02	1.2
VDD(V)	1.1-2	0.6-1.1	3-4.5	1.1-2	0.9-1.8
Energy Efficiency(uW/MHz)	2.06	4.4	160	5.72	0.68
Power(uW)	11.64	56.2	160	5.72	0.82
Temperature Range(°C)	-40-125	-30-120	-40-125	-40-125	-25-125
TC(ppm/°C)	22	53	32	59	100
Startup time	1 cycle	N.A.	1 cycle	1 cycle	3 cycles

*simulation result

the OSC core consumes 8.05uA from 1V supply. From Fig.4 (a), TCs of V_{LDO} and Vref are 19.2ppm/°C and 19.3ppm/°C, respectively. Improved delay technique, together with the stable supply voltage and reference voltage, ensures the oscillation frequency to be temperature-insensitive. Fig.4 (b) elaborates that TC of frequency is only 21.7ppm/°C. From Fig. 4 (c) and (d), when supply voltage varies from 1.1V to 2V, Vref changes for 0.002% and VLDO changes for 0.015%. However, channel length modulation causes frequency and Idc to vary for 13% and 16% respectively. Table I summarizes and compares the proposed oscillator with some state-of-the-art works.

IV. CONCLUSION

This paper presents a high power-efficiency relaxation oscillator with low sensitivity to temperature and supply voltage. This work cuts energy-per-cycle into 2.06μW/MHz and simplifies the control circuit by reducing the required number of comparators from two to one and optimizing the comparator delays. The design can provide a stable 5.05MHz square wave signal with very low energy consumption, thus it would be economical and useful for fully-on-chip oscillator applications.

REFERENCES

[1] J. Wang et al., "A 12.77-MHz on-chip relaxation oscillator with digital compensation for loop delay variation," 2015 IEEE A-SSCC, 2015, pp. 1-4.

[2] J. Mikulić et al., "A 1-MHz relaxation oscillator core employing a selfcompensating chopped comparator pair," in IEEE Trans. on Circuits and Systems I: Regular Papers, vol. 66, no. 5, pp. 1728-1736, May 2019.

[3] C. Chen et al., "A 5.72-μW, 1.02-MHz, -40 125°C, 1s-Startup Time Relaxation Oscillation with Fully-on-Chip Voltage Reference and LDO Regulator," 2021 IEEE ICTA, 2021, pp. 237-238.

[4] A. Savanth et al., "A Sub-nW/kHz Relaxation Oscillator With Ratioed Reference and Sub-Clock Power Gated Comparator", IEEE J. of Solid-State Circuits, vol. 54, no. 11, pp. 3097-3106, Nov. 2019.

A 4 to 5GHz Digitally Controlled Ring Oscillator with 100kHz Resolution using Noise Cancellation Technology in 40nm CMOS

Shuyue Fang[1], Jinrui Hu[1], Haigang Feng[1], Xinpeng Xing[1], Han Wang[2], Lei Yang[2]

[1] Shenzhen International Graduate School, Tsinghua University, Shenzhen 518055, China
[2] Megahunt Technologies Inc., China

Abstract—Herein a digitally controlled ring oscillator (DCRO) using noise cancellation technology is presented for an all digital phase-locked loops (ADPLLs) system. The design introduced a noise insensitive current source combined with current digital-to-analog converter (DAC) to achieve high resolution with wide tuning range, and better supply noise immunization. Meanwhile, regulated cascode topology is utilized to ensure the equality of drain source voltage of current mirror arrays under various current injecting into DCRO alleviating the channel-length modulation effect. The proposed design was implemented in 40 nm CMOS process operating from 4 to 5GHz with 100kHz resolution. Simulation results show that the supply sensitivity of current source can averagely reach -151.5dB and DCRO achieves static and dynamic supply noise immunity of 0.021%-fout/1%-VDD and 0.011%-fout/1%-VDD respectively at 4.5GHz. The overall power dissipation is 0.84mW from a 1.2V supply.

Keywords—ADPLLs, DCRO, noise cancellation, regulated cascode

I. INTRODUCTION

Nowadays, all digital phase-locked loops (ADPLLs) are more competitive than their analog counterparts in some application fields and thus are widely used. Digitally controlled oscillator (DCO) is a core building block in ADPLLs, among which digitally controlled ring oscillator (DCRO) receives considerable attention recently due to its wide frequency tuning range, compact layout and multi-phase output. The current mode DCRO (CM-DCRO) combines current digital-to-analog converter (DAC) with ring voltage controlled oscillator (VCO) to fulfill frequency tuning. Although easily realized, typical DCRO system exists the following drawbacks. First of all, DCRO is sensitive to the supply voltage noise and the different current injecting into DCRO will affect the replication accuracy of the current mirror because of inequality of drain source voltage of MOSFETs[1]. Both of these will cause frequency deviation from desired value and deteriorate linearity. In addition, it is difficult to achieve both high-precision replication and wide replication range grounded on a single current source at the same time [2].

To address the aforementioned issues, this work introduced a reference current circuitry using noise cancellation technology to provide noise insensitive current for DCRO. A regulated cascode topology is adopted for the sake of fixed value of V_{ds} of current mirror arrays to suppress the second-order effect of MOSFETs under various operating frequency. This movement results in the high linearity of frequency tuning.

By taking all these efforts, the proposed CM-DCRO core for a ADPLLs system demonstrates -151.5dB in average for the supply sensitivity ($\Delta I_{out}/\Delta V_{DD}$). Its static and dynamic supply noise immunity are 0.021%-fout/1%-VDD and 0.011%-fout/1%-VDD, respectively, which gets better performance than [3-4]. The DCRO operates from 4GHz to 5GHz with 100kHz resolution and consumes only 0.84mW at 4.5GHz.

The rest of paper is organized as follows. The proposed CM-DCRO core inside ADPLLs system is described specifically in Section II. The simulation results are illustrated in Section III followed by conclusions drawn in Section IV.

II. CM-DCRO DESIGN

A. CM-DCRO Core and Frequency Tuning Scheme

As shown in Fig. 1, the proposed CM-DCRO core inside ADPLLs system consists of ring DCO, self-biased current source using noise cancellation technology, current mirror arrays and regulated cascode topology. The frequency of ring oscillator (f_{osc}) can be expressed by the formula (1), where I is current injecting into the DCRO, n is the stage of delay cell, C is the output capacitor of delay cell, and U is the voltage of DCRO. Thus, the expected frequency can be obtained by adjusting the value of output capacitor and current.

$$f_{osc} = \frac{I}{nCU} \quad (1)$$

CM-DCRO receives oscillator tuning word (OTW) from digital loop filter (DLF). The OTW is divided into coarse-tuning and fine-tuning parts. Coarse-tuning word (CTW) is composed of 5 bit separating the entire band of 4 to 5GHz into five small bands (4-4.25GHz, 4.2-4.45GHz, 4.4-4.65GHz, 4.6-4.85GHz, 4.8-5.05GHz) tuned by capacitor arrays. Fine-tuning word (FTW), which has a total of 12 bits, is grouped as 1 bit of always-on unit, 6 bits of multiplication unit and 5 bit of division unit. Under the fine-tuning scheme, each of five small bands can be fully covered and the least significant bit (LSB) determines the 100kHz frequency resolution.

The self-biased current source and its mirror arrays set up the operation condition for DCRO. Such noise cancellation technology combined with regulated cascode topology reduces the interference from other circuits for the current mirror arrays. They will be discussed in detail in the following sections.

B. Current Source using Noise Cancellation Technology

The simplified schematic of the self-biased current source using a novel noise cancellation technology is demonstrated in Fig. 2. (a).

Compared with the conventional current source, it adds two extra modules, noise sensing stage and negative feedback loop.

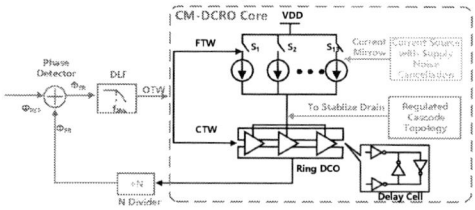

Fig. 1. Concept of proposed CM-DCRO core inside ADPLLs system

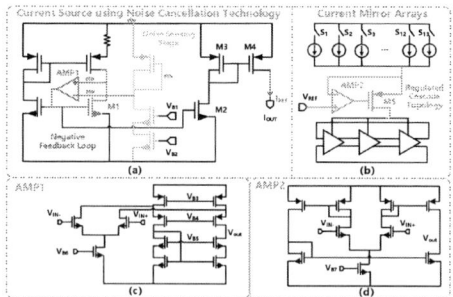

Fig. 2. Simplified schematic of the current source using noise cancellation technology (a) and regulated cascode topology (b) inside current mirror

978-1-6654-9270-6/22 $31.00 © 2022 IEEE

The noise sensing stage transfers the noise voltage from the supply to the negative input of the amplifier (AMP1). The positive input of AMP1 serves two purposes. The first is to constitute a negative feedback loop with M1 to stabilize the gate terminal voltage of M1, thereby generating a relative DC current. Another one is to receive the noise from power supply, so as to obtain approximately equal common mode noise voltage with the negative input. The common mode noise is eliminated because of the differential input of amp and therefore there is nearly no power supply noise component in the gate of M1. As a result, the current mirrored from M1 almost does not contain the noise current generated by the power supply noise. Similarly, the current I_{DC} mirrored by M3 to M4 does not contain noise component ΔI_n and the DC current injected into the CM-DCRO is nearly eliminated. It is noted that some compensation capacitors of negative feedback loop are not displayed in Fig. 2.

C. Regulated Cascode Topology

Regulated cascode topology consisted of amplifier (AMP2) and M5 is illustrated in Fig. 2. (b). It not only tackles the problem of inequality of V_{ds} of current mirror arrays, but also ensures a constant value of drain voltage under different working current of DCRO. At the same time, the voltage fluctuation generated by DCRO is fairly small with decoupling capacitor, which hardly affects dc current from current mirror arrays hence improving the linearity of DCRO.

D. Current Mirror Arrays

The current mirror arrays in Fig. 1 are replicated by the current source using noise cancellation technology. It is necessary to generate a wide range and high precision current in order to cover a fine tuning frequency band of 250MHz while achieving resolution within 100kHz. Nevertheless, if the MOSFETs of current mirror are solely used in parallel type to generate current of $2^{-5}uA \sim 2^6uA$, it will occupy a large area concerning circuit layout implement, which is unacceptable in actual design. Consequently, MOSFETs applied in current mirror are employed in the type of series and parallel concurrently and divided into always-on, multiplication and division units. The always-on unit decides the basic oscillatory frequency of DCRO. The multiplication and division unit are exploited to cover the current arrays of $2^{-5}uA \sim 2^{-1}uA$ and $2^0uA \sim 2^5uA$ respectively.

III. SIMULATION RESULTS

The proposed DCRO system is designed and implemented in a generic 40nm CMOS process with a 1.2V voltage supply. It has both wide frequency tuning of 4 to 5GHz and relative high resolution of 100kHz. The novel noise cancellation technology applied in current source shows that its power supply sensitivity ($\Delta I_{out}/\Delta V_{DD}$) can reach -151.5dB. Fig. 3 displays the Mente Carlo simulation results (Left) and comparison (Right) of supply sensitivity of current source concerning whether using noise cancellation technology or not. Its comprehensive performance increases more than 40dB compared with the current source in [1]. To characterize supply noise rejection performance, the static sensitivity is simulated with ±10% VDD deviation and the results are illustrated in the left of Fig. 4. With the noise cancellation technology, the frequency drift caused by ±10% VDD deviation is only 21.89MHz, which suggests that the static sensitivity of DCRO is 0.021%-fout/1%-VDD. The right of Fig. 4 shows the frequency of DCRO under a 0.012Vpp sinusoidal signal of 10MHz interference. To conclude the dynamic sensitivity of DCRO is 0.011%-fout/1%-VDD.

Table I summarizes the performance of the proposed DCRO, and compares it to other works. The power dissipation is

obviously less than other works while achieving better performance against supply noise.

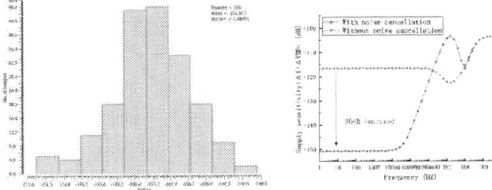

Fig. 3. Mente Carlo simulation results (Left) and supply sensitivity comparison (Right)

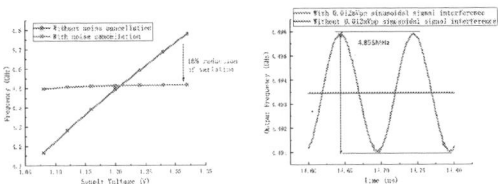

Fig. 4. Static (Left) and dynamic (Right) supply sensitivity of DCRO

TABLE I. PERFORMANCE AND COMPARISON

	ISSCC'14 [1]	ISCAS'11 [3]	APCCAS '16 [4]	This work
Tuning Type	Digital	Analog	Hybrid	Digital
Output frequency	1.6GHz	N/A	3GHz	4.5GHz
Tuning Range(GHZ)	0.025-1.6	N/A	1.5/3	4-5
Supply sensitivity $\Delta I_{OUT}/\Delta V_{DD}$(dB)	-104	N/A	N/A	-151.5
Static Sensitivity (%-f_{out}/1%-V_{DD})	N/A	0.013	0.03	0.021
Dynamic Sensitivity (%-f_{out}/1%-V_{DD})	1.2**	0.08 @10MHz	2.54	0.011 @10MHz
Jitter(RMS) (ps)	28	N/A	N/A	20.42
Power (mW)	3.1*	19.56	5.2	0.84
Technology	20nm	0.13um	0.18um	40nm

* With other PLL building blocks; ** Calculated from deterministic jitter.

IV. CONCLUSION

In this paper, a 4 to 5GHz DCRO with 100kHz resolution using noise cancellation technology has been presented. At 4.5GHz, the overall circuit consumes 0.84mW power and the RMS jitter is 20.42ps. The static and dynamic supply noise immunity of CM-DCRO are 0.021%-fout/1%-VDD and 0.011%-fout/1%-VDD, respectively.

ACKNOWLEDGMENT

This work is funded by Guangdong Key Area Research and Development Plan under Grant 2019B010143003.

REFERENCES

[1] J. Liu et al., "15.2 A 0.012mm2 3.1mW bang-bang digital fractional-N PLL with a power-supply-noise cancellation technique and a walking-one-phase-selection fractional frequency divider," 2014 IEEE International Solid-State Circuits Conference Digest of Technical Papers (ISSCC), 2014.

[2] S. Toprak et al, High accuracy potentiostat with wide dynamic range and linearity, AEU-International Journal of Electronics and Communications, Volume 142, 2021.

[3] Y. Park and W. Choi, "Supply noise insensitive ring VCO with on-chip adaptive bias-current and voltage-swing control," 2011 IEEE International Symposium of Circuits and Systems (ISCAS), 2011, pp. 229-232.

[4] V. Sharma et al, "Design of a hybrid ring oscillator at 1.5/3.0 GHz with low power supply sensitivity," 2016 IEEE Asia Pacific Conference on Circuits and Systems (APCCAS), 2016, pp. 567-570.

An IMPLY-based Memristive Multiplier for Computing-in-Memory Systems with Weight-Stationary CNN Acceleration

Wenhui Liang[1], Jiarui Xu[1], Yuansheng Zhao[1], Zixuan Shen[1], Guoyi Yu[1,2], Yuhui He[1,2], Chao Wang[1,2]

[1] School of Optical and Electronic Information, Huazhong University of Science and Technology, Wuhan, China
[2] Wuhan National Laboratory of Optoelectronics, Huazhong University of Science and Technology, Wuhan, China

Abstract— Adders and multipliers based on memristive Material Implication (IMPLY) logic are widely used in primary building blocks of Arithmetic Logic Unit (ALU). To solve the issue that the existing IMPLY-based multipliers cannot protect the input operands, this paper presents a novel data non-destructive memristive IMPLY-based semi-parallel multiplier for Computing-in-Memory (CIM) systems, by assigning function-specific memristors for data-protection and introducing additional switches for higher parallelism. Simulation results show that the proposed multiplier can achieve 30% faster than conventional semi-parallel design and 9.1% less memristors against the state-of-art semi-serial design for 4-bit multiplication, while preventing the input weight from destruction as required by CNN weight reuse.

Keywords—Memristor, Material Implication (IMPLY), Multiplier, Computing-in-Memory (CIM), Convolution Neural Network (CNN)

I. INTRODUCTION

Memristor-based Computing-in-Memory application is a promising candidate to overcome Memory-wall bottleneck thanks to its excellent properties of non-volatile processing, high density, and ultra-low power, etc. Material Implication (IMPLY) and FALSE logic are fundamental logic operations for constituting a complete logic system, which is suitable memristor-based CIM implementation, as depicted by Fig. 1 (a) and (b) [1]. Although IMPLY-based multipliers as core computing blocks have been widely discussed in [1]-[3], the mismatch between the requirement of weight reuse for efficient CNN acceleration and the input operands destruction of existing multiplier designs is still a major challenge that has not been addressed. This paper presents a data non-destructive IMPLY-based multiplier to protect the input operands, while keeping the operation steps and hardware overhead comparable with the state-of-art designs, which is suitable for the weight-stationary dataflow of IMPLY-based CIM CNN acceleration.

	Input (P)	Input (Q)	Output (Q')
Case 1	0	0	1
Case 2	0	1	1
Case 3	1	0	0
Case 4	1	1	1

Fig.1. (a) Truth table of IMPLY logic and (b) its memristive implementation: IMPLY operation between P and Q can be denoted by $Q'=P\rightarrow Q$. Note: FALSE logic always yields "0", which is Q'= FALSE Q.

II. PROPOSED SEMI-PARALLEL MULTIPLIER

A. Introduction of IMPLY-based Multiplication Realization

Fig. 2 shows the basic principle of 2-bit multiplication between multiplicand A (A_1, A_0) and multiplier B (B_1, B_0), where the product is P (P_2, P_1, P_0). The multiplication consists of two phases, i.e., partial-product calculation (i.e., A_0B_0, A_1B_0, A_0B_1, A_1B_1.) and partial-product summation, respectively.

$$a_kb_j=(a_k\rightarrow(b_j\rightarrow 0))\rightarrow 0 \quad (1)$$

$$S=[((\sim a)\rightarrow b)\rightarrow((a\rightarrow(\sim b))\rightarrow c)]\rightarrow\sim(a\oplus b)\rightarrow(\sim c)) \quad (2)$$

$$C_{out}=\sim[((\sim a)\rightarrow b)\rightarrow\sim((a\rightarrow(\sim b))\rightarrow c)] \quad (3)$$

Fig. 2. Principle of 2-bit multiplication. Note: 0 is implemented by FALSE logic.

More specifically, A one-bit partial-product calculation is indeed a logical AND operation, which can be realized by IMPLY logic and FALSE logic as described by the first equation in Fig. 2. Similarly, the summation of partial-production can also be realized by IMPLY logic and FALSE logic as described by the rest two equations in Fig. 2.

B. Existing Memristive IMPLY-based Multiplier Design

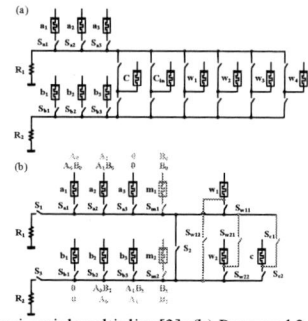

Fig.3. (a) A 2-bit semi-serial multiplier [3]; (b) Proposed 2-bit semi-parallel multiplier. Note: initial input states are denoted by green color, and the states after partial-product calculation are denoted by purple color.

Fig.3. (a) shows the 2-bit IMPLY-based multiplier using semi-serial adder [3], where memristors a_i and b_i (i=1,2,3) are used for input operands storage, memristors C and C_{in} are used for carry-out, while memristors w_k (k=1, 2, 3, 4) are used for intermediate data transformation. Note that the input B stored in the memristors is destructed during partial-product summation phase, this feature makes the semi-serial multiplier difficult to satisfy the requirement of weight reuse for CNN application. The data non-destructive full adder proposed in [4] firstly introduce the concept of input operands protection, which is only suitable for the weight protection of binarized neural networks, due to the different function of the specific memristors that need to be protected between addition and multiplication tasks.

C. Proposed Semi-parallel Multiplier Design

To address the issues of data destruction and applicability for multiplication, we propose a data non-destructive IMPLY-based semi-parallel multiplier for input protection as shown in Fig.3. (b). For a 2-bit multiplier, only two extra multiplier memristors (i.e., m_1 and m_2) and three switches (i.e., S_{w12}, S_{w21} and S_{c1}) are induced against conventional semi-parallel full adder, for input operands storage and parallelism improvement, respectively.

step	operation		state after operation							
	1st line	2nd line	a_1	a_2	a_3	m_1	b_1	b_2	b_3	m_2
init state				A_0	A_1	0	B_0	0	A_0	A_1	B_1
1	$m_1\rightarrow a_3$	$m_2\rightarrow b_1$	A_0	A_1	$\sim B_0$	B_0	$\sim B_1$	A_0	A_1	B_1
2	$a_1\rightarrow a_3$	$b_2\rightarrow b_1$	A_0	A_1	$A_0\rightarrow\sim B_0$	B_0	$A_0\rightarrow\sim B_1$	0	A_1	B_1
3	FALSE a_1	FALSE b_2	0	A_1	$A_0\rightarrow\sim B_0$	B_0	$A_0\rightarrow\sim B_1$	0	A_1	B_1
4	$a_3\rightarrow a_1$	$b_1\rightarrow b_2$	A_0B_0	A_1	$A_0\rightarrow\sim B_0$	B_0	$A_0\rightarrow\sim B_1$	A_0B_1	A_1	B_1
5	FALSE a_3	FALSE b_1	A_0B_0	A_1	0	B_0	0	A_0B_1	A_1	B_1
6	$m_1\rightarrow a_3$	$me_2\rightarrow b_1$	A_0B_0	A_1	$\sim B_0$	B_0	$\sim B_1$	A_0B_1	A_1	B_1
7	$a_2\rightarrow a_3$	$b_3\rightarrow b_1$	A_0B_0	A_1	$A_1\rightarrow\sim B_0$	B_0	$A_1\rightarrow\sim B_1$	A_0B_1	A_1	B_1
8	FALSE a_2	FALSE b_3	A_0B_0	0	$A_1\rightarrow\sim B_0$	B_0	$A_1\rightarrow\sim B_1$	A_0B_1	0	B_1
9	$a_3\rightarrow a_2$	$b_1\rightarrow b_3$	A_0B_0	A_1B_0	$A_1\rightarrow\sim B_0$	B_0	$A_1\rightarrow\sim B_1$	A_0B_1	A_1B_1	B_1
10	FALSE a_3	FALSE b_1	A_0B_0	A_1B_0	0	B_0	0	A_0B_1	A_1B_1	B_1

TABLE I. Execution steps of partial product calculation in 2×2bit multiplication. Note: Multiplier B is in green color, and the partial-products are in purple color.

Table I shows the detailed operation steps during 2-bit multiplication phase 1, i.e., partial-product calculation, according to Fig.2. As shown in Fig. 3. (b), the initial multiplicand A (A_1, A_0) are duplicated into the memristors (a_2,

---

This work was partially supported by National Natural Science Foundation of China (61974053). Wenhui Liang and Jiarui Xu equally contributed to this paper. #Corresponding email: chao_wang_me@hust.edu.cn.

978-1-6654-9270-6/22 $31.00 © 2022 IEEE

a_1) and (b_3, b_2), while the multiplier B (B_1, B_0) are stored into the memristors (m_2, m_1), respectively, for calculating (A_0B_0, A_1B_0) and (A_0B_1, A_1B_1) on the separated lines simultaneously. During the phase 1, the state of multiplier memristors that store B remain unchanged for stationary weight. The 2-bit partial-product calculation can be finished in 10 steps with the highest parallelism based on the semi-parallel design.

In the Phase 2 of multiplication, i.e., partial-product summation, the addition of the updated partial-products stored in memristors a_i and b_i (purple color in Fig. 3. (b)) will be performed by the proposed addition scheme. Table II. (b) shows the optimized operation steps for 1-bit addition. Notably, the operands B would be still stored in the memristors m_1 and m_2. The product P (P_2, P_1, P_0) would be stored in the memristors ($a3$, $a2$, $a1$) and the carry-out would be stored in the memristor c.

In general, once partial-product calculation and once 3-bit partial-product addition would be performed for a 2-bit multiplication by the IMPLY-based multiplier. The proposed semi-parallel multiplier can finish the calculation with the highest parallelism and effectively keep the input weight stationary, which can achieve up to 30% faster than the conventional semi-parallel design and 9.1% less memristor than the semi-serial design.

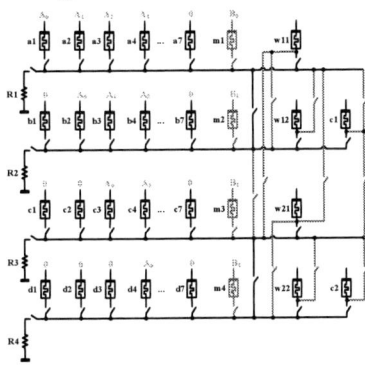

Fig.4. Proposed 4-bit data non-destructive semi-parallel multiplier.

Fig.4 shows the topology of the proposed non-destructive semi-parallel multiplier in 4-bit implementation, consisting of two semi-parallel adders and four additional switches to connect adders. The two adders can calculate the lower 2-bit partial-product and higher 2-bit partial-product in parallel, respectively. Then the results of partial-product calculation stored in the 1st and 3rd line can be added by the reconstructed adder topology consisting of the memristors and switches in these two lines, without disrupting the input operands B.

III. SIMULATION AND DISCUSSION

The proposed data non-destructive multiplier is designed and simulated with a UNIMORE Verilog-A model [5] and a 40-nm CMOS on Cadence Virtuoso by using the parameters in [4].

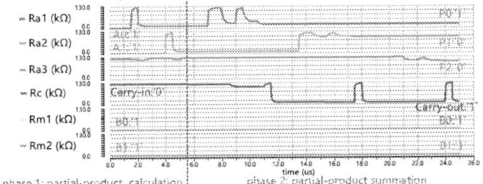

Fig.5. Simulation of 2-bit multiplier with A=11, B=11 and Carry-in=0. The calculated product is in c&a3−1 = 1001 after 49 execution steps.

Fig. 5 as an example shows that proposed multiplier design can perform 2-bit multiplication operation correctly. All the input cases of 2-bit multiplication have been tested on the

proposed multiplier design. TABLE II compares the steps of the multiplier based on the conventional semi-parallel design and the proposed data non-destructive design during partial-product summation phase. By executing more steps in parallel with the additional switches shown in Fig.2. (b), i.e., the steps with blue color in TABLE II. (b)，the proposed multiplier can achieve 30.0% faster than the conventional semi-parallel multiplier.

step	operation			step	operation		
	1st line	2nd line	between lines		1st line	2nd line	between lines
1	FALSE w_1	FALSE w_2		1	FASLE w_1	FASLE w_2	
2	a → w_1	b → w_2		2	a→w_1	b→w_2	
3			w_1 → b	3	a→w_2	w_1→b	
4			a → w_2	4	FALSE a	FALSE w_1	
5	FALSE a			5			b→a
6			b → a	6	w_2→a	c→w_1	
7			w_1 → a	7	a→w_1	w_2→c	
8	FASLE w_1			8	FALSE a	FALSE w_2	
9			c → w_1	9	w_1→a	c→w_2	
10	a → w_1	w_2 → c		10			b→c
11	FASLE a	FALSE w_2		11	c→a	b→w_2	
12	w_1 → a	c → w_2		12	FALSE c		
13		b → w_2		13	w_2→c		
14		b → c		14			
15			c → a	15			
16		FASLE c		16			
17			w_2 → c	17			

TABLE II. Execution steps of partial-product addition in 1 bit: (a) conventional semi-parallel multiplier and (b) proposed semi-parallel multiplier. Note: Operations in green color are executed in parallel as compared to the corresponding operations in red color executed in serial from conventional design.

TABLE III compares the performance and hardware overhead between the proposed design and the state-of-art designs. We re-evaluate the critical metrics by approximately calculating the average bit-width of the adders when considering N×N-bit multiplication. The proposed multiplier can realize excellent operands protection features, with an about 10% reduction in hardware overhead and a 30% increase in calculation speed, against the semi-serial and semi-parallel methods, respectively.

Designs	Memristors	Imp.*	Transistors	Imp.*	Number of steps	Imp.*	Protect
Semi-serial [3]	$\lceil\frac{N}{2}\rceil(4N+4)$	9.1%	$12\lceil\frac{N}{2}\rceil+\lfloor\frac{N-1}{2}\rfloor+2N^2-N$	10.4%	$\lceil\log_2 N\rceil(15N+2)+4N+2$	-20.2%	No
Semi-parallel[2]	$\lceil\frac{N}{2}\rceil(4N+1)$	-18.2%	$3\lceil\frac{N}{2}\rceil+\lfloor\frac{N-1}{2}\rfloor+2N^2-N+3(\lceil\log_2 N\rceil-1)$	-20.8%	$\lceil\log_2 N\rceil(\frac{51}{2}N)+5N+2$	30.0%	No
Proposed	$\lceil\frac{N}{2}\rceil(4N+3)$	-	$8\lceil\frac{N}{2}\rceil+\lfloor\frac{N-1}{2}\rfloor+2N^2-N+3(\lceil\log_2 N\rceil-1)$	-	$\lceil\log_2 N\rceil(\frac{39}{2}N)+5N+2$	-	Yes

TABLE III. Comparison of N×N-bit multiplication among the semi-serial multiplier, re-implemented semi-parallel multiplier and the proposed design. *Percentage of improvement (Imp.) is calculated based on N=4.

IV. CONCLUSION

In this paper, a novel data non-destructive memristive IMPLY-based semi-parallel multiplier is proposed for CIM systems. This design has achieved 9.1% less memristors and 10.4% less transistors than the existing semi-serial multiplier, which can also protect the input operands. This is of great significance for the data reuse of weight-stationary CNN dataflow in CIM-based acceleration.

REFERENCES

[1] J. Xu et al., "In Situ Aging-Aware Error Monitoring Scheme for IMPLY-Based Memristive Computing-in-Memory Systems," in *IEEE Trans. Circts. and Systs. I: Reg. Papers*, vol. 69, no. 1, pp. 309-321, Jan. 2022

[2] S. Ganjeheizadeh Rohani *et al.*, "A Semiparallel Full-2Adder in IMPLY Logic," in *IEEE Trans. Very Large Scale Integr. (VLSI) Systs.*, vol. 28, no. 1, pp. 297-301, Jan. 2020.

[3] D. Radakovits, *et al.*, "A Memristive Multiplier Using Semi-Serial IMPLY-Based Adder," in *IEEE Trans. Circtss and Systs. Part I Reg. Papers*, vol. 67, no. 5, pp. 1495-1506, May 2020.

[4] X. Hu *et al.*, "A Data Non-destructive IMPLY-based Memristive Semi-parallel Full-Adder for Computing-in-memory Systems," in *Proc. IEEE Intl. Conf. on Integr. Circts., Techs. and Apps*, 2021, pp. 212-213.

[5] F. M. Puglisi *et al.*, "Unimore Resistive Random Access Memory (RRAM) Verilog—A Model," in *Proc. nanoHUB*, 2019.

A Novel Concept of using Double Threshold Voltage Coupling to Improve the linearity of AlGaN/GaN HEMTs for millimeter-wave applications

Pengfei Wang [1,2], Minhan Mi [1,2], Sirui An [1,2], Xiang Du [1,2]
Xiaohua Ma [1,2], Yue Hao [1,2]

[1] School of Microelectronics, Xidian University, Xi'an 710071, China
[2] State Key Discipline Laboratory of Wide Bandgap Semiconductor Technology, Xidian University, Xi'an 710071, China

Abstract—In this letter, we demonstrate an AlGaN/GaN HEMT fabricated by synthesizing recess and planar devices along the gate width and incorporating N_2O plasma treatment to form an oxide layer at the gate electrode of the proposed HEMT. The transconductance curve of the fabricated device has a plateau region larger than 7 V, with a flattened response curve of f_T/f_{max} with respect to the gate bias voltage. At the operating frequency of 30 GHz, the maximum power-added efficiency (PAE) is 41%, the value of the power density (P_{out}) is 5.3 W/mm, and the associated 1dB compression point (P_{1dB}) is 28 dBm. The device presented in this article has excellent potential for millimeter-wave applications where high linearity is essential.

Keywords— AlGaN/GaN HEMT, linearity, 1dB compression point

I. INTRODUCTION

High electron mobility transistors (HEMTs) based on gallium nitride (GaN) are considered the most promising components in power amplification and telecommunication due to their excellent material properties [1]. However, the conventional AlGaN/GaN HEMTs have unwanted nonlinearities and a bell-shaped transconductance (G_m) curve due to several physical phenomena [2]. Several techniques have been utilized to figure out the G_m non-linearity [3].

A novel concept of using threshold coupling improves the device's linearity. It is possible to synthesize such devices with a dual threshold voltage (V_{th}). In our earlier work, a device with a highly linear transconductance with a plateau region and gate voltage swing (GVS) of ~6 V was achieved by utilizing the double-V_{th} coupling (DVC) [3]. However, the off-state current induced by etching severely harms the reliability of the device and deteriorates power-added efficiency (PAE) [4]. It has been found that the N_2O plasma treatment can effectively reduce the leakage current [5].

This letter demonstrates an AlGaN/GaN HEMT fabricated by synthesizing the recess and planar devices along the gate width. Moreover, N_2O plasma treatment was incorporated to form an oxide layer at the gate electrode of the metal-insulator-semiconductor HEMT (MISHEMT) with DVC. The proposed DVC-MISHEMT exhibits a GVS-G_m >7 V and a fixed f_T/f_{max} of 40 /60 GHz over a wide gate voltage range. For the 30GHz operating frequency, a maximum power-added efficiency (PAE) of 41% was observed along with the power density (P_{out}) of 5.3 W/mm at V_{ds} = 20V and the associated 1dB compression point (P_{1dB}) of 28 dBm. Thus, the DVC-MISHEMT is an attractive alternative for millimeter-wave applications requiring high linearity and efficiency.

II. DEVICE STRUCTURE AND FABRICATION

Fig. 1(a) shows the schematic diagram of the respective device. The epilayers include a 1.3 μm GaN buffer layer, a 1 nm AlN interlayer, and a 24 nm AlGaN barrier. The 2-dimensional electron gas mobility and density after measurement are 2096 $cm_2 \cdot V^{-1} \cdot s^{-1}$ and $1.04 \times 10^{13} cm^{-2}$, respectively. Fabrication of the device starts with the source and formation of the ohmic contact using the Ti/Al/Ni/Au evaporation process. An ohmic contact resistance (R_c) of 0.36 $\Omega \cdot mm$ was achieved. A 120 nm SiN layer was deposited by plasma-enhanced chemical vapor deposition (PECVD). An electron beam lithography (EBL) was used to define the gate legs. The detailed fabrication process of DVC-HEMT is described in our previous work [3].

The SEM image of the DVC structure is shown in Fig. 1 (b). After removal of SiN from the gate area, an EBL was employed to define the recess element. Afterwards, the gate area was treated with nitrous oxide (N_2O) plasma in the plasma-enhanced chemical vapor deposition (PECVD) system to form an oxide layer, as shown in Fig.1 (c). Finally, Ni/Au metal stacks deposited by beam evaporation was designed to form gate cap. All devices have the same gate length (L_g) of 200nm and a source-drain distance (L_{sd}) of 3 μm.

III. RESULTS AND DISCUSSION

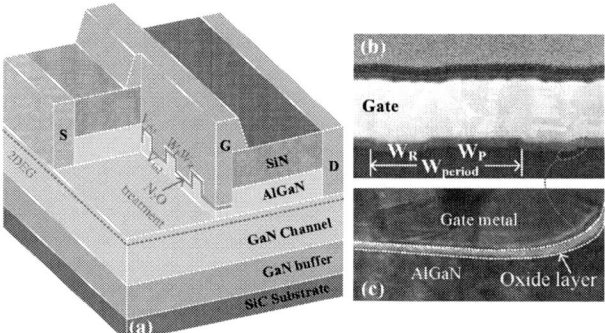

Fig. 1. (a) The structural diagram of DVC-MISHEMT. (b) The SEM image of the DVC structure. (c) The TEM image of the oxide layer formed by N_2O treatment.

A. DC Characterization

As shown in Figs. 2 (a) and (b), the DVC-MISHEMT has a dramatically wider G_m range with GVS-G_m of ~7 V. The DVC-HEMT, on the other hand, has a GVS-G_m of 5.7 V. Compared to the DVC-HEMT, the DVC-MISHEMT possesses a lower G_m' of ~0.3 S/mm·V, and G_m'' of ~0.6 S/mm·V². Fig. 2(c) compares Schottky characteristics of DVC-HEMT and DVC-MISHEMT. It can be found that the DVC-MISHEMT shows a gate leakage current that is 4 orders of magnitude lower than the gate leakage current of the DVC-HEMT. In short, due to the insertion of a thin oxide gate insulator between the gate and the channel, the DVC-MISHEMT exhibits a flatter G_m curve and lower gate leakage current. As a result, lower levels of RF distortions are observed, and an enhancement in P_{out} and PAE of AlGaN/GaN HEMT can be seen at high power levels.

B. Small-signal Characterization

As shown in Fig. 2 (d), the maximum current-gain cut-off frequency (f_T) and the maximum power gain cut-off frequency (f_{max}) of the DVC-MISHEMT are 40 and 60GHz, respectively. The f_T and f_{max} have an associated GVS-f_T/f_{max} of ~5 V, which is broader than the conventional device discussed in Ref. [3]. The excellent GVS-f_T/f_{max} of DVC-MISHEMT indicates a

978-1-6654-9270-6/22 $31.00 © 2022 IEEE

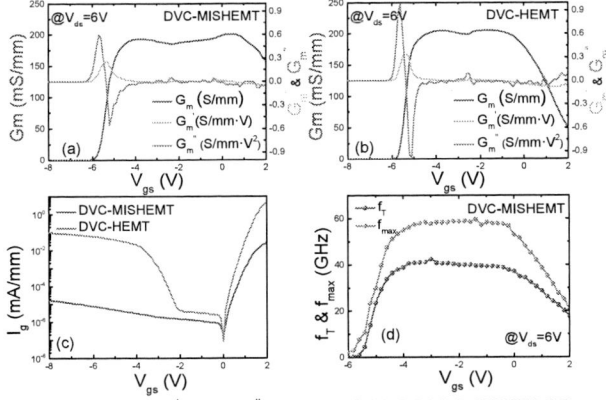

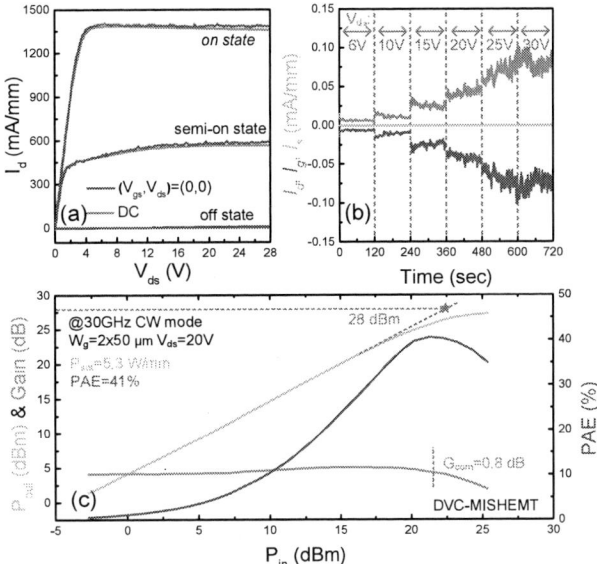

Fig. 2. (a) G_m, G_m' and G_m'' *vs.* V_{gs} of (a) DVC-MISHEMT, and (b) DVC-HEMT. (c) Comparison of Schottky characteristics of DVC- HEMT, and DVC-MISHEMT. (d) f_T/f_{max} with respect to gate voltage for DVC-MISHEMT.

better linear operation. The respective load line covers a broad V_{gs} range, and the linearity is improved by suppressing the gain compression (G_{com}) factor. Thus, the relatively flat cut-off frequency curve of the DVC-MISHEMT reflects the enhanced RF linearity of the device [6,7].

C. Robustness Assessment

In order to evaluate how robust the device is, the I-V characteristics and stress measurements in the pulsed mode were carried out, as shown in Figs. 3 (a) and (b). Slight fluctuations in the current can be observed in Fig. 3 (a). Fig. 3 (b) shows the applied negative gate stress of V_{gs}= -6 V and the drain stresses of V_{ds}= 6, 10, 15, 20, 25, and 30 V biased for 120 sec. The maximum leakage current is below 0.1 mA/mm. Thus, by ignoring the self-heating, the fabricated high linearity device is suitable for large-signal measurements due to the outstanding stress performance.

D. Large-signal performance

The power measurement of HEMT at 30 GHz was performed in continuous wave (CW) mode to check the device's power characteristics. Fig. 3 (c) illustrates the P_{out}, the power gain, and the PAE as a function of the input power (P_{in}) of the DVC-MISHEMT. The device was characterized at deep-class AB operation at V_{ds} = 20 V. Because of the excellent characteristics, minimum gate current, and negligible current fluctuations, a P_{out} of 5.3 W/mm and a maximum PAE of 41% are notably observed. The device exhibits a flattened gain curve with the gain compression (G_{com}) as low as 0.8 dB. Due to the suppression of G_{com}, a P_{1dB} of 28 dBm is obtained by the DVC-MISHEMT, having RF linearity with slight distortions. The improvement in RF linearity is mainly due to the flatter Gm profile. Hence, the proposed architecture has potential in highly linear millimeter-wave applications.

IV. CONCLUSION

This letter presents AlGaN/GaN HEMT with a highly linear operation obtained by integrating the recess and planar elements parallel to the gate width with N_2O plasma treatment on the gate region. The fabricated device achieved a steady G_m profile with a GVS of 7 V and a constant f_T/f_{max} of 40/60. Furthermore, AlGaN/GaN HEMT features exceptional linearity with P_{1dB} of 28 dBm at a frequency of 30 GHz and V_{ds} = 20 V. The

proposed architecture has excellent potential in millimeter-wave applications requiring good linearity and high efficiency.

ACKNOWLEDGMENT

This work was supported by the National Key R&D Program of China under Grant No. 2020YFB1804902; in part by the National Natural Science Foundation of China under Grant Nos. 61904135, 62090014, 11690042, and 62188102;in part by the Key R&D Program of Guangzhou under Grant No. 202103020002; in part by Wuhu and Xidian University special fund for Industry-University-Research Cooperation under Grant No. XWYCXY-012021014-HT; in part by the Fundamental Research Funds for the Central Universities under Grant Nos.QTZX22022 and XJS221110; and in part by the Innovation Fund of Xidian University under Grant No. YJS2213

Fig. 3. (a) Comparison of DC and pulsed output characteristics of DVC-MISHEMT. (b) The off-state leakage currents (I_d, Ig, I_s) of DVC-MISHEMT under different stress conditions versus stress time. (c) The load-pull power characteristics at 30 GHz with the biased voltage V_{ds} of 20 V.

REFERENCES

[1] P. Choi, U. Radhakrishna, and D. Antoniadis, "Linearity Enhancement of a Fully Integrated 6-GHz GaN Power Amplifier," *IEEE Microw. Wireless Compon. Lett.*, vol. 27, pp. 927-929, Oct. 2017.

[2] K. R. Bagnall *et al.*, "Experimental Characterization of the Thermal Time Constants of GaN HEMTs Via Micro-Raman Thermometry," *IEEE Trans. Electron Devices*, vol. 64, pp. 2121-2128, May 2017.

[3] P. Wang *et al.*, "Influence of Fin-Like Configuration Parameters on the Linearity of AlGaN/GaN HEMTs," *IEEE Trans. Electron Devices*, vol. 68, pp. 1563-1569, April 2021.

[4] X. Hu et al., "Si₃N₄/AlGaN/GaN-metal-insulator-semiconductor heterostructure field-effect transistors," *Appl. Phys. Lett.*, vol. 79, pp. 2832, Oct. 2001.

[5] M. Mi et al., "Millimeter-wave power AlGaN/GaN HEMT using surface plasma treatment of access region," *IEEE Trans. Electron Devices*, vol. 64, pp. 4875–4881, Dec. 2017.

[6] M. Nagahara *et al.*, "Improved intermodulation distortion profile of AlGaN/GaN HEMT at high drain bias voltage" in *IEDM Tech. Dig.*, pp. 693-696, 2002.

[7] W. Song, Z. Zheng, T. Chen, J. Wei, L. Yuan and K. J. Chen, "RF Linearity Enhancement of GaN-on-Si HEMTs With a Closely Coupled Double-Channel Structure," *IEEE Electron Device Lett.*, vol. 42,pp. 1116-1119, Aug. 2021.

Photoresponses and Memory Effects in Optoelectronic Synaptic Devices Based on CdSe Quantum Dots and Poly(3-hexylthiophene)

Zhicheng Li, Zhulu Song, Zhaojin Wang, Jiayun Sun, Kai Wang

Department of Electronic and Electrical Engineering, Southern University of Science and Technology, Shenzhen, China

Abstract—The optical responses and memory effects of photoelectric synaptic devices based on CdSe quantum dots (QDs) and poly(3-hexylthiophene) (P3HT) are studied in this work. Compared with devices only incorporating CdSe QDs, the devices based on CdSe QDs and P3HT exhibit higher photocurrents because the heterojunction formed by CdSe QDs and P3HT enhances the separation of photogenerated excitons, and the loss of excitons in the QDs reduces. In addition, due to the effect of the surface defect trapping charge of CdSe QDs, the photocurrent of the device can still be maintained for more than 100 seconds under the condition of zero gate voltage. Finally, the device can perform each synaptic activity with a low power consumption of 12.9 pJ by adjusting the concentration of QDs.

Keywords—optoelectronic synaptic device, poly(3-hexylthiophene), quantum dots, heterojunction

I. INTRODUCTION

Inspired by the biological nervous system, neuromorphic computing is expected to be applied to solve the problems of insufficient computing power and low energy efficiency of traditional computers based on von Neumann architecture [1]. As an important stage in the development of neuromorphic computing, the design and fabrication of artificial synaptic devices have attracted much attention. In recent years, devices including phase-change memory, memristor and field effect transistor have been explored as electronic synaptic devices [2-4]. Because it is difficult for electronic synaptic devices to take into account several factors such as bandwidth, connection and density, optoelectronic synaptic devices that can simultaneously process electrical and optical signals become a new research hotspot benefiting for their characteristics of high bandwidth, low crosstalk and low power. Among the many materials used to prepare photoelectric synaptic devices, semiconductor colloidal quantum dots (QDs) have attracted more attentions for their excellent electronic / optical properties and structural stability. At the same time, they can also meet the needs of low-cost, large-area, solution-based manufacturing processes, and have the potential for practical production and application [5]. Nowadays, some reported synaptic devices are used for handwritten digit recognition and face recognition [6-7], which can show the characteristics of low energy consumption and high accuracy. However, the memory ability of synaptic devices is often ignored, which determines the application potential of synaptic devices, and is also an important feature that distinguishes synaptic devices from photodetectors.

In this paper, we report an optoelectronic synaptic device based on CdSe QDs and poly(3-hexylthiophene) (P3HT), which exhibit low energy consumption and long retention time. The Fig. 1 (a) shows the equivalent model diagram of optoelectronic synaptic device in which the optical signal represents the input signal of the presynaptic membrane, and the electrical signal represents the output signal of the postsynaptic membrane. The

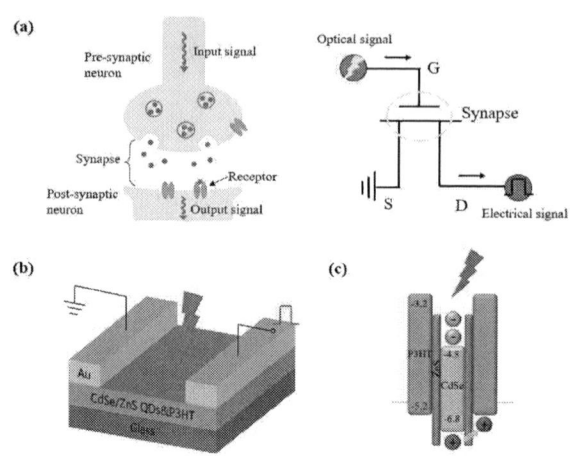

Fig. 1. (a) The equivalent model diagram of optoelectronic synaptic device; (b) The model diagram of device structure; (c) The energy band structure diagram of the device.

Fig. 1 (b) shows the model diagram of device structure. The Fig. 1 (c) shows the energy band structure diagram of the device. Nowadays some nanomaterials such as silicon nanocrystals have been tried to be used in optoelectronic synaptic devices [6]. However, there are few examples of CdSe QDs and P3HT mixtures applied to optoelectronic synaptic devices. Due to the effect of heterojunction formed by CdSe QDs and P3HT separating excitons and the surface defect which can trap charges in CdSe QDs, the device mentioned above exhibits good photocurrent response and memory ability.

II. FABRICATION AND CHARACTERIZATION

A. Fabrication of Synaptic Transistors

CdSe QDs and P3HT with weight ratio of 1:5 for QD to P3HT were prepared in a brown bottle, and then an appropriate volume of toluene was added to the bottle to prepare the hybrid solutions containing CdSe QDs (1 mg mL^{-1} in toluene) and P3HT (5 mg mL^{-1} in toluene). The hybrid solutions were pretreated with ultrasonication for 15 min. After stirring for 12 h, the hybrid solutions were spin-coated onto cleaned glass with a rate of 1500 rpm for 45 s under nitrogen. After the spin coating was completed, the device was placed on the hot plate and annealed for 15 min. Subsequently, the source and drain electrodes were thermally deposited with 100 nm-thick gold through a regular shadow mask with channel length (L) and width (W) of 100 μm and 2000 μm in the vacuum evaporation equipment.

B. Absorption and Steady-State Fluorescence Spectra

Fig. 2 (a) shows the absorption spectra of CdSe QDs and P3HT respectively. As can be seen from Fig. 2 (a), when lasers at the wavelength of 656.5nm are used, P3HT hardly absorbs light. It is also found in our experiments that P3HT-only devices have a very small photocurrent response. We assume that CdSe QDs are the main contributors of photogenerated carriers in the test.

Fig. 2 (b) shows the steady-state fluorescence spectra of CdSe QDs, P3HT and their mixtures excited by a 450 nm light source. It can be seen from Fig. 2 (b) that the composite films of CdSe QDs and P3HT exhibit fluorescence peaks at 666 nm and 716 nm. Referring to the fluorescence spectrum of CdSe QDs, we can know that the peak at 666 nm is from the influence of the radiative recombination of excitons in CdSe QDs. And we

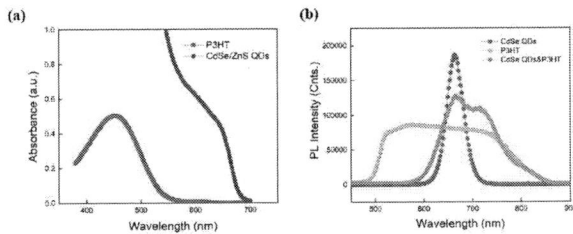

Fig. 2. (a) The absorption spectra of CdSe QDs and P3HT respectively; (b) The steady-state fluorescence spectra of CdSe QDs, P3HT and their mixtures.

TABLE I. FITTING SUMMARY AND COMPARISON

Materials (QDs:P3HT)	τ_1 (s)	τ_2 (s)	$a_1 = A_1/(A_1+A_2) \times 100\%$	$a_2 = A_2/(A_1+A_2) \times 100\%$	$\tau_{average}$ (s)
1:5	9.8	64.6	34.8%	65.2%	45.5
3:5	4.6	49.2	33.9%	66.1%	34.1
5:5	5.2	44.5	33.5%	66.5%	31.3

speculate that the 716 nm peak originates from the process of radiative recombination of P3HT which is enhanced by the charge transfer between CdSe QDs and P3HT. By analyzing the steady-state fluorescence spectra, we understand the possible recombination pathways of carriers after illumination better.

C. Optical Responses and Memory Effects

Fig. 3 (a) shows the time-resolved photoluminescence (TRPL) spectra of CdSe QDs and the mixture of CdSe QDs and P3HT. From Fig. 3 (a), we can see that compared with CdSe QDs, the fluorescence lifetime of the composite films is significantly decreased. This indicates that the photogenerated excitons in the QDs rapidly separate due to the heterojunction between CdSe QDs and P3HT after illumination. Fig. 3 (b) shows the path of carrier movement in CdSe QDs and P3HT hybrid films after illumination.

Specifically, the generated holes can be transferred to P3HT due to the mismatching between the highest occupied molecular orbital (HOMO) of P3HT and the conduction band (CB) of CdSe QDs, while the electrons will be trapped in CdSe QDs.

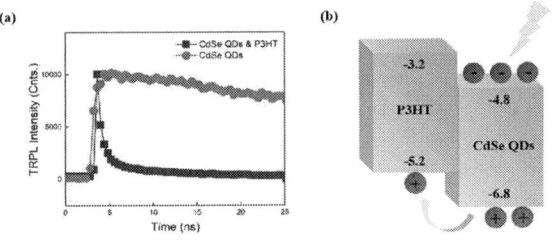

Fig. 3. (a) The TRPL spectra of CdSe QDs and the mixtures of CdSe QDs and P3HT; (b) The diagram of carrier movement after illumination.

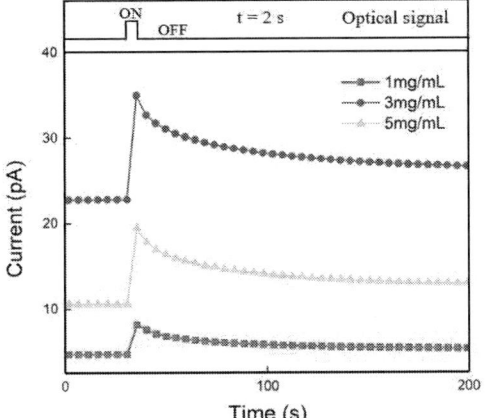

Fig. 4. The optical response curves of composite films with different QDs concentrations.

Some electrons are also trapped by defects on the surface of CdSe QDs simultaneously.

Fig. 4 shows the optical response curves of composite films with different QDs concentrations. For optoelectronic synaptic devices, the increased drain current due to optical signal input represents the response of the postsynaptic membrane, which is called excitatory postsynaptic current (EPSC). When CdSe QDs are incorporated into P3HT, the defect energy and defect density in P3HT decrease, and the existence of QDs also increases the scattering probability of carriers [8]. The two factors lead to the difference of dark current in different QDs concentrations. It can be seen from Fig. 4 that the device prepared from 1 mg mL^{-1} CdSe QDs and 5 mg mL^{-1} P3HT has the minimum dark current, and the energy consumption can be calculated as in:

$$\text{Energy consumption} = \int_{t_0}^{t_1} V \cdot I(t) dt \qquad (1)$$

where t_0 and t_1 is the time when the photocurrent starts to increase and reaches the peak respectively, and V is the constant drain voltage of the transistor whose value is 1 V. The lowest energy consumption in this work is about 12.9 pJ. When the size of our devices is reduced to the nanometer scale, the energy consumption can approach the level of biological synapses (~10 fJ per synaptic event) [9].

The photocurrent attenuation curve can be fitted by biexponential decay curves as in:

$$I = I_0 + A_1 e^{-t/\tau_1} + A_2 e^{-t/\tau_2} \qquad (2)$$

where I_0 is the dark current, A_1 and A_2 are the corresponding weight factors, τ_1 is the fast decay component that is related to the radiative recombination of free carriers, τ_2 is the slow decay component that is attributed to the process of slow release of charge trapped by defects and the potential nonradiative recombination [10]. The fitting results are shown in Table I where $\tau_{average}$ is calculated as in:

$$\tau_{average} = a_1 \tau_1 + a_2 \tau_2 \qquad (3)$$

With the increase of QDs concentrations, the decrease of $\tau_{average}$ may mean that the proportion of nonradiative recombination assisted by surface defects in QDs increases.

III. CONCLUSIONS

This work states the mechanism of optical responses and memory effects of optoelectronic synaptic devices based on CdSe QDs and P3HT. Meanwhile, by adjusting the concentration of QDs, we prove that the device can perform a synaptic activity with a low energy consumption of 12.9 pJ.

REFERENCES

[1] Furber, Steve. "Large-scale neuromorphic computing systems." Journal of neural engineering 13.5 (2016): 051001.

[2] Tuma, Tomas, et al. "Stochastic phase-change neurons." Nature Nanotechnology 11.8 (2016): 693-699.

[3] Yin, Jun, et al. "Adaptive crystallite kinetics in homogenous bilayer oxide memristor for emulating diverse synaptic plasticity." Advanced Functional Materials 28.19 (2018): 1706927.

[4] Tian, He, et al. "Anisotropic black phosphorus synaptic device for neuromorphic applications." Advanced Materials 28.25 (2016): 4991-4997.

[5] Lv, Ziyu, et al. "Semiconductor quantum dots for memories and neuromorphic computing systems." Chemical Reviews 120.9 (2020): 3941-4006.

[6] Yin, Lei, et al. "Synaptic silicon-nanocrystal phototransistors for neuromorphic computing." Nano Energy 63 (2019): 103859.

[7] Wang, Yue, et al. "Dual - Modal Optoelectronic Synaptic Devices with Versatile Synaptic Plasticity." Advanced Functional Materials 32.1 (2022): 2107973.

[8] Kumari, Kusum, et al. "Effect of CdSe quantum dots on hole transport in poly (3-hexylthiophene) thin films." Applied Physics Letters 92.26 (2008): 263504.

[9] Kuzum, Duygu, Shimeng Yu, and HS Philip Wong. "Synaptic electronics: materials, devices and applications." Nanotechnology 24.38 (2013): 382001.

[10] Dong, Yifan, et al. "Solution processed hybrid polymer: HgTe quantum dot phototransistor with high sensitivity and fast infrared response up to 2400 nm at room temperature." Advanced Science 7.12 (2020): 2000068.

A Current Domain Computing-in-Memory SRAM Macro with Hybrid IAF-SAR ADC for Signal Margin Enhancement

Tianqi Xu, Shumeng Li, Fukun Su, Xian Tang

Shenzhen International Graduate School, Tsinghua University, Shenzhen 518055, China

Abstract—This paper presents a computing-in-memory (CIM) SRAM macro utilizing current domain computing for multiply-and-accumulate (MAC) operations. A 32x64 8T bitcell array is used to store the weight data. This design adopts the modulated word line pulse-width method to convert 4-bit digital input data to analog domain. The MAC operation of weight and input data is accomplished through bitwise multiplication and the result is transformed to current which accumulates on the reading bit line (RBL). In order to improve the signal margin without sacrificing the readout precision, a hybrid integrate-and-fire (IAF)-SAR ADC is proposed to convert the computing result into digital domain. The presented design is implemented in a standard 65nm CMOS process and simulation results indicate the proposed 32x64 CIM macro achieves energy efficiency of 18.76 TOPS/W and peak throughput of 10.24 GOPS with 4-bit inputs and 4-bit weights.

Keywords—Computing in Memory, Analog Computing, SRAM

I. INTRODUCTION

The conventional Von Neumann architecture suffers from the frequent and heavy data movement between memory and computing units, limiting the computation throughput and efficiency. This issue is even more critical in the Convolution Neural Network (CNN) inference in the edge computing scene, which is computation intensive and energy limited. Computing-in-memory is recognized as a promising solution to relieve the data moving burden. To further improve the computing efficiency, recent papers [1-4] incorporate the analog computing into the CIM architecture featuring a high computing efficiency at a cost of little inference accuracy loss.

From the aspect of computing domain, analog computing can be classified into current domain [2], charge domain [3] and time domain [4]. In the analog CIM design, the signal margin of readout data and the readout precision is in an inverse proportion when the full voltage range ($V_{max} - V_{min}$) is a constant, as is illustrated in (1).

$$Signal\ Margin = \frac{V_{max} - V_{min}}{2^N} \qquad (1)$$

Here the signal margin refers to the minimum voltage difference between two consecutive MAC values (MACVs) [5] and N equals the bits-number of the ADC. A higher readout

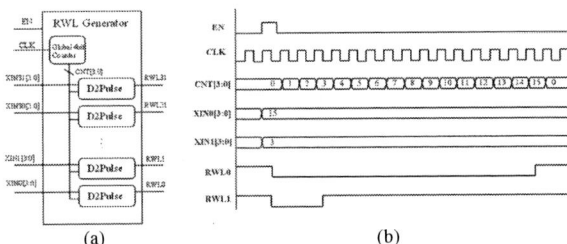

Fig. 2. (a) Architecture of the RWL generator, (b) waveform of the RWL generator

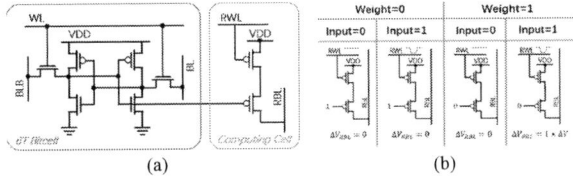

Fig. 3. (a) Proposed 8T bitcell, (b) bitwise multiplication concept

precision comes with a smaller signal margin, which gives a stringent restriction on the input offset and noise of ADC. To overcome the problem above, we propose a hybrid IAF-SAR ADC composed of IAF ADC's coarse-conversion and SAR ADC's fine-conversion in the analog CIM to expand the equivalent full voltage range, which can improve the signal margin without sacrificing the readout precision.

This paper presents the analog CIM macro designed in 65nm technology, and is organized as follows. Section II illustrates the macro design architecture and implementations. In section III, the simulation results of the proposed design are presented with further discussion. At last, a conclusion is drawn in Section IV.

II. CIM MACRO DESIGN DETAILS

The high-level block diagram of the CIM macro is shown in Fig. 1. The principal part of the macro is a 32x64 8T Array. The basic read and write operation of SRAM is the same as the classical method through read/write interface and corresponding control module. The computing peripherals include a reading word line (RWL) generator module for the conversion of digital input to a width modulated pulse signal, the hybrid IAF-SAR ADC for the conversion of computing results to digital domain and finally digital bitwise accumulation module for the weighted summation of the bitwise MAC results.

A. RWL generator

The architecture of the RWL generator is shown in Fig. 2(a). A group of 32 4-bit inputs is sent to the RWL generator and every 4-bit digital input is converted to a pulse signal with pulse-width proportional to the data value. For a low energy cost, a global 4-bit counter is adopted in the design to offer the count number for every Digital-to-pulse (D2Pulse) block. The conversion of input data is accomplished by corresponding D2Pulse block, which can be synthesized through EDA tools. The waveform of the RWL Generator is presented in Fig. 2(b). It is apparent that the unit pulse width time is equal to the CLK period.

B. Bitwise multiplication and current summation

The proposed 8T cell is presented in Fig. 3(a). It is composed of a standard 6T SRAM cell for fundamental data storing and a 2T computing cell for bitwise multiplication. As is described in Fig. 3(b), every PMOS gate voltage has two optional values, which is high for the drain to source disconnecting and low for drain to source connecting. Besides, the cascode structure of the computing cell keeps a highly linear current, which improves the

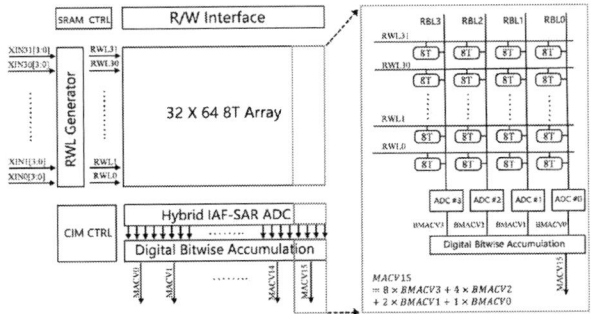

Fig. 1. Block diagram of the CIM macro

978-1-6654-9270-6/22 $31.00 © 2022 IEEE

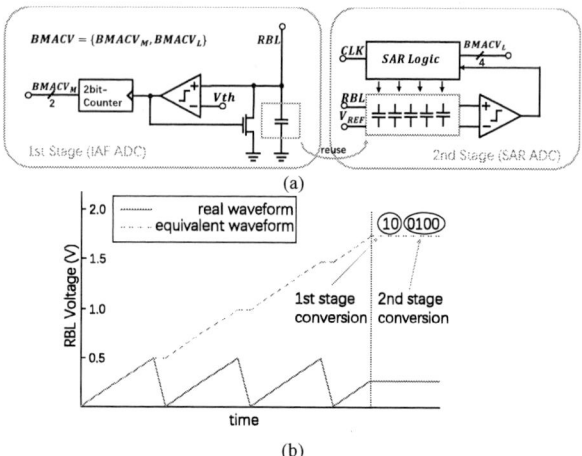

(a)

(b)

Fig. 4. (a) Architecture of the hybrid IAF-SAR ADC, (b) waveform of the hybrid IAF-SAR ADC

linearity of the computing results. The charge contributed by every 8T cell in the computing period is proportional to the multiplication of 1-bit weight stored in the 6T SRAM cell and the input 4-bit data, which finally appears as the voltage rise on the RBL.

C. Proposed hybrid IAF-SAR ADC

The architecture and waveform of proposed hybrid IAF-SAR ADC are presented in Fig. 4. We incorporate an IAF ADC [6] to execute the coarse conversion as the first stage, which is accomplished by integrating current on the RBL capacitor and resetting the capacitor when the capacitor voltage reaches the threshold. The reset frequency is counted by the counter and recognized as the first stage conversion result. After that, the residue voltage on the capacitor will be sent to the second stage and a 4-bit result is converted by the SAR ADC. The two results got from the two stages above are combined to get a final 6-bit bitwise-MAC value (BMACV). Besides, the capacitor in the first stage can be reused by the second conversion stage. As the introducing of the IAF ADC, the final signal margin of the SAR ADC is 4 times as large as the conventional design with only 4-bit SAR ADC.

D. Digital accumulation

To accumulate the BMACVs, we adopted the weighted summation, combining BMACVs of every 4 columns' BMACVs to get 9-bit results of 4-bit inputs and 4-bit weights' MAC values, as is illustrated in Fig. 1. This module can be synthesized through EDA tools.

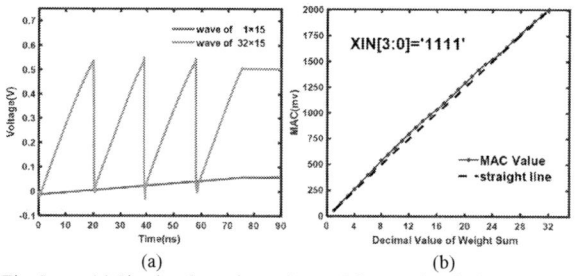

(a)   (b)

Fig. 5.   (a) Simulated transient voltage of the capacitor in CIM macro, (b) linearity of the computing results

TABLE I.   PERFORMANCE AND COMPARISON

	This work[1]	JSSC' 2020[2]	L-SSC' 2019[3]	ISSCC' 2019[4]
Technology	65nm	55nm	65nm	28nm
Computing domain	Current	Current	Charge	Time
Memory capacity	2Kb	3.75Kb	16Kb	156Kb
Input/weight/output precision(bit)	4/4/9	1~4/2~5/ 3~7	1/1/5	8/1/-
Signal margin x Readout precision(V)	2	-	0.6	-
Throughput[2](GOPS)	10.24	10.6~33.7	1638	-
Energy efficiency (TOPS/W)	18.76	18.37~ 72.03	671.5	12.8~ 119.7

[1] Simulation results

[2] Normalized to array size of 32x64

III. SIMULATION RESULT AND DISCUSSION

The proposed CIM macro has been designed and simulated in 65nm CMOS. Fig. 5 shows the simulated transient voltage and linearity of the CIM macro. It can be seen the speed of the comparator has little impact on the linearity of the computation results but can be further calibrated by some post processing. Performance and some comparisons are shown in Table I. The simulated results show a power consumption of 53.29 fJ/operation at supply voltage of 1V, corresponding to an energy efficiency of 18.76 TOPS/W. And the compute throughput is equal to 10.24 GOPS.

IV. CONCLUSION

We propose a current domain computing-in-memory SRAM macro with hybrid IAF-SAR ADC in this paper, featuring signal margin enhancement. The signal margin of the second stage SAR ADC in our design is 4 times as large as the conventional design with only one stage, which largely relaxes the input offset and noise restriction of the ADC design.

ACKNOWLEDGMENT

This work is supported by Guangdong Province Natural Science Foundation (No. 2022A1515011934) and Shenzhen Basic Research Foundation (No. WDZC20200819152116001 and JCYJ20200109143003935). (Corresponding author: Xian Tang.)

REFERENCES

[1] M. E. Sinangil et al., "A 7-nm Compute-in-Memory SRAM Macro Supporting Multi-Bit Input, Weight and Output and Achieving 351 TOPS/W and 372.4 GOPS," in IEEE Journal of Solid-State Circuits, vol. 56, no. 1, pp. 188-198, Jan. 2021.

[2] X. Si et al., "A Twin-8T SRAM Computation-in-Memory Unit-Macro for Multibit CNN-Based AI Edge Processors," in IEEE Journal of Solid-State Circuits, vol. 55, no. 1, pp. 189-202, Jan. 2020.

[3] Z. Jiang, S. Yin, J. -S. Seo and M. Seok, "C3SRAM: In-Memory-Computing SRAM Macro Based on Capacitive-Coupling Computing," in IEEE Solid-State Circuits Letters, vol. 2, no. 9, pp. 131-134, Sept. 2019.

[4] J. Yang et al., "24.4 Sandwich-RAM: An Energy-Efficient In-Memory BWN Architecture with Pulse-Width Modulation," 2019 IEEE International Solid- State Circuits Conference - (ISSCC), 2019, pp. 394-396.

[5] C. -J. Jhang, C. -X. Xue, J. -M. Hung, F. -C. Chang and M. -F. Chang, "Challenges and Trends of SRAM-Based Computing-In-Memory for AI Edge Devices," in IEEE Transactions on Circuits and Systems I: Regular Papers, vol. 68, no. 5, pp. 1773-1786, May 2021.

[6] C. Liu et al., "A spiking neuromorphic design with resistive crossbar," 2015 52nd ACM/EDAC/IEEE Design Automation Conference (DAC), 2015, pp. 1-6

A Low Frequency Drift LC-DCO with Wide Temperature Range

Jinrui Hu[1], Shuyue Fang[1], Haigang Feng[1], Xinpeng Xing[1], Han Wang[2], Lei Yang[2]

[1] Shenzhen International Graduate School, Tsinghua University, Shenzhen 518055, China
[2] Megahunt Technologies Inc., China

Abstract—A 2.4GHz LC-DCO with low frequency drift is presented to support frequency synthesizer under narrow band system like BLE. In the LC-DCO, a comprehensive temperature compensation scheme, which includes a Proportional To Absolute Temperature (PTAT) current bias and the varactor arrays varying linearly with voltage, is proposed to reduces the frequency drift of LC-DCO as a result of temperature fluctuations. By applying the circuit, frequency drift is reduced from 31MHz (without compensation) to 6MHz within the temperature from -40°C to 120°C. And the results show that no extra in-band noise is added to LC tank. It consumes 860uW from a 0.9V supply in 40nm CMOS process technology.

Keywords—Digital Control Oscillator, temperature compensation, PTAT current, compensation capacitance.

I. INTRODUCTION

Digital Control Oscillator (DCO) is sensitive to temperature. The frequency drift caused by temperature fluctuations could lead to big performance degradation in communication systems, particularly in narrow band system like BLE. Adding redundancy circuit into LC-tank is a conventional way to reduce the frequency drift[1] with extra noise penalty.

In this paper, a comprehensive compensation scheme to reduce the frequency drift is proposed after analyzing the effect of capacitance and resistance under the change of temperature and amplitude of the oscillator. The compensation scheme consists of the PTAT current bias plus the varactor arrays as secondary compensation topology. By applying the circuit, frequency drift is reduced from 31MHz (without compensation) to 6MHz within the temperature from -40°C to 120°C.

II. CIRCUIT ANALYSIS AND DESIGN

A. Theoretical Analysis of Frequency Drift

Fig.1 shows the schematic of a LC-DCO with static bias voltage, Metal-Oxide-Metal (MOM) capacitors tank, switch arrays, PMOS cross-coupled devices and the inductor which the center tap is connected to ground. The amplitude of the DCO can be expressed as

$$A_{sig} = \frac{2}{\pi} I_{bias} R_t \qquad (1)$$

Where I_{bias} is the tail current and R_t is the equivalent parallel resistance of the resonant tank. The value of R_t at high frequency is given by (2), where T is temperature, ω is angular frequency, and L is inductance of the resonant tank [2]. The amplitude decrease of LC-DCO leads to the degradation of phase noise of DCO at a high temperature.

$$R_t = \frac{\omega}{l} (\omega L)^2 \sqrt{\frac{2}{\mu \omega \rho_0 T}} \qquad (2)$$

As shown in Fig. 1(b), M2 is in saturation region when $-V_{TH} < V_P - V_N (V_{PN}) < 0$; M2 is in triode region when $V_{PN} < -V_{TH}$. C_{EQ1} and C_{EQ2} are given by (3) and (4) respectively.

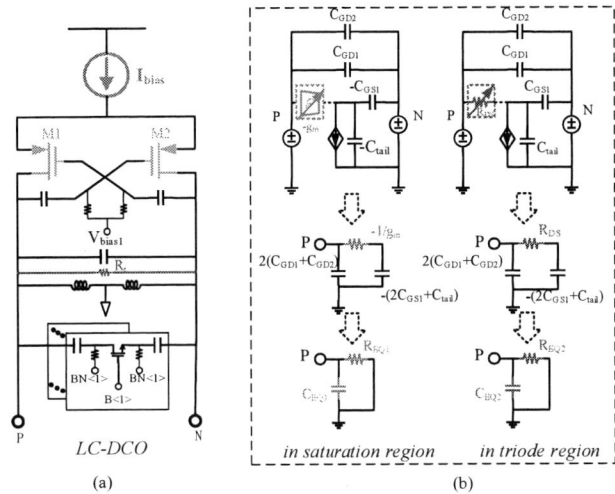

Fig. 1. (a) Schematic of LC-DCO   (b) Small-signal model of M2

$$C_{EQ1} = \frac{-(2C_{GS1} + C_{tail})}{1 + \omega^2 \left(1/g_m\right)^2 (2C_{GS1} + C_{tail})^2} + 2(C_{GD1} + C_{GD2}) \qquad (3)$$

$$C_{EQ2} = \frac{(2C_{GS1} + C_{tail})}{1 + \omega^2 (R_{DS})^2 (2C_{GS1} + C_{tail})^2} + 2(C_{GD1} + C_{GD2}) \qquad (4)$$

$$f = \frac{\omega}{2\pi} = \frac{1}{2\pi\sqrt{LC}} \qquad (5)$$

With the raising of temperature, the value of C_{EQ1} will increase and C_{EQ2} will change inversely. The reason is that $1/g_m$ and R_{DS} are proportional to temperature. C_{EQ1} is therefore proportional to temperature when the amplitude of the differential nodes (V_{PN}) is smaller than V_{TH} ; when V_{PN} is larger than V_{TH}, C_{EQ2} is inversely proportional to temperature. At the same time, the process also has an effect on the capacitance by affecting the value of V_{TH}. With the decrease of V_{TH}, M2 will have longer time in the triode region in a cycle. The equivalent parasitic capacitance of M2 is C_{EQ2} instead of C_{EQ1}.

According to the analysis above, a comprehensive compensation scheme is proposed. By using the property of PTAT, the loss of amplitude with Rt reducing is compensated. At the same time, the varactor arrays varying linearly with voltage is applied to compensate the capacitance.

B. Circuit Design

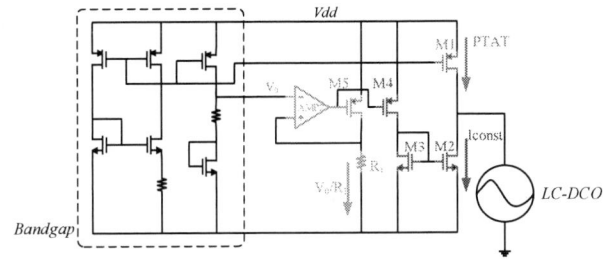

Fig. 2. Optimized PTAT current bias

Fig. 2 shows the schematic of PTAT current bias with adjustable temperature coefficient. The first part generates a bandgap voltage reference V_0. The constant current in the second part is generated by AMP_2, M_5 and R_1. Next, the current mirror made of M_4, M_3 and M_2 mirrors the current from M_5 to M_2. The current that eventually flows into the LC-DCO is the delta current between PTAT current of M_1 and constant current of M_2.

978-1-6654-9270-6/22 $31.00 © 2022 IEEE

The circuit can obtain the PTAT current bias with desired proportional temperature coefficient.

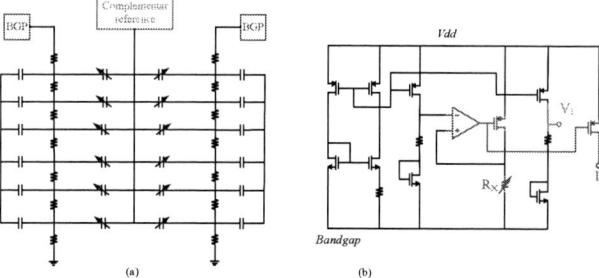

Fig. 3.   (a) Schematic of compensation capacitance   (b) Reference circuit

A schematic of compensation capacitance is shown in Fig. 3(a). Current bias provides the reference voltage of the varactor G terminal by resistor arrays and complementary reference is added to the S terminal of the varactor. The capacitance varying linearly with V_S can be obtained by superimposing varactors with various voltages at the G terminal. Fig. 3(b) shows a reference circuit, where V_1 and I_1 represent complementary reference and current bias respectively.

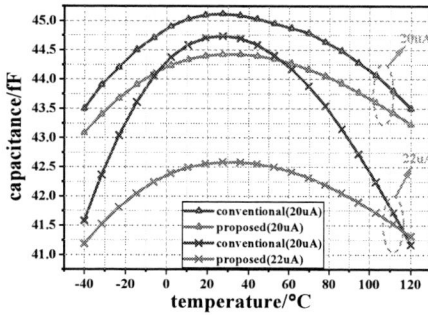

Fig. 4.   Compensation capacitance

As shown in Fig. 4, the desired compensation capacitance can be obtained by the aforementioned circuit. The proposed compensation capacitance still satisfies the desired compensation curves (red curves) when the value of reference current varies from 20uA to 22uA. A stable capacitance curve is an effective way to prevent compensation failure caused by process sensitivity.

III. SIMULATION RESULTS

The proposed comprehensive compensation scheme with a very small value of frequency drift is designed in 40nm CMOS process. Fig. 5 shows frequency curves under different corners. By adjusting the value of current bias and PTAT current, the frequency drift of DCO can reach extremely small under the corner of ff, tt and ss. The value of temperature coefficient decreases from 79 ppm/℃ without compensation to 15 ppm/℃ under corner of tt.

The design only deteriorates the noise at low frequency offset as depicted in Fig. 6 and the low frequency noise of the DCO is suppressed by the loop bandwidth in a PLL system. For the power consumption, the LC-DCO reaches 860uW. The phase noise of LC-DCO reaches -115dBC/Hz at offset 1MHz of 2.4GHz.

Table I compares the results for several temperature compensation schemes of previous studies. The proposed scheme reaches extremely small frequency drift with wide temperature range.

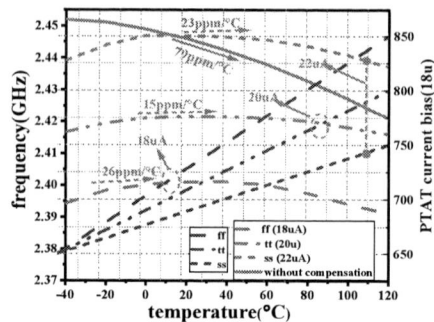

Fig. 5.   Temperature compensation under different corners

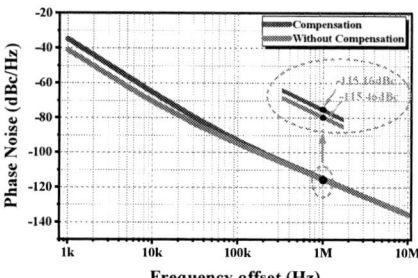

Fig. 6.   Phase noise comparison

TABLE I.   PERFORMANCE COMPARATION

	RFIC 2014 [3]	ISCAS 2018 [4]	This work
Process	130nm	130nm	40nm
P_{DC} (mW)	50.88	17.7	0.86
Freq (GHz)	6	13.1	2.4
PN*(dBc/Hz)	-108@1MHz	-120.6@1MHz	-115@1MHz
Frequency drift (Hz)	18M	7.6M	6M
Temperature Range	[-40℃,85℃]	[-40℃,125℃]	[-40℃,120℃]

IV. CONCLUSION

A LC-DCO with temperature compensation technique is presented. A PTAT current bias with adjustable temperature coefficient and the varactor arrays varying linearly with voltage are proposed and adopted. The design reduces frequency drift down to 6 MHz over a temperature range of - 40°C to 120°C. The overall power dissipation is 860uW from a 0.9V supply in 40nm CMOS process technology.

ACKNOWLEDGMENT

This work was supported by Guangdong Key Area Research and Development Plan under Grant 2019B010143003.

REFERENCES

[1] H. Akima, A. Dec, T. Merkin and K. Suyama, "A 10 GHz frequency-drift temperature compensated LC VCO with fast-settling low-noise voltage regulator in 0.13 μm CMOS," IEEE CICC 2010, pp. 1-4.

[2] P. C. Maulik and P. W. Lai, "Frequency Tuning of Wide Temperature Range CMOS LC VCOs," in IEEE Journal of Solid-State Circuits, vol. 46, no. 9, pp. 2033-2040, Sept. 2011.

[3] Y. You, D. Huang, J. Chen and S. Chakraborty, "A 12GHz 67% tuning range 0.37pS RJrms PLL with LC-VCO temperature compensation scheme in 0.13μm CMOS," 2014 IEEE Radio Frequency Integrated Circuits Symposium, 2014, pp. 101-10.

[4] P. Mirajkar, J. Chand, S. Aniruddhan and S. Theertham, "Low Phase Noise Ku-Band VCO with Reduced Frequency Drift Across Temperature," 2018 IEEE International Symposium on Circuits and Systems (ISCAS), 2018, pp. 1-5.

A High PSR and Fast Transient Response Output-Capacitorless LDO using Gm-Boosting and Capacitive Bulk-Driven Feed-Forward Technique in 22nm CMOS

Heng Liu, Dongxu Li, Xian Tang

Shenzhen International Graduate School, Tsinghua University, Shenzhen 518055, China

Abstract—This paper presents an output-capacitorless low-dropout regulator (OCL-LDO) using capacitive bulk-driven feed-forward (CBDFF) technique and an adaptive-biasing error amplifier with gm-boosting to enhance the power supply rejection (PSR) and the transient response. The proposed OCL-LDO has been implemented in a 22nm CMOS technology. It consumes a quiescent current of 49 μA from a power supply of 1.05-1.25 V and has a dropout voltage of 200 mV. The OCL-LDO achieves -84 dB PSR at low frequency and -69 dB PSR at 1 MHz for the load current of 20 mA. It achieves a line regulation of 0.18 mV/V, a load regulation of 0.77 μV/mA, and a settling time of 135 ns.

Keywords—power supply rejection, capacitorless, gm-boosting, adaptive-biasing, bulk-driven feed-forward, low-dropout regulator

I. INTRODUCTION

Usually, in a system-on-chips (SoCs) solution, DC-DC converters are always followed by low-dropout regulators (LDOs) to eliminate power supply noise for noise-sensitive circuits. Output-capacitorless low-dropout regulators (OCL-LDOs) are better architecture solutions for SoCs to enhance integration. However, due to the absence of the off-chip capacitor, the transient response and the power supply rejection (PSR) will degrade significantly in OCL-LDOs. Therefore, OCL-LDOs have many design challenges. To address those challenges, lots of work have been done. In [1], a multistage scheme is proposed to broaden the unity-gain bandwidth (UGB) of LDO above 100 MHz, which obtains ultrafast transient response and a relatively good PSR, but it consumes a quiescent current of 112 μA and the PSR is degraded to -31 dB at 5 MHz. In [2] and [3], the bulk-driven feed-forward (BDFF) technique is adopted to extend PSR bandwidth, but it is sensitive to PVT conditions in [3] and those are not suitable for applications with different power supply voltages.

In this paper, firstly, we propose a capacitive bulk-driven feed-forward (CBDFF) technique to enhance the BDFF circuit's robustness in different PVT conditions while improving the PSR at high frequency. Secondly, a low-power amplifier with high DC gain and ultra-wide bandwidth is applied to increase the low-frequency PSR and the transient response.

The paper is organized as follows. In section II, the proposed high PSR and fast transient response OCL-LDO is described in detail. In section III, simulation results are presented. In section IV, the conclusion is shown.

II. PROPOSED OUTPUT-CAPACITORLESS LDO

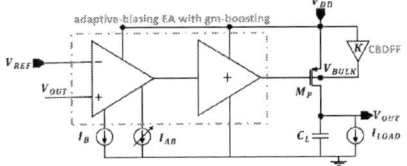

Fig. 1. Simplified block diagram of the proposed OCL-LDO.

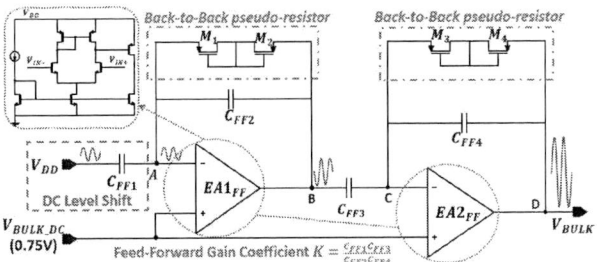

Fig. 2. Schematic of the capacitive bulk-driven feed-forward circuit.

Fig. 1 is a simplified block diagram of the proposed OCL-LDO, it consists of a power transistor (M_P) with bulk bias, an adaptive-biasing error amplifier (EA) with gm-boosting, and an on-chip output capacitor C_L. C_L represents the parasitic capacitance of power line and load circuits, and it is not used to stabilize the OCL-LDOs' loop. The CBDFF circuit and EA are the main blocks of the LDO, which will be described in detail as follows.

A. Capacitive Bulk-driven Feed-forward circuit

Previous work on the BDFF circuit in [3] used a resistor (R_B) and a constant current source (I_C) to generate a fixed DC level shift from V_{DD}, and a non-inverting amplifier with a resistive feedback network to achieve K times AC gain of the power supply noise. The level shift voltage generated in [3] is given by

$$V_{shift} = V_{DD} - I_C R_B \qquad (1)$$

In (1), the resistor R_B is very sensitive to PVT conditions and V_{shift} will be changed as V_{DD} changes. The offset of V_{shift} will be further amplified by the non-inverting amplifier, which may cause the bulk terminal voltage of M_P to be too low, and the bulk-source parasitic diode will turn on. To solve this problem, we propose a CBDFF circuit to guarantee that the bulk-source parasitic diode never turns on and ensure the feed-forward path's performance in terms of PSR.

In Fig. 1, the CBDFF circuit provides DC bias and the feed-forward signal which is a weighted version of V_{DD} multiplied by a feed-forward coefficient K to M_P's bulk terminal. Through [2], low-frequency PSR or PSR bandwidth can be optimized by adjusting K, but a trade-off is required. In this design, a fixed K of 4.4 is implemented for extending PSR bandwidth. V_{th} of M_P could be reduced by decreasing the bulk voltage according to the body effect of the PMOS transistor. To prevent the forward turn-on of the bulk-source parasitic diode, improve M_P's current driving capability, and reduce M_P's size, the bulk voltage is set to 750mV in this design.

Fig. 2 is the schematic of the proposed CBDFF circuit. The CBDFF circuit contains two inverting AC amplifiers and two back-to-back pseudo-resistors. The equivalent resistance of the back-to-back pseudo-resistor is up to hundreds of Giga ohms [4]. The back-to-back pseudo-resistor is used to define DC bias points. In Fig. 2, $V_{shift}=V_A=V_B=V_C=V_D=V_{BULK}=V_{BULK\_DC}$. The DC level shift implemented in this way is not sensitive to PVT variation because V_{BULK} is mainly related to amplifier mismatch. Furthermore, V_{BULK} is always equal to the preset bulk-bias voltage V_{BULK_DC} for different supply voltages. C_{FF1} is used instead of a resistor to sense the power supply noise. The Feed-Forward ac gain coefficient K of the CBDFF circuit is ($C_{FF1}C_{FF3}/C_{FF2}C_{FF4}$). To set K=4.4, C_{FF1} and C_{FF3} are 500 *fF*, C_{FF2} and C_{FF4} are 228 *fF* in this design.

As shown in Fig. 2, EA_{FF1} and EA_{FF2} are implemented by a two-stage amplifier to achieve high gain and large UGB. High gain can reduce OTA's mismatch. Because the selected feed-forward capacitor and the parasitic capacitor of M_P's bulk

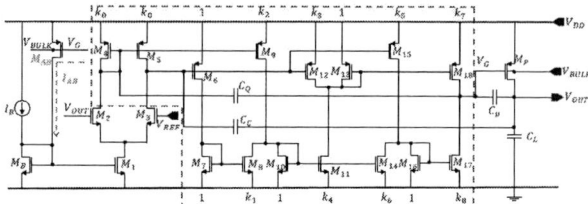

Fig. 3. Schematic of adaptive-biasing error amplifier with gm-boosting.

terminal are enough large to set amplifiers' dominant pole at their output, the Miller compensation is not used and the OTA could achieve larger UGB with less current consumption.

B. Adaptive-biasing EA with Gm-boosting

As shown in Fig. 3, we adopt an adaptive-biasing EA with gm-boosting that uses enhanced multipath nested miller compensation (EMNMC) with embedded feed-forward path in [1]. The feed-forward path brings a left-half-plane zero to stabilize LDO's loop. M1 to M5 form a simple five-transistor differential amplifier as the first stage of the amplifier. M4 to M18 form EMNMC block with an embedded feed-forward path, which could be considered as the second stage of the amplifier. The effective gm of the second stage is given by

$$G_{m2,eff} = (k_1 k_6 k_8 + k_1 k_4 k_7 + k_5 k_8 + k_3 k_7)g_{m6} \quad (2)$$

We set $k_0 = 2$, $k_1 = k_3 = k_6 = k_7 = k_8 = 3$, and $k_2 = k_4 = k_5 = 4$ in this design referring to [1]. So $G_{m2,eff} = 84g_{m6}$, EMNMC block boosts gm of the second stage extremely and the loop gain of LDO will increase greatly. A high dc gain will not only improve the line regulation and load regulation of LDO but also ensure a high low-frequency PSR of LDO. The amplifier uses a combination of fixed bias I_B and adaptive bias I_{AB}. I_{AB} is adjusted according to I_{LOAD} by M_{AB} which is a scaled replica of the power transistor with a ratio of 1:2000. The adaptive-biasing technique can effectively reduce the quiescent current of the amplifier and the settling time from heavy load to light load. C_C and C_P are miller compensation capacitors and C_Q is used for Q-reduction.

III. SIMULATION RESULTS

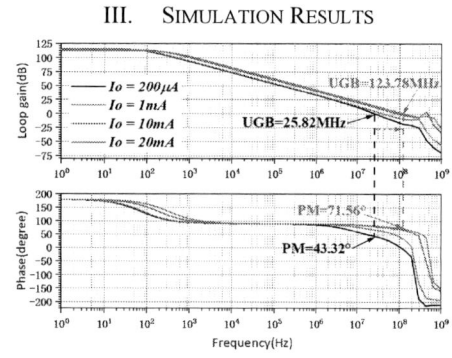

Fig. 4. Bode plots of the proposed LDO from light load to heavy load. $V_{DD} = 1.2V$, $V_{OUT} = 1V$, $C_L = 20pF$.

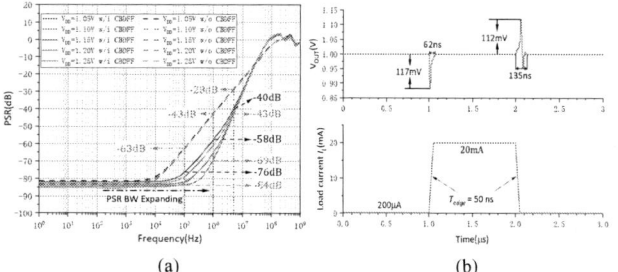

Fig. 5. (a) PSR at $I_L = 20$mA. (b) Transient response of the proposed LDO.

TABLE I. PERFORMANCE SUMMARY AND COMPARISON

	This work*	TPE 2018[1]	TPE 2019[2]	JSSC 2022[4]
Process(nm)	22	130	130	180
V_{DD}(V)	1.05-1.25	1-1.4	1.2	1.8-2.5
V_{DO}(mV)	200	200	200	200
$I_{L,MAX}$(mA)	20	25	50	200
I_Q(µA)	49	112	42	0.9
C_L	20pF	25pF	400pF	1µF**
Load regulation(µV/mA)	0.77	173	10	55
Line regulation(mV/V)	0.18	2.25	0.3	4.86
PSR (dB@MHz)	-84@0.1 -69@1 -43@5	-65@0.1 -57@1 -31@5	-90@0.1 -64@1 -38@5	-57@0.1 -40.5@1 -40.5@5
Ts(µs)	0.135	0.19	0.3	>10
$\triangle V_{OUT}$(mV)/T_{edge}(ns)	117/50	284@0.3	140/100	78/100

*: Simulation results at $V_{DD} = 1.2$ V   **: Off-chip

Fig. 4 shows the simulated bode plots of the proposed LDO from light load to heavy load. From the results of frequency response, the minimum DC gain is 112.4 dB when the load current (I_L) is 20mA. The UGB and phase margin (PM) of the LDO will improve with the load increase. When I_L is 200 µA, the UGB is 25.82 MHz and the PM is 43.32°. When I_L is 20 mA, the UGB is 123.78 MHz and the PM is 71.56°.

The proposed OCL-LDO is designed in a 22nm CMOS technology. The input voltage range of LDO is 1.05-1.25 V and the dropout voltage of LDO is 200 mV. When V_{DD} is 1.2 V, the quiescent current is 49 µA, and the PSR is -84 dB@100kHz, -69 dB@1MHz, and -43 dB@5MHz as it is shown in Fig. 5 (a). The PSR in the worst case is -76 dB@100kHz, -58 dB@1MHz, and -40 dB@5MHz when V_{DD} is 1.05 V. Fig. 5 (b) shows the load transient response of the proposed LDO with a 50ns edge time (T_{edge}) when V_{DD} is 1.2 V. The undershoot and overshoot voltages are 117 mV and 112 mV, respectively, with a settling time (T_S) of 65 ns and 135 ns, respectively. In addition, the proposed LDO achieves the best line regulation and load regulation under comparison in Table I.

IV. CONCLUSION

An OCL-LDO with gm-boosting and CBDFF technique for high PSR and fast transient response was introduced. The PSR performance has a 26 dB improvement at 1 MHz by the proposed CBDFF technique. The settling time is 135 ns at a load step from 200 µA to 20 mA in 50 ns. Moreover, the proposed OCL-LDO has excellent line regulation and load regulation.

ACKNOWLEDGMENT

This work is supported by Guangdong Province Natural Science Foundation (No. 2022A1515011934) and Shenzhen Basic Research Foundation (No. WDZC20200819152116001 and JCYJ20200109143003935). (Corresponding author: Xian Tang.)

REFERENCES

[1] S. Bu, J. Guo and K. N. Leung, "A 200-ps-Response-Time Output-Capacitorless Low-Dropout Regulator With Unity-Gain Bandwidth >100 MHz in 130-nm CMOS," in IEEE Transactions on Power Electronics, vol. 33, no. 4, pp. 3232-3246, April 2018, doi: 10.1109/TPEL.2017.2711017.

[2] F. Lavalle-Aviles, J. Torres and E. Sánchez-Sinencio, "A High Power Supply Rejection and Fast Settling Time Capacitor-Less LDO," in IEEE Transactions on Power Electronics, vol. 34, no. 1, pp. 474-484, Jan. 2019, doi: 10.1109/TPEL.2018.2826922.

[3] W. Wang and B. Chi, "A Wideband High PSRR Capacitor-Less LDO with Adaptive DC Level Shift and Bulk-Driven Feed-Forward Techniques in 28nm CMOS," 2019 IEEE International Symposium on Circuits and Systems (ISCAS), 2019, pp. 1-5, doi: 10.1109/ISCAS.2019.8702501.

[4] T. Guo, W. Kang and J. Roh, "A 0.9-µA Quiescent Current High PSRR Low Dropout Regulator Using a Capacitive Feed-Forward Ripple Cancellation Technique," in IEEE Journal of Solid-State Circuits, doi: 10.1109/JSSC.2022.3161014.

ROPY-SLAM: a Heterogeneous CPU-FPGA System for Simultaneous Localization and Mapping

Weiyi Zhang, Liting Niu, Chaoyang Ding, Yiyang Wang, Fasih Ud Din Farrukh, Chun Zhang*

School of Integrated Circuits, Tsinghua University

Abstract—Simultaneous localization and mapping (SLAM) is an emerging robotic technology enabling autonomous robots to self-localize and map the surrounding environments. However, SLAM system is time and resource consuming, making it hard to implement on mobile devices. In this work, ROPY-SLAM, a novel heterogeneous framework for SLAM is proposed to take advantage of various devices such as personal computer and FPGA. The proposed system reduces the execution time by 28.6% and resource utilization by at most 71.2%. Moreover, ROPY-SLAM supports more flexible hardware and software implementations, providing a solution for distributed calculation of robotic applications.

Keywords—robotic, SLAM, FPGA, heterogeneous system

I. INTRODUCTION

Autonomous robots have been widely used in different areas. Simultaneous localization and mapping (SLAM) is an emerging technology enabling robots in self-localization and constructing the map of surrounding environments. However, SLAM frameworks are hard to implement on lightweight devices due to limited hardware resources. Some FPGA-based accelerators have been proposed [1]. However, current works focus on implementing the system on a single device via tools such as Petalinux. The programable logic (PL) was used to accelerate specific modules, while the processing system (PS) was used to execute other tasks. Both the augmentability of the system and the overall performance suffer. In addition, the operating system for the evaluation board has to be generated and burned into the SD card of the board when the design is updated, resulting in more developing labor and time. In this work, we propose ROPY-SLAM (ros-pynq-SLAM), fully leveraging both the powerful processor of personal computer (PC) and the efficient PL of the evaluation board. The proposed design supports the updating of hardware design in a hot swap method. We also propose one optimized Gaussian filter for feature extraction, achieving 1.75x improvement of throughput and at most 33.2% reduction of hardware resources. The overall system reduces the execution time by 28.6% and resource utilization by at most 71.2% compared with the reference work.

II. BACKGROUND KNOWLEDGE

A. SLAM

SLAM is an emerging technology, which consists of two main tasks, constructing the map in a new environment and estimating the state of the moving robot in it. A typical SLAM framework is mainly composed of the following parts: a) reading of sensor information, b) frontend visual odometry, c) backend optimization, d) loop closure detection, e) mapping. In this work, the structure of PTAM [2] is used as the backbone. The feature extraction of frontend visual odometry is accelerated by FPGA.

B. ROS

ROS (Robot Operating System) is a set of operating system architecture designed specifically for the development of robot software. It provides services including hardware abstract

Supported by the National Natural Science Foundation of China (No.U20A20220).

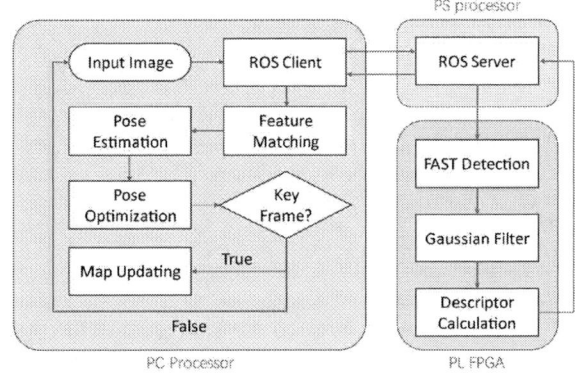

Fig.1. General framework of ROPY-SLAM

description, driver management, message transmission between programs, and program distribution package management. It also provides some tools and libraries for acquiring, establishing, writing and executing programs of multi-computer integration.

C. PYNQ

PYNQ (Python productivity for Zynq) is an open-source framework for Xilinx Zynq All Programmable SoCs, which runs an Ubuntu operating system on the processing system (PS) side and is able to control the (PL) by python scripts. These two properties enable the evaluation board to run as one ROS node and interact with other devices such as PC.

III. PROPOSED DESIGN

A. General framework of ROPY-SLAM

ROPY-SLAM aims to leverage the advantages of all accessible devices. ROPY-SLAM consists of three major parts: host PC, evaluation board, and connector. The processor of the host PC is powerful to execute complex tasks, thus the control of the whole system and the un-accelerated functions of SLAM are assigned to be executed on the PC. PC has a larger memory system which is suitable to run large-scale applications. With the development of artificial neural networks, GPU can be configured to extract semantic information from the input images. In mainstream applications on Zynq architecture, PS is used as the major controller and processing element. However, the PS is used as coprocessor to control the PL and communicate with the host PC in our proposed work. The connector is realized by ROS at the software level and interfaces such as Ethernet at the hardware level. The workflow of ROPY-SLAM is shown in Fig.1. The ROPY-SLAM is initialized by three steps: creating one ROS host on PC, creating one ROS node as a service server on the evaluation board, and creating one ROS node as a service client on PC. Then the system can process the input images serially. When the input image arrives and the system is idle, the tracking thread firstly wraps the RGB image as one service message. Then the calculation request and the message will be transferred to the PS of the evaluation board. The service server receives the request and image, then controls the PL to extract the keypoints and corresponding descriptors. The result will be returned and stored as an $N \times 10$ array as the type of 32-bit unsigned integer, where N is the max number of keypoints. For each line of the result, the first eight elements constitute a 256-bit ORB descriptor and the last two elements are respectively the x-coordinate and y-coordinate of the keypoints. The array is then wrapped as a ROS message and transferred to the PC for other processes including the matching of the feature points, pose estimation, pose optimization, and map updating.

B. Hardware modules of ROPY-SLAM

The framework of ROPY-SLAM supports various devices and functions. In this work, feature extraction is implemented on PL while other calculations are performed on PC. The feature extraction comprises three modules: keypoint detection, Gaussian filter, and descriptor calculation. All the modules are connected by the AXI-stream interface and the architecture is pipelined. The keypoint detection module scans the image and selects the FAST (Features from Accelerated Segment test) [3] corners as keypoints. The keypoint detection module has one NMS (non-maximum suppression) submodule, which filters the keypoints, and only the keypoint with the highest score is reserved in one neighborhood space. For each reserved keypoint, the moments on both x-axis and y-axis are calculated representing the orientation of the keypoints. The descriptor calculation module calculates the BRIEF [4] (Binary Robust Independent Elementary Features) based on the image filtered by Gaussian filter, which comprises 256 Boolean values.

C. Optimization of Gaussian filter

The Gaussian filter with size 7×7 is used before the descriptor calculation to improve the robustness of the descriptors. An efficient Gaussian filter is designed leveraging sparsity. Given a Gaussian filter with size 3×3 shown in Fig.2, there are 9 multiplications. However, the window has only 3 different values. The kernel is decomposed into 3 sub-kernels with only element values 0 and 1. For each sub-kernel, the pixel values of non-zero area are summed up and then multiplied by the weight. Then results from all sub-kernels are summed up to get the final result. Thus, the multiplication is reduced to 3 for each pixel calculation. In the implemented Gaussian filter, the kernel size is 7×7 with 7 weights. Thus, 7 multiplications are needed for each pixel.

IV. EXPERIMENT RESULTS

The proposed ROPY-SLAM is implemented on a PC with intel i7 processor at 2.2GHz, and a ZCU104 evaluation board with a Cortex-A53 processor at 1.5GHz. The PL part runs at the frequency of 200 MHz. The TUM dataset is used to validate the functionality of the system. The visualized SLAM result is shown in Fig.3. The tracking accuracy drops slightly compared with only PC execution, due to the error from the fixed-point calculation. The resource and time consumption of the optimized Gaussian kernel are shown in Fig.4, where the Gaussian filter is used from Xilinx xfopencv library. The optimized kernel reduces 29.7% flip-flops (FFs), 33.2% look-up tables (LUTs), and one DSP. In the Meanwhile, the throughput is improved to 1.75x.

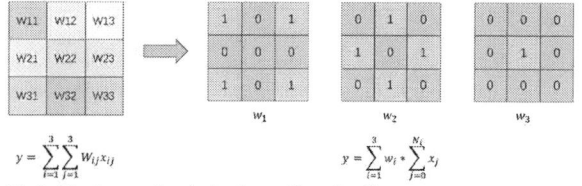

$$y = \sum_{i=1}^{3}\sum_{j=1}^{3} W_{ij}x_{ij} \qquad y = \sum_{i=1}^{3} w_i * \sum_{j=0}^{N_i} x_j$$

Fig.2. The theory of optimization to Gaussian filter

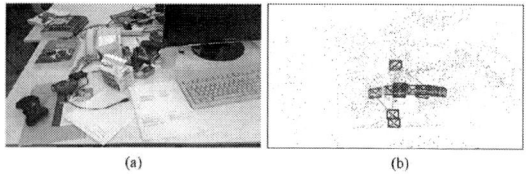

Fig.3. Visualization of ROPY-SLAM

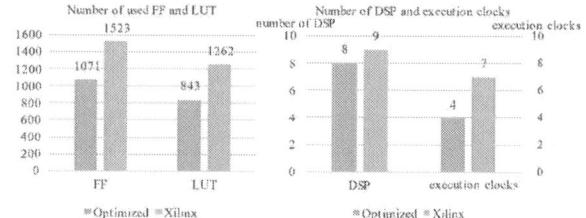

Fig.4. Resource and time consumption of optimized Gaussian filter

TABLE I.   TIME CONSUMPTION OF ROPY-SLAM

	Ours	eSLAM [1]	ARM [1]
Feature Extraction	5.4ms	9.1ms	291.6ms
Feature Matching	13.4ms	4.0ms	246.2ms
Others	10.4ms	27.8ms	27.8ms
Total	29.2ms	40.9ms	565.6ms

TABLE II.   RESOURCE CONSUMPTION OF ROPY-SLAM

	FF	LUT	DSP	BRAM
Ours	38639	36399	32	45.5
eSLAM [1]	67809	56954	111	78

The image size is 480×640 and the max number of keypoints is set as 1000. The time consumption of ROPY-SLAM is shown in Table.1, where eSLAM and the execution on pure ARM core are compared [1]. The feature extraction time is reduced by 40.7%. The feature matching is accelerated by PL in eSLAM, while executed by PC in ROPY-SLAM. Thus, the feature matching costs more time. The time consumption of other calculations is largely reduced compared with eSLAM and the ARM core. The total time consumption is reduced respectively by 28.6% and 94.8% compared with eSLAM and ARM core. The proposed design largely reduces the resource consumption as shown in Table.2, where DSP utilization is reduced by 71.2%.

V. CONCLUSION

In this work, we proposed ROPY-SLAM, a heterogeneous CPU-FPGA system for SLAM, which takes advantage of different hardware and provides a new solution for distributed calculation. To improve the hardware efficiency, an optimized Gaussian filter is proposed leveraging the sparsity of the Gaussian kernel. Both the time and resource consumption are largely reduced. To improve flexibility, ROPY-SLAM supports hardware updating in a hot swap way, which can be achieved by reloading the bitstream and re-run the server node in the PS of the evaluation board. The distributiveness of the ROPY-SLAM brings compatibility for different hardware and software, which is promising for future research.

REFERENCES

[1] Liu, Runze, et al. "eslam: An energy-efficient accelerator for real-time orb-slam on fpga platform." Proceedings of the 56th Annual Design Automation Conference 2019. 2019.

[2] Pire, Taihú, et al. "S-PTAM: Stereo parallel tracking and mapping." Robotics and Autonomous Systems 93 (2017): 27-42.

[3] Viswanathan, Deepak Geetha. "Features from accelerated segment test (fast)." Proceedings of the 10th workshop on image analysis for multimedia interactive services, London, UK. 2009.

[4] Calonder, Michael, et al. "BRIEF: Binary Robust Independent Elementary Features." European Conference on Computer Vision Springer, Berlin, Heidelberg, 2010.

A Single-input Dual-output Three-level Buck Converter for SoC Applications

Zhuoneng Li, Zhongming Xue*, Chenglong Liang, Yongchao Zhang, Mengqi Duan, Shangzhou Zhao, Xihao Liu, Zhuoqi Guo, Li Geng*

School of Microelectronics, Xi'an Jiaotong University

Abstract—This paper proposes a single-input dual-output (SIDO) three-level buck converter to meet the requirements of multi-voltage domain and high voltage stress for SoC applications. The topology is based on three-level buck with standard 1.8V devices, where only an additional power switch is added. By using this structure, the second output can be powered by a fly capacitor, and V_{CF} calibration is achieved by the control loop for reliability issues. Hence the efficiency and the power density are enhanced with standard devices and reducing the number of power switches. Moreover, the control loop with error processor and driver module is demonstrated based on the topology analysis. Ultimately, the proposed converter is designed and fabricated with 0.18μm CMOS process, which handles the input range of 3.3-2.8V and dual-output of 1.8V and 1.2V with 96.9% peak efficiency. The power density is 2.557W/mm², and the active area is only 0.49 mm².

Keywords—SIDO three-level buck, high efficiency, high power density, error processor, SoC applications

I. INTRODUCTION

The fast development of microelectronics process prompts the system on chip (SoC) with larger scale and higher performance. However, the decrease of breakdown voltage for the upgrade of process poses a challenge to the on-chip power management for SoC applications. Meantime, the improvement of SoC complexity makes multi-voltage supply become very necessary.

In recent years, some improved schemes have been presented to deal with high voltage stress or dual-outputs. For instance, a three-level buck using 2.5V I/O devices under 5V input voltage is demonstrated, in which the calibration of fly capacitor voltage V_{CF} is designed for reliability issues [1][2]. Besides, a traditional SIDO three-level buck is proposed with two additional power switches for dual-outputs [3], but the voltage V_{CF} isn't calibrated with high voltage devices. Therefore, it remains difficulties for buck converter with high voltage and dual-outputs.

In this paper, a SIDO three-level buck converter for SoC applications is proposed to settle the issues of high voltage stress and multi-voltage supply. Both dual-outputs and V_{CF} calibration are realized by one power MOS transistor and control loop, aiming to enhance the reliability, high efficiency and high power density with standard 1.8V devices and less power switches. Moreover, the topology is analyzed, and the control loop with error processor and driver module is designed for the stability and reliability of the loop.

II. PROPOSED SIDO THREE-LEVEL BUCK CONVERTER

A. System architecture

Fig. 1 shows the architecture of the proposed SIDO three-level buck converter. Based on three-level buck structure for one output, the power topology adopts a fly capacitor to reduce the voltage stress on power switches. Moreover, the additional switch S_5 is added at V_X node, and provides the other output by the fly capacitor. To control these outputs and calibrate the V_{CF}

simultaneously, the control loop with pulse width modulation (PWM) is designed. In the control loop, the error processor and multi-type compensation are employed for the steady of control loop and the decouple between two branches of control loop. The driver module is presented with level shifter to reduce the voltage stress of the power switches.

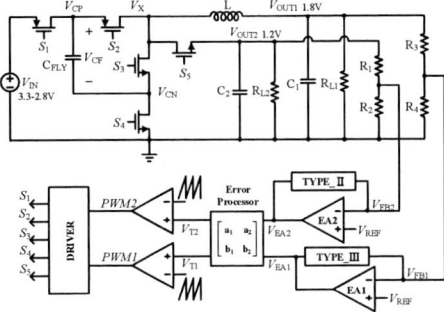

Fig. 1. Architecture of the proposed SIDO three-level buck converter

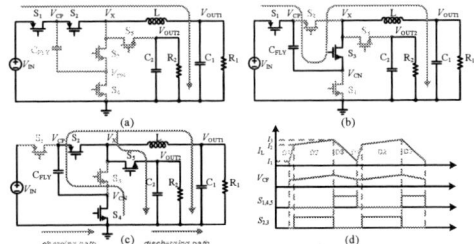

Fig. 2. Operation of the proposed converter: (a)phase 1 (b)phase 2 (c)phase 3 and (d) waveforms of time diagram

Consequently, there are three advantages for the presented SIDO three-level buck converter. a) Compared with the previous SIDO three-level buck, one power switch can be saved to increase the power density of the converter. b) By utilizing the fly capacitor and the switch S_5, V_{CF} calibration is realized by control loop, which settles the breakdown issue and enables the standard 1.8V devices in power stage. c) The conversion efficiency is also improved due to the smaller parasitic capacitance and on-resistance of standard 1.8V devices.

B. Topology analysis

Fig. 2 shows the operation and waveforms of the proposed converter. In phase 1, S_1 and S_2 turn on. Hence the inductor L is magnetized, and the capacitor C_{fly} is hanged. In phase 2, S_1 and S_3 turn on, and the capacitor C_{fly} is charged by inductor current. The inductor L is magnetized or demagnetized according to the input and output voltages. In phase 3, only S_1 and S_3 turn off. The capacitor C_{fly} is discharged, and the inductor L is demagnetized when the V_{out2} is lower than V_{out1}. In Fig. 2(d), the waveforms of inductor current I_L and capacitor voltage V_{CF} is shown with the switch state of S_{1-5}.

The small signal analysis is also accomplished with state space averaging method. Assuming that the load current of V_{out2} is relatively less than the load current of V_{out1}, and the ripple of inductor current is adequately small, the small signal model of this power stage is expressed by (1)(2), which establishes the base for control loop design.

$$v_{out1} = [V_{out2}d_1 + (V_{in} - 2V_{out2})d_2]\frac{1}{1 + sL/R_1 + s^2LC_1} \tag{1}$$

$$v_{out2} = \frac{I_{load1}(2d_2 - d_1)R_2}{1 + s(C_{fly} + C_2)R_2} \tag{2}$$

$$
\begin{aligned}
v_{ea1} &\propto V_{out2}d_1 + (V_{in} - 2V_{out2})d_2 \\
v_{ea2} &\propto 2d_2 - d_1
\end{aligned} \tag{3}
$$

$$
\begin{pmatrix} v_{T1} \\ v_{T2} \end{pmatrix} = \begin{pmatrix} \frac{5}{8} & -\frac{5}{8} \\ \frac{1}{4} & \frac{3}{4} \end{pmatrix} \begin{pmatrix} v_{EA1} \\ v_{EA2} \end{pmatrix} \tag{4}
$$

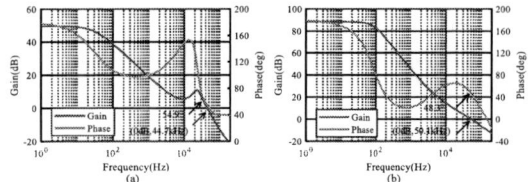

Fig. 3. Bode diagram for (a) loop of V_{out1} and (b) loop of V_{out2}

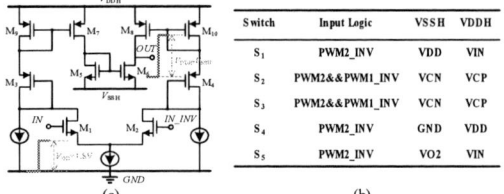

Fig. 4. (a) Level shifter and (b) its parameters for driver module

S witch	Input Logic	VSSH	VDDH
S_1	PWM2_INV	VDD	VIN
S_2	PWM2&&PWM1_INV	VCN	VCP
S_3	PWM2&&PWM1_INV	VCN	VCP
S_4	PWM2_INV	GND	VDD
S_5	PWM2_INV	VO2	VIN

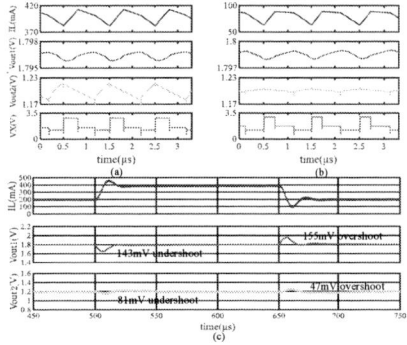

Fig. 5. Transient simulation results for (a) V_{in}=3.0V V_{out1}=1.8V V_{out2}=1.2V I_{load1}=400mA, (b) V_{in}=3.0V V_{out1}=1.8V V_{out2}=1.2V I_{load1}=80mA, and (c) I_{load1}=200-400-200mA V_{in}=3.0V V_{out1}=1.8V V_{out2}=1.2V

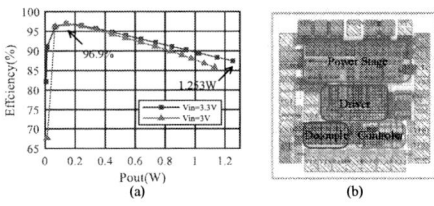

Fig. 6. (a) Efficiency changing with output power and (b) layout of converter

C. Control loop design

1) Error processor

For the sake of the steady and decouple of control loop, the relationship between error signals v_{ea1}, v_{ea2} and duty cycles d_1, d_2 is set as expression (3) according to the small signal model. Therefore, the error processor is proposed in expression (4), which adopts adder and subtractor to translate the error signals $v_{EA1,2}$ to the input signals of PWM comparator $v_{T1,2}$ [4]. Thus, the type III and type II compensation network are designed for the error amplifiers of V_{out1} and V_{out2}, respectively. By multi-type compensation and error processor, the loop of V_{out1} and the loop of V_{out2} obtain the crossover frequency of 44.7kHz and 50.1kHz, respectively, while the phase margins are 54.9° and 48.3°, respectively, as shown in Fig. 3.

2) Driver module

To drive power switches with high voltage stress tolerance, the level shifter and its parameters in driver module are shown in Fig. 4. The level shifter with standard 1.8V devices utilizes the gate-source voltage of M_3 and M_4 to protect M_9 and M_{10} away from the breakdown risk and adopts current comparator of M_{5-8} for rail-to-rail output from V_{SSH} to V_{DDH}. Fig. 4(b) shows the detailed parameters of the level shifters for each power switch, such as input logic, the voltage setting of V_{SSH} and V_{DDH}.

TABLE I. COMPARISON WITH PRIOR WORKS

	JSSC[1]	VLSI[2]	IFEEC[3]		This work	
Process	65nm 2.5V I/O device	250nm 2.5V I/O device	350nm 5V device		180nm 1.8V device	
Frequency/MHz	50	4.4-11.8	0.2		1	
Input voltage/V	5	2.6-5	5		3.3-2.8	
Output voltage/V	0.6-4.2	0.34-4.5	1.5	1.2	1.8	1.2
Structure	3level	3level	3level SIDO		**3level SIDO**	
No. of switches	4	4	6		**5**	
V_{CF} calibration	yes	yes	no		**yes**	
Control loop	PWM	AOOT	PWM		**PWM with error processor**	
Peak efficiency/%	90	92.8	N/A		**96.9**	
Max. Pout/W	3.000	1.320	2.430		1.253	
Power density/(W/mm²)	5.882	0.367	0.426		**2.557**	
Area/mm²	2.3*/0.51**	5.8*/3.6**	5.7**		0.64*/0.49**	

* With integrated decouple and fly capacitor ** Without integrated decouple and fly capacitor

III. SIMULATION RESULTS AND ANALYSIS

Fig. 5 shows the transient simulation results of the proposed converter. The output voltages of 1.8 and 1.2V under 3.3-2.8V input voltage can be obtained by automatically adjusting the duty cycle of three phases for different loads. And due to error processor and multi-type compensation, when I_{load1} varies between 400mA and 200mA, the undershoot and overshoot voltages for V_{out1} and V_{out2} are 143/155mV and 81/61mV, respectively, which presents a stable transient response, as shown in Fig. 5(c).

Fig. 6 gives the efficiency for different output power and layout of the proposed converter. The converter is designed and fabricated with 0.18μm CMOS process only using 1.8V standard devices, and the active area is 0.49mm². The peak efficiency is 96.9%, and the maximum output power is 1.253W. As a result, the power density of 2.557W/mm² is achieved. Table I shows the comparison with other prior works.

IV. CONCLUSION

In this paper, a single-input dual-output three-level buck converter is proposed for the issues of multi-voltage supply and high voltage stress. The V_{CF} calibration and dual-outputs are realized with just only five standard power devices for excellent features of high reliability, 96.9% peak efficiency and 2.557W/mm² power density. Moreover, based on the topology analysis and the control loop with error processor and driver module, the better steady and transient operation are obtained. This work provides a good power supply solution for SoC applications.

ACKNOWLEDGMENTS

This work was supported by National Natural Science Foundation of China (62004159), National Key Research and Development Program of China (2019YFB2204700), and the Key Project of "two chains" Integration (2021LL-JB-05).

REFERENCES

[1] X. Liu, C. Huang and P. K. T. Mok, "A High-Frequency Three-Level Buck Converter With Real-Time Calibration and Wide Output Range for Fast-DVS," JSCC, vol. 53, no. 2, pp. 582-595, Feb. 2018.

[2] Y. Karasawa, T. Fukuoka and K. Miyaji, "A 92.8% Efficiency Adaptive-On/Off-Time Control 3-Level Buck Converter for Wide Conversion Ratio with Shared Charge Pump Intermediate Voltage Regulator," VLSI, 2018, pp. 227-228.

[3] Y. -C. Hsu, J. -Y. Lin, P. -C. Shih and R. -L. Lee, "Single Inductor Dual Output Three Level Buck Converter," IFEEC, 2021, pp. 1-6.

[4] M. Belloni et al., "A 4-Output Single-Inductor DC-DC Buck Converter with Self-Boosted Switch Drivers and 1.2A Total Output Current," ISSCC, 2008, pp. 444-626.

Dynamically Reconfigurable Memory Address Mapping for General-Purpose Graphics Processing Unit

Weiliang Chen[1,2], Zhaoshi Li[2], Leibo Liu[1], Shaojun Wei[1]

[1] Tsinghua University   [2] Metax Tech Inc.

Abstract—**GPGPUs utilize multi-dimensional memory subsystems to provide the bandwidth needed by their multi-dimensional parallelism. However, an unfavorable address mapping leads to imbalanced memory request distribution across the memory resources, causing degraded performance and poor power efficiency. The optimal mapping is both application- and hardware-dependent. This paper provides a software-hardware co-design to dynamically reconfigure the address mapping according to the trace of the targeted application. First, a circuit to sample the entropy of address bits is proposed to capture the optimal address mapping. Second, a dynamic reconfiguration mechanism is designed to apply the optimal address mapping. Simulation results show up to 45% performance improvement over fixed address mappings.**

Keywords—GPGPU, reconfigurable architecture, DRAM

I. INTRODUCTION

To cope with the bandwidth demand of the General-Purpose Graphics Processing Unit (GPGPU), high-bandwidth DRAM solutions such as GDDR5 and HBM are utilized. To fully exploit their bandwidth, GPGPU applications need to choose an optimal address mapping.

Fig.1 shows an example DRAM organization and a naïve address mapping. A DRAM is organized in a hierarchical structure of channels, banks, rows, and columns. To maximize bandwidth, applications should exploit the channel/bank-level parallelism and row-buffer locality of DRAM. The row bit should vary frequently to avoid closing an open row, whereas the channel bit and bank bit should vary as infrequently as possible to spread accesses across channels and banks. Typically, CPUs map row bits to the Most-Significant Bit (MSB), column bits to the Least-Significant Bit (LSB), and channel/bank to middle bits, on the assumption that consecutive accesses to arrays tend to change less frequent from LSB to MSB. Modern CPUs use permutation-based address mapping to further reduce row conflicts [1].

However, on GPGPU the optimal address mapping is highly dependent on both application memory access patterns and hardware configurations. A GPGPU manages its massive parallelism by dividing applications into grids, which are further divided into Thread Blocks (TB). All threads of a TB are allocated on one Streaming Multiprocessor (SM) at runtime.

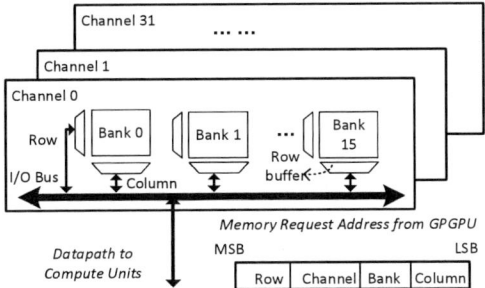

Figure 1. An Address Mapping to Multi-dimensional DRAM

Req. Seq.	TB_0	TB_1	Row / Ch
1	···000 00···	···001 00···	Mapping-1 ···000 10···
2	···000 01···	···001 01···	Row, ch = (0, 3)
3	···000 10···	···001 10···	Mapping-2 ···000 10···
4	···000 11···	···001 11···	Row, ch = (1, 0)

(a) Memory Requests of Two TBs   (b) Two Address Mappings

$TB_0 \rightarrow TB_1$ @ Mapping-1 : *(0,0), (0,1), (0,2), (0,3)* => uniformly distributed√

$TB_0 \rightarrow TB_1$ @ Mapping-2 : *(0,0), (0,1), (1,0), (1,1)* => channel conflict

$TB_0 \mid TB_1$ @ Mapping-1 : *(0,0), (1,0), (0,1), (1,1)* => channel conflict

$TB_0 \mid TB_1$ @ Mapping-2 : *(0,0), (0,1), (0,2), (0,3)* => uniformly distributed√

(c) Different schedules favor different mappings.

Figure 2 On GPGPU the optimal address mapping is decided by both application and hardware status.

Fig.2 demonstrates how different TB allocations of the same application favor different address mappings. On GPGPU threads within a TB calculates their memory addresses from dimensional indexing of threads and TBs. In this simplified scenario, TB0 and TB1 send sequences of memory requests shown in Fig.2 (a). For brevity, only the bits related to row and channel bits are shown. Fig.2 (b) shows two address mappings to the row and channel indexes. The first mapping is the same with the fixed mapping of Fig.1, whereas the second one exchange one bit. These two mappings result in different DRAM performances for different TB scheduling. As shown in Fig.2 (c), sequentially executing TB0 and TB1 on the same SM (denoted as TB0->TB1) favors the first mapping due to uniform channel distribution. In contrast, concurrent execution of TB0 and TB1 (denoted as TB0 | TB1) prefers the second mapping.

To cope with the high variance of application and hardware resource availability of GPGPU, this paper proposes a software-hardware co-design to dynamically reconfigure the address mapping according to the trace of applications. To capture the optimal address mapping, section II designs a circuit to sample the entropy of address bits. Section III designs a dynamic reconfiguration mechanism to apply the optimal address mapping. Simulation results show up to 35% performance improvement over fixed address mappings.

II. SAMPLING THE ADDRESS ENTROPY AT RUNTIME

Entropy measures the frequency of changes in a sequence of values. To unlock the performance potential of hierarchical DRAM solutions, address mappings of GPGPUs should harvest entropy by concentrating the entropy into the address bits used for channel and bank selection, leaving low entropy address bits for rows. Y. Liu et al. [2] statistically calculates the entropy of various GPGPU workloads and observes that many workloads exhibit the *entropy valley* distributed throughout the lower bits of memory addresses, which may result in many channel and bank conflicts in traditional mappings. The exact positions of entropy valleys are dependent on application and hardware scheduling, justifying our motivation for dynamically reconfigurable address mappings on GPGPUs. However, their approach requires knowledge of all memory request addresses of an application before executing the application, which is only applicable for offline profile-guided optimizations.

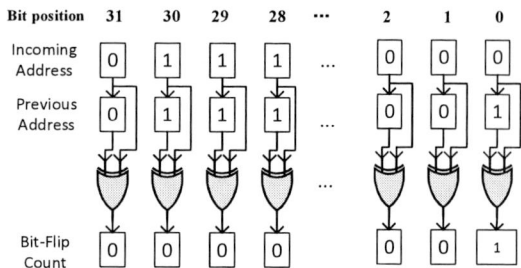

Figure 3 Circuit design for window-based entropy sampling

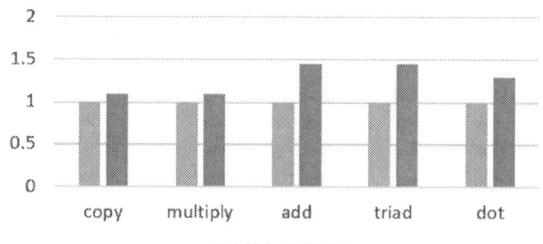

Figure 5 Performance speedup of the proposed scheme over the fixed address mapping of Fig. 1, with varying number of Stream Multiprocessors.

This section proposes a dynamically (online) sampling-guided approach to optimal address mappings. To calculate the entropy of address bits, instead of counting the number of 0s and 1s for each bit position of all addresses as previous works [2][3] did, our approach designs an XOR-circuit to count the number of bit-flips during selected address sampling windows.

As shown in Fig. 3, when the memory request addresses stream in the circuit, the incoming address is bitwise-XORed with the previous address to detect the bit changes of consecutive requests. The changes in bit positions are accumulated in an array of bit-flip counters. The higher the number of changes, the higher entropy is observed for the corresponding bit position.

To limit the width of the bit-flip counters, the sampling is done in a window fashion among memory requests. The window size is decided by the position of the sampling circuit, which will be discussed in section III.

III. RECONFIGURING ADDRESS MAPPING ON GPGPU

A. Address-Mapping Estimation

The sampler circuit is integrated into GPGPUs so that runtime entropy is obtained. Since the memory request of all TBs will be serviced by a shared DRAM controller, both intra- and inter-TB entropy is needed to estimate the optimal address mapping. As shown in Fig. 4, mainstream GPGPUs feature a per-SM L1 cache and a shared L2 cache. Per-thread memory requests of TBs are coalesced in the L1 cache, whereas inter-TB data reuse is fulfilled by the L2 cache. Thus, to sample the intra-TB entropy, a sampler is added to each L1 cache, whose window size is equal to the number of outstanding memory requests supported by the L1 cache. Similarly, the inter-TB entropy is sampled at the L2 cache, with its window size set to be the size of the request reordering buffer in the DRAM controller.

After gathering the entropy information, a simple heuristic is used to decide the optimal address mapping. Address bits with the higher to lower inter-TB entropy are mapped to channel, bank, and row, respectively. If a tie is met, the intra-TB entropy is used similarly. The channel is chosen to be mapped to higher entropy because channel-level parallelism has a larger impact on performance than bank-level parallelism.

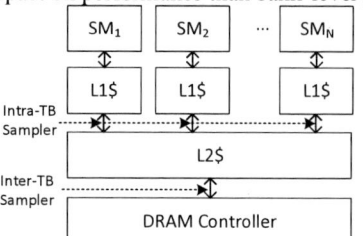

Figure 4 Intra-TB and Inter-TB entropy sampler is integrated to GPGPUs

B. Reducing Data Relocation Overhead of Reconfiguration

However, naïvely relocating data after reconfiguring the address mapping at runtime needs copying all data in DRAM to accommodate the new addressing mode. This is why previous works on CPU [3] limit changing of address mapping to offline configuration.

To enable dynamic reconfiguration of address mappings, the proposed design leverages the fact that the DRAM of GPGPU is managed by a host CPU in page granularity. The estimated optimal address mapping can be encoded in the Page Table Entry (PTE) so that the following memory access to that page will be addressed by the new mapping.

To constrain the width of address mapping configuration bits, several predefined mappings are built into the memory controllers. Each memory request will choose one mapping indicated by its PTE.

IV. EVALUATION

We evaluate the design by running the BabelStream benchmark [4] on an in-house GPGPU simulator. The fixed address mapping pattern of Fig.1 is used as the baseline. The result is shown in Fig. 5. We evaluate different hardware configurations by varying the number of SMs. For 1 SM case, performance is not bound at the DRAM interface. For memory-bound workload, i.e. add and triad, the proposed scheme on 32 SM gets up to 1.45x speedup over baseline. It shows the proposed design is beneficial for memory-bound scenarios.

V. CONCLUSION

This paper introduced a dynamically reconfigurable address mapping scheme on GPGPU. Based on the observation that the optimal address mapping is decided both by application and hardware architecture, a runtime sampling approach to estimate the optimal mapping is proposed, along with a dynamically reconfigurable scheme to mitigate overheads.

REFERENCES

[1] D. Kaseridis, J. Stuecheli, and L. K. John, "Minimalist open-page: A DRAM page-mode scheduling policy for the many-core era," in 2011 44th Annual IEEE/ACM International Symposium on Microarchitecture (MICRO), Dec. 2011, pp. 24–35.

[2] Y. Liu et al., "Get out of the valley: power-efficient address mapping for GPUs," in Proceedings of the 45th Annual International Symposium on Computer Architecture, Los Angeles, California, Jun. 2018, pp. 166–179.

[3] S. Adavally and K. Kavi, "Towards Application-Specific Address Mapping for Emerging Memory Devices," in The International Symposium on Memory Systems, Washington DC USA, Sep. 2020, pp. 105–113.

[4] Deakin T, Price J, Martineau M, McIntosh-Smith S. Evaluating attainable memory bandwidth of parallel programming models via BabelStream. International Journal of Computational Science and Engineering. Special issue. Vol. 17, No. 3, pp. 247–262. 2018

Strain-regulated flexible molecular sensors enabled by 2D PtTe$_2$

Zhehan Wang[1,2#], Dingxuan Kang[1,2#], Jiayi Chen[1,2], Xiao Xu[3], Xu Jing[1,2*], Li Tao[1,2*]

[1] School of Materials Science and Engineering, Southeast University, China.
[2] Center of 2D Materials, Southeast University, China.
[3] Department of Electrical and Computer Engineering, The University of Texas at Austin, United States.

Abstract—Two-dimensional transition metal dichalcogenide (TMDs) holds great promise for future wearable technologies. The large-scale synthesis of TMDs and the investigation of their molecular sensing properties are current research hotspots. Herein, we construct flexible PtTe₂ molecular sensors directly on polyimide, followed by the exploration on ammonia sensing properties and stability under strain. The sensor shows an excellent ammonia response of 0.4% ppm⁻¹ surpassing most reported semimetal-based sensors. In-situ sensing testing reveals 500% enhancement in response after controlled strain engineering, which suggests the potential to expand limit of detection and linearity range.

Keywords—PtTe₂, flexible devices, molecular sensors, strain engineering, IoTs.

I. INTRODUCTION

In Internet of Things (IoTs) devices, flexible molecular sensors are one of the key components, providing a variety of effective information about surroundings that are imperceptible to humans [1-3]. Ammonia is widely used in the production of refrigerants, fertilizers, nitric acid, etc. However, leaked ammonia may result in a multitude of diseases which pose a threat to human health. A highly reliable and easily deployable ammonia sensor is required to sound the alarm and prevent people from danger. Layered 2D transition metal dichalcogenide materials have tunable band gap, rich surface chemical states, and strong air stability, making them a promising material for wearable technology [4, 5]. Despite the rapid development of 2D TMDs materials in the field of molecular sensing, problems and challenges remain in large-area material synthesis, sensing performance enhancement and stability under deformation. Herein, we report a flexible molecular sensor based on 2D semimetal PtTe₂ and investigate its sensing response to ammonia when operating under bending deformation.

II. EXPERIMENT

A. Sensor Fabrication and Characterization

Different thicknesses of Pt seed films are evaporated on polyimide (PI) substrates by e-beam evaporation (VZS-600 Pro), and thermal assisted conversion (TAC) is performed in a tube furnace to prepare 2D PtTe₂. Raman spectrum (Witec Alpha 300) is collected with a 532 nm laser to confirm the quality of as-grown films. Surface topography and thickness of PtTe₂ are carried out by using Atomic Force Microscope (AFM, Dimension ICON). Ring-shaped electrodes (5 nm Ti and 45 nm Au) are patterned by photolithography (JB-VIII Mask Aligner) and deposited by e-beam evaporation for the construction of the PtTe₂-based sensor devices.

B. Sensing Test

A Polytetrafluoroethylene chamber with a gas inlet and several windows is designed. The PtTe₂ molecular sensors are placed inside the chamber and connected to an electrochemical workstation with electrical cables. During the sensing test, a specific amount of ammonia is introduced into the chamber filled with high purity nitrogen (N₂). The corresponding ammonia concentration is calculated by volume ratio and is also calibrated by the commercial ammonia detector (AR8500). Last, in-situ tests are performed to obtain current-time curves for the characterization of the response of the sensors.

III. RESULTS AND DISCUSSION

A. PtTe₂ and Ammonia Sensor

The thickness of PtTe₂ can be controlled by evaporating different thicknesses of Pt. In this work, 1 nm Pt film is evaporated on the polyimide, and the thickness of as-grown PtTe₂ is about 6 nm, as shown in Fig. 1.a. The results show that the surface roughness of our prepared film is small with Ra= 2 nm and Rq= 2.7 nm, respectively. In the Raman spectrum, there are two dominant peaks located at ~113.2 and ~156.8 cm⁻¹, corresponding to the in-plane and out-of-plane vibrational motions of Te atoms, which are consistent with the previous study [6], as shown in Fig. 1.b.

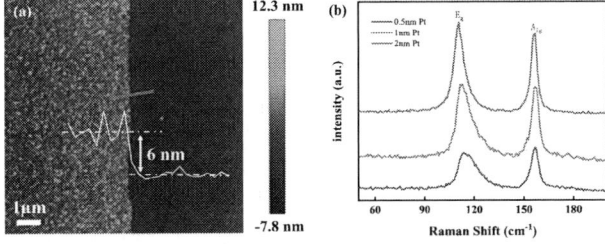

Fig. 1. (a) AFM topography map for PtTe₂ film and substrate. (b) Raman spectrums of 2D PtSe₂ synthesized by various thicknesses of Pt films.

Since holes are the dominant carriers in PtTe₂ synthesized by the TAC method, the process of trapping ammonia molecules in the film can be regarded as an n-type doping process, which results in the rise of the sensor resistance. The response of the sensor can be evaluated by (1).

$$Response = \frac{R - R_0}{R_0} = \frac{V/_I - V/_{I_0}}{V/_{I_0}} \times 100\% \qquad (1)$$

The parameter R_0 and R refers to the resistance before and after ammonia adsorption of the PtTe₂ film. V is the fixed operating voltage of 0.05 V. I and I_0 correspond to the currents in the presence or absence of ammonia.

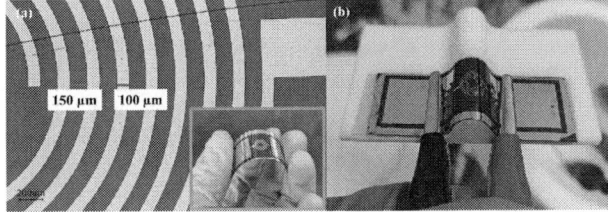

Fig. 2. (a) Microscope image of PtTe₂ sensor on flexible polyimide. (b) In-situ measurement setup of PtTe₂ sensor via customized platform.

To enhance the stability and ensure a high response, ring-shaped electrodes with a width of 100 um are prepared for the sensor devices, whose trench width is 150 um (see Fig. 2.a). Inset shows the overall picture of the device. Fig. 2.b illustrates the in-situ measurement setup for the PtTe₂ sensor *via* customized platform.

B. Response to Ammonia

The real-time response curves of the sensors for both flat and 6 mm strain curvature conditions are shown in Fig. 3.a. Under flat condition, our flexible molecular sensors exhibit a good gradient response to ammonia molecules with a response of 2.5%

at a concentration of 25 ppm and 8.2% at 2000 ppm concentration, which surpasses most single component semimetal materials, shown in Table I. Besides, the response has a 200% improvement at 6 mm strain curvature, which is related to the strain-induced decrease in gas adsorption energy. Similar to semiconductors, the increasing trend of $PtTe_2$ response decreases with increasing ammonia concentration (Fig.3.b), and how to further expand the linear region is a significant challenge for 2D semimetal $PtTe_2$ sensors.

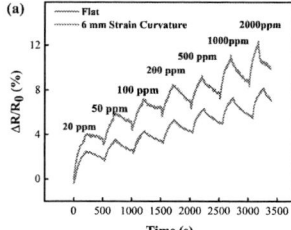 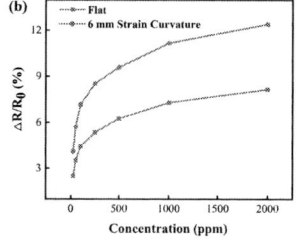

Fig. 3. In-plane sensing responses of the devices. (a) Real-time response test under gradient concentrations (b) The response tendency with increasing concentration, which is similar to that of semiconductors.

TABLE I. COMPARISON OF SEMIMETAL-BASED AMMONIA SENSORS

| Active materials | Sensing performance | | | Substrate | Ref. |
	Sensing range (ppm)	Sensitivity (% ppm^{-1})	LOD (ppm)		
GO	5–40	0.03	5	PET	[7]
Ag–rGO	5–100	0.65	5	PET	[8]
MWCNTs	100–500	NA	100	Nylon	[9]
MoTe$_2$	10-1000	0.048	10	SiO$_2$/Si	[10]
PtTe$_2$	5-2000	0.4	5	PI	This work

The strain platforms, which have curvatures radius of 6, 7, 8, 9, 10 mm, are designed to induce various deformations on the $PtTe_2$ sensors during the testing process. At various ammonia concentrations, the device's response increases with the larger strain (curvatures radius reduced from 10 mm to 6 mm), as shown in Figure 4.a. It is assumed that under a greater degree of strain, the outer layer of the $PtTe_2$ will be subjected to tensile stress resulting in distortion of its lattice. On the one hand, this distortion reduces the adsorption energy of ammonia on $PtTe_2$ according to the first-principal calculation [11]. While on the other hand it exposes more adsorption sites on its surface, which is manifested in the enhanced response of the device. Accordingly, controlled strain engineering may be able to further extend molecular sensors' linearity range and detection limits, which is also of value for other semimetal sensors.

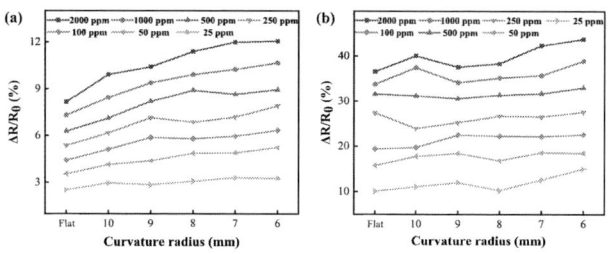

Fig. 4. (a) Strain-enhanced molecular sensing performance. Nearly linear increase in response with strain curvature vary from 10 to 6 mm. (b) Post-deformation response test. At each strain, the response is about 500% of what it was before the large deformation. Although there are some fluctuations in response at certain curvatures, the overall remains stable.

To verify the stability we perform several tests on the device after severe deformation treatment. After the device experienced a large strain with a radius of curvature of 5 mm, there was an irreversible increase in the response. Fig. 4.b shows the sensing response of the device to ammonia under in-situ strain after a large strain bending. The response is still approximately one order of magnitude higher than the previously measured data when returning to the flat and low strain condition. Comparison with data in Fig. 4.a reveals that the sensor gets 400% to 500% improvement in response in each case while the stability decreases. There are fluctuates in the response for each concentration, and the tendency to increase with strain is ambiguous relative to the data in Fig. 4.a., which may be caused by irreversible damage to the material due to large strains. To sum, this means that large strains are beneficial for improving device performance, but may affect sensing stability when re-strained.

IV. CONCLUSION

We report flexible $PtTe_2$-based resistive molecular sensors on polyimide. Gradient concentration measurements show that $PtTe_2$ outperforms other semimetal in terms of ammonia sensing with a high sensitivity. Additionally, our sensors are 500% more sensitive to ammonia molecules after experiencing a large strain with radius of curvature of 5 mm while ensuring acceptable stability. This means that controlled strain engineering can further broaden the linearity range and limits of detection of current sensors. Our work provides a promising approach for realizing flexible molecular sensors with high performance, and the potential for practical use in IoTs, electronic skins and other applications.

ACKNOWLEDGEMENT

The authors acknowledge the funding support from NSFC (92164102), Innovation Talent Program of Jiangsu Province (2019). This research is also supported by Postdoctoral Science Foundation of China (2021M700774) the Fundamental Research Funds for the Central Universities (2242022R20047), Jiangsu Funding Program for Excellent Postdoctoral Talent and Postdoctoral Research Start-up Funding of Southeast University.

REFERENCES

[1] E. Singh, et al., "Flexible Graphene-Based Wearable Gas and Chemical Sensors," ACS Appl. Mater. Interfaces, vol. 9, no. 40, pp. 34544–34586, Oct. 2017.

[2] G. T. Chandran, et al., "Electrically Transduced Sensors Based on Nanomaterials (2012–2016)," Anal. Chem., vol. 89, no. 1, pp. 249–275, Jan. 2017.

[3] S. S. Varghese, et al., "Recent Advances in Graphene Based Gas Sensors," Sensors and Actuators B: Chemical, vol. 218, pp. 160–183, Oct. 2015.

[4] A. Bag and N.-E. Lee, "Gas Sensing with Heterostructures Based on Two-Dimensional Nanostructured Materials: a Review," J. Mater. Chem. C, vol. 7, no. 43, pp. 13367–13383, 2019.

[5] D. Jariwala, et al., "Emerging Device Applications for Semiconducting Two-Dimensional Transition Metal Dichalcogenides," ACS Nano, vol. 8, no. 2, pp. 1102–1120, Feb. 2014.

[6] X.-W. Tong et al., "Direct Tellurization of Pt to Synthesize 2D PtTe$_2$ for High-Performance Broadband Photodetectors and NIR Image Sensors," ACS Appl. Mater. Interfaces, vol. 12, no. 48, pp. 53921–53931, Dec. 2020.

[7] L. T. Duy et al., "Flexible Transparent Reduced Graphene Oxide Sensor Coupled with Organic Dye Molecules for Rapid Dual-Mode Ammonia Gas Detection," Adv. Funct. Mater., vol. 26, no. 24, pp. 4329–4338, Jun. 2016.

[8] L. Zhang, et al., "Highly Sensitive NH$_3$ Wireless Sensor Based on Ag-RGO Composite Operated at Room-Temperature," Sci Rep, vol. 9, no. 1, pp. 9942, Dec. 2019.

[9] Z. Gao et al., "Fiber Gas Sensor-integrated Smart Face Mask for Room-Temperature Distinguishing of Target Gases," Nano Res., vol. 11, no. 1, pp. 511–519, Jan. 2018.

[10] X. Chen, et al., "Gas Sensitive Characteristics of Polyaniline Decorated with Molybdenum Ditelluride Nanosheets," Chemosensors, vol. 10, no. 7, pp. 264, Jul. 2022.

[11] J. Du et al., "Elastic, Electronic and Optical Properties of the Two-Dimensional PtX$_2$ (X = S, Se, and Te) Monolayer," Applied Surface Science, vol. 435, pp. 476–482, Mar. 2018.

BaFe$_{12}$O$_{19}$ based Ferroelectric Memristor for Applications of True Random Number Generator

Ziyang Chen[1,2], Miaocheng Zhang[1,2], Zixuan Ding[1], Aoze Han[1], Xingyu Chen[1], Xinpeng Wang[2], Lei Wang[1]*, Hao Zhang[2]*, Yi Tong[1,2]*

[1] College of Integrated Circuit Science and Engineering, Nanjing University of Posts and Telecommunications, Nanjing 210023, China
[2] Gusu Laboratory of Materials, Suzhou 215123, China

Abstract—**Ferroelectric memristors are in principle a promising candidate for realizing effective computing in memory, for their advantages of multi-bit storage, ultra-fast switching behavior, and low power consumption. Here, we successfully fabricated the Cu/BaFe$_{12}$O$_{19}$/Pt ferroelectric memristive device with multi-resistance state (2 bits), reliable reproducibility (>10^2), and desired on/off ratio (10^3). The conductive mechanism of the device is attributed to the variation of ferroelectric barrier, which is verified by first-principle calculations. In addition, based on the randomness of the SET voltage of the devices, we innovatively propose a schematic diagram of true random number generator (TRNG) circuit. This work may pave the way for next-generation ferroelectric memristors and further enable a broad range of multifunctional applications.**

Keywords—*ferroelectric memristor, BaFe$_{12}$O$_{19}$, multi-bits storage, TRNG*

I. INTRODUCTION

With the advent of post-Moore Era, the traditional Von Neumann computer architecture faces the issue of storage wall, where data communication between memory and processor incurs high power consumption and slow operation speed [1]. As one of the candidates for surpassing traditional logic devices, memristors possess excellent performance of high speed, low power consumption, CMOS compatibility, and the application potential of computing in memory (CIM) [2-3]. Recently, memristors based on ferrielectric materials have attracted great attention to implement non-volatile storage owing to their stable resistance states, excellent endurance, fast read/write speed, and high-integration density [4]. In previous reports, some randomness inherent of resistance switching in the physical process of ferroelectric memristors tended to be ignored [5]. However, these stochastic phenomena may be used as a random source in the application of true random number generator (TRNG), which are widely used in the fields of image encryption, secure communication, and radar waveform design [6]. BaFe$_{12}$O$_{19}$ (BFO), a new kind of ferroelectric material, possesses unsurpassed low cost, excellent chemical stability, and corrosion resistance [7]. In addition, compared with the tradition ferroelectric materials, BFO owns weaker ferroelectric behavior with Pr of 11.8 μC/cm^2, which may be beneficial for the realization of stochastic phenomena. Therefore, the introduction of BFO in ferroelectric memristor and research its electrical characteristics in TRNG is meaningful.

In this work, the BFO ferroelectric memristor has been demonstrated. Specifically, the device exhibits stable non-volatile RS properties *i.e.*, high ON/OFF ratio, stable switching between the high (HRS) and low resistance (LRS) states, and long retention characteristics. Moreover, by setting various magnitude of current compliances (CCs), the device can realize multi-state of resistance. In order to investigate the bipolar switching process of the device, density functional theory (DFT) calculations and Schottky emission (SE) model fitting were performed to demonstrate the atomic structure of BFO and carrier migration mechanism. In addition, a novel true random number generator can be constructed by a random source of memristors with the dispersive set voltage parameters, providing a new idea for TRNG of memristive devices.

II. EXPERIMENTS

To fabricate the Cu/BFO/Pt memristors, the silicon (Si) wafers were ultrasonic cleaned by absolute ethanol. By utilization of physical vapor deposition (PVD) assisting with magnetron sputtering, the 100 nm Pt bottom electrode, 60 nm BaFe$_{12}$O$_{19}$ resistive layer, and 80 nm Cu top electrode was deposited on the Si substrate with shadow mask, respectively. The schematic diagram of 8 × 8 crossbar array of Cu/BFO/Pt ferroelectric memristor is shown in Fig. 1(d). The planar morphology of the device was measured by the metallurgical microscope (Leica DM6000M). The composition of the resistive-switching material BFO was verified by X-ray diffraction (XRD) (PANalytical X'Pert3Powder). The ferroelectric properties of BFO film were characterized by PFM (Burker Dimension ICON). Additionally, the electrical characteristics of the device were tested by semiconductor parameter analyzer (Keithley4200A-SCS) with probe station (Cascade Microtech M150).

III. RESULTS

The XRD result in Fig. 1(a) confirms that the component of the film was BaFe$_{12}$O$_{19}$ (PDF#39-1433: BaFe$_{12}$O$_{19}$). The surficial uniformity of BFO film was characterized by AFM as shown in Fig. 1(b). The surface roughness is 0.881 nm with the scanning area of 10 μm × 10 μm. Fig. 1(c) displays the PFM phase image after poling the film under ±5 V bias voltage, along with 0 bias voltage. The ferroelectric domain in BFO film can be completely inverted, which verifies the ferroelectric characteristic of BFO film. The ferroelectric properties of the BFO film were characterized by the ferroelectric analyzer and PFM. The representative RS behaviors of BFO based memristor are illustrated in Fig. 2. The BFO memristor exhibits stable bipolar RS characteristics. As shown in Fig. 2(a), with the voltage sweeping from - 4 V to + 5 V with 0.1 A of CC, the device can be switched for at least 100 cycles. The typical *I-V* curve has been highlighted in red. The transition process from HRS to LRS is defined as SET. Next, as the positive voltage sweeps back from + 5 V to 0 V, the current decreases proportionally. When the negative voltage sweep was applied to the device, the resistance state returned to HRS. This process is defined as RESET. Similarly, the device exhibits non-volatile behavior when the voltage sweeps back from -4 V to 0 V. To explore the multiple-bit behavior of the device, the *I-V* characteristics under different CCs (0.1 mA, 1 mA, 10 mA, 100 mA) have been measured in Fig. 2(b). It can be observed that higher CC corresponds to higher conductance of device. As the CCs decrease, the resistance of the device in LRS increases gradually. To investigate the RS mechanism of BFO ferroelectric memristors, the fitting of typical *I-V* curve was plotted on a log-log scale, as shown in Fig. 3(a). For the HRS, charge transport function can be classified into two regions: the Ohmic region (linear) and the Schottky emission region (nonlinear). These results indicate that the carrier transport mechanism of LRS is dominated by ohmic conduction, whereas HRS obeys the Schottky emission mode (ln I ~ V$^{1/2}$), as shown in Fig. 3(b) [8]. The possible mechanism may be concluded as followed. When the orientation of ferroelectric polarization to the BFO interface under positive voltage signal, the height of the junction barrier of BFO will be lower. On the contrary, the ferroelectric barrier will be raised when the negative voltage is applied to the memristor [9]. The transmission of carriers is affected by diversion of the ferroelectric barrier. The process was investigated by DFT calculations. The schematic diagram of the sub-circuit to generate true random bits based on the BFO memristor, one switch, and one comparator is demonstrated in Fig. 4. The internal mechanism is dependent on the cumulative probability of SET voltages during 100 cycles in Figure 4a. During SET process, the SET voltage is distinguished by the median value 3.725 V, which could be utilized to generate true random numbers. The timing diagram to generate TRNG is illustrated in Fig. 4(c). The device was initially in HRS. When the applied voltage (V_m) was connected to the devices, the probability of the device which switches to LRS ("1") or HRS ("0") was 50%. The resistance state of the device can be detected by the read voltage (V_{read}) connected to the switch connected to the comparator.

IV. CONCLUSION

In summary, the ferroelectric memristor based on novel $BaFe_{12}O_{19}$ has been successfully fabricated, which exhibits excellent bipolar resistive-switching properties and multi-state behavior (2 bits). The physical characteristic of BFO devices is analyzed by XRD, AFM, and PFM. The RS behavior of the devices may be attributed the variations of the ferroelectric barrier. In addition, TRNG circuit using a random source is proposed. This work demonstrates a promising way for fabrication of the ferroelectric memristors that can help give insight into practical application.

ACKNOWLEDGMENT

This work was supported by the 2030 Major Project of the Chinese Ministry of Science and Technology (2021ZD0201200), the National Natural Science Foundation of China (61964012), and Science and Technology Department of Jiangsu Province (BK20211273). Ziyang Chen and Miaocheng Zhang equally contributed to this paper. (Corresponding author: Lei Wang, email: leiwang1980@njupt.edu.cn; Hao Zhang, email: zhanghao2021@gusulab.ac.cn; Yi Tong, email: tongyi@njupt.edu.cn.)

REFERENCES

[1] Sebastian, Abu, et al, "Memory devices and applications for in-memory computing," *Nature nanotechnology*, vol. 15, (7), pp. 529-544, 2020.

[2] Li, Can, et al, "Efficient and self-adaptive in-situ learning in multilayer memristor neural networks," *Nature communications*, vol. 9, (1), pp. 1-8, 2018.

[3] Wang, Xiao-Yuan, et al, "FPGA synthesis of ternary memristor-CMOS decoders for active matrix microdisplays," *IEEE Transactions on Circuits and Systems I: Regular Papers*, 2022.

[4] Dongale, Tukaram D., et al, "Recent Progress in Selector and Self-Rectifying Devices for Resistive Random-Access Memory Application," *physica status solidi (RRL)–Rapid Research Letters*, vol. 15, (9), 2021.

[5] Qin, Qi, et al, "Fabrication and investigation of ferroelectric memristors with various synaptic plasticities," *Chinese Physics B*, vol. 31, (7), 2022.

[6] Sun, Bai, et al, "A True Random Number Generator Based on Ionic Liquid Modulated Memristors," *ACS Applied Electronic Materials*, vol. 3, (5), pp. 2380-2388, 2021.

[7] Li, Xue, and Guo-Long Tan, "Multiferroic and magnetoelectronic polarizations in $BaFe_{12}O_{19}$ system," *Journal of Alloys and Compounds*, vol. 858, 2021.

[8] Tsai, Shu - Chin, et al, "Structural Analysis and Performance in a Dual - Mechanism Conductive Filament Memristor," *Advanced Electronic Materials*, vol. 7, (10), 2021.

[9] Ma, Chao, et al, "Sub-nanosecond memristor based on ferroelectric tunnel junction," *Nature communications*, vol. 11, (1), pp. 1-9, 2020.

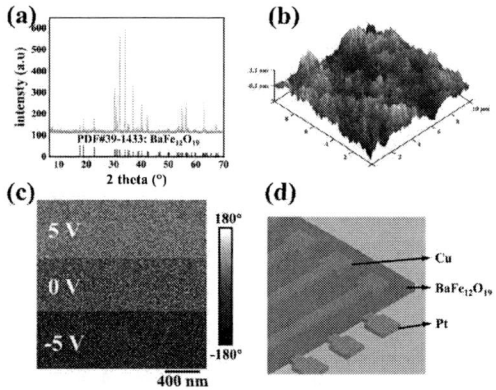

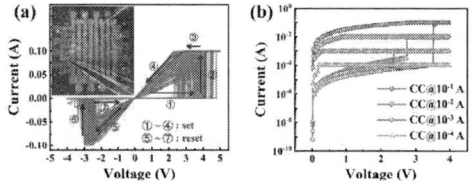

Fig. 1. (a) The analysis of the composition of BFO flakes by XRD. (b) The AFM image of BFO film. The roughness (Rq) is 0.881 nm. (c) PFM image of BFO film. (d) The structure of Cu/BFO/Pt memristor devices.

Fig. 2. (a) Typical DC *I-V* characteristics of the Cu/BFO/Pt memristor device. (b) *I-V* curves under different I_{CCS}.

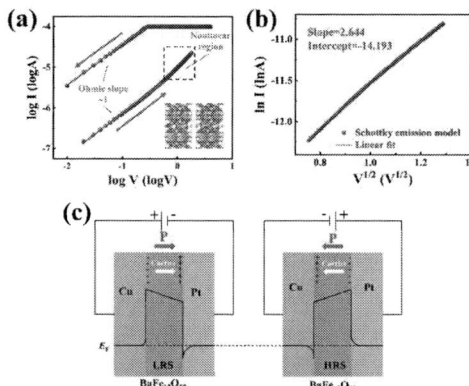

Fig. 3. (a)(b) Schottky emission model fitting for positive region in the SET process. The internal mechanism was verified by first-principle calculations. (c) Schematic illustrations of the switching mechanism.

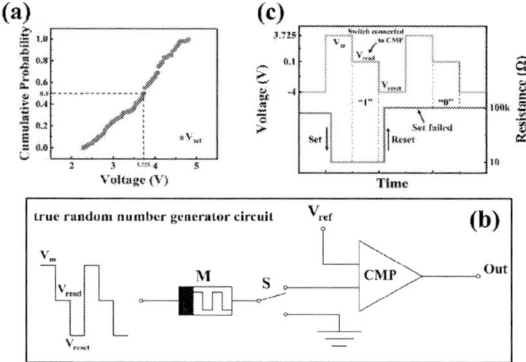

Fig. 4. Schematic circuit of TRNG.

Memristor-based Digital Circuits for Realizing the Pavlov's Associative Neural Network

Yu Wang[1], Yi Liu[1], Jiayu Bao[1], Yu Yan[1], Ertao Hu[1], Xiang Wan[1], Rongqing Xu[1*], Hao Zhang[2*], Yi Tong[1,2*]

[1] College of Electronic and Optical Engineering&College of Flexible Electronics (Future Technology), Nanjing University of Posts and Telecommunications, Nanjing, China
[2] Gusu Laboratory of Materials, Suzhou, China
* Email: tongyi@njupt.edu.cn

Abstract—Memristors have sparked substantial interest in the hardware implementation of brain-inspired neuromorphic devices and systems. In this work, we propose a digital circuit to emulate the Pavlov's associative memory experiments based on fabricated Ag/TiO₂/Pt memristors. Memristors operate as a logical signal processing unit in conjunction with the register to implement the emulation. The design of digital circuitry substantially increases the frequency of the system and reduces its power consumption and cost.

Keywords—memristor, digital circuit, associative memory

I. INTRODUCTION

Inspired by biological neural network (BNN), the technology of brain-inspired devices and systems is advancing rapidly and has become a focus of academic and commercial research [1-4]. As manufactured by Hewlett-Packard in 2008, memristors have been shown to mimic brain-inspired processes such as the Pavlov's associative learning [5, 6]. The Pavlov's associative memory network is a classical experiment that describes the conditional reflex, including learning principles [7]. However, almost all other memristor-based Pavlov's associative memory networks are analogue circuitry at the circuit or system level, resulting in low accuracy and slow speed of the circuitry [8].

In this work, we fabricated the 12×12 Ag/TiO2/Pt crossbarmemristors with a 150-μm feature size, exhibits ~0.50 Vthreshold voltage, and good stability of resistance. Based on thefitting of the VTEAM model, we have constructed a memristor-based digital circuitry for implementing the Pavlov'sassociative memory networks. This whole digital circuitryrequires just four flip-flop triggers, and sixteen memristors,exhibiting an operating frequency of several megahertz and apower consumption of microwatts.

II. EXPERIMENTS

The 12×12 Ag/TiO₂/Pt crossbar memristors were fabricated by magnetron sputtering. Firstly, the 80-nm Pt bottom electrode (BE) was sputtered on the 300-μm-thick SiO₂/Si wafer by the first mask. Then, a 100-nm TiO₂ switching layer was deposited subsequently on the BE using the second mask. Finally, Ag (100 nm) as top electrodes were deposited on the TiO₂ film by magnetron sputtering through the third mask. All the electrical characteristics were measured by Keithley 4200A-SCS semiconductor characterization analyzer at room temperature.

III. RESULTS

Fig. 1. (a) illustrates the structure schematic of the 12×12 Ag/TiO₂/Pt crossbar memristors. The metallographic microscopy crossbar array picture in Fig. 1. (b) depicts the 150-μm feature size of memristors. With the application of positive direct current (DC) sweeping voltage (0 - 0.5 - 0 V), the device sharply switches from a high resistance state (HRS) to a low resistance state (LRS) at a threshold voltage (V_{set}) of ~ 0.5 V, This process is known as the "SET" process. For the negative DC bias (0 - -0.4 - 0 V), this device abruptly shifts from the LRS to the original HRS at a voltage (V_{reset}) of ~ -0.4 V, a process known as "RESET". According to Fig. 1. (c), the measured *I-V* curves (red lines and balls) demonstrate the non-volatility of the Ag/TiO₂/Pt memristors, which could be better fitted (blue dashed lines) by the model of the voltage threshold adaptive memristor (VTEAM). Fig. 1. (c) and (e) plot the retention and its cumulative probability distribution of ON and OFF resistance states, respectively.

Fig. 2 sketches the process of the Pavlov's associative memory network. In Pavlov's experiments, the dog initially started to salivate when it saw food. The ring alone did not cause the dog to salivate. The dog salivated when the food and rings were applied simultaneously. After several training sessions, the dog could salivate when the ring rang on its own. As a result, the dog has learned to associate the conditions of food and ringing bell.

The designed circuit of the Pavlov's associative memory network is plotted in Fig. 3, which contains a 4-bit shift register and the module of memristors. This shift register could move signals of DATA to its output (Q1, Q2, Q3, and Q4) once every clock cycle. Following this, pairs of parallel connected memristors perform AND or OR logical operations depending on whether the signal comes in through the BE or TE of the device, respectively.

The output results are displayed in Fig. 4, from where we can see that the signals are divided into Bell, Food, DATA, and salivation, which are clock cycle, periodic square wave, constant high-level voltage, and output signals. The high-level and low-level voltages are set at 0.4 and 0.1 V respectively. VOUT1 and VOUT4 are respectively designated as the least significant bit (LSB) and the most significant bit (MSB), where the "1111" indicates that the dog starts to salivate. And the parallel VOUT4, VOUT3, VOUT2, and VOUT1 signal of "0000", "0001", and "0111" means that the dog does not salivate as summarized in Table I. At the first clock, the output logical state is "0000". The output of the second clock is "1111", whereas the Food and Bell are both high-level voltages, which is the training for Pavlov's dog. The output logical state switches to "0001" on the subsequent clock cycle. After two training sessions, the output is "1111", whereas the Bell has a high-level voltage and the Food has a low value. It indicates that the dog has associated the rings with food.

Table II lists the simulated performance of this digital circuits. The supply voltage is 2.4 / 0.1 V. The operating frequency can reach 7.6 MHz. The circuit power consumption, including the 4-bit shift register and all memristors, is 26.04 μW. The area is simulated as 455.94 μm².

IV. CONCLUSION

In this paper, Ag/TiO₂/Pt 12×12 crossbar memristors are prepared and fitted better by the VTEAM model. On this basis, a memristor-based digital circuitry for implementing the Pavlov's associative memory networks has been proposed and exhibits high frequency and quantitative associative learning. This memristor-based digital realization of analog circuits may provide a method for achieving large-scale integration in brain-inspired circuits and systems.

ACKNOWLEDGMENT

This work was supported in part by the 2030 Major Project of the Chinese Ministry of Science and Technology (Grant No 2021ZD0201200).

REFERENCES

[1] B. Gao, Y. Zhou, Q. Zhang, S. Zhang, P. Yao, Y. Xi, Q. Liu, M. Zhao, W. Zhang, and Z. J. N. c. Liu, "Memristor-based analogue computing for brain-inspired sound localization with in situ training," *Nature Communications*, vol. 13, no. 1, pp. 1-8, 2022.

[2] Y. Zhang, P. Qu, Y. Ji, W. Zhang, G. Gao, G. Wang, S. Song, G. Li, W. Chen, W. Zheng, F. Chen, J. Pei, R. Zhao, M. Zhao, and L. Shi, "A system hierarchy for brain-inspired computing," *Nature*, vol. 586, no. 7829, pp. 378-384, 2020.

[3] P. Yao, H. Wu, B. Gao, J. Tang, Q. Zhang, W. Zhang, J. J. Yang, and H. Qian, "Fully hardware-implemented memristor convolutional neural network," *Nature*, vol. 577, no. 7792, pp. 641-646, 2020.

[4] H. Zhou, J. Chen, Y. Wang, S. Liu, Y. Li, Q. Li, Q. Liu, Z. Wang, Y. He, H. Xu, and X. Miao, "Energy‐Efficient Memristive Euclidean Distance Engine for Brain‐Inspired Competitive Learning," *Advanced Intelligent Systems*, vol. 3, no. 11, 2021.

[5] D. B. Strukov, G. S. Snider, D. R. Stewart, and R. S. Williams, "The missing memristor found," *Nature*, vol. 453, no. 7191, pp. 80-3, 2008.

[6] S. G. Hu, Y. Liu, Z. Liu, T. P. Chen, Q. Yu, L. J. Deng, Y. Yin, and S. Hosaka, "Synaptic long-term potentiation realized in Pavlov's dog model based on a NiOx-based memristor," *Journal of Applied Physics*, vol. 116, no. 21, 2014.

[7] S. Maren, "Synaptic mechanisms of associative memory in the amygdala," *Neuron*, vol. 47, no. 6, pp. 783-6, 2005.

[8] Z. Wang and X. Wang, "A Novel Memristor-Based Circuit Implementation of Full-Function Pavlov Associative Memory Accorded With Biological Feature," *IEEE Transactions on Circuits and Systems I: Regular Papers*, vol. 65, no. 7, pp. 2210-2220, 2018.

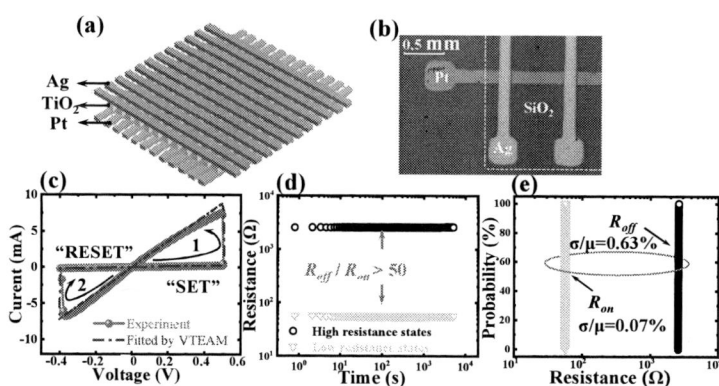

Fig. 1. (a) The structure schematic, (b) The optical crossbar array image, (c) DC *I-V* measured and fitted curves of Ag/TiO₂/Pt memristors, (d) Retention and (e) Cumulative probability distribution of ON and OFF resistance states.

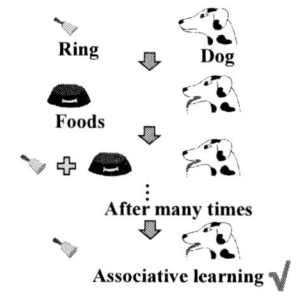

Fig. 2 Biological behaviors of the Pavlov's experiments.

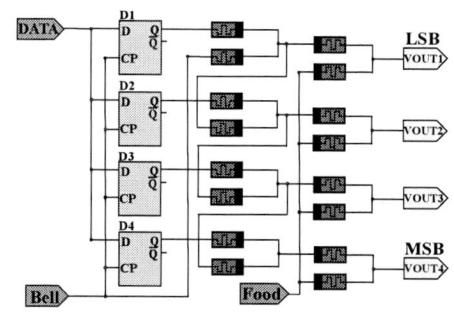

Fig. 3 Schematic circuits of the Pavlov's associative neural networks.

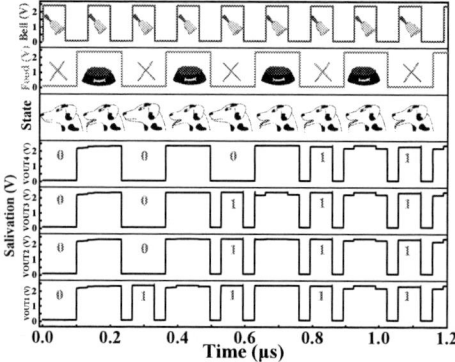

Fig. 4 Simulated results of the Pavlov's associative neural networks.

TABLE I.   SALIVATION TRUTH TABLE

Output Results				
VOUT1	VOUT2	VOUT3	VOUT4	Salivate
0	0	0	0	No
1	0	0	0	No
1	1	1	0	No
1	1	1	1	Yes

TABLE II.   PERFORMANCE

Supply Voltage	Frequency	Power	Area
2.4 / 0.1 V	7.6 MHz	26.04 μW	455.94 μm²

A Split-Ring Resonator with Interdigitated Electrodes Aimed at the Dielectric Characterization of Liquid Mixtures (Invited Paper)

Giovanni Gugliandolo[1], Xiue Bao[2,3], Haoyun Yuan[2], Jinkai Li[3], Juncheng Bao[4], Giovanni Crupi[5], Nicola Donato[1]
[1] Department of Engineering, University of Messina, 98166 Messina, Italy
[2] School of Integrated Circuits and Electronics, Beijing Institute of Technology Beijing, 100081 Beijing, China
[3] Tangshan Research Institute of BIT, 063000 Tangshan, China
[4] Div. ESAT-WAVECORE, KU Leuven, 3001 Leuven, Belgium
[5] BIOMORF Department, University of Messina, 98125 Messina, Italy

Abstract—This paper is focused on a microwave sensor for the evaluation of the dielectric properties of binary liquid mixtures at RF/microwave frequencies. The sensor consists of a split ring resonator (SRR), built using the microstrip technology. Interdigitated electrodes are integrated into the ring as a sensing element for liquid detection. A proper extraction procedure has been proposed for the accurate evaluation of the resonant frequency of the developed prototype. The resonant extraction procedure is based on the analysis of the frequency-dependent behavior of the complex forward transmission coefficient (S_{21}) that is accurately modeled locally around the resonance by using a fitting function. According to the tests carried out with water-isopropanol liquid mixtures at various volume fractions, the studied device is more sensitive than the more conventional SRR sensor.

Keywords—sensor, interdigitated electrodes, liquid mixtures, split-ring resonator (SRR), dielectric characterization, permittivity, modeling.

I. INTRODUCTION

Dielectric spectroscopy is a prominent and widespread method used for materials characterization. It measures the material dielectric response to electromagnetic fields [1], enabling the analysis of the frequency-dependent dielectric properties of various biological samples, such as tissues, blood, proteins, viruses, and cultured cells [2]–[6]. The main advantages of the spectroscopy characterization technique are the immediate real-time quantification of the material properties and its easy application in the laboratory (thus simple usage). Non-invasive, real-time measurements also imply an easy application. These features make such technique even more attractive for materials characterization. In recent years, this technique has become more and more attractive in biological and chemical applications due to the combination of permittivity measurement techniques with microfluidic devices. This is because the technology has several benefits, such as high sensitivity, robustness, and reduced manufacturing cost [7], [8].

The presented research activity is focused on the effective permittivity evaluation of a liquid binary mixture at microwave frequencies. The proposed sensor is fabricated in microstrip technology and is based on the conventional split-ring resonator (SRR) combined with interdigitated electrodes (IDE) that acts as sensing elements. The classic SRR design includes a simple gap that acts as a capacitance [9]. In the proposed IDE-SRR sensor, an interdigitated capacitor made by IDEs is placed in place of the traditional gap, thus forming the sensing element used to evaluate the dielectric properties of liquid mixtures with relatively high sensitivity. A two-dimensional (2D) sketch of the

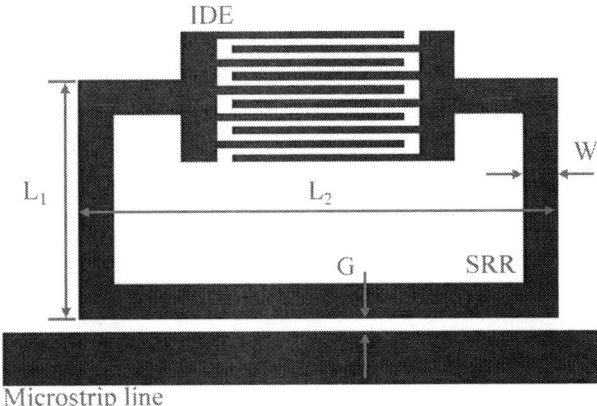

Fig. 1: 2D sketch of the IDE-SRR. $L_1 = 7$ mm; $L_2 = 14$ mm; $W = 1$ mm; $G = 0.4$ mm.

IDE-SRR is reported in Fig.1

This geometry has been successfully used in [10]. It inherits the SRR's high sensitivity and the IDE's linear relationship between its exhibited capacitance and the permittivity of the surrounding material. A comparison of the state-of-the-art microwave sensors aimed at the dielectric characterization of liquid mixtures is reported in [10].

The interdigitated capacitor is a widely employed device in RF/microwave field [11], [12]. It has been used as a transducer in gas sensing applications [13]–[15], as a sensing element for remote relative humidity measurement [16], and for dielectric materials characterization [17], [18].

In the present study, the behavior of the complex forward transmission coefficient (S_{21}) as a function of the frequency is straightforwardly and faithfully modeled locally around the resonance through a fitting process, which is carried out in Python using the Levenberg–Marquardt algorithm. The achieved results of the fitting procedure are exploited for obtaining an accurate evaluation of the resonant frequency (f_r) of the studied sensor. As a case study, a water-isopropanol liquid mixtures at various volume fractions in considered. Depending on the isopropanol content in water, a variation of f_r from 443.43 MHz to 892.07 MHz is observed, implying that a variation of the isopropanol content can be detected as a variation of f_r.

The paper is organized as follows. In the next section more details on the sensor design and fabrication are given. Moreover, the experimental characterization is described. Next, the proposed methodology for resonant frequency extraction is presented and discussed. The main achieved results are reported in Section III, while conclusive remarks are given in Section IV.

II. SENSOR DESIGN AND CHARACTERIZATION

A. Sensor Design

The sensor is designed based on a microstrip line, on one side of which a coplanar split ring is designed. The distance between the ring and the microstrip line is G; the lengths of the ring in the direction of perpendicular and parallel to the line are L_1 and L_2, respectively; and the width of the ring is W. A borofloat 33 glass is used as the substrate. This substrate is 1 mm thick and is characterized by a dielectric constant $\varepsilon'_r = 4.6$ and a

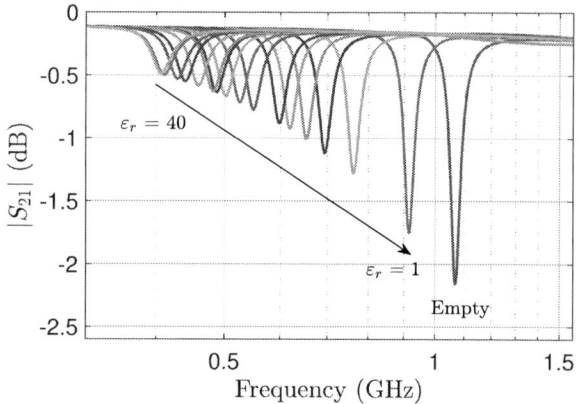

(a)

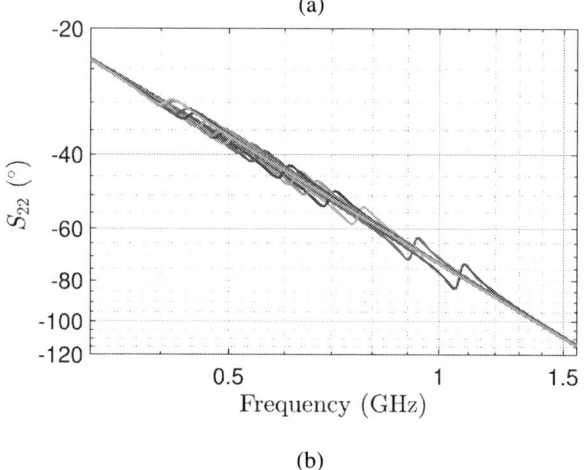

(b)

Fig. 2: (a) Magnitude and (b) phase of the simulated scattering parameter S_{21} for the IDE-SRR loaded with a group of materials, whose real part permittivity is decreasing from 40 to 1 while the imaginary part permittivity is 0.

loss tangent $\tan\delta = 0.0037$ (at 1 MHz and 25 °C). Additional information of the prototype design can be found in [10].

To ensure liquid confinement, a Polydimethylsiloxane (PDMS) ring is attached to the surface of the IDE-SRR device. This allows the liquid under test (LUT) to be loaded on top of the sensing region. Moreover, in order to guarantee that the measurement of the liquid mixture permittivity is independent of the LUT volume, a thickness of 3 mm is chosen for the PDMS ring [19].

B. Resonance Modeling and Resonant Frequency Extraction

A series of simulations are performed on the designed IDE-SRR, which is loaded with a material whose real part permittivity is set at 1, 2.5, 5, ..., 40, whereas the imaginary part permittivity is set at 0. Fig. 2 presents the simulated scattering parameters. In the traditional method, the resonance frequency and quality factor (Q) are read directly from the magnitude of S_{21}. The read values are presented in Table I. However, this method is complicated, time-consuming and may results in a low accurate estimation of the resonant parameters (i.e., f_r and Q).

In general, the accurate estimation of the resonant frequency of a microwave resonator is not a simple task. Noise in the

acquired scattering (S-) parameters, limited number of points or resonator non-idealities may bring to a low accurate evaluation of f_r. The literature proposes alternative methods based on the determination of the complex scattering coefficients S_{ij} of the two-port resonator, and their fitting on the complex plane [20], [21]. This approach is useful to compensate the device non-idealities which alter the sensor response affecting the accurate evaluation of the resonant parameters. In this work, the complex S_{21} parameter has been modeled as:

$$S_{21}(f) = 1 - \frac{Q/S_c}{1 + 2iQ(\frac{f - f_r}{f_r})} + \sum_{k=0}^{N} c_k (f - f_r)^k \quad (1)$$

The first term is used to describe the resonant dip [22], while the second term is used to model the background signal [23]. In this work an $N = 2$ was enough to describe the background. Q and f_r are the sensor quality factor and resonant frequency, respectively. S_c and c_k are two complex parameters. The fitting process is carried out in Python using the *lmfit* package [24] and selecting the Levenberg–Marquardt algorithm. The Python script takes the complex transmission coefficient S_{21} as input parameter, performs the fitting procedure, and provides as output the model function that best describes the input parameter. From the model, it is possible to have an estimation of both f_r and Q. The higher is the fitting model goodness (coefficient of determination $R^2 \approx 1$) the more accurate is the estimation of the resonant parameters. The fitting results for the simulation with $\varepsilon'_r = 1.0$ are reported in Fig. 3. In this case, the model provides a $R^2 > 0.999$ as proof of a good fitting. Moreover, the simulation and fitting tracks in Figs. 3a and 3b are completely overlapped.

The fitting procedure has been carried out for each ε'_r considered in this study, and the extracted resonance frequencies at different ε'_r settings are presented in Table I. It can be seen that the extracted resonance frequency points by using the proposed method are in good agreement with the read-out ones with differences lower than 5 MHz.

TABLE I. Comparison of the extracted parameters using two different methods. f'_r is the resonance frequency value evaluated with the read-out method; f_r is the resonance frequency extracted using the proposed model, Δf_r is defined as $f'_r - f_r$.

ε'_r	f'_r (MHz)	f_r (MHz)	Δf_r (MHz)
1.0	1070.0	1067.63	2.37
2.5	910.0	914.78	-4.78
5.0	760.0	761.52	-1.52
7.5	690.0	693.32	-3.32
10.0	650.0	652.55	-2.55
12.5	620.0	617.78	2.22
15.0	600.0	596.92	3.08
17.5	550.0	548.20	1.80
20.0	520.0	524.43	-4.43
22.5	500.0	500.53	-0.53
25.0	490.0	485.57	4.43
27.5	480.0	480.28	-0.28
30.0	460.0	457.56	2.44
32.5	440.0	438.04	1.96
35.0	430.0	427.23	2.77
37.5	410.0	408.90	1.10
40.0	410.0	405.23	4.77

The ε'_r values are also reported in Fig. 4 as a function of $1/f_r^2$ as well as a linear regression.

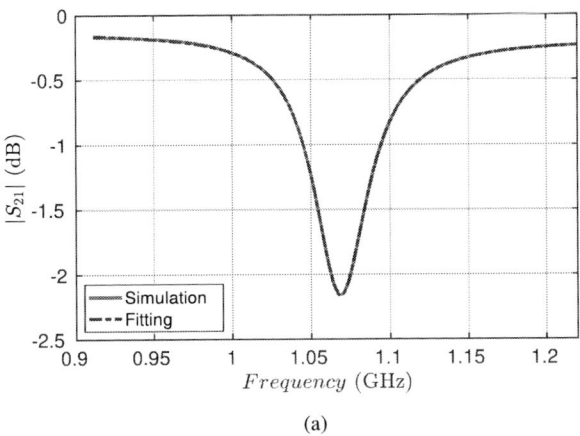

(a)

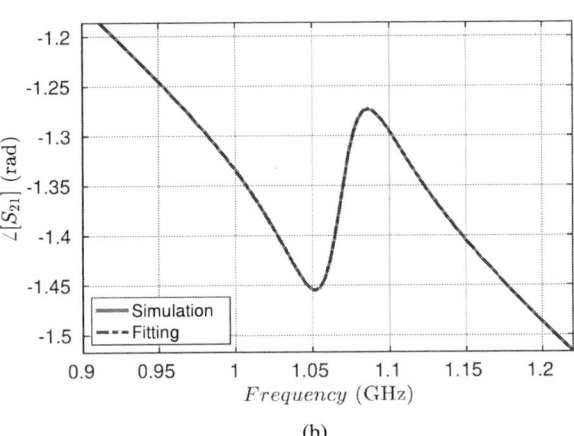

(b)

Fig. 3: Comparison between simulations (red) and fitting (blue) of the frequency-dependent behavior of S_{21} for the studied sensor: (a) magnitude and (b) phase.

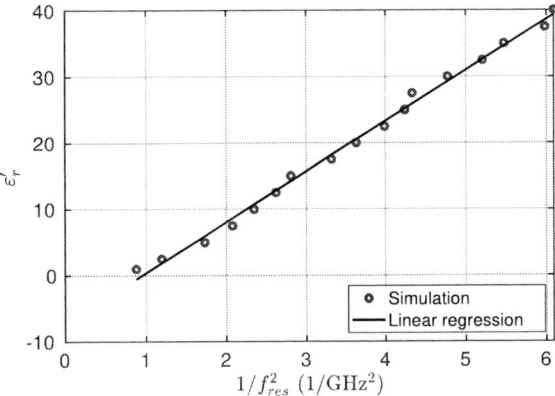

Fig. 4: Simulations (blue symbols) and fitting (black line) of the real part of the complex permittivity as a function of $1/f_r^2$.

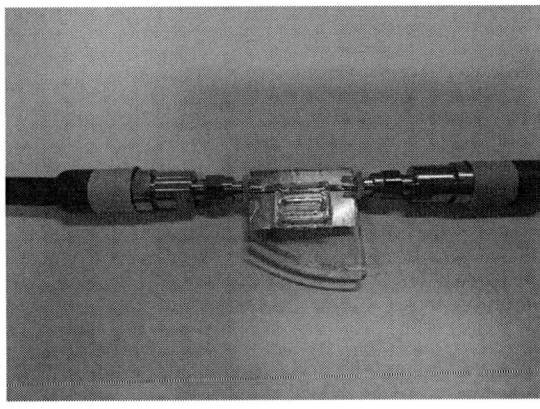

Fig. 5: A schematic diagram of practically fabricated IDE-SRR sensor for mixture characterization.

III. MEASUREMENT AND RESULTS

The sensor is fabricated using the standard photolithography and chemical etching techniques on a Borofloat 33 glass wafer. A photo of the fabricated device is presented in Fig. 5. Measurements on the IDE-SRR through liquid mixtures are performed by means of the Keysight M9375A PXIe Vector Network Analyzer (VNA). A two-port short-open-load-thru (SOLT) calibration procedure is carried out prior to any tests in order to get rid of contribution of cables and connectors between the VNA and the sensor under test. To accomplish this task, the commercial Agilent 85052B 3.5 mm calibration kit is used. For each measurement, 1601 frequency points spanning from 300 kHz to 3 GHz are recorded.

Tests are conducted considering a binary mixture of isopropanol (99.9% purity) and deionized water at different volume fractions, from 0% of isopropanol (i.e., pure water) to 100% with steps of 5%, for a total 21 measurements.

The fitting procedure has been exploited for the evaluation of the sensor resonant frequency at each isopropanol volume fraction considered in this study. The unloaded practical IDE-SRR is characterized by a resonant frequency f_r of 1.056 GHz ($Q = 16.3$) that results in a dip on the magnitude of the forward transmission coefficient (S_{21}). When it is loaded with the water-isopropanol mixture, f_r varies from 443.43 MHz ($Q = 8.1$) to 892.07 MHz ($Q = 3.9$), depending on the isopropanol content in water.

The dependence of the IDE-SRR resonant frequency on the isopropanol volume fraction (ϕ_{isp}) is reported in Fig. 6a. The resonant frequency increases with ϕ_{isp}, this results in a decrease of the quantity $1/f_r^2$. For the sake of completeness, the Q factor is reported in Fig. 6b as a function of ϕ_{isp}. It is characterized by a clear downward trend even if the signal-to-noise ration is lower in this case.

The results, combined with the permittivity measurements carried out with the interdigital capacitor sensor [25], are shown in Fig. 7. Here the real part of the complex permittivity (i.e., ε_r') of the liquid mixture is plotted against $1/f_r^2$ and exhibits a pretty linear trend:

$$\varepsilon_r' = 17.835 \times 1/f_r^2 - 14.858 \qquad (2)$$

where ε_r' is evaluated from the measurement using the microstrip interdigital capacitor technique and f_r (expressed in GHz) is extracted with the fitting procedure from the proposed IDE-SRR sensor. The linear fitting describes pretty well the variation of the ε_r' with $1/f_r^2$ ($R^2 = 0.98$). At low $1/f_r^2$ values, some deviations can be observed between the linear trend and the measurements. This can be ascribed to the fact that the isopropanol at high concentrations is highly volatile thus, during

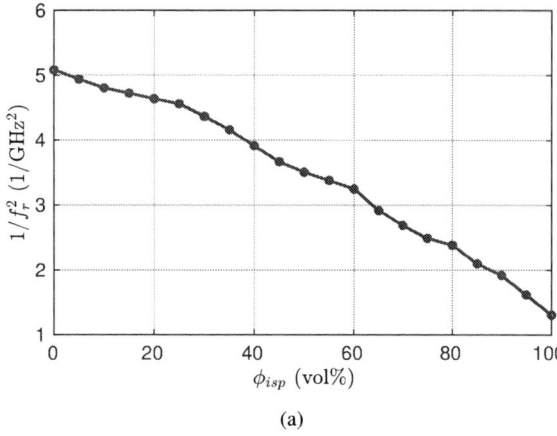

(a)

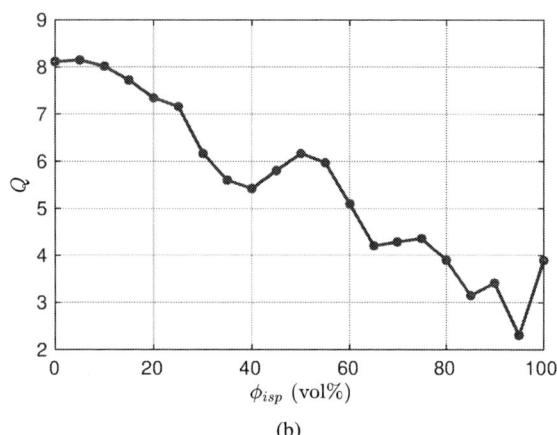

(b)

Fig. 6: IDE-SRR resonant frequency (a) and Q factor (b) as a function of the isopropanol concentration in water (ϕ_{isp}).

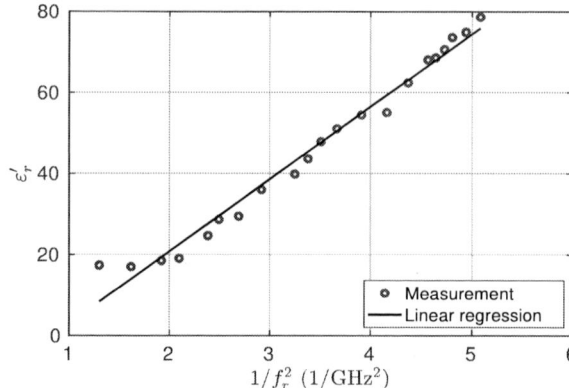

Fig. 7: Measurement (blue symbols) and fitting (black line) of the real part of the complex permittivity of the liquid mixture as a function of $1/f_r^2$.

the measurements, the local temperature might have changed, resulting in the change of the measured ε'_r.

It is worth noting that the linear relationship is different between the simulation and the measurement (see Figs. 4 and 7). This can be explained by the effects of the SMA connectors which are not considered in the simulation. Moreover, the dielectric properties of PDMS set in the simulation are also different from the practical values in the measurements. However, this is not a limitation for this study since Fig. 4 is mainly used to validate the linear relationship between ε'_r and $1/f_r^2$ while Fig. 7 represents the actual relationship used for the sensor.

IV. CONCLUSION

In this paper a microwave sensor for the evaluation of the dielectric properties of binary liquid mixtures is investigated. The device is an improved version of the classic SRR, with interdigitated electrodes used as a sensing element. A resonant frequency extraction procedure has been proposed and the sensor has been validated considering a water-isopropanol liquid mixtures as case study with promising results. As a future perspective, the device will be used with additional liquid mixture for their dielectric characterization.

REFERENCES

[1] F. Kremer, "Dielectric spectroscopy–yesterday, today and tomorrow," *J. Non-Cryst. Solids*, vol. 305, no. 1, pp. 1–9, 2002.
[2] C. Gabriel, S. Gabriel, and Y. Corthout, "The dielectric properties of biological tissues: I. literature survey," *Physics in medicine & biology*, vol. 41, no. 11, p. 2231, 1996.
[3] G. Facer, D. Notterman, and L. Sohn, "Dielectric spectroscopy for bioanalysis: From 40 hz to 26.5 ghz in a microfabricated wave guide," *Appl. Phys. Lett.*, vol. 78, no. 7, pp. 996–998, 2001.
[4] K. Yokoyama, T. Kamei, H. Minami, and M. Suzuki, "Hydration study of globular proteins by microwave dielectric spectroscopy," *The Journal of Physical Chemistry B*, vol. 105, no. 50, pp. 12 622–12 627, 2001.
[5] S.-C. Yang, H.-C. Lin, T.-M. Liu, J.-T. Lu, W.-T. Hung, Y.-R. Huang, Y.-C. Tsai, C.-L. Kao, S.-Y. Chen, and C.-K. Sun, "Efficient structure resonance energy transfer from microwaves to confined acoustic vibrations in viruses," *Scientific reports*, vol. 5, no. 1, pp. 1–10, 2015.
[6] K. Asami, "Characterization of biological cells by dielectric spectroscopy," *J. Non-Cryst. Solids*, vol. 305, no. 1, pp. 268–277, July 2002.
[7] J. C. Booth, N. D. Orloff, J. Mateu, M. Janezic, M. Rinehart, and J. A. Beall, "Quantitative permittivity measurements of nanoliter liquid volumes in microfluidic channels to 40 GHz," *IEEE Trans. Instrum. Meas.*, vol. 59, no. 12, pp. 3279–3288, Dec. 2010.
[8] X. Bao, I. Ocket, J. Bao, J. Doijen, J. Zheng, D. Kil, Z. Liu, B. Puers, D. Schreurs, and B. Nauwelaers, "Broadband dielectric spectroscopy of cell cultures," *IEEE Trans. Microw. Theory Techn.*, vol. 66, no. 12, pp. 5750–5759, Oct. 2018.
[9] G. Gugliandolo, G. Vermiglio, G. Cutroneo, G. Campobello, Giuseppe amd Crupi, and N. Donato, "Inkjet-printed capacitive coupled ring resonators aimed at the characterization of cell cultures," in *2022 IEEE International Symposium on Medical Measurements and Applications (MeMeA)*. IEEE, 2022, pp. 1–5.
[10] X. Bao, M. Zhang, I. Ocket, J. Bao, D. Kil, Z. Liu, R. Puers, D. Schreurs, and B. Nauwelaers, "Integration of interdigitated electrodes in split-ring resonator for detecting liquid mixtures," *IEEE Transactions on Microwave Theory and Techniques*, vol. 68, no. 6, pp. 2080–2089, 2020.
[11] G. Gugliandolo, Z. Marinković, A. Quattrocchi, G. Crupi, and N. Donato, "Development of an inkjet-printed interdigitated device: Cad, fabrication, and testing," in *2021 IEEE International Conference on Integrated Circuits, Technologies and Applications (ICTA)*, 2021, pp. 153–154.
[12] G. Gugliandolo, K. Naishadham, N. Donato, G. Neri, and V. Fernicola, "Sensor-integrated aperture coupled patch antenna," in *2019 IEEE International Symposium on Measurements & Networking (M&N)*. IEEE, 2019, pp. 1–5.
[13] G. Gugliandolo, K. Naishadham, G. Crupi, and N. Donato, "Design and characterization of a microwave transducer for gas sensing applications," *Chemosensors*, vol. 10, no. 4, p. 127, 2022.
[14] Z. Marinković, G. Gugliandolo, M. Latino, G. Campobello, G. Crupi, and N. Donato, "Characterization and neural modeling of a microwave gas sensor for oxygen detection aimed at healthcare applications," *Sensors*, vol. 20, no. 24, p. 7150, 2020.
[15] G. Gugliandolo, D. Aloisio, S. Leonardi, G. Campobello, and N. Donato, "Resonant devices and gas sensing: From low frequencies to microwave range," in *2019 14th International Conference on Advanced Technologies,*

978-1-6654-9270-6/22 $31.00 © 2022 IEEE

Systems and Services in Telecommunications (TELSIKS). IEEE, 2019, pp. 21–28.

[16] G. Gugliandolo, K. Naishadham, G. Neri, V. C. Fernicola, and N. Donato, "A novel sensor-integrated aperture coupled microwave patch resonator for humidity detection," *IEEE Transactions on Instrumentation and Measurement*, vol. 70, pp. 1–11, 2021.

[17] X. Bao, I. Ocket, G. Crupi, D. Schreurs, J. Bao, D. Kil, B. Puers, and B. Nauwelaers, "A planar one-port microwave microfluidic sensor for microliter liquids characterization," *IEEE J. Electromagn. RF Microw. Med. Biol.*, vol. 2, no. 1, pp. 10–17, Feb. 2018.

[18] X. Bao, G. Crupi, I. Ocket, J. Bao, F. Ceyssens, M. Kraft, B. Nauwelaers, and D. Schreurs, "Numerical modeling of two microwave sensors for biomedical applications," *International Journal of Numerical Modelling: Electronic Networks, Devices and Fields*, vol. 34, no. 1, p. e2810, 2021.

[19] X. Bao, I. Ocket, J. Bao, Z. Liu, B. Puers, D. M.-P. Schreurs, and B. Nauwelaers, "Modeling of coplanar interdigital capacitor for microwave microfluidic application," *IEEE Trans. Microw. Theory Techn.*, vol. 67, no. 7, pp. 2674–2683, Jun. 2019.

[20] N. Pompeo, K. Torokhtii, F. Leccese, A. Scorza, S. Sciuto, and E. Silva, "Fitting strategy of resonance curves from microwave resonators with non-idealities," in *2017 IEEE International Instrumentation and Measurement Technology Conference (I2MTC).* IEEE, 2017, pp. 1–6.

[21] G. Gugliandolo, S. Tabandeh, L. Rosso, D. Smorgon, and V. Fernicola, "Whispering gallery mode resonators for precision temperature metrology applications," *Sensors*, vol. 21, no. 8, p. 2844, 2021.

[22] M. S. Khalil, M. Stoutimore, F. Wellstood, and K. Osborn, "An analysis method for asymmetric resonator transmission applied to superconducting devices," *Journal of Applied Physics*, vol. 111, no. 5, p. 054510, 2012.

[23] J. B. Mehl, "Analysis of resonance standing-wave measurements," *The Journal of the Acoustical Society of America*, vol. 64, no. 5, pp. 1523–1525, 1978.

[24] M. Newville, T. Stensitzki, D. B. Allen, M. Rawlik, A. Ingargiola, and A. Nelson, "Lmfit: Non-linear least-square minimization and curve-fitting for python," *Astrophysics Source Code Library*, pp. ascl–1606, 2016.

[25] X. Bao, I. Ocket, M. Zhang, J. Bao, D. Schreurs, and B. Nauwelaers, "Microwave characterization of liquid mixtures with a miniaturized interdigital sensor," in *2019 IEEE MTT-S Int. Microwave Workshop Series on Advanced Materials and Processes for RF and THz Applications (IMWS-AMP).* Bochum, Germany, 16-18 Jul. 2019, pp. 1–3.

Two-bit multi-level spin orbit torque MRAM with the fully one-step write operation

Chenyi Wang, Min Wang, Zhaohao Wang*, Weisheng Zhao

Fert Beijing Institute, MIIT Key Laboratory of Spintronics, School of Integrated Circuit Science and Engineering, Beihang University, Beijing 100191, China.
* Email: zhaohao.wang@buaa.edu.cn

Abstract—As an emerging non-volatile memory technology, the spin orbit torque magnetic random access memory (SOT-MRAM) has attracted intensive research interest due to its advanced performance. However, the binary storage feature of the SOT-MRAM has become one of the obstacles. In this paper, we present a study of two-bit multi-level SOT-MRAM where two canted in-plane-anisotropy magnetic tunnel junctions (MTJs) store a pair of data. Compared with the previous schemes of multi-level SOT-MRAMs, our proposal enables fully one-step writing without the need of the preset operation. Micromagnetic simulation is performed to validate the functionality of the proposed multi-level cell (MLC) SOT-MRAM, meanwhile, the details of magnetization switching are clearly shown. Simulation results also demonstrate that the device could accomplish the magnetization switching at the sub-nanosecond speed and continuously decreasing power consumption with the size scaling down. In addition, the dipolar field between two cells has little influence on the switching process.

Keywords—SOT-MRAM, Multi-level cell (MLC), Fully one-step write operation, Micromagnetic simulation

I. INTRODUCTION

Spin orbit torque magnetic random access memory (SOT-MRAM) has shown great advantages such as non-volatility, sub-nanosecond switching speed, low power, high endurance, etc. [1-3]. However, SOT-MRAM is generally used for binary memory due to the rareness of intrinsic multi-state features. As a result, the storage capacity and application scope of the SOT-MRAM are limited. Although multi-level cell (MLC) SOT-MRAM could be implemented by series-connected or parallel-connected multiple cells [4, 5], it requires additional preset operation while writing '01' or '10' state, leading to speed and power overhead. We have proposed an MLC SOT-MRAM where each state can be deterministically written with the fully one-step operation [6], promising to simplify the signal sequence and periphery circuits. In this paper, we will reveal more details about the proposed one-step-writing MLC SOT-MRAM, including the behaviors of magnetization switching, the scaling study of the device, and the influence of the dipolar field. Our work will provide guidance for the physical realization of MLC SOT-MRAM.

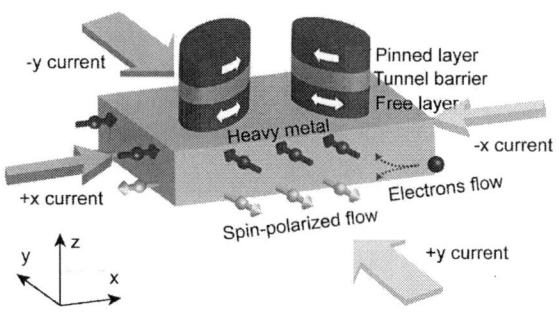

Fig. 1 Schematic of the proposed MLC SOT-MRAM structure.

II. DEVICE MODEL

The schematic of the proposed MLC SOT-MRAM is illustrated in Fig. 1, where two canted in-plane-anisotropy magnetic tunnel junctions (MTJs) [7] with different tilted angles are deposited above a heavy metal. The MTJ, whose core structure consists of a tunnel barrier sandwiched between pinned and free layers, shows low or high resistance state when the magnetization directions of free and pinned layers are parallel or antiparallel. Thus four states (2-bit data) could be represented therein. The magnetization direction of pinned layer is fixed, whereas that of free layer could be switched by an injected spin-polarized current, which is induced by a charge current flowing through the heavy metal, as explained by SOT mechanism [1-3]. The principle of MLC writing is explained in Table I. It is seen that the orientation of spin polarization is dependent on the direction of the charge current. Furthermore, the tilted angles of two MTJs are specifically designed in order that the four combinations of left and right magnetization vectors are achieved by four current directions, which is attributed to the interaction between spin polarization and easy axis. In this way, no additional preset operation is required while writing 2-bit data.

III. RESULTS AND DISCUSSION

A. Behaviors of magnetization switching

Micromagnetic simulations were performed with OOMMF platform for showing the behaviors of magnetization switching. Two MTJs are symmetrically placed with respect to the y-axis. The main parameters are set as follows: the exchange constant is 1×10^{-11} J/m, saturation magnetization is 1.1×10^6 A/m , damping constant is 0.01, size is 90 nm×40 nm×π/4. Typical simulation results can be seen in Fig. 2, where the deterministic magnetization switching is clearly validated. Obviously, the magnetization precession occurs during the switching process, which is the feature of in-plane-anisotropy magnetization switching. In addition, the symmetry can be observed in Fig. 2.

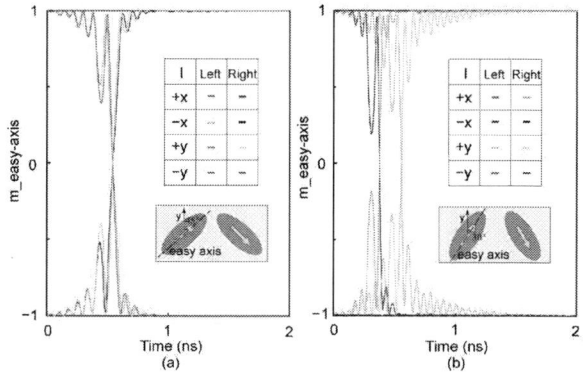

Fig. 2 Time-varying easy axis component of the free layer magnetization vectors (m_easy-axis) of the two cells under the different current directions. The tilted angle is set as 45° (a) and 30° (b), respectively. The current is applied at 0 ns and removed at 1 ns.

TABLE I   OPERATION PRINCIPLE

Current Direction	Spin Polarization	Final State (parallel/antiparallel)				Data State
		Left Cell		Right Cell		
+x	+y	P	↗	P	↖	{1,1}
−x	−y	AP	↙	AP	↘	{0,0}
+y	−x	AP	↙	P	↖	{0,1}
−y	+x	P	↗	AP	↘	{1,0}

For instance, the curves of {−*Ix*, Left cell} and {+*Ix*, Right cell} are symmetric with respect to the line 'm_easy-axis = 0'. Furthermore, for the tilted angle of 45°, the curves of {+*Ix*, Right cell} and {+*Iy*, Right cell} are almost identical, but it is not true for the tilted angle of 30°. These features could be explained by the relative orientation between spin polarization and easy axis (see Table I). It is worth noting that, for the tilted angle of 45°, the little difference between curves of {+*Ix*, Right cell} and {+*Iy*, Right cell} is induced by the inevitable edge effects.

B. Scaling study

Below we study the switching speed and power consumption of the proposed MLC device with the size scaling down. To maintain thermal stability, the size is set according to the following model [8].

$$\Delta \propto t^2 w(AR - 1) \qquad (1)$$

where Δ, t, w and AR represent the thermal stability, thickness, width and aspect ratio of the free layer, respectively. The thickness and width are adjusted under the constant AR. The design rule for the scaling study is illustrated by Fig. 3(a).

As shown in Fig. 3(b), the switching delay increases and power consumption decreases with the size scaling down. Overall, the switching delay is kept at sub-nanosecond order, meanwhile the power consumption is continuously reduced as the device is scaled. Therefore the proposed device shows good scalability. However, the switching delay varies more significantly at the smaller size. This phenomenon is partly attributed to the rise of in-plane anisotropy field (H_k), because H_k in common is related to the thickness and width of the free layer. Thus while holding the thermal stability, more effort is required to improve magnetization switching speed of the device at the smaller size.

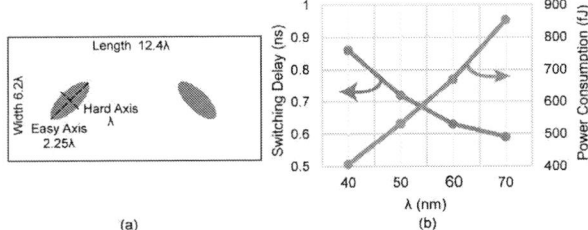

(a) (b)

Fig. 3 (a) The design rule of the device size in the simulation (two cells have the same size). (b) Switching delay and power consumption as a function of the width of the free layer. The applied current density is set to 8×10^{11} A/m².

C. Influence of the dipolar field

The dipolar field exists between two MTJs and needs to be considered when it comes to the practical MLC device. Here, using micromagnetic simulation, we preliminarily study the influence of dipolar field on the magnetization switching process. Only the free layers are considered for the sake of simplicity. Besides the MLC device, another single-level cell (SLC) with only one canted MTJ is studied for comparison.

Simulation results show that generally the dipolar field has little influence on the magnetization switching (not shown here). However, while the applied current density is close to the threshold value, the influence of the dipolar field may be dramatic. As shown in Fig. 4(a), the dipolar field prohibits the magnetization switching of the MLC device. But in Fig. 4(b), the

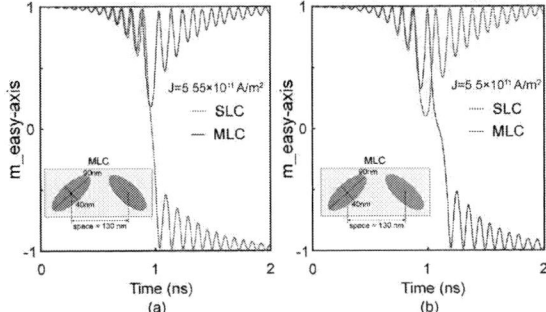

Fig. 4 Time-varying easy axis component of the left MTJ magnetization vector (m_easy-axis) at SLC and MLC, of which the space is set as the illustration. The current is applied at 0 ns and removed at 1 ns.

influence becomes positive. In real applications, the pinned layer and synthetic antiferromagnetic layer (SAF) can control the dipolar field through well-designed fabrication.

IV. CONCLUSION

We have studied an MLC SOT-MRAM device that features the fully one-step write operation. Through micromagnetic simulation, we found that the relative orientation between spin polarization and easy axis plays an important role in the magnetization switching process. Moreover, the scalability of the proposed device is explored in terms of switching delay and power consumption. In addition, the influence of the dipolar field between two cells is complicated around the threshold current density. This work provides a potential solution to design MLC SOT-MRAM, and helps in expanding the application scope of SOT-MRAM to multi-bit storage or neural network.

ACKNOWLEDGMENT

This work was supported by the National Natural Science Foundation of China under Grant 62171013, the National Key Research and Development Program of China (Nos. 2021YFB3601303, 2021YFB3601304, 2021YFB3601300), and the Fundamental Research Funds for the Central Universities.

REFERENCES

[1] Z. Guo, et al. "Spintronics for Energy- Efficient Computing: An Overview and Outlook," *Proc. IEEE.*, 2021.

[2] B. Dieny, et al. "Opportunities and challenges for spintronics in the microelectronics industry," *Nat. Electron.*, vol. 3, no. 8, pp. 446-459, 2020.

[3] X. Han, et al. "Spin-orbit torques: Materials, physics, and devices," *Appl. Phys. Lett.*, vol. 118, no. 12, pp. 120502, 2021.

[4] S. Dhull, et al. "SOT and STT Based Four-Bit Parallel MRAM Cell for High-Density Applications," *IEEE Trans. Nanotechnol.*, vol. 20, pp. 653-661, 2021.

[5] Y. Kim, et al. "Multilevel spin-orbit torque MRAMs," *IEEE Trans. Electron Devices*, vol. 62, no. 7, pp. 561-568, 2014.

[6] M. Wang, et al. "Design of a 2-bit Multi-Level Cell with Fully One-Step Writing Mode Based on Spin-Orbit Torque," in *Abstract Book of The 15th Joint MMM–Intermag Conference*, Abstract No. GOL-07, pp. 307, Submitted July 2021, Presented January 2022.

[7] H. Honjo, et al. "First demonstration of field-free SOT-MRAM with 0.35 ns write speed and 70 thermal stability under 400 degrees C thermal tolerance by canted SOT structure and its advanced patterning/SOT channel technology," *IEDM.*, pp. 28.5.1-28.5.4, 2019.

[8] D. Apalkov, et al. "Comparison of Scaling of In-Plane and Perpendicular Spin Transfer Switching Technologies by Micromagnetic Simulation," *IEEE Trans. Magn.*, vol. 46, no. 6, pp. 2240-2243, 2010.

2022 IEEE International Conference on
Integrated Circuits, Technologies and Applications

3D-NWA: A Nested-Winograd Accelerator for 3D CNNs

Huafeng Ye[1], Huipeng Deng[2], Jian Wang[2], Mingyu Wang[1,*], Zhiyi Yu[1]
[1]School of Microelectronics Science and Technology, Sun Yat-sen University
[2]School of Electronics and Information Technology, Sun Yat-sen University
[1,*]wangmingyu@mail.sysu.edu.cn

Abstract—**3D Convolutional neural networks (3D CNNs) perform better in some scenarios, such as video understanding and 3D medical image diagnosis. With the increase in the dimension and size of the convolution kernel, CNN's computational complexity and implementation difficulty increase severely. Winograd transformation can significantly reduce the number of multiplications in convolution operations. However, large convolution filters will bring numerical instability. In this article, we presented a novel method called 3D nested Winograd algorithm to address the problem. Compared with the state-of-art OLA-Winograd algorithm, the proposed algorithm reduces the multiplications by 1.72 to 5.83× for computing $5 \times 5 \times 5$ to $9 \times 9 \times 9$ convolutions. Finally, we demonstrate the efficiency of 3D-NWA on the FPGA platform (Xilinx VCU118) and achieve highest DSP efficiency up to 4.67× compared with the state-of-art accelerators.**

Keywords—**3D CNN, Winograd algorithm, accelerator, large filters**

I. INTRODUCTION

Convolutional Neural Networks (CNNs) have shown excellent performance in many fields, such as voice recognition, image processing, and video understanding. However, the enormous computational complexity of CNNs limits their implementation on hardware accelerators. The Winograd algorithm can replace some of the multiplications with additions which can conserve DSP resources and improve computation throughput. The Winograd algorithm's acceleration effect will improve with increasing the convolution kernel's dimension.

Equation (1) shows the 3D Winograd transformation algorithm, and Equation (2) is for F(2,2). Note that $F(m^k, r^k)$ are used to represent the k dimensions Winograd algorithms with filter size r and partial sum size m.

$$Y = \left[A^T \left[\left(GgG^T \right)^R G^T \odot \left(B^T dB \right)^R B \right] A \right]^R A \quad (1)$$

$$A^T, B^T, G = \begin{vmatrix} 1 & 0 \\ 0 & 1 \\ 0 & -1 \end{vmatrix} \begin{vmatrix} 1 & 0 & -1 \\ 0 & 1 & 1 \\ 0 & -1 & 1 \end{vmatrix} \begin{vmatrix} 1 & 0 \\ 1/2 & 1/2 \\ 1/2 & -1/2 \end{vmatrix} \quad (2)$$

However, using the Winograd algorithm to accelerate convolution with large filters is numerically unstable[1]. Therefore, most of the current work is aimed at small-sized convolution kernels. Notably, there is currently less research on how to speed up 3D convolution with large kernel sizes effectively. Work [2] studied the nested Winograd algorithm with large kernels but only aimed at 2D convolution. Our previous works[5] extended the OLA-Winograd algorithm to 3D convolution and proposed computing units that support different dimensions and two Winograd units. Still, the low efficiency of the OLA-Winograd algorithm computing unit is more evident in the 3D convolution operation.

This work, based on the OLA-Winograd algorithm and the nested Winograd algorithm, contributes a novel 3D nested

This work was supported in part by the Key-Area Research and Development Program of Guangdong Province 2021B0101410004, in part by the Guangdong Basic and Applied Basic Research Foundation 2022A1515011708

Winograd implementation approach that overcomes the shortcomings of the previous Winograd implementations in that 3D CNNs with large filters can be accelerated more efficiently and more succinctly.

Overall, this paper makes the following contributions:

- A 3D nested Winograd algorithm is proposed to accelerate the computation of large-scale 3D convolutions. For $5 \times 5 \times 5$, $7 \times 7 \times 7$, and $9 \times 9 \times 9$ convolution kernels, our algorithm can reduce the computational complexity by up to 82.86%, 93.75%, and 97.06%, respectively.

- A novel hardware accelerator compatible with the 3D nested Winograd algorithm and the OLA-Winograd algorithm is proposed to accelerate 3D CNNs. DSP efficiency achieves 4.67× compared with SOTA works.

II. ALGORITHM DETAILS

Winograd algorithm is an efficient method to replace some of the expensive multiplications with cheap operations such as additions, which can significantly reduce the computational cost for CNNs.Nevertheless, the traditional Winograd algorithm needs to perform complex splits when dealing with large convolution kernels, which leads to the low efficiency of the computing unit. This work proposed a 3D nested Winograd algorithm, which can significantly reduce the processing and computational complexity of large convolution kernels.

In this section, we give details on the theoretical derivation of the 3D Nested Winograd algorithm when the kernel W, the input X, and the output W sizes are 3x3x3, 6x6x6, and 4x4x4, respectively. The convolution can be written as $X * W = Y$. Using the 2D nested Winograd algorithm to perform 3D transformation on each 2D matrix, the convolution above can be converted into Equation (3)

$$\begin{bmatrix} A_0 & B_0 & C_0 \\ A_1 & B_1 & C_1 \end{bmatrix} * \begin{bmatrix} a_0 & b_0 & c_0 & d_0 & e_0 & f_0 \\ a_1 & b_1 & c_1 & d_1 & e_1 & f_1 \\ a_2 & b_2 & c_2 & d_2 & e_2 & f_2 \end{bmatrix} = \begin{bmatrix} y_0 & y_1 & y_2 & y_3 \\ y_4 & y_5 & y_6 & y_7 \end{bmatrix} \quad (3)$$

Equation (3) represents a 2D convolution. Each element in the matrix is a 3D matrix. At this step, we have nested a 3D convolution inside a 2D convolution. We then convert this 2D convolution to a 3D convolution.

$$\begin{bmatrix} A_0 & B_0 \\ C_0 & 0 \end{bmatrix} * \begin{bmatrix} a_0 & b_0 & c_0 \\ c_0 & d_0 & e_0 \\ e_0 & f_0 & 0 \end{bmatrix} = \begin{bmatrix} y_0 & y_1 \\ y_2 & y_3 \end{bmatrix}$$

$$\begin{bmatrix} A_1 & B_1 \\ C_1 & 0 \end{bmatrix} \quad \begin{bmatrix} a_1 & b_1 & c_1 \\ c_1 & d_1 & e_1 \\ e_1 & f_1 & 0 \end{bmatrix} \quad \begin{bmatrix} y_4 & y_5 \\ y_6 & y_7 \end{bmatrix} \quad (4)$$

$$\begin{bmatrix} a_2 & b_2 & c_2 \\ c_2 & d_2 & e_2 \\ e_2 & f_2 & 0 \end{bmatrix}$$

Equation (4) represents 3D convolution, and each element is a 3D matrix, so the original 3D convolution is converted into a 3D convolution nested in another 3D convolution.

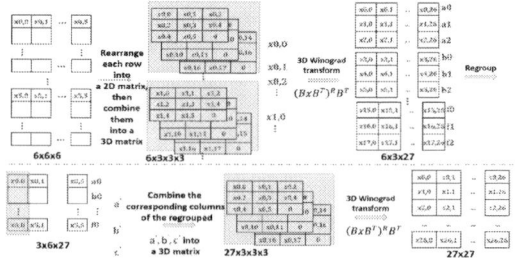

Fig. 1: the input transform

As shown in Fig.1, 2D Winograd transformation is processed on each input matrix row. The first row is converted into three 3x3x3 matrices. Then we perform 3D Winograd transformation on these matrices into three new 3D matrices and convert them into three 1D matrices after tiling. We do the same for the remaining 2D matrices to get a 1D matrix. All these 1D matrices are reassembled into three 2D matrices. Afterward, we convert the columns of the three matrices into a 3D matrix. After transformation, we send the input matrix with the preprocessed filter to do element-wise Winograd multiplication (EWMM).

III. HARDWARE ARCHITECTURE

In this section, we provide details of our Nested Winograd based uniform accelerator. We can obtain input, filter, and output transformation by shifting and adding on the electric circuit. Fig.2 depicts a partial circuit diagram of the input transformation.

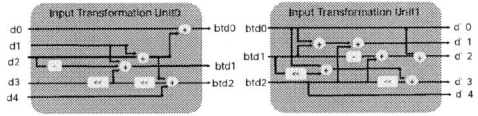

Fig. 2: the input transformation module for F(3,3)

Based on the 3D Nested Winograd Algorithm, we proposed a uniform Winograd architecture capable of supporting F(3,3) OLA-Winograd and Nested-Winograd units. Winograd transformation processes for different dimensions are mapped into several PATHs chosen by the MUX. For example, PATH 1.1-1.3 represents transformation for different dimension input (1D, 2D, or 3D).

Our 3D nested algorithm includes a two-level Winograd transformation. The data after the first transformation needs to be transposed. So we added three buffers called Ibuff, Wbuff, and Obuff to reorganize the data before serving EWMM in the Winograd domain. We can bypass these buffers to be compatible with the traditional OLA-Winograd algorithm. Besides, we add a buffer called Abuff to gather the data of all channels before performing the output transformation. This buffer can dramatically reduce the bandwidth requirements of output transformation.

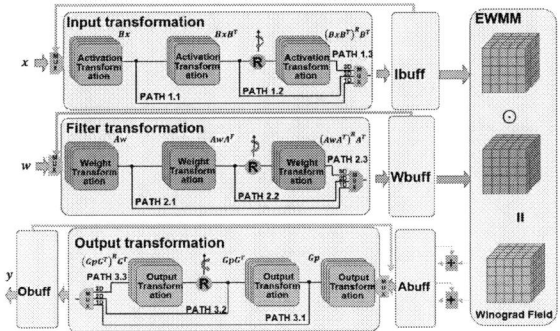

Fig. 3: the nested Winograd hardware architecture

IV. IMPLEMENTATION RESULTS

We set the Winograd unit as F(3,3) during the simulation and implementation, which can be compatible with the 3x3x3 filter. Simulations are made to compare the computational complexity with different filter sizes ranging from $5 \times 5 \times 5$ to $9 \times 9 \times 9$. Fig.4 summarizes the comparison results between direct convolution[3], WRA[4] and 3D nested-Winograd(this work).

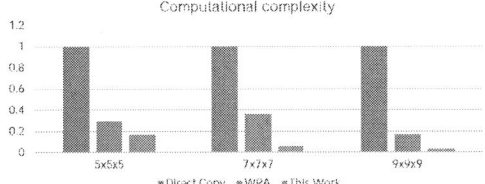

Fig. 4: Computational complexity comparison

The proposed accelerator 3D-NWA is implemented on the Xilinx FPGA VCU118 evaluation board, using 31500kb more BRAMs and 600 digital signal processors(DSPs). Our FPGA implementation is operated at 200MHz. The throughput achieves up to 4570 GOPS for $9 \times 9 \times 9$ convolution. Fig.5 depicts the DSP efficiency (GOPS/DSPs) of different types of Winograd units in processing commonly used convolutional layers. The NWA unit has the highest DSP efficiency, reaching $4.67\times$ that of the WHD unit[4] and $3.3\times$ that of the Dimension Fusion unit[5].

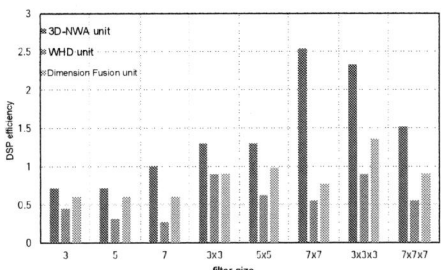

Fig. 5: DSP efficiency comparison

V. CONCLUSION

In conclusion, we propose a novel 3D nested Winograd algorithm to accelerate 3D CNNs with large kernel sizes. The computational complexity can be reduced by up to $5.83\times$ and $34.01\times$ compared with OLA-Winograd(WRA) and direct convolution, respectively. Besides, we implement the 3D-NWA on the Xilinx VCU118 evaluation board FPGA and process at a clock frequency of 200MHz.The evaluation results show that 3D-NWA achieves the highest DSP efficiency up to $4.67\times$ compared with the state-of-art accelerators.

REFERENCES

[1] K. Vincent, K. Stephano, M. Frumkin, B. Ginsburg, and J. Demouth,"On improving the numerical stability of winograd convolutions," 5th Int.Conf. Learn.Represent.ICLR 2017-Work. Track Proc.,no.1, pp.1-4, 2019.

[2] J.Jiang, X.Chen and C.Tsui, "A Reconfigurable Winograd CNN Accelerator with Nesting Decomposition Algorithm for Computing Convolution with Large Filters",2021, [Online]. Available: https://arxiv.org/abs/2102.13272

[3] K. Hegde, R. Agrawal, Y. Yao and C. W. Fletcher, "Morph: Flexible Acceleration for 3D CNN-Based Video Understanding," 2018 51st Annual IEEE/ACM International Symposium on Microarchitecture (MICRO), 2018, pp. 933-946, doi: 10.1109/MICRO.2018.00080.

[4] C. Yang, Y. Wang, X. Wang and L. Geng, "A Stride-Based Convolution Decomposition Method to Stretch CNN Acceleration Algorithms for Efficient and Flexible Hardware Implementation," in IEEE Transactions on Circuits and Systems I: Regular Papers, vol. 67, no. 9, pp. 3007-3020, Sept. 2020, doi: 10.1109/TCSI.2020.2985727.

[5] H. Deng, J. Wang, H. Ye, S. Xiao, and Z. Yu,:" Dimension fusion: Dimension-level dynamically composable accelerator for convolutional neural networks". IEICE Electron. Express 18(24): 20210491 (2021)

Long Short-Term Memory Networks for Behavioral Modeling of A GaN Sequential Power Amplifier

Peng Chen[1*], Yucheng Yu[1], Chao Yu[1,2]

[1] State Key Laboratory of Millimeter Waves, Southeast University, Nanjing, China
[2] Purple Mountain Laboratory, Nanjing, China
*pchen@seu.edu.cn

Abstract—this paper investigates wideband behavioral modeling of Gallium Nitride (GaN) power amplifiers (PAs) using long short-term memory (LSTM) networks. Due to the memory mechanisms used in LSTM networks, they have the capability of accurately capturing both the short term and long term memory effects presenting in GaN PAs. The LSTM network-based model is verified experimentally on a GaN sequential power amplifier (SPA) under wideband multi-channel modulated signals, with showing good alignment between the modeled and measured data.

Keywords—behavioral modeling, long short-term memory networks, memory effects, power amplifier.

I. INTRODUCTION

When operating with wideband modulated signals, memory effects are of great concern in Gallium Nitride (GaN)-based power amplifiers (PAs). It has been recognized that the long-term memory effects are of paramount importance on the dynamic nonlinearity, driving the need for building nonlinear models with high fidelity. The Conventional Volterra series-based models are widely used to characterize both the short-term and long-term memory effects, achieving substantial improvement in wideband PA linearization [1]. As the polynomials are not amplitude-bounded in Volterra series-based models, there is a probability of catastrophic error degradation that happened in the parameter extraction procedure [2]. Unlike the Volterra series-based models, artificial neural networks (ANNs) adopt bounded activation functions to train the behavioral models, avoiding the problem of catastrophic error degradation. Due to ANNs' distinct features, they have attracted a growing interest in PA modeling and linearization, such as time-delay neural networks (TDNN) [2], polynomial-assisted neural networks [3], recurrent neural networks [4], and long short-term memory (LSTM) networks [5].

This paper expands our prior work [5] to investigate wideband behavioral modeling of GaN PAs. For verification, a 12-carrier 60 MHz Universal Mobile Telecommunication System (UMTS) signal and a 5-carrier 100 MHz LTE signal were used for extracting the behavioral models of a GaN-based sequential power amplifier (SPA).

II. BEHAVIORAL MODELING OF POWER AMPLIFIER

A. Mechanisms of Long Short-Term Memory Networks

LSTM networks employ three types of gates to update the long-term memory state c_t and the short-term memory state h_t, at the expense of extra parameters. As shown in Fig 1, the forget gate, input gate and output gate, are used to implement the functionality of updating, writing and reading memory. When calculating the gradient flow from c_t to c_{t-1}, it only involves the multiplication by f_t, without the weight matrix W, preventing the vanishing or exploding gradients. The mechanisms of LSTM networks are presented as follows:

1) When the short-term memory state h_{t-1} and the current input x_t are fed into the LSTM networks, the forget gate,

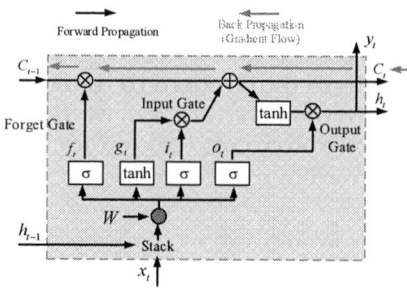

Fig. 1. A diagram of LSTM networks. The symbols of "$\oplus$" and "$\otimes$" denote the operation of addition, element-wise multiplication, respectively.

controlled by the parameter $f_t \in [0,1]$, is firstly used to decide what information to discard from the memory state, i.e., fading the memory,

$$f_t = \sigma(W_f \cdot [h_{t-1}, x_t] + b_f) \quad (1)$$

where $\sigma(\cdot)$ denotes the sigmoid function, and W_f and b_f are a weight matrix and a bias vector with regards to f_t. The forget gate will completely remove the short-term memory state h_{t-1} when $f_t = 0$, while it will remains h_{t-1} when $f_t = 1$.

2) The input gate is used to decide what new information to store in the memory state, controlled by the parameters i_t and g_t,

$$i_t = \sigma(W_i \cdot [h_{t-1}, x_t] + b_i) \quad (2)$$

$$g_t = \tanh(W_g \cdot [h_{t-1}, x_t] + b_g) \quad (3)$$

where W_i and W_g are weight matrixes, and b_i and b_g are bias vectors. Here, i_t decides which values will be updated, while g_t creates new memory information that will be added.

3) After knowing what information will be forgotten or remembered in the last two steps, we update the old long-term memory state c_{t-1} into the new long-term memory state c_t

$$c_t = f_t \otimes c_{t-1} + i_t \otimes g_t \quad (4)$$

4) Finally, the output gate decides what to output as the current short-term memory state h_t, calculated by

$$o_t = \sigma(W_o \cdot [h_{t-1}, x_t] + b_o) \quad (5)$$

$$h_t = o_t \otimes \tanh(c_t) \quad (6)$$

Similarly, W_o and b_o are a weight matrix and a bias vector, respectively. The parameter o_t determines what parts of the scaling long-term memory state that will be outputted. In this way, LSTM networks have the capability of learning useful features and storing them in the long-term memory state, albeit they are deployed in modeling very long sequence data.

B. Behavioral Modeling

Finding an appropriate size of LSTM networks is a process of trial-and-error experimentation. It is advised to start with fewer layers and neurons, and then gradually increase the number until the validation losses reach the minimum value. In this paper, we selected two layers of LSTM networks and a fully connected linear layer, with nine neurons for each layer. The loss function was calculated in mean square error (MSE)

$$L(y, \bar{y}) = \frac{1}{T} \sum_{t=0}^{T} [I_t(\bar{y}) - I_t(y)]^2 + [Q_t(\bar{y}) - Q_t(y)]^2 \quad (7)$$

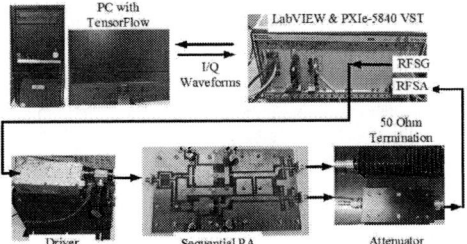

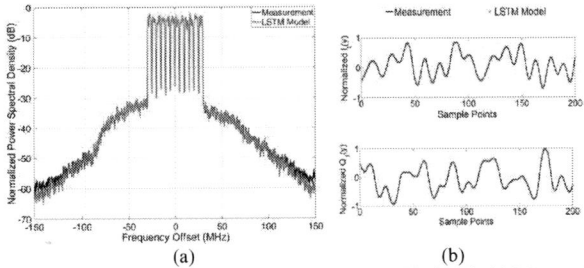

Fig. 2. Experimental platform.

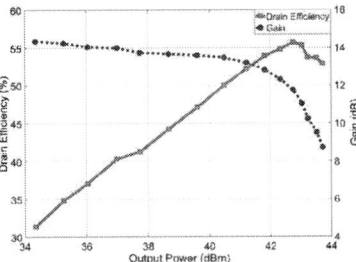

Fig. 3. Measured results of the SPA at 2.2 GHz using CW signal.

(a)

(b)

Fig. 4. Verification with a 12-carrier 60 MHz UMTS signal. (a) Measured and modeled output spectra. (b) Normalized I/Q components of SPA output.

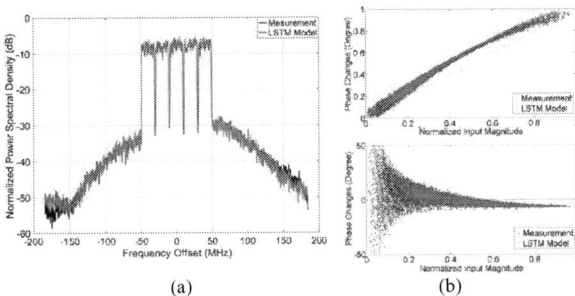

(a)

(b)

Fig. 5. Verification with a 5-carrier 100 MHz UMTS signal. (a) Measured and modeled output spectra. (b) AM/AM and AM/PM of SPA output.

The Adam Optimizer with a decayed learning rate starting from 0.01, was used to search the optimum coefficients by minimizing (7). The training data were the input and output I/Q data, which were fed into the LSTM networks at each iteration. When the limited tolerance or iteration number was reached, the behavioral model could be used for new prediction.

III. EXPERIMENTAL VALIDATION

As a wideband architecture of interest for achieving high efficiency, a GaN sequential PA, comprising a main (Cree CGH40010F) and a peaking (Cree CGH40025F) devices, was used in the experiments, as shown in Fig. 2. The main device was biased in class-AB (I_{DSQ}=89 mA), while the peaking PA was biased in class-C at -6 V, with both drain supply voltages at 28 V. Fig. 3 shows that under continuous-wave (CW) excitation at 2.2 GHz, the SPA achieved a drain efficiency higher than 40 % at 6.5 dB power back off, with a gain varying from 14.0 to 8.7 dB. After that, a 12-carrier 60 MHz UMTS signal with PAPR=6.5 dB and a 5-carrier 100 MHz LTE signal with PAPR=7.8 dB, were employed in the test, respectively. The behavioral model was built using LSTM networks with the baseband input and output I/Q data of the SPA.

The input and output I/Q data were acquired when the SPA was tested at 2.2 GHz under the excitation of the two modulated signals, respectively. The LSTM networks-based models were trained using 10,000 pairs of input and output I/Q data, while a further, different 5,000 pairs of data points were used for the model validation. Fig. 4(a) shows the output spectra from the measurement and the LSTM model when using a 12-carrier 60 MHz UMTS signal with an average output power of 37.2 dBm. The measured and modeled output spectra showed good agreement from the carrier channel to the second offset-channels, albeit there was little deviation in the third offset-channels. Fig. 4(b) compared a sample of the time-domain output envelope waveforms, showing precise alignment between the measured and the modeled values. Similarly, the SPA was tested at an average output power of 35.9 dBm with a 5-carrier 100 MHz LTE signal, which was very likely to cause more severe memory effects. Fig. 5 shows the output spectra and the AM/AM and AM/PM characteristics from the

measurement and the LSTM model, respectively. The measured data coincided with the predicted values, validating the accuracy of the proposed LSTM model.

IV. CONLUSIONS

In this paper, LSTM networks were applied to investigate the wideband behavioral modeling of a GaN SPA. The LSTM network-based model was validated on the SPA using two wideband multi-channel modulated signals. Owing to the nature of memory mechanisms involved, LSTM networks can accurately present the useful pattern of the baseband I/Q components in the time-domain, indicating its potential in wideband PA linearization.

ACKNOWLEDGMENT

This work was supported in part by the National Natural Science Foundation of China under Grant 62101116, Grant 62022025, in part by the Natural Science Foundation of Jiangsu Province under Grant BK20200065, and in part by the Jiangsu Shuangchuang Talent Program under Grant JSSCBS20210062.

REFERENCES

[1] C. Yu, Q. Lu, H. Yin, J. Cai, J. Chen, X. -W. Zhu, and W. Hong, "Linear-Decomposition Digital Predistortion of Power Amplifiers for 5G Ultrabroadband Applications," *IEEE Trans. Microw. Theory Techn.*, vol. 68, no. 7, pp. 2833-2844, July 2020

[2] E. G. Lima, T. R. Cunha, and J. C. Pedro, "A physically meaningful neural network behavioral model for wireless transmitters exhibiting PM-AM/PM-PM distortions," *IEEE Trans. Microw. Theory Techn.*, vol. 59, no. 12, pp. 3512–3521, Dec. 2011.

[3] Y. Yu, J. Cai, X. -W. Zhu, P. Chen and C. Yu, "Self-Sensing Digital Predistortion of RF Power Amplifiers for 6G Intelligent Radio," *IEEE Microw. Wireless Compon. Lett.*, vol. 32, no. 5, pp. 475-478, May 2022.

[4] B. O'Brien, J. Dooley, and T. J. Brazil, "RF power amplifier behavioral modeling using a globally recurrent neural network," in *IEEE MTT-S Int. Microw. Symp. Dig.*, San Francisco, CA, Jun. 2006, pp. 1089–1092.

[5] P. Chen, S. Alsahali, A. Alt, J. Lees, and P. J. Tasker, "Behavioral Modeling of GaN Power Amplifiers Using Long Short-Term Memory Networks," in *Proc. IEEE Int. Integr. Nonlinear Microw. Millim.-Wave Circuits Workshop*, Brive-La-Gaillarde, France, Jul. 2018, pp. 1–3.

A Cap-Less High PSR and Low Output Noise Low-Dropout Regulator for Cryogenic Applications

Lingyun Liu, Chenglong Liang, Zhuoqi Guo, Zhongming Xue, Li Geng*

School of Microelectronics, Xi'an Jiaotong University

Corresponding Author Email: gengli@xjtu.edu.cn

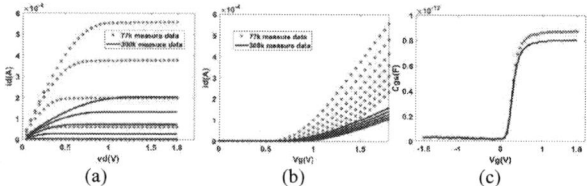

Fig. 1. Measurement results of 10μm/10μm NMOS at 77 K and 300 K. (a) I_{ds} versus V_{ds}, (b) I_{ds} versus V_{gs}, and (c) C_{gs} versus V_{gs}.

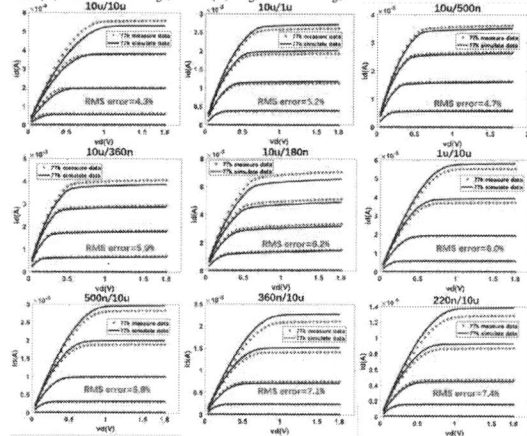

Fig. 2. Comparison between simulation with cryogenic model and measurement results

***Abstract*—Low dropout (LDO) voltage regulators are essential for noise-sensitive circuit systems in cryogenic temperature environments. This paper characterized and modeled a full-scale BSIM4-based 180nm MOSFET at cryogenic temperature. Then, a high PSR low output noise cap-less LDO is implemented with the cascade and feed-forward current technology for cryogenic applications. At 77K, simulation results show that PSR is -98dB at 10kHz and -78dB at 100kHz, and the integrated noise is 0.82 μVrms among the frequency from 100Hz to 100kHz.**

Keywords—cryogenic, low-dropout (LDO) regulator, power supply rejection (PSR), output noise

I. INTRODUCTION

Cryogenic electronics has played an indispensable role in several applications, mainly for spacecraft, quantum computer, and radio astronomy. These applications generally operate in cryogenic temperature environments. Their power supplies have strict requirements for power supply rejection and output noise. Also, it needs to integrate on a chip[1].

Low dropout (LDO) voltage regulator is widely used in power management systems because of its high PSR and low output noise, and it can provide a stable voltage even at cryogenic temperature[2]. When LDO works in cryogenic environment, thermal noise can further reduce with the decrease of temperature, which is conducive to the design of low output noise[3]. However, due to the change in MOSFET characteristics, loop stability of the LDO is greatly affected, resulting in the deterioration of PSR. Adding off-chip capacitors can improve PSR performance to a better extent, but it is adverse to the application in cryogenic environment [4]. In this paper, a full-scale cryogenic model based on BSIM4 was established by analyzing the characteristics of MOSFETs at 77K. This model can accurately simulate the circuit performance changes at cryogenic temperature to ensure the feasibility of the designed circuit. Then, a cap-less LDO structure with cascade and feed-forward current is proposed to simultaneously achieve high PSR and low output noise performance.

II. CIRCUIT DESIGN

A. Modeling Of Cryogenic MOSFET

A group of MOSFETs with full size using a standard 180nm CMOS process is characterized at 77K. Fig. 1. shows the comparison of the measurement result between 300K and 77K. When the temperature drops, the threshold voltage will increase. In the subthreshold region, the leakage current increases, increasing the subthreshold slope. At cryogenic temperature, carrier lattice scattering dominates, and lattice scattering weakens with the decreased temperature, so electron mobility and electron saturation rate increase at 77K. As a result, the drain current and the parasitic capacitance of MOSFET increase. These changes make it hard for a circuit designed for normal temperature to be used in a cryogenic temperature. To get the excellent performance of the device in 77K, an accurate simulation model of cryogenic MOSFET is necessary.

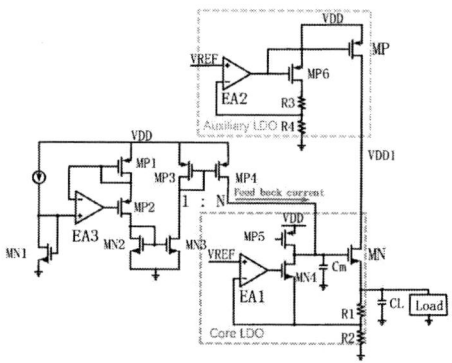

Fig. 3. Circuit schematic of designed LDO

By using the measurement data, the DC related parameters were extracted from the BSIM4 default model. Ultimately, a MOSFET model that can be accurately simulated at 77k and is suitable for full-scale devices is implemented. Fig. 2. shows the root mean square (RMS) error comparison between the model simulation results and measure values. The RMS erros of all sizes NMOS and PMOS are about 5%. The error range is considered suitable for further cryogenic circuit design.

B. Design of High PSR and Low Output Noise LDO

At cryogenic temperature, the needed gate voltage of the power transistor in the LDOs with the same output increases. Moreover, g_m of the power transistor will increase, which leads to the greater influence of the output ripple and deteriorates the performance of PSR. Reducing the ripples at the gate of the power transistor is the key to improve the PSR performance of LDO. In addition, it is necessary to increase the amplifier gain to ensure the stability because the parasitic capacitance increases, which may worsen the stability of the control loop.

Fig. 3. shows the topology of the designed LDO. The improvement of PSR performance depends on two parts. The first part is the cascade structure. Compared with the traditional cascade LDO, the auxiliary LDO uses PMOS as the power

978-1-6654-9270-6/22 $31.00 © 2022 IEEE

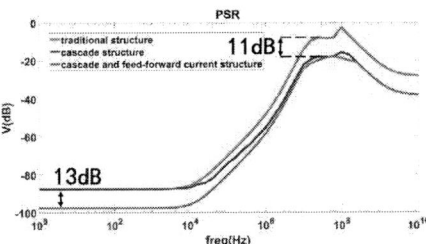

Fig. 4. PSR performance of different LDO structures

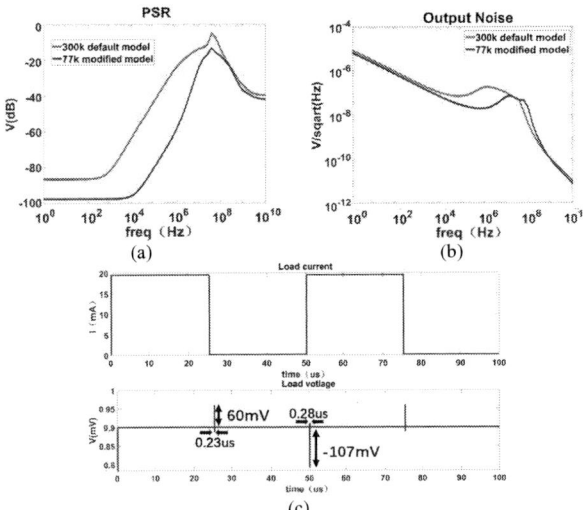

Fig. 5. (a) PSR performance of LDO at 77 K and 300 K (b) Output noise of LDO at 77 K and 300 K (c) Transient responses when the load current changes from 0.1 mA to 20 mA and from 20 mA to 0.1 mA at 77 K.

transistor instead of NMOS with complicated structure of charge pump, which reduces the noise due to the extra components. This structure can effectively improve the PSR at high frequencies.

On the other hand, feed-forward current circuit is used to improve LDO's PSR performance. The feed-forward current amplifies part of the ripple caused by the power supply VDD to act on the gate of the main LDO power transistor to offset part of the influence of VDD on *MN*. However, there is no change in the output voltage of LDO, so that the PSR performance can be effectively improved. Finally, the PSR performance of the LDO circuit at high and low frequencies is

$$PSR_{low} \approx \frac{sC_{gdN} + sC_{gdN}g_{mn}r_{on}}{sC_{gdN} + sC_m + sC_{gdN}g_{mn}r_{on}}$$
$$\times \frac{1}{EA_1(s)} \times \frac{1}{1 + Ng_{mn}r_{on}EA_3(s)} \quad (1)$$

$$PSR_{high} \approx \frac{sC_{gdN} + sC_{gdN}g_{mn}r_{on}}{sC_{gdN} + sC_m + sC_{gdN}g_{mn}/g_{mp}}$$
$$\times g_{mn}\left(r_{on} \parallel R_{Load} \parallel \frac{1}{sC_L}\right) \quad (2)$$

where C_{gdN} and C_{gdP} are the capacitance of *MN* and *MP*, g_{mp}, r_{op} are the transconductance and the output resistance of *MP*. $EA_1(s)$ and $EA_3(s)$ stand for the transfer function of the error amplifier. At low frequencies, $N \times EA_3(s)$ is large, thus the feed-forward current improves PSR. At high frequency, $EA_3(s)$ rolls off to close to zero, and PSR can be improved by g_{mp} of *MP*. Fig. 4. shows the change in PSR caused by the improvement of the structure.

LDO's output noise can be divided into flicker noise and thermal noise. One of the factors affecting thermal noise is temperature, which can be reduced at 77K. The flicker noise mainly comes from the EA circuit, which can be reduced by proportionally increasing the channel width and length of the MOS transistor. Therefore, the low output noise performance can be achieved by reducing the noise of the EA circuit.

III. SIMULATION RESULTS

The prototype cap-less LDO with high PSR and low noise is designed in a standard 180nm CMOS process. The output voltage of the LDO is 0.9 V. The maximum load current is 20mA, the total capacitance on the chip is only 20 pF. Fig. 5. shows the simulation results of LDO working at 77 K and 300 K. The PSR performance of LDO at 77 K is better than that at 300 K, also it is below -20dB in the whole band at 77 K when the load current is 20mA. The PSR is -98dB at 10 Hz and -78 dB at 100 kHz when the LDO works at 77K. The output noise of LDO at 77 K is generally lower than that at 300 K. The integrated noise from 100 Hz to 100 kHz is 0.82μVrms. Fig. 5(c) is the transient response when the load current changes from 0.1 mA to 20 mA at 77K. The undershoot is 60mV and the response is 0.23 μs. The overshoot is 107 mV and the response is 0.28 μs. Table I summarizes the performance of recently published cryogenic

TABLE I. PERFORMANCE SUMMARY

Reference	[3] (77K)	[4] (77K)	[5] (300K)	This work (77K)
Process	65nm	180nm	55nm	180nm
VDD	1.5V	2V	1.6V	1.8V
VOUT	1.2V	0.9V	1.2V	0.9V
IOUT (mA)	150	100	12	20
PSR	-54.6dB (@10HZ) -55.1dB (@100kHZ)	-81dB (@1kHZ)	-23dB (@50MHZ)	-98dB (@10HZ) -78dB (@100kHZ)
RMS Noise	0.987μV	NA	NA	0.82μV
capacitance	off chop	On chip	On chip	On chip

LDOs. Compared with other works, the proposed LDO has higher PSR and lower output noise.

IV. SUMMARY

A full-scale cryogenic MOSFET model based on BSIM4 is established through the measurement of cryogenic MOS transistors. Then, a cap-less LDO applied to cryogenic temperature is designed, which adopts cascade and feed-forward current structure to achieve high PSR and low output noise. Simulation results show the LDO can work at 77 K. The PSR is increased by 11~13dB in wide frequency range when it operates at cryogenic case. The output noise is quite lower than that at 300 K.

REFERENCES

[1] N. Sahin and M. B. Yelten, "A 0.18 μm CMOS X-Band Low Noise Amplifier for Space Applications," 2017 New Generation of CAS (NGCAS), pp. 205-208, 2017.12.

[2] D. Andrade-Miceli et al., "Cryogenic Low-Drop-Out Regulators Fully Integrated with Quantum Dot Array in 22-nm FD-SOI CMOS," 2021 IEEE MTT-S International Microwave Symposium (IMS), pp. 635-637.

[3] W. Hou, S. Li, G. D. Geronimo and M. Stanaćević, "An Ultra-Low-Noise LDO Regulator in 65 nm for Analog Front-End ASICs in Cryogenic Environment," 2018 IEEE Nuclear Science Symposium and Medical Imaging Conference Proceedings (NSS/MIC), pp. 1-4, 2018.

[4] H. İ. Kayıhan, A. Kabaoğlu and M. B. Yelten, "A Cryogenic CMOS Low Dropout Voltage Regulator Design for Space Applications," 2019 11th International Conference on Electrical and Electronics Engineering (ELECO), pp. 392-396, 2019.

[5] C. Zhan and W. Ki, "Analysis and Design of Output-Capacitor-Free Low-Dropout Regulators With Low Quiescent Current and High Power Supply Rejection," in IEEE Transactions on Circuits and Systems I: Regular Papers, vol. 61, no. 2, pp. 625-636, Feb. 2014.

2022 IEEE International Conference on
Integrated Circuits, Technologies and Applications

All-Digital Full-Precision In-SRAM Computing with Reduction Tree for Energy-Efficient MAC Operations

Dengfeng Wang[#], Zhi Li[#], Chengjun Chang, Weifeng He, Yanan Sun*
[#]These authors contributed equally to this work

The Department of Micro-Nano Electronics, Shanghai Jiao Tong University, Shanghai, China

Abstract—This paper proposes an all-digital full-precision static random-access memory based computing-in-memory (SRAM-CIM) macro with compressor-based reduction tree (CRT) for energy-efficient multiplication-and-accumulation (MAC) operations. The proposed CRT composed of hybrid 28T/18T/14T 3-2 compressors (full adders, FAs) and 18T half adders (HAs) with lower supply voltage consumes lower power compared to conventional binary adder tree (BAT). The experimental results show that the power and area of the proposed CRT are reduced by up to 56.15% and 28.11%, respectively, as compared to BAT. The proposed SRAM-CIM macro with CRT achieves 78.07% higher energy efficiency per unit area, compared to previous all-digital full-precision SRAM-CIM macro with BAT.

Keywords—static random-access memory, all-digital, computing-in memory, compressor-based reduction tree, full-precision

I. INTRODUCTION

SRAM-CIM is a promising solution to alleviate the memory bottleneck of conventional Von Neuman architecture. SRAM-CIM works can be primarily categorized into analog-domain [1]-[2] and digital-domain [3]-[4] for MAC operations. The analog-domain SRAM-CIM works with either current-based [1] or charge-based [2] mechanism are susceptible to process variations with requirement of energy-hungry analog-to-digital convertor (ADC). In contrast, the digital-domain SRAM-CIM designs eliminate ADCs and possess high robustness against process variations with good scalability of CMOS technology.

Recently, an all-digital full-precision SRAM-CIM macro is presented with hybrid 28T-14T BAT [3] for high-performance and energy-efficient NN acceleration. The bit-wise multiplications of input activations and weights stored in SRAM-CIM cells are implemented with embedded NOR2 gates which are then accumulated by BAT [3]. However, BAT occupies significant portions for both power and area in SRAM-CIM macro [3]. Approximate computing is employed in BAT with reduced hardware cost for BNN while the accuracy loss is suppressed by approximation-aware training algorithm [4]. However, computation errors can be significantly accumulated for complex DNN models with large datasets, increasing difficulties to maintain sufficient accuracy in [4].

In this paper, an all-digital full-precision SRAM-CIM macro with compressor-based reduction tree based on hybrid 28T/18T/14T FAs and 18T HAs is proposed for energy-efficient NN accelerations. With higher compression efficiency, the CRT operating at lower supply voltage reduces the number of required adders, yielding higher energy and area efficiency compared to BAT. The proposed SRAM-CIM macro achieves 78.07% higher energy efficiency per unit area, compared to the previous all-digital full-precision SRAM-CIM work in [3].

This work was supported by National Natural Science Foundation of China (NSFC) under Grant No. 62174110 and National Key R&D Program of China under Grant No. 2021YFA0717400. (*Corresponding author: Yanan Sun, email: sunyanan@sjtu.edu.cn*)

II. PROPOSED SRAM-CIM MACRO

A. SRAM-CIM Macro Overview

The proposed SRAM-CIM macro powered by normal supply V_{DD} and lower supply (V_{DDL}) based on CRT is shown in Fig. 1(a). The macro is composed of 64 sub-CIMs which share the same input activations and word lines, enabling 4b/4b activation/weight CNN model where the 4-bit inputs are serialized in by four cycles. Each sub-CIM consists of 256×4-bit SRAM-CIM array, SRAM I/O periphery, CRT (powered by V_{DDL}), and shift-and-add units (powered by V_{DDL}). In each computing cycle, SRAM-CIM array generates 256 4-bit multiplication results which are then fed in CRT to get 12-bit results. The shift-and-add units further accumulate the results of CRT for four cycles to get 16-bit outputs in each sub-CIM.

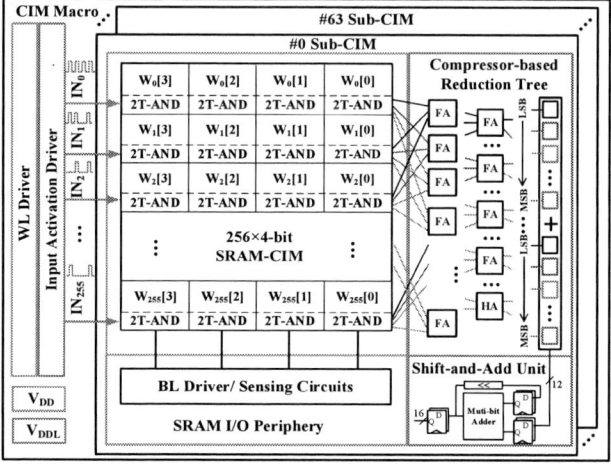

(a)

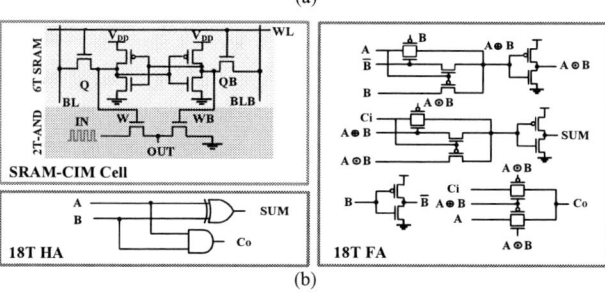

(b)

Fig. 1. (a) Architecture of the proposed all-digital full-precision SRAM-CIM macro. (b) Schematics of SRAM-CIM cell, 18T HA, and 18T FA.

B. MAC Circuits

The SRAM-CIM cell of proposed marco is composed of a standard 6T SRAM and a 2T-AND. As shown in Fig. 1(b), the weight W (weight_bar WB) stored in the Q (QB) node of each SRAM-CIM cell drives the gate terminal of NMOS transistor in each side of 2T-AND which generates the outputs (OUT) of bit-wise multiplications between input IN and weight W. The voltage of OUT denponds on the minimum voltage of V_{DDL} and ($V_{DD}-V_{tn}$), where V_{tn} is the threshold of NMOS. The input drivers, FAs, and HAs are powered by V_{DDL} for improving power, performance and area (PPA) of adder tree. The OUT signals of SRAM-CIM cells are directly sent to reduction tree which is composed of hybrid 28T/18T/14T FAs and 18T HAs. Note that, the structure of 28T and 14T FAs are the same as [3]. The 18T FA is newly proposed to enable a stronger sum cascaded propogation by employing additional output driver, as shown in the right of Fig. 1(b).

978-1-6654-9270-6/22 $31.00 © 2022 IEEE

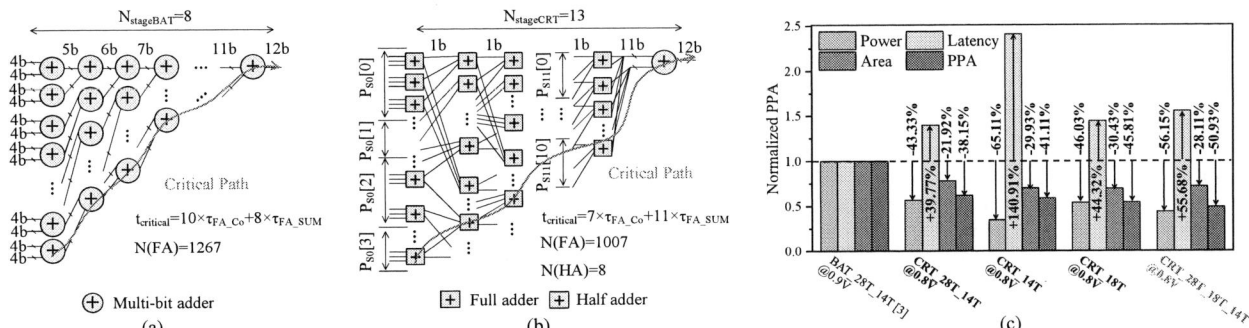

Fig. 2. (a) Conventional BAT. (b) Proposed CRT. (c) Results of different adder tree designs assuming 28nm technology normalized to interleaved 28T/14T BAT [3] considering power, latency, area, and product of PPA. The BAT is powered by 0.9V V_{DD}, and the CRTs are powered by 0.8V V_{DDL}.

C. CRT Design

Conventional BAT used in [3] occupies a significant portion in the area and energy of CIM macro. To alleviate the hardware overhead of BAT, CRT is proposed in this paper. The conventional BAT and the proposed CRT for 256 4-bit data summation are shown in Figs. 2(a) and (b), respectively. The BAT has total eight stages. For each stage, multi-bit adders are used to reduce inputs to half. The proposed CRT has total thirteen stages. For each stage, every three data bits that share the same significance, are reduced by one FA, while the HA is used only when the remaining data bits are two.

Moreover, hybrid FAs are widely used in those designs. In [3], BAT is composed of interleaved 28T and 14T FAs. In this work, two different CRT structures are explored, including interleaved 28T/14T FAs, and hybrid 28T/18T/14T FAs. For hybrid CRT, in the first three stages, which occupies over 70% number of FAs of CRT, 14T FAs are used to reduce the area and power. In the 4th stage, 28T FAs are used as buffer stage to recover cascaded signals. In the 5th to the 12th stage, 18T FAs are used to reduce power and area overhead. In the final stage, multi-bit carry ripple adder is implemented with 28T FAs for enhanced performance.

III. EVALUATIONS ON THE PROPOSED SRAM-CIM MACRO

A. Evaluations of Different Adder Trees

Five different adder trees for 256 4-bit data summation are evaluated in this section. The evaluations are based on TSMC 28nm standard cell library with Design Compiler for synthesis. The 18T and 14T FAs are customized by Candace Virtuoso and extracted to lib file. The 28T FA and 18T HA are based on the standard cell library. The power, latency, area, and product of PPA of different adder trees are shown in Fig. 2(c).

The interleaved 28T/14T CRT achieves 43.33% lower power and 21.92% less area with performance degradation compared to previous BAT [3], due to low supply voltage and fewer number of adders in CRT. The 14T CRT reduces the power by up to 65.11%, achieving lowest power among all the designs. However, the latency of 14T CRT is increased by up to 140.91%, due to the weak driving strength of 14T FAs. The product of PPA of 18T CRT is reduced by up to 45.81% compared to the previous BAT. 18T CRT also outperforms 14T CRT in the product of PPA due to the additional output drivers used in 18T FAs to trade off between performance and power. Finally, the proposed hybrid 28T/18T/14T CRT reduces the power and area by up to 56.15% and 28.11%, respectively, achieving the best product of PPA, due to the massive usage of 14T/18T FAs for significant power and area saving with tolerable performance degradation compared to BAT [3].

B. System Level Evaluations

The system level evaluations on [3] and this work are listed in TABLE I, assuming 28nm technology. With the area-efficient and energy-efficient CRT, the proposed design achieves 18.54% less area and 45.05% higher energy efficiency than [3]. The throughput of proposed design has 19.68% degradation compared to [3] primarily due to the lower supply voltage for CRT to achieve higher energy efficiency. Finally, the energy efficiency per unit area of proposed design is achieved at 321.91 TOPS/W/mm² which is 78.07% higher than [3].

TABLE I. PERFORMANCE COMPARISON

SRAM-CIM Design	21'ISSCC[3]	This work*
Technology	28nm	28nm
MAC operation	Digital	Digital
Power supply(V)	0.9	0.9&0.8
Array Size	64Kb	64Kb
Macro Size (mm²)	0.23	0.19
Energy Efficiency (TOPS/W)	42.40(4b/4b)	61.50(4b/4b)
Throughput (GOPS)	3276.80(4b/4b)	2631.97(4b/4b)
Energy Efficiency Per Unit Area (TOPS/W/mm²)	180.78	321.91

IV. CONCLUSION

In this paper, an all-digital full-precision SRAM-CIM macro based on compressor-based reduction tree with lower supply voltage is proposed for high energy-efficient NN accelerations. The power and area of the proposed compressor-based reduction tree are reduced by up to 56.15% and 28.11%, respectively, as compared to binary adder tree. The area, energy efficiency, and energy efficiency per unit area of proposed design are reduced by up to 18.54%, 45.05%, and 78.07%, respectively, compared to the previous all-digital full-precision SRAM-CIM work based on binary adder tree.

V. REFERENCES

[1] Z. Lin et al., "Cascade current mirror to improve linearity and consistency in SRAM in-memory computing," *IEEE Journal of Solid-State Circuits*, Vol. 56, No. 8, pp. 2550-2562, August 2021.

[2] Z. Chen et al., "CAP-RAM: a charge-domain in-memory computing 6T-SRAM for accurate and precision-programmable CNN inference," *IEEE Journal of Solid-State Circuits*, Vol. 56, No. 6, pp. 1924-1935, June 2021.

[3] Y. -D. Chih et al., "An 89TOPS/W and 16.3TOPS/mm² all-digital SRAM-based full-precision compute-in memory macro in 22nm for machine-learning edge applications," *IEEE International Solid- State Circuits Conference (ISSCC)*, pp. 252-254, February 2021.

[4] D. Wang et al., "DIMC: 2219TOPS/W 2569F²/b digital in-memory computing macro in 28nm based on approximate arithmetic hardware," *IEEE International Solid- State Circuits Conference (ISSCC)*, pp. 2376-8606, February 2022.

A Compact 60 GHz LNA with 22.7-dB Gain and 4.4-dB NF in 40nm CMOS

Jiacong Ke, Guangyin Feng, Yanjie Wang*

School of Microelectronics, South China University of Technology
*wangyanjie@scut.edu.cn

Abstract—This paper presents a compact low-noise amplifier (LNA) design for 60 GHz phased-array applications. Utilizing single transformer-based 4th-order resonators for input and inter-stage matching, a compact core area of only 0.08mm² is achieved. A unit transistor-cell layout design technique for millimeter-wave (mm-Wave) circuit design is adopted to reduce the uncertain high-frequency coupling effects, leading a peak gain of 22.7dB. A 4.4-dB noise figure (NF) and a 3-dB bandwidth from 54 to 63 GHz are achieved based on the post layout simulation results, with a total power consumption of 29.9 mW.

Keywords—low noise amplifier, millimeter-wave, phased-array, matching network, transformer

I. INTRODUCTION

With the rapid increasing requirements of short-range wireless communication and 3D sensing in recent years, mm-wave phased-array attracted the academia and the industry as a highly accurate object sensor and point-to-point wideband transceiver. Since the application of most mm-wave bands are list on designated uses in most regions, the unlicensed 60 GHz band shows sufficient application potential. As the first active module of the receiver at the antenna end, LNA's performance determines the overall noise figure (NF) and link budget of the receiver. Moreover, due to the large number of receiver front-end in large-scale phase-array, the chip size of low-noise amplifier is critical to the overall chip area and cost. Therefore, LNAs with compact size, high gain, and low NF are desired for large-scale phased-array. This paper presents an ultra-compact 60 GHz LNA topology using single transformer-based 4th-order resonators for input and inter-stage matching.

II. COMPACT LNA CIRCUIT DESIGN

A. Pad-Transformer 4th-Order Resonator

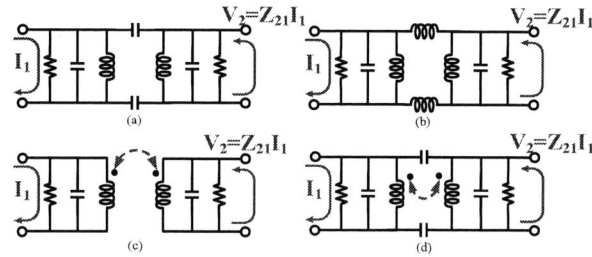

Fig. 1. (a) Capacitively coupled resonators. (b) Inductively coupled resonators. (c) Magnetically coupled resonators. (d) Magnetically and capacitively coupled resonators.

The inter-stage matching network of high-frequency amplifiers are divided into four types as shown in Fig.1. The 4th-order resonator based on magnetic coupling shown in Fig. 1(c) achieves an optimal in-band flatness of normalized high transimpedance Z_{21}, which is critical for common-source (CS) amplifier since it is seen by a G_m stage [1]. Meanwhile, the most compact input matching network based on magnetic coupling is also composed of this 4th-order resonator, which is constructed by absorbing the parasitic of input pads and transistors, as shown in Fig.2(a). To illustrate the conjugate matching capability of the pad-transformer 4th-order resonator, a T-model is utilized to deemed the mutual inductance of transformer, leading to the trajectory extraction on smith chart, as shown in Fig.2(b). By tuning component values in desired

networks, the available matching range for a typical 50-ohm input matching requirement is obtained, which is represented by the shaded region as shown in Fig. 2(c). Considering the large input parallel resistance R_P of CS amplifiers, the region covers all matching requirements of general size transistor. Therefore, a most compact magnetic coupling-based LNA is composed of several G_m stages and single transformer-based 4th-order resonators, which enables input matching and obtains optimal in-band flatness.

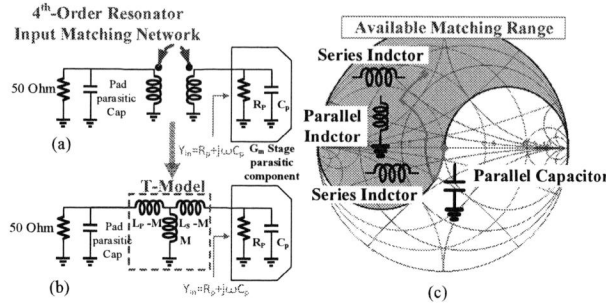

Fig. 2. (a) 4th-order resonator input matching network. (b) 4th-order resonator input matching network with T model. (c) available matching range of this 4th-order resonator.

B. The Proposed Compact LNA Circuit Topology

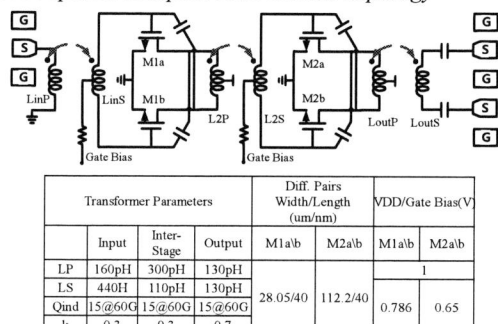

Transformer Parameters				Diff. Pairs Width/Length (um/nm)		VDD/Gate Bias(V)	
	Input	Inter-Stage	Output	M1a\b	M2a\b	M1a\b	M2a\b
LP	160pH	300pH	130pH	28.05/40	112.2/40	1	
LS	440H	110pH	130pH			0.786	0.65
Qind	15@60G	15@60G	15@60G				
k	0.3	0.3	0.7				

Fig. 3. The proposed LNA topology and the component values.

As shown in Fig.3, the proposed LNA circuit topology is composed of two pseudo differential CS amplifiers instead of fully differential topology to relax the voltage head room issue, and the neutralization capacitors are used to eliminate the coupling effect of the gate-drain capacitor C_{gd} at mm-wave with improved stability and power gain. Both the input and inter-stage matching networks consist of single passive transformer, which absorbs the parasitic components of pads and transistors to form the 4th-order resonators, satisfying the input matching requirements and enhancing the overall power gain, as shown in Fig.4. The LNA's minimum noise figure (NF_{min}) value is optimized by biasing the transistors of G_m-amplifiers at an optimal current density of 0.2 mA/um [4]. A transistor's finger width of 850nm is chosen to improve the compatibility between input conjugated matching and noise matching [4].

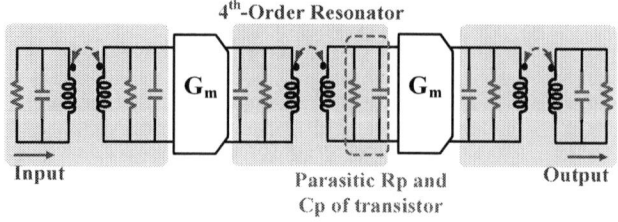

Fig. 4. Two stages compact LNA composed of three 4th-order resonators and two G_m stages.

C. Compact LNA Layout Optimazation

Due to inaccuracy of transistor's RF model provided by foundries and unpredictable large loss of passive components at target frequency, Fig.5 shows the layout method of a unit transistor-cell used in this design to reduce the extra parasitic effect and loss. The widths of transistor M1a/b and M2a/b are 28.05um and 112.2um, respectively. Each transistor was divided into a parallel connection of multiple unit-cell with a finger width of 850nm and 32 fingers. Fig.5 (a) shows the 3D view of the layout of CS transistor unit-cell. The signal travels vertically inside the unit-cells while they are connected laterally for a low output power loss [5]. Since the metal ground thickness is approaching the skin depth of cooper at the desired frequency, the lowest two metal layers Matel-1 and Matel-2 are used as the ground plane and the source-terminal of the unit-cell surrounding the transistor, which reduces the magnetics loss, improving the quality factor of the inductors, and form an approximate ideal connection to common-ground.

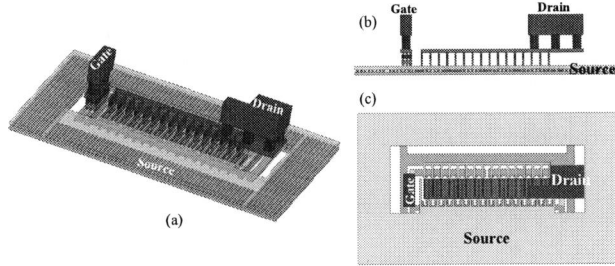

Fig. 5. (a) The 3D view of a unit transistor-cell layout. (b) Side view of a unit transistor-cell layout. (c) Top view of a unit transistor-cell layout.

III. LNA SIMULATION RESULTS

The proposed LNA was implemented in TSMC 40nm LP bulk CMOS process. Fig.6 (a) shows the proposed LNA layout, with a core area of 0.20mm*0.40mm. Fig.6 (b) shows EM simulation model. Fig.7 shows that a 22.7 dB peak gain at 58 GHz, and a 9 GHz 3-dB bandwidth from 54 to 63 GHz are achieved in the post layout simulation results. Fig.8 (a) illustrates a 4.4 dB NF achieved at 60 GHz. An average IP1dB of -16 dBm in operating frequency band is shown in Fig.8 (b).

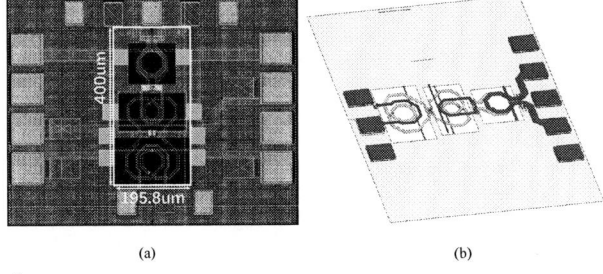

(a)   (b)

Fig. 6. (a) The proposed LNA layout. (b) EM model of the proposed LNA.

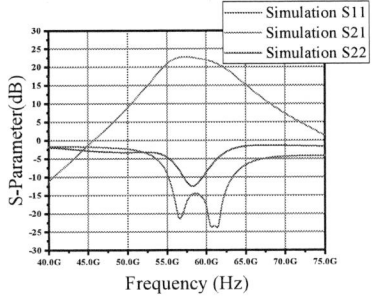

Fig. 7. Simulation results of s-parameter.

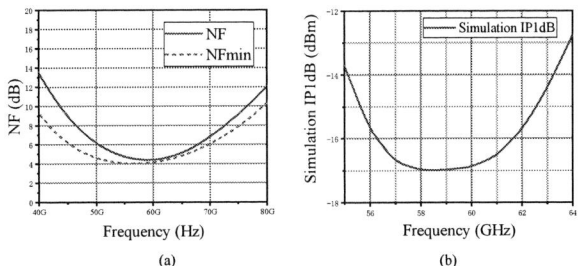

Fig. 8. (a) NF and NF$_{min}$ simulation result. (b) Average IP1dB of -16 dBm in the operating frequency band.

TABLE I.   COMPARISON TABLE

Ref.	PROCESS	Freq (GHz)	NF (dB)	Gain (dB)	DC Power (mW)	IP1dB (dBm)	3-dB BW (GHz)
[6] ASSCC '08	65nm CMOS	60	<6	16.7	31	-15.1	9
[2] RFIC' 20	65nm CMOS	60	4	25	47	-22	7.5
[3] IWS' 18	40nm CMOS	60	5.13	6.67	12.1	N/A	9
THIS WORK	40nm CMOS	60	4.4*	22.7*	29.9*	-16.1*	9*

*post layout simulation result

IV. CONCLUTION

This paper presents an ultra-compact 60 GHz LNA implemented in TSMC 40nm LP bulk CMOS. Utilizing single transformer-based 4th-order resonators for input and inter-stage matching, the proposed LNA achieves a core area of only 0.08 mm^2. Moreover, a 4.4 dB NF, a 22.7 dB peak gain and a 3-dB bandwidth from 54 to 63 GHz are achieved in the post layout simulation result, with a total power consumption of 29.9 mW.

ACKNOWLEDGMENT

This work is supported by the Fundamental Research Funds for the Central Universities under Grant 2020ZYGXZR067 and Natural Science Foundation of Guangdong Province under Grant No. 2021A1515011677.

REFERENCES

[1] M. Vigilante and P. Reynaert, "20.10 A 68.1-to-96.4GHz variable-gain low-noise amplifier in 28nm CMOS," 2016 IEEE International Solid-State Circuits Conference (ISSCC), 2016, pp. 360-362, doi: 10.1109/ISSCC.2016.7418056.

[2] D. Bierbuesse and R. Negra, "60 GHz variable Gain & Linearity Enhancement LNA in 65 nm CMOS," 2020 IEEE Radio Frequency Integrated Circuits Symposium (RFIC), 2020, pp. 163-166, doi: 10.1109/RFIC49505.2020.9218337.

[3] B. Wang, H. Gao, R. van Dommele, M. K. Matters-Kammerer and P. G. M. Baltus, "A 60 GHz low noise variable gain amplifier with small noise figure and IIP3 variation in a 40-nm CMOS technology," 2018 IEEE MTT-S International Wireless Symposium (IWS), 2018, pp. 1-4, doi: 10.1109/IEEE-IWS.2018.8400880.

[4] G. Feng et al., "Pole-Converging Intrastage Bandwidth Extension Technique for Wideband Amplifiers," in IEEE Journal of Solid-State Circuits, vol. 52, no. 3, pp. 769-780, March 2017, doi: 10.1109/JSSC.2016.2641459.

[5] D. Zhao and P. Reynaert, "A 60-GHz Dual-Mode Class AB Power Amplifier in 40-nm CMOS," in IEEE Journal of Solid-State Circuits, vol. 48, no. 10, pp. 2323-2337, Oct. 2013, doi: 10.1109/JSSC.2013.2275662.

A Fully-integrated 110-GHz Wideband Direct-detection Imaging Unit MMIC Integrating Balanced Power Detector and Log-periodic Antenna

Jinyu Xie[1], Xiaojun Bi[1]

[1] School of Optical and Electronic Information,
Huazhong University of Science & Technology,
Wuhan 430074, China
E-mail: bixj@hust.edu.cn

Abstract—this paper presents a 110-GHz wideband direct-detection imaging unit integrating balanced power detector and log-periodic antenna in 0.25 μm GaAs pHEMT technology. The detector mainly consists of a 90° bridge, two identical transistors for detection, and a low-pass filter. Coplanar waveguides (CPW) are deployed to implement transmission lines in a compact size. In addition, an on-chip log-periodic antenna is co-designed to connect with the detector. Simulation results demonstrate that the peak responsivity of the detector is 8 kV/W at 98 GHz with a 3 dB bandwidth from 85.9 GHz to 128.9 GHz and a low in-band noise equivalent power (NEP) of 0.87-1.5 pW/Hz$^{1/2}$. The simulated S_{11} is less than -7.6 dB at 75-140 GHz. The overall size of the chip is 2.7 × 1 mm^2 including pads.

Keywords—detector, direct-detection, log-periodic antenna, NEP

I. INTRODUCTION

The rapid growth of wireless communications up to 110 GHz and beyond has driven the emergence of millimeter-wave systems for various applications such as radar, passive imaging sensing, short-range high-rate communications, etc. Particularly, the detector is the key component in millimeter-wave passive imaging systems, which plays the role of extracting the information carried in the high-frequency carrier wave for processing at the back end. In recent published works, more and more on-chip millimeter-wave detector operating at W-band, D-band or beyond [1-6] are reported. In addition to the traditional sperate detector design, there are also works where the on-chip antenna is integrated with the detector thus realizing a direct-detection receiver unit [4][6]. However, such works are usually integrated with LNA or assembled with silicon lens to enhance the responsivity of the system, which makes the detection unit cumbersome and disables large imaging array. In this paper, a co-design of a 110 GHz wideband power detector and on-chip antenna in 0.25 μm GaAs pHEMT technology is presented.

II. DIRECT-DETECTION UNIT DESIGN

A. Wideband Balanced Power Detector

Fig. 1 illustrates the schematic of the proposed 110 GHz wideband detector. It can be seen that in order to obtain transmission lines with the desired characteristic impedance on a thick (100 μm) GaAs substrate with reasonable dimensions, coplanar waveguide is deployed instead of microstrip line.

The characteristic impedance of TL_1 and TL_3 is 35 Ω, and the characteristic impedance of TL_2 and TL_4 is 50 Ω, their electrical lengths are $\lambda/4$, thus forming a branch line coupler (BLC). As long as the transistors connected to the two output ports have the same characteristics, two reflected waves are in-phase and out-of-phase at the input and isolated ports respectively, then a good input matching can be obtained.

As shown in Fig. 1, two 2×25 μm transistors are used as nonlinear components to generate DC component. The bias

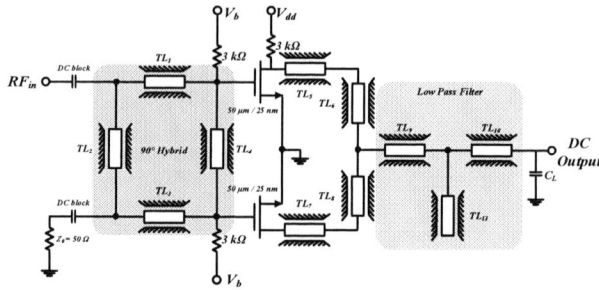

Fig. 1. Schematic of the proposed wideband balanced power detector.

voltage V_b is optimized to around 0.42 V, slightly higher than the threshold voltage V_{th} of the E-mode pHEMT, in order to obtain a large enough second-order transconductance. The DC voltage output ultimately depends on the supply voltage V_{dd} and bias resistor R_b. After optimizing the responsivity (i.e., the ratio of DC voltage to microwave input power), V_{dd} of 2.7 V and R_b of 3 kΩ are used, which results in a drain current of 160 μA and a drain voltage of 1.6 V for the transistor. Note that although increasing the resistance and supply voltage can obtain a greater responsivity, the noise also deteriorates at the same time. According to [5], the NEP of the circuit can be expressed as:

$$NEP = \frac{\sqrt{4k_B TR}}{R_v} \qquad (1)$$

where, k_B is the boltzmann constant, T is the temperature, R is the load resistance, and R_v is the responsivity.

Since the detector needs to convert the microwave signal into a DC signal, a high frequency component are not expected at the output. As shown in Fig. 1, a low-pass filter is designed with two series lines and a shunt stub (TL_9-TL_{11}), which achieves a relative low cut-off frequency.

B. Log-periodic Antenna

In order to obtain a compact wideband direct-detection unit, a log-periodic antenna is co-designed with the detector, which is a frequency-independent broadband antenna capable of increasing the bandwidth without increasing the size. The structure is shown in Fig. 2. The antenna consists of three folded dipoles with different lengths and spacings, and the relation between these physical parameters is geometric ratio. The same as the BLC of the detector, the antenna is fed with a CPW of 25 μm line width and 10 μm gap width. In this way, the antenna and the detector can be connected compactly without using additional matching networks. In order to excite such a dipole array in the balanced mode, a transition from CPW to coplanar stripline (CPS) is used. The detailed configuration can be observed in the layout, where a crisscross structure is formed with vias and bottom metal layer. The propagation mode changes during the combination of the left ground and the right ground. At the same time, the directivity of the antenna is enhanced because the truncated ground from the detector plays the role of the reflector of the antenna, and an end-fire radiation pattern can be obtained as shown in Fig. 2.

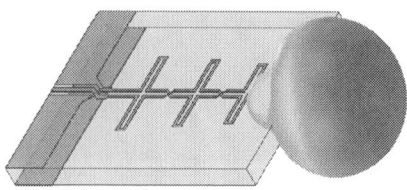

Fig. 2. On-chip log-periodic antenna structure with 3D radiation pattern.

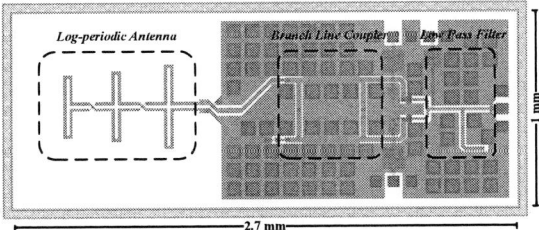

Fig. 3. Layout of the proposed wideband direct-detection unit.

III. SIMULATION RESULTS

Fig. 3 demonstrates the layout of the proposed wideband direct-detection unit with a chip area of 2.7×1 mm² including the DC bias and output GSG pads. Fig. 4 shows a impedance matching of the detector in the range of 75-140 GHz with S_{11} < -7.6 dB. Fig. 5(a) presents the output voltage and responsivity of the detector at 98 GHz for different input power levels, the responsivity is completely linear when the input power is below -30 dBm, and it starts to compress at input power above -30 dBm and is completely saturated until at around 0 dBm. Fig. 5(b) shows the responsivity and NEP versus frequency when the input power is -30 dBm, the peak responsivity is 8.175 kV/W @ 98 GHz with a 3 dB bandwidth of 85.9-128.9 GHz, and an in-band NEP of 0.87-1.5 pW/Hz$^{1/2}$. Fig. 6-7 are the simulation results of the antenna. The radiation gain is greater than 0 dBi in operating band of the detector with a maximum gain of 5.8 dBi at 110 GHz and the cross-polarization rejection level is higher than 20 dB.

IV. CONCLUSION

A 110 GHz wideband direct-detection unit has been demonstrated. A branch line coupler structure has been used to obtain good input matching, resulting in a 3 dB fractional bandwidth of 40% and a peak responsivity of 8.175 kV/W at 98 GHz. Additionally, an broadband log-periodic antenna is co-designed with enhanced area-efficiency, thus forming a direct-detection imaging unit without other modules such as LNA or off-chip configurations. It shows potential for large array or passive imaging applications. The performance comparison between this work and some published detectors is shown in Table I. It can be seen that the proposed direct-detection unit demonstrates competitive performance using a relative unadvanced technology, providing a very substantial operating bandwidth and good noise performance while ensuring high responsivity.

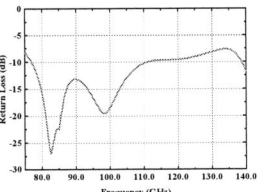

Fig. 4. Simulated S_{11} of the proposed wideband detector.

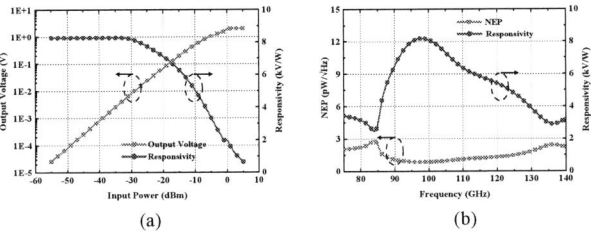

(a)   (b)

Fig. 5. (a) Simulated output voltage and responsivity versus input power at 98 GHz; (b) Simulated responsivity and NEP versus frequency at P_{in} = -30 dBm.

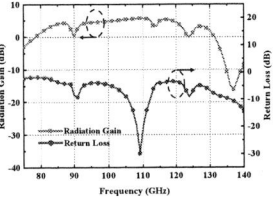

Fig. 6. Simulated Radiation Gain and S_{11} of the proposed antenna.

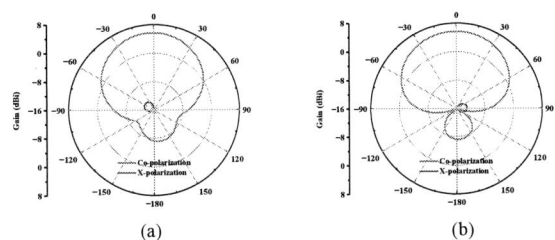

(a)   (b)

Fig. 7. Radiation patterns by simulation results: (a) E-plane; (b) H-plane.

TABLE I. PERFORMANCE COMPARISON WITH THE PUBLISHED WORKS

Parameter	Reference					
	[2]	[3]	[4]	[5]	[6]	This work
Technology	55 nm SiGe BiCMOS	45 nm CMOS SOI	0.25 μm SiGe BiCMOS	0.18 μm SiGe BiCMOS	0.13 μm GaAs pHEMT	0.25 μm GaAs pHEMT
f_T (GHz)	300	200	110	200	100	57
Antenna Integration	No	No	Yes	No	Yes	Yes
R_v (kV/W)	2	3	5	11	0.22	8
NEP (pW/Hz$^{1/2}$)	4.56-5.53	10-20	10-16	5-10	25	0.87-1.5
3 dB Bandwidth (GHz)	140-220 (44%)	150-195 (26%)	83-94 (12%)	75-107 (35%)	225-280 (22%)	86-129 (40%)

ACKNOWLEDGMENT

This work was supported by the National Key Research and Development Program of China under Grant 2018YFB2201503.

REFERENCES

[1] X. Bi, M. A. Arasu, Y. Zhu and M. Je, "A Low Switching-Loss W-Band Radiometer Utilizing a Single-Pole-Double-Throw Distributed Amplifier in 0.13-μm SiGe BiCMOS," in *IEEE Transactions on Microwave Theory and Techniques*, vol. 64, no. 1, pp. 226-238, 2016.

[2] I. Alaji, S. Lépilliet, D. Gloria, G. Ducournau and C. Gaquière, "Design and Characterization of (140–220) GHz Frequency Compensated Power Detector," in *IEEE Transactions on Microwave Theory and Techniques*, vol. 69, no. 4, pp. 2352-2356, 2021.

[3] M. Uzunkol and G. M. Rebeiz, "A Low-Noise 150–210 GHz Detector in 45 nm CMOS SOI," in *IEEE Microwave and Wireless Components Letters*, vol. 23, no. 6, pp. 309-311, 2013.

[4] D. Dancila et al., "Wide band on-chip slot antenna with back-side etched trench for W-band sensing applications," *2013 7th European Conference on Antennas and Propagation (EuCAP)*, 2013, pp. 1576-1579.

[5] L. Gilreath, V. Jain and P. Heydari, "Design and Analysis of a W-Band SiGe Direct-Detection-Based Passive Imaging Receiver," in *IEEE Journal of Solid-State Circuits*, vol. 46, no. 10, pp. 2240-2252, 2011.

[6] S. Nahar, A. Muraviev, D. But and M. M. Hella, "A 220 V/W, 25 pW/√Hz NEP bow-tie antenna-coupled pHEMT detector at 250 GHz," *2016 41st International Conference on Infrared, Millimeter, and Terahertz waves (IRMMW-THz)*, 2016, pp. 1-2.

978-1-6654-9270-6/22 $31.00 © 2022 IEEE

A 250Mbps 100kV/μs CMTI On-Chip Double-Isolated Transformer-Based Digital Isolator

Zhiyong Xiong, Dongfang Pan, Guolong Li, Lin Cheng

University of Science and Technology of China (USTC), Hefei, China

Email: pdf1992@ustc.edu.cn   eecheng@ustc.edu.cn

Abstract—This paper presents a double-isolated on-chip transformer-based digital isolator exploiting integrated double SiO_2 galvanic barriers to achieve an isolation rate of 4kV. The isolator employs pulse polarity modulation to reduce the power consumption while maintaining a high data rate and high common mode transient immunity (CMTI). Simulation results show that the proposed isolator achieves a 250Mbps maximum data rate with a 5ns propagation delay and a 100kV/μs CMTI performance. With a load of 15pF, TX and RX consume 0.32mA and 3.4mA currents at a 1Mbps data rate and 11.7mA and 3.5mA currents at a 250Mbps data rate, respectively.

Keywords—digital isolator, double-isolated, transformer, pulse polarity, transmitter (TX), receiver (RX)

I. INTRODUCTION

Digital isolators are widely used in electrical vehicles, communications, medical equipment and other industrial fields to guarantee safety and reliability of data transmission between the high-voltage domain and the low-voltage domain in these systems. CMOS-based digital isolators are replacing the traditional optocoupler isolators in many applications for the higher data rates, lower delay, better reliability, and lower cost.

On-chip transformers using SiO_2 as isolation dielectric with high noise immunity are an attractive candidate for digital isolators in a CMOS process, providing high dielectric isolation strength of >500V/μm. However, the thickness of SiO_2 (i.e., 3-4μm) provided by the standard 5-layer metal 0.18μm CMOS process is not sufficient to achieve an isolation rate >3kV [1-4]. Basic isolation can be achieved using only one isolation barrier. For an enhanced level of safety, double isolation is needed. In [5] and [6], capacitor-based digital isolators are reported. These isolators doubled the breakdown voltage by connecting two capacitors in series at the transmitter (TX) chip and receiver (RX) chip, and the isolation capability is improved. However, the capacitor-based isolator adopts on-off keying (OOK) modulation, and its power consumption will increase because the TX is always in an oscillation state.

In this paper, an on-chip double-isolated transformer-based digital isolator is presented to improve the isolation rate. The isolator utilizes the pulse polarity modulation scheme to achieve 250Mbps data rate and 100kV/μs common mode transient immunity (CMTI) with low power dissipation.

II. CIRCUIT DESIGN

A. System Architecture

Fig. 1 shows the circuit schematic of the proposed digital isolator. The isolator consists of a TX chip and an RX chip and adopts a pulse polarity modulation scheme [3]. The double isolation barrier is performed by two on-chip transformers on the TX and the RX, respectively, in a series connection. In addition, the series transformer can reduce the parasitic capacitance on both sides of the isolation barrier and thus further benefit CMTI. The TX chip encodes the rising and falling edges of the input signal into positive and negative narrow pulse signals V_{PULSE} with a signal width of 300ps, respectively. Then, the pulse signal is filtered and shaped to restore the data signal through a hysteretic comparator in the RX chip.

Fig. 2 shows the structure and performance of the on-chip transformer. The 4.7-μm-thick top metal-5 and 0.53-μm-thick metal-2, both with 6-μm width and 6 turns of windings, are used to form the primary and the secondary coil, respectively, and metal-1 is used for the interconnections among the coils. As a result, the single on-chip isolation barrier is implemented with a total thickness of 4μm multilayer of SiO_2 between the primary and secondary coils, providing a high dielectric insulation strength (i.e., ~500V_{RMS}/μm). As shown in Fig. 2, the electromagnetic-simulated inductance/Q value of the primary and the secondary coils are 16.2nH/15.1 and 16.1nH/2.2 at 2GHz, respectively, with a mutual inductance/k of 15.1nH/0.9.

B. Transmitter design

The TX first detects the rising and falling edges of the input signal IN, and then the short pulse signals S_1 and S_2 are used to control the driver transistors M_{P1}, M_{N1}, M_{P2} and M_{N2} to drive the coils to generate pulse signals V_{PULSE}.

The operation principle diagram of the TX is shown in Fig 3(a). Taking the digital input signal IN from low to high as an example, the gate voltage of M_{P1} becomes low, and the short pulse S_2 generated by the rising edge detector turns on the driving transistors M_{N2}. At this time, a sufficiently large instantaneous current I_L is generated on the primary coil to generate a positive narrow pulse representing the rising edge. After the end of short pulse S_2, the current I_L is gradually cut off. To avoid the high swing counter pulse generated by the sharp falling edge of I_L, the RX is prevented from distinguishing between the V_{PULSE} and the counter pulse. The falling edge slope of the M_{N1} and M_{N2} gate voltages is reduced by the current source I_B so that the counter pulse swing can be suppressed. The magnitude of V_{PULSE} is proportional to MdI_L/dt, where M is the mutual inductance of the transformer. In this design, the typical swing and width of the V_{PULSE} with two transformers coupled are 1.4V and 300ps, respectively.

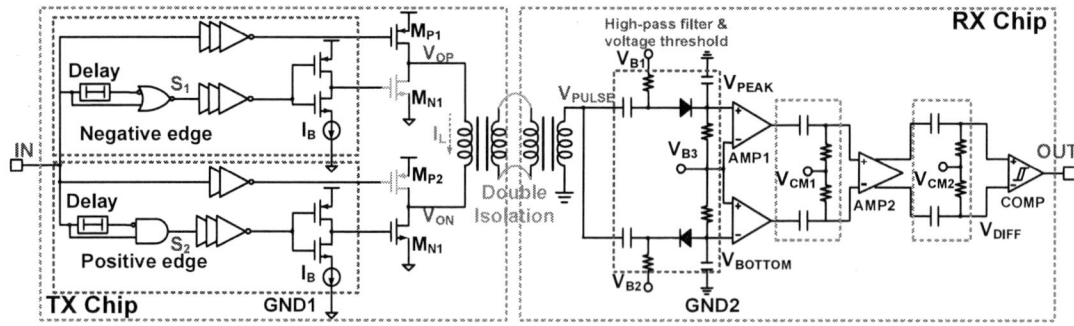

Fig. 1. The circuit schematic of the proposed double-isolated digital isolator.

978-1-6654-9270-6/22 $31.00 © 2022 IEEE

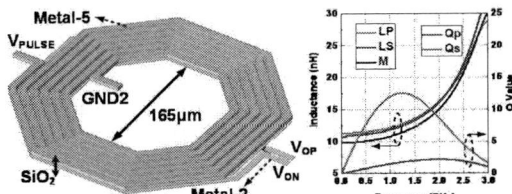

Fig. 2. EM-model and performance of the on-chip transformer.

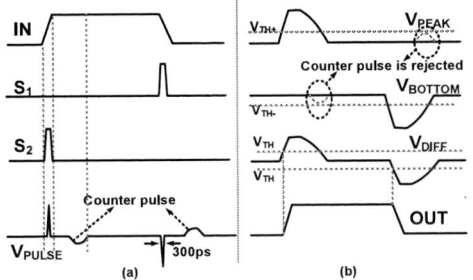

Fig. 3. (a) Operation diagram of the TX. (b) Operation diagram of the RX.

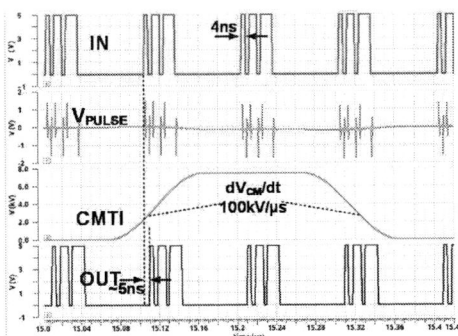

Fig. 4. Simulated key waveforms of the digital isolator at 1/100/250Mbps.

TABLE I. PERFORMANCE SUMMARY

	This work	[3]	[4]	[5]
Isolation Barrier	Transformer	Transformer	Transformer	Capacitor
Process	0.18μm CMOS	0.5μm CMOS	0.18μm CMOS	0.18μm CMOS
Isolation rating	4kV$_{RMS}$	2.5kV$_{RMS}$	N/A	3kV$_{RMS}$
Modulation Architecture	Pulse polarity	Pulse polarity	OOK	OOK
Maximum Data Rate	250Mbps	250Mbps*	1Mbps*	200Mbps
Propagation Delay	5ns	5.5ns*	11ns*	6ns
CMTI	100kV/μs	35kV/μs	200kV/μs*	100kV/μs
Maximum I$_{DD}$@5V	11mA @100Mbps	14.3 mA @100Mbps	2.8mA@ 1Mbps	6.4mA @100Mbps

*: Measurement results.

C. Receiver design

The RX must detect the narrow pulses V$_{PULSE}$ and filter out low swing counter pulses and noise before recovering the input signal IN. Fig. 3(b) shows the operation diagram of the RX. The input stage of the RX consists of a high-pass filter and diodes D$_{1,2}$ used as a filter and shaper for V$_{PULSE}$, suppressing low-swing counter pulses by its threshold, which can be adjusted by changing the bias voltage V$_{B1-3}$(the voltage threshold V$_{TH+}$=0.7-(V$_{B1}$-V$_{B2}$) and V$_{TH-}$=0.7-(V$_{B2}$-V$_{B3}$)). After filtering and shaping, the V$_{PEAK}$ and V$_{BOTTOM}$ signals are amplified in two stages, and finally, the input signal is restored through a hysteresis comparator. Moreover, a common-mode DC point is set among each amplifier stage and the comparator to improve the robustness of the fully differential amplifier and to reduce mismatch.

III. SIMULATION RESULTS

The proposed double-isolated on-chip transformer digital isolator is simulated in a 0.18μm CMOS technology. Under 5V power supply and 15pF load, the TX and the RX consume 0.32mA and 3.4mA at 1Mbps data rate and 11.7mA and 3.5mA currents at 250Mbps data rate, respectively.

Fig. 4 shows the simulated key waveforms of both TX and RX. The input signal with an amplitude of 5V is simulated by a pseudo random sequence generator to generate signals of different data rates (up to 250Mbps). These waveforms verify the encoding and decoding capabilities of the isolator at different data rates. The isolator is capable of transmitting a maximum data rate of 250Mbps with only 5ns propagation delay. Fig. 4 also shows the simulated CMTI performance of the isolator. When a high voltage signal with a 100kV/μs edge is added between the TX and RX ground, the isolator still operates normally. Table I summarizes and compares the performance of the proposed digital isolator with state-of-the-art designs. The proposed double-isolated isolator achieves high isolation rate and high data rate.

IV. CONCLUSIONS

An on-chip transformer-based digital isolator with double galvanic isolation is presented to improve the isolation rate. Double galvanic isolation provides twice the thickness of the SiO$_2$ dielectric through two series-connected on-chip transformers placed at TX and RX, and this on-chip transformer is suitable for the integration of multiple isolation channels. The isolator uses pulse polarity modulation to achieve low propagation delay and shorten the current duration, improving the data rate and power performance. The simulation results show that the isolator achieves a 100kV/μs CMTI performance and 250Mbps data rate with a 5ns propagation delay. Under a load of 15pF, TX and RX consume 11.7mA and 3.5mA currents at a 250Mbps data rate, respectively.

ACKNOWLEDGMENTS

This work was supported in part by the National Natural Science Foundation of China under Grant 62104220. The authors would like to thank the Information Science Experiment Center of University of Science and Technology of China for the EDA tools support.

REFERENCES

[1] D. Pan et al., A 1.2W 51%-Peak-Efficiency Isolated DC-DC Converter with a Cross-Coupled Shoot-Through-Free Class-D Oscillator Meeting the CISPR-32 Class-B EMI Standard, IEEE International Solid- State Circuits Conference (ISSCC), San Francisco, CA, USA, 2022. pp. 240-242.

[2] D. Pan et al., 33.5 A 1.25W 46.5%-Peak-Efficiency Transformer-in-Package Isolated DC-DC Converter Using Glass-Based Fan-Out Wafer-Level Packaging Achieving 50mW/mm2 Power Density, IEEE International Solid- State Circuits Conference (ISSCC), San Francisco, CA, USA, 2021, pp. 468-470.

[3] S. Kaeriyama et al, "A 2.5 kV isolation 35 kV/us CMR 250 Mbps digital isolator in standard CMOS with a small transformer driving technique," IEEE J. Solid-State Circuits, vol. 47, no. 2, pp. 435– 443, Feb. 2012.

[4] R. Yun, et al., "A transformer-based digital isolator with 20kV$_{PK}$ surge capability and > 200kV/μs Common Mode Transient Immunity," IEEE Symposium on VLSI Circuits, Honolulu, HI, USA, 2016, pp. 1-2.

[5] G. Shi et al., "A Compact 6 ns Propagation Delay 200 Mbps 100 kV/μs CMR Capacitively Coupled Direction Configurable 4-Channel Digital Isolator in Standard CMOS," IEEE International Conference on Electronics, Circuits and Systems, Bordeaux, France, 2018, pp. 721-724.

[6] F. Miao et al., " A 250-Mbps 2.6-ns Propagation Delay Capacitive Digital Isolator with Adaptive Frequency Control," IEEE International Midwest Symposium on Circuits and Systems, Fukuoka, Japan, 2022, pp. 1-4.

Implementation of CNN Heterogeneous Scheme Based on Domestic FPGA with RISC-V Soft Core CPU

Hailong Wu[1], Jindong Li [1], Xiang Chen[1]

[1] School of Electronics and Information Technology, Sun Yat-sen University, China

Abstract—**Field Programmable Gate Array (FPGA) has the characteristics of low power consumption, high performance and flexibility. Research on FPGA neural network acceleration is emerging, but most of the researches are based on foreign FPGA devices. In order to improve the current situation of domestic FPGA, a novel Convolutional neural networks (CNNs) accelerator for domestic FPGA equipped with lightweight RISC-V soft core is proposed. The peak performance of the proposed accelerator reaches 153.6 GOP/s, occupying only 14K LUTs (Look-Up-Table), 32 DRMs (Dedicated RAM Modules) and 208 APMs (Arithmetic Process Modules). The proposed accelerator has enough computing power for most of the Edge-AI applications and embedded systems, providing a possible AI inference acceleration solution for domestic FPGA.**

Keywords—*FPGA, CNN, Accelerator , RISC-V, Domestic*

I. INTRODUCTION

Convolutional neural networks have become more and more popular in machine vision tasks, including image classification and object detection. How to give full play to the maximum performance of FPGA under limited conditions is the main direction of researchers. Nowadays most CCNs use foreign FPGA devices, such as AMD (Xilinx) and Intel (Altera). Due to the late start of domestic FPGA, its related development tools and devices lag behind other foreign manufacturers. Therefore, building high performance CNN on domestic FPGA and replacing the existing mature heterogeneous schemes is a challenging task.

The first in-depth analysis and exploration of the data sharing and parallelism in inference with convolutional networks were conducted by Zhang[1] in 2015. A accelerator proposed by Guo[2] reached a peak performance of 84.3 GOP/s under 214MHz. In 2016, Qiu [3] explored more deeply on accelerator using the line buffer. In this paper, a more efficient and general convolution accelerator is proposed. The peak performance of our accelerator reaches 153.6GOP/s, occupying only 14K LUTs, 32 DRMs and 208 APMs. The chapters of this paper are arranged as follows, the detailed design of our proposed accelerator and the control scheduling scheme of accelerator implementation based on RISC-V is introduced in Sec.2. The experimental results is show in Sec. 3.

II. SYSTEM DESIGN

The whole RISC-V on chip system design is shown in Fig. 1. The system mainly consists of RISC-V Soft Core CPU, Instruction/data memory, bus bridge, peripherals, DMA(Direct Memory Access) and convolution accelerator.

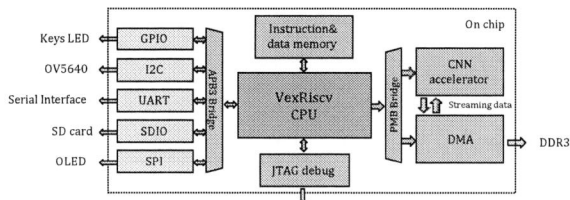

Fig. 1. RISC-V System On Chip Structure

This work was financially supported by the Guangdong R&D Project in Key Areas under Grant 2019B010158001, Industry-University-Research Coopera-

---

Our work mainly lies in three aspects. First, we use soft core CPU as the main control of on-chip system, control peripherals, DMA, CNN accelerator to implement data scheduling and operation. Second, the 1D (One Dimension) accelerator was designed to change the buffer mechanism.Third, a DMA IP was designed for the FPGA device of PangoMicro for the application of convolution acceleration.

A. RISC-V Soft core CPU Structure

1) Soft Core. Using RISC-V soft core VexRiscv instead of Ibex[4] to build RISC-V on-chip system and software-oriented approach can make VexRiscv highly flexible and extensible.

2) Interface. Peripherals such as I2C and SPI are connected to the RISC-V soft core through APB3 bus. DMA and accelerator are connected to the RISC-V soft core by PMB bus.

3) Instruction and Data Memory. The program is cross-compiled to obtain a specific file, which is burned to the on-chip instruction/data memory by JTAG.

B. CNN accelerator Structure

1) Input buffer. Using Ping-Pong cache to implement the buffer, and throughput can be effectively increased.

2) Weight Cache. The weight cache module is composed of a series of distributed RAM and serial-to-parallel units.

3) Convolution. The 1D convolution module in Fig.2 is divided into four group which contains four 1D convolution units. Each unit responsible for one channel of 1D convolution.

4) Integration. The Intergration module has four groups of adder tree. Each group of adder tree adds up the results of each group of convolution operation units to get the output result one way.

5) Accumulation. There are four sets of FIFOs and four adders in accumulation module. The accelerator can only receive four channels of input feature map data at a time.

6) Quantization. This Quantization module consists of multiplication units and shift units. It retransforms the 24-bit accumulation result to 8-bit[5] by means of a scale transformation.

7) Activation. The activation functions are implemented by looking up the table consists of a series of distributed RAMs.It store the INT8 function tables of the ReLu, Leaky ReLu and sigmoid function.

8) Pooling. Determine whether the current convolutional layer is cascaded with the pooling layer and then decide whether to use the pooling module to complete the pooling operation.

9) Output Buffer. The output buffer is implemented by FIFO instead of a ping-pong cache. The output cache FIFO stores the result back to off-chip memory as input for the next convolutional layer.

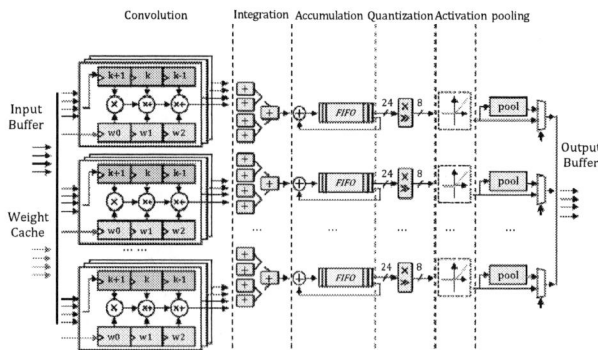

Fig. 2. Implementation of CNN accelerator

C. DMA Structure

Neural network not only has a high demand for computing power, but also has a large demand for memory. Medium and low-end FPGA often requires DDR SRAM (Double Data-Rate Synchronous Dynamic Random-Access Memory) to carry the weight of the entire neural network and all the intermediate operation results. DDR3 memory driver IP of the FPGA created by PangoMicro provides the user with the memory access interface of Simplified AXI4 bus.

Because of the differences Standard between Simplied AXI and AXI, a new DMA design is required. The DMA is designed as follows. The read and write address channels are directly controlled by the RISC-V soft core. The FIFO of the read and write data channel serves as the buffer of the convolution accelerator and DDR3 driver IP to complete the port conversion.

D. Implementation Details

1) Array of 1D Convolution units Design

The convolution operation module of the accelerator is based on the line units. Multiple 1D convolutional units form an array of 1D convolution operation to achieve parallel acceleration, as shown in Fig.3. The Fig.4 shows the hardware design diagram of 1D convolution units. The weight data is stored in L independent registers and fixed until the current convolution is complete. The input vector flows through a chain of shift registers of length L, each of which output taps. The weight data is multiplied with the input vector obtained by buffering to obtain the output vector of the 1D convolution.

The design of convolution units based on 1D convolution is an improvement of the line buffer[6] mechanism. It only expands the inner product operation part and uses the lateral spatial correlation of convolution operation to achieve the balance of generality and efficiency. In addition, the accelerator solves the problem of overlapping operation due to feature map chunking and eliminates the step of data rearrangement during the inference process.

2) Convolution accelerator controller

In this paper, a design based on instruction queue is proposed to reduce the response delay of DMA and accelerator in RISC-V soft core. Instead of waiting for feedback from DMA and accelerator, RISC-V CPU can continuously send multiple memory read and write request instructions and multiple operation scheduling control instructions. DMA and accelerator get instructions from the queue, and take out the next instruction directly from the queue after the task is completed, without waiting for the corresponding CPU, so as to achieve low delay scheduling.

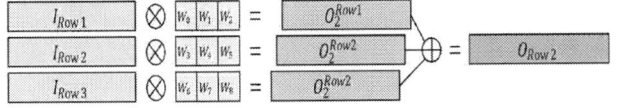

Fig. 3. Schematic of 1X3 1D convolution operation

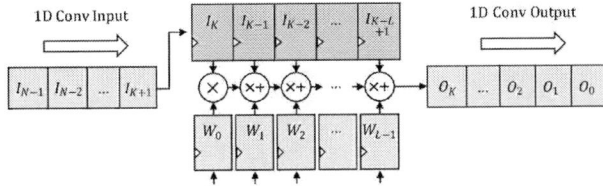

Fig. 4. One-dimensional convolution units

III. EXPERIMENTAL RESULTS AND REMARKS

Through the implementation of a CNN accelerator with the same configuration on domestic FPGA PG2L100H and Xilinx Zynq series X7Z020.The performance test of the CNN accelerator is completed and the feasibility of domestic FPGA CNN acceleration scheme is verified. The resource consumption and performance of the accelerator are shown in Table I and Table II.

TABLE I. RESOURCE UTILIZATION

Resource	Pango PG2L100H	Xilinx X7020
LUT	14354/66600	13110/53200
FF	29867/133200	14834/106400
APM/DSP48	208/240	22.5/140
DPM/BRAM	31.5/155	208/220

The resource consumption in PG2L100H and X7Z020 is similar. PG2L100H requires additional logical resources to build a VexRiscv CPU, while X7Z020 uses more logical resources for AXI DMA IP. In terms of accelerator performance, it can be seen from Table II. Because of the differences in FPGA device architecture, the convolution operation of the accelerator can only achieve better convergence at 200MHz on PG2L100H compared with X7Z020. RISC-V soft core can only achieve timing convergence at 100MHz.

TABLE II. PERFORMANCE COMPARISON

Performance	Pango PG2L100H	Xilinx X7020
Clock/MHz	200/100	250
GOP/s	153.6	192.0
Memory BW GByte/s	1.6	2.0

We propose a novel design based on 1D convolution operation with RISC-V. The implementation and deployment of the accelerator on a domestic FPGA is completed, and the performance is comparable to that of Xilinx FPGAs with the same scale of hardware resources.

This paper demonstrates the feasibility of CNN heterogeneous scheme based on domestic FPGA and the research is a rare attempt in the field of CNN acceleration in the application ecology of domestic FPGA.

REFERENCES

[1] Zhang. C, et al. "Optimizing FPGA-based Accelerator Design for Deep Convolutional Neural Networks. " the 2015 ACM/SIGDA International Symposium ACM, 2015.

[2] K. Guo et al., "Angel-Eye: A Complete Design Flow for Mapping CNN Onto Embedded FPGA," in IEEE Transactions on Computer-Aided Design of Integrated Circuits and Systems, vol. 37, no. 1, pp. 35-47, Jan. 2018.

[3] Qiu.J, et al. "Going Deeper with Embedded FPGA Platform for Convolutional Neural Network." the 2016 ACM/SIGDA International Symposium ACM, 2016.

[4] E. Gholizadehazari, T. Ayhan and B. Ors, "An FPGA Implementation of a RISC-V Based SoC System for Image Processing Applications," 2021 29th Signal Processing and Communications Applications Conference (SIU), 2021, pp. 1-4.

[5] B. Jacob et al., "Quantization and Training of Neural Networks for Efficient Integer-Arithmetic-Only Inference," 2018 IEEE/CVF Conference on Computer Vision and Pattern Recognition, 2018, pp. 2704-2713.

[6] B. Bosi, G. Bois and Y. Savaria, "Reconfigurable pipelined 2-D convolvers for fast digital signal processing," in IEEE Transactions on Very Large Scale Integration (VLSI) Systems, vol. 7, no. 3, pp. 299-308, Sept. 1999.

A TWN Inspired Speaker Verification Processor with Hardware-friendly Weight Quantization

Xuanhao Zhang[1], Haige Wu[1], Renyuan Zhang[2], Zihang Xu[3], Hao Zhang[4], Bo Liu[1], Hao Cai[1]*

[1] School of Electronic Science and Engineering, Southeast University, Nanjing, China
[2] School of Integrated Circuits, Southeast University, Wuxi, China
[3] Chien-Shiung Wu College, Southeast University, Nanjing, China
[4] Nanjing Research Institute of Electronics Technology, Nanjing, China
* Email: hao.cai@seu.edu.cn

Abstract—Speaker verification (SV) is not only a convenient biometric recognition technology but also an important method to ensure information security. Since SV systems are often deployed in mobile terminals, this places a higher demand on the trade-off between ensuring recognition accuracy and reducing system power consumption. Thus, this paper proposes an implementation of a speaker verification system based on a ternary weight network (TWN). First, we design a TWN structure for the SV system. Then a weight quantization scheme to reduce hardware storage overhead is adopted. After that, the hardware of the SV system is designed and simulated. The recognition accuracy of the proposed TWN is tested to be 83.3%@5dB, 87.9%@15dB, and 93.1%@clean, respectively. Using an industry of 22nm ULL process, the overall power consumption of the system is 16.3μW.

Keywords—Speaker Verification, Ternary weight network, Weight Quantization.

I. INTRODUCTION

With the continuous development of the Internet of Things (IoT), the demand for human-computer interaction is growing. The convenience of automatic speech recognition (ASR) technology has made it an important means of human-computer interaction. While traditional ASR technologies are mainly keyword spotting (KWS) technologies, in recent years, speaker verification (SV) has also made rapid progress. SV considers only the information related to the speaker in the speech, without considering the specific meaning of a word in it. Since voice input devices (e.g. microphones) tend to be cheap to manufacture, the use of voiceprints for identification is a simpler and more economical method.

Traditional SV is mainly implemented by statistics and signal processing methods, such as Gaussian Mixture Models (GMM), but they are more demanding to train and compute and are more affected by noises. Recently, solutions have emerged to implement speaker verification using Deep Neural Networks (DNN). This approach can learn highly abstract phoneme features from a large number of samples for classification and is highly immune to noise. In mobile devices, weight compression is often used to make neural networks lightweight. A common weight compression method is binary weight networks (BWN). Ternary weight network (TWN) improves recognition accuracy and robustness compared to BWN, although it needs to store higher bit-width weights. Besides, TWN is more advantageous in wide signal-to-noise ratio and multi-classification voice recognition tasks.

In this paper, we propose a speaker verification system based on a neural network for multiclassification in the wide SNR range, including the design of a hardware-friendly ternary weight network and the hardware design of a speaker verification system, which includes MFCC (Mel Frequency Cepstrum Coefficient)-based feature extraction and TWN-based feature recognition. Finally, the system is verified by functional simulation and synthesis.

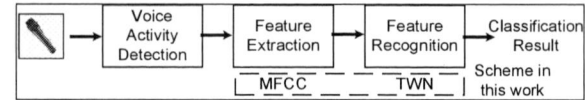

Fig. 1. SV system structure

II. TERNARY WEIGHT NETWORK DESIGN FOR SV

Neural network-based feature recognition is the mainstream scheme in the feature recognition module of advanced SV systems [1]. Traditional deep neural networks (e.g. AlexNet) have large sizes and a large number of parameters, which can cause large computation and storage overheads. Therefore, in this paper, a convolutional neural network structure with four convolutional layers and two fully connected layers with ternary weight is proposed for speaker verification in order to achieve a trade-off between recognition accuracy and hardware overhead.

The DARPA TIMIT Acoustic-Phonetic Continuous Speech Corpus (TIMIT) speech dataset has a speech sampling frequency of 16 kHz and contains a total of 6300 sentences, with 630 individuals from eight major dialect regions of the United States, each uttering 10 sentences. To characterize the effect of environmental noises on the recognition accuracy of the convolutional neural network, five of the speakers were selected for training with mixed signal-to-noise ratios (SNRs) of 5dB, 15dB, and clean (no background noise). The background noises are a mixture of 10 noises selected from the NoiseX-92 library. The neural network training is based on Python 3.6 and Adam optimizer.

During training, the quantized weights are not really ternary values (0, +1, -1), but contain a full-precision scaling factor Wg. Wg is initialized at the beginning of training, as are the full precision weights W. At each epoch of training, when the weights are updated to ternary values, Wg is also updated with its partial derivatives according to the loss function. The final ternary weight Wt is the product of Wg and the ternary values (0, +1, -1). In hardware, since each convolutional layer (and the first fully connected layer) is followed by a batch normalization (BN) layer, Wg can be considered in the BN linear operation and the weights only need to be stored by ternary values, thus reducing the storage expense while improving the complexity of CNN. The recognition accuracy of the proposed TWN is tested to be 83.3%@5dB, 87.9%@15dB, and 93.1%@clean with 10 mixed noises, respectively.

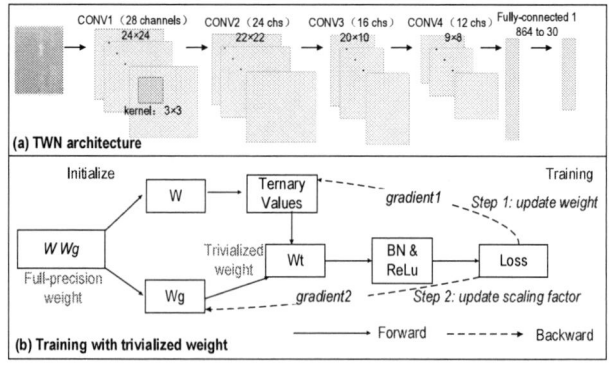

Fig. 2. (a) TWN architecture. (b) Training with trivialized weight.

III. HARDWARE IMPLEMENTATION OF SV SYSTEM

The overall hardware architecture of the TWN-based low-power SV system is shown in Figure 3(a). The system mainly consists of MFCC feature extraction and TWN accelerator. MFCC feature extraction receives external speech input and converts it into Mel speech features in the format of 26×49, which are used as input to the TWN accelerator. The TWN accelerator consists of three main modules: computation, storage, and control. During system initialization, TWN training parameters need to be loaded into the storage module. The control module includes layer control logic and computational control logic. The memory includes the data storage SRAM (Data SRAM) and the parameter storage SRAM (Weight SRAM). The operation module includes the array of processing elements (6×5 PEs), parallel to serial, batch normalization module, and activation module (ReLU).

The detailed structure of the MFCC feature extraction module is shown in Figure 3(b). The pre-emphasis module is a high-pass filter to weigh the high-frequency part of the speech signal. Assuming that the discrete time domain speech signal input is $s(n)$ and the output of the pre-emphasis module is $y(n)$, we have equation (1):

$$y(n) = s(n) - a \times s(n-1) \qquad (1)$$

, where a is a constant less than but close to 1. Here, for hardware implementation, a=0.9375 is taken. Then the half-frame long interleaved speech signal is obtained and stitched together. The fast Fourier transform uses a 512-point FFT scheme based on the radix-2 of the butterfly cell, while the Modulus-taking module is based on a 16-bit coordinate rotation digital computation method (CORDIC). The final Mel filter module is implemented by solidification, storing the information of the Meier triangular filter in ROM, and multiplying the value in ROM with the modulus value of the corresponding frequency to obtain the result of the Meier filter.

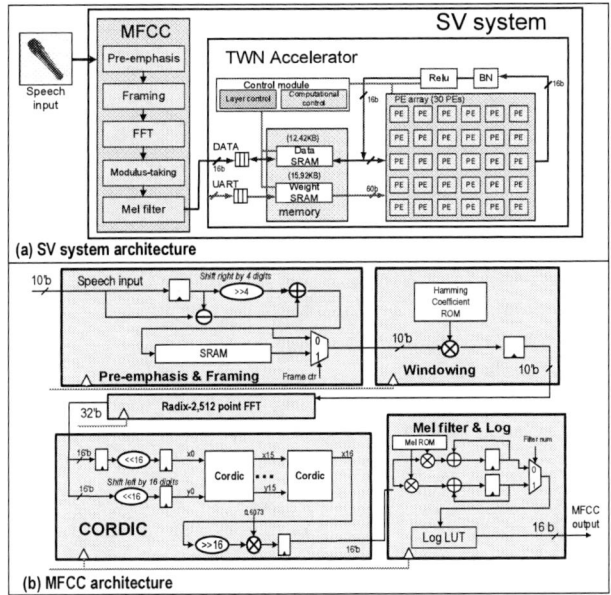

Fig. 3. (a) connection in the computing cluster. (b) accuracy rate in origin adder and hybrid tree.

The core of the TWN accelerator is the computing module, where the convolutional operation in the forward inference of the CNN can be processed efficiently. The Processing Element (PE) is the basic component of the computing module, which is based on the principle of holding, taking the opposite, or

bypassing the input data according to different ternary weights. There are 30 PEs in the PE array because the first fully-connected layer has 30 output channels, which is the largest number among all layers of the neural network. The kernel size is changed from 3×3 to a size comparable to that of a Feature Map. Therefore, to save the overhead in the hardware area and realize the reuse of hardware resources, the proposed PE array is the computation of both convolutional and fully connected layers. For the last layer computation of the neural network, batch normalization and ReLU computation will not be required, and the output of the computing module is the result of the current speech classification.

To evaluate the power consumption of the proposed SV system, the prototype system is synthesized with the logic supply voltage of 0.6 V, and the clock frequency is 250 kHz. The overall power consumption of the system is 16.3 µW on an industry of 22 nm ULL process HVT technology. Table 1 shows the comparison of this paper with previous state-of-art automatic speech recognition ASICs. Compared with work in [2] and [3], this work can hold out a higher SNR range. Compared with work in [4], power consumption can be reduced by 88.4%. And the recognition accuracy of the proposed TWN is tested to be 83.3%@5dB, 87.9%@15dB, and 93.1%@clean with 10 mixed noises, respectively.

TABLE I.   COMPARISONS WITH THE STATE-OF-THE-ART WORKS

Compare Items	TVLSI 2014[2]	JSSC 2020[3]	VLSI 2018[4]	This work
Technology	90 nm	65 nm	28nm	22 nm
Voltage/V	0.9	0.6	0.57-0.9	0.6
Bitwidth	NA	8bits	1bits	2bits
Main Accelerator	LPCC SVM	MFCC LSTM	Filter bank BCNN	MFCC TWN
Power	8.12mW	14.95µW	141µW	16.3µW
SNR range	clean	clean	5db - clean	5db - clean
Accuracy@clean	91.73	99.5%	95%	93.1%

IV. CONCLUSION

In this paper, we propose a TWN structure for SV, where each weight is hardware-friendly and quantized into ternary values. The proposed TWN is trained using a speech dataset with mixed SNR, and the recognition accuracy is 83.3%@5dB, 87.9%@15dB, and 93.1%@clean SNR. A hardware architecture for the SV system is proposed. Using an industry of 22nm ULL process, the system power consumption is 16.3µW by synthesis. Compared with similar work, it reveals a good trade-off between accuracy and power consumption.

ACKNOWLEDGMENT

This work was supported by the National Key R&D Program of China (Grant No. 2018YFB2202102).

REFERENCES

[1] J. S. Yue, Y, P. Liu, and Z, Yuan, "A 3.77TOPS/W Convolutional Neural Network Processor With Priority-Driven Kernel Optimization", *IEEE Transactions on Circuits and Systems II: Express Briefs*, 2015, vol. 66, no. 2, pp. 277-281, Feb. 2019.

[2] J. C. Wang, L. X. Lian, and Y. Y. Lin, "VLSI Design for SVM-Based Speaker Verification System", *IEEE transactions on very large scale integration (VLSI) systems*, 2015, vol. 23, no. 7, pp. 1355-1359, July 2015.

[3] J. S. P. Giraldo, S. Lauwereins, K. Badami and M. Verhelst, "Vocell: A 65-nm Speech-Triggered Wake-Up SoC for 10-µW Keyword Spotting and Speaker Verification", *IEEE Journal of Solid-State Circuits*, 2020, vol. 55, no. 4, pp. 868-878, April 2020.

[4] S. Yin P. Ouyang, and S. Zheng., "A 141 UW, 2.46 PJ/Neuron Binarized Convolutional Neural Network Based Self-Learning Speech Recognition Processor in 28NM CMOS," *2018 IEEE Symposium on VLSI Circuits*, 2018, pp. 139-140.

A 64Gb/s PAM-4 Digital Equalizer With Tap-Configurable FFE and Partially Unrolled DFE in 28nm CMOS

Xinjie Feng, Yongzhen Chen[*], Youzhi Gu, Jiangfeng Wu

College of Electronics and Information Engineering,
Tongji University, Shanghai 201804, China

Abstract—This paper presents a high-performance digital equalizer with four-level pulse amplitude modulation (PAM-4) for 64Gb/s backplane I/Os. The digital equalizer consists of a tap-configurable feed-forward equalizer (FFE) and a partially unrolled decision-feedback equalizer (DFE). The first two post-cursor is covered by DFE and then FFE follows, which can largely reduce the influence of noise and crosstalk. The configurable FFE taps enable better adaption for different kind of channels. In order to optimize the internal algorithm, the look-up table (LUT) is used in both FFE and DFE. And the DFE is unrolled for timing closing using a new architecture introduced in this paper. Fabricated in 28nm CMOS, the digital equalizer operates at 64Gb/s with only 5pJ/bit power consumption at 1V.

Keywords—*Digital equalizer, FFE, DFE, partially unrolled, tap-configurable*

I. INTRODUCTION

New technology industries such as the Internet of Things (IoT), blockchain and cloud computing have put forward higher requirements for data bandwidth. However, when the data rate comes to over 50Gb/s, even in middle reach (MR) channels, none return to zero (NRZ) signaling suffers from high insertion loss (IL) in its Nyquist which could cause serious intersymbol interference (ISI) and then degrade the data transmission performance. Four-level pulse amplitude modulation (PAM-4) signaling with half Nyquist of NRZ is a better choice in high speed transceiver. Many PAM-4 receivers use analog equalization technology to handle ISI [1], [2]. Although analog equalizers show good performance in area and power consumption, they are greatly limited by process, voltage and temperature (PVT). In contrast, digital equalizers with an analog-to-digital converter (ADC) in front end is more robust in PVT, and they have greater potential to compensate ISI over channels with severe IL [3]. The biggest challenge of designing a high-speed ADC-based digital equalizer is how to optimize its power consumption while meeting the bit error rate (BER) requirement.

In this paper, a 64Gb/s PAM-4 digital equalizer with low-power structure is proposed. Tap configurable feature of FFE gives flexibility to the digital equalizer, making it more adaptive to different kind of channels. And the proposed partially unrolled architecture solves the timing problem of DFE at a relatively small area and power cost.

II. PAM-4 DSP ARCHITECTURE

Fig. 1 shows the block diagram of the DSP. Before the data sending to ADC, it will be preprocessed by the CTLE in analog front end, which is not shown in the diagram. Then ADC samples the data in parallel to get 32 channels' 8-bit values. Given the transmission rate in channel is 64 Gb/s, the samples will be demux to 128 channels for further parallelism. Such approach lowers the clock frequency of the DSP so as to relieve timing pressure in synthesis. The digital equalization is realized by combining the tap-configurable FFE and the partially unrolled DFE. For the sake of flexibility and better adaption,

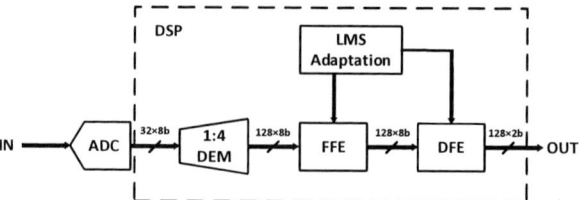

Fig.1. DSP block diagram

the tap distribution of FFE is designed in a configurable way. However, because of the linear nature, noise and crosstalk enhancement is inevitable in FFE. So DFE, as a nonlinear equalization method, is adopted to make up for the deficiency of FFE. In order to solve the timing problem of DFE in high-speed receiver with minimum power and area cost, a new unrolling structure is introduced in this paper. With this architecture, the unrolled branch needed in one tap will be reduced by half. Finally, both FFE and DFE coefficients are updated by sign-to-sign least mean square (LMS) algorithm.

III. TAP-CONFIGURABLE FFE

FFE is essentially a transversal linear filter which superposes a finite number of waveform responses to compensate ISI at the sampling time. Its input-output relationship can be expressed as

$$y_i = \sum c_n x_{i-n} \qquad (1)$$

In this proposed design, x_{i-n} is the 8-bit parallel input of FFE, c_n represents the tap coefficient, and y_i is the output. It is not difficult to learn that the characteristics of the filter depend on the tap coefficient c_n. So the equalization in different channels can be achieved by adjusting c_n. Theoretically, an infinite-length transversal filter is able to completely eliminate ISI at the sampling time. It is apparently impossible to realize in practice, so FFE has 12 configurable taps in this paper. The c_n has a bit width of 8 for ensuring calculation accuracy, where 1 bit is the symbol bit, 1 bit is the integer bit, and the remaining 6 bits are the decimal bits. Implementing 8-bit multipliers in a traditional way is convenient and straightforward on a small scale. However, for high data parallelism, such approach will undoubtedly increase the chip area and power consumption greatly. In order to optimize the internal multiplications of FFE, this paper introduces LUT to simplify the operation.

The coefficients in FFE are updated iteratively by ss-LMS algorithm. The formula is

$$w_m(n+1) = w_m(n) + \alpha \cdot x(n+m-1) \cdot e(n) \qquad (2)$$

where $w_m(n)$ and $w_m(n+1)$ are the coefficients before and after update, α is the iteration step size which can be set in the module, and $e(n)$ is the error between equalization result and target value. At the same time, each tap has an initial value before updating. By adjusting these values, we can change the position of the main tap. Besides, all the 12 taps can be individually configured into open or close mode. When dealing with the channels of low insertion loss, we can turn off some far-end taps to save power. In contrast, coping with high IL channels would need to turn on all taps and use the maximum equalization ability to compensate ISI.

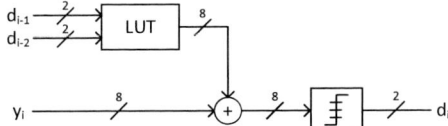

Fig. 1. DFE slice

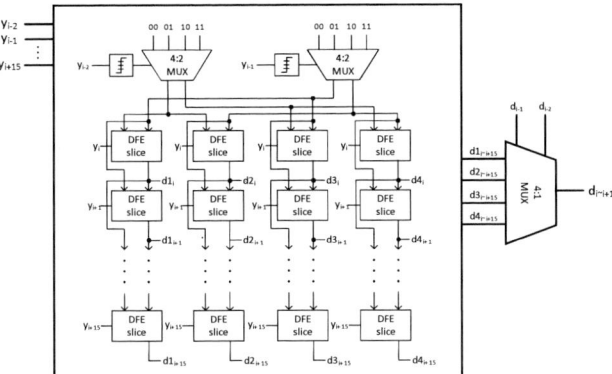

Fig. 3. PU-DFE architecture

IV. PARTIALLY UNROLLED DFE

Fig. 3 shows the architecture of the partially unrolled DFE (PU-DFE). It is well known that three reference voltages are required to determine the signal in PAM-4. Suppose they are V_3, V_2 and V_1 from high to low, respectively. Though there is a high probability of error if the equalization results from FFE are used to make decision directly, they can be utilized to estimate the range of the signals. When the equalization result of FFE is greater than $(V_3 + V_2) / 2$, the decision is probably 11 or 10. When the equalization result of FFE is between $(V_3 + V_2) / 2$ and $(V_1 + V_2) / 2$, the decision is probably 10 or 01. When the equalization result of FFE is smaller than $(V_1 + V_2) / 2$, the decision is probably 01 or 00. As we can see, the first two channels y_{i-1} and y_{i-2} are utilized to make hard decisions. Based on either one, we can choose two most likely decisions from 00, 01, 10 and 11. By pre-determining the range of the signal, the branches needed for loop unrolling can be reduced by half. Hence, two-tap PAM-4 PU-DFE only requires 4 branches in total. Compared with the traditional one which needs 16 branches, it saves a lot of area and power. The 16-channel decisions under four branches are $d1$, $d2$, $d3$ and $d4$ respectively. The correct branch is selected according to the actual post decisions d_{i-1} and d_{i-2}, and then d_{i+14} and d_{i+15} continue to server as the selection basis for the next PU-DFE block.

The internal structure of DFE slice is shown in Fig. 2. The equalization process of DFE can be expressed as follow

$$d_i = f\left(y_i + \sum c_n d_{i-n}\right) \quad (3)$$

where c_n is the tap value of the DFE. There are four possible outcomes for $c_n d_{i-n}$, i.e., $-c_n$, $+c_n$, $-0.33\,c_n$ and $+0.33\,c_n$. Similar to FFE, the LUT in PU-DFE stores all possible combinations of $c_1 d_{i-1} + c_2 d_{i-2}$ in advance. By using d_{i-1} and d_{i-2} as the address, the result of $c_1 d_{i-1} + c_2 d_{i-2}$ can be extracted. Then it adds to y_i to get a new 8-bit equalization value which will be sent to comparator finally for the 2-bit decision. On account of the LUT generated outside of DFE slice, the internal structure is simplified, further relieving the timing pressure in the critical path. The updating algorithm of DFE taps is similar to that of FFE, so the details are omitted here.

V. RESULT AND CONCLUSION

The digital equalizer that operates at 64Gb/s is fabricated in 28nm CMOS. For the purpose of power saving, the RTL is coded in a special style to increase the clock gating rate in the automatic synthesis. The overall chip performance and comparison are summarized in Table 1. As shown in Fig. 4, after a period of time, the tap coefficients of FFE converge to fix values, which hints the accomplishment of adaption. Fig. 5 displays the eye in different cases. The first one (a) shows the eye before equalization. For a middle-reach channel with 23 dB loss at 16 Gb/s, a good equalization performance can be achieved with 3 pre-cursor taps and 3 post-cursor taps configured in FFE. Its eye measurement result is displayed in Fig. 5 (d). Here also gives Fig. 5 (b) and Fig. 5 (c), corresponding to 6 pre-cursor taps configuration and 6 post-cursor taps configuration respectively, for comparison. From the simulation, the bit error rate (BER) measures lower than 10^{-4}, which means the BER requirement of PAM-4 signaling is met.

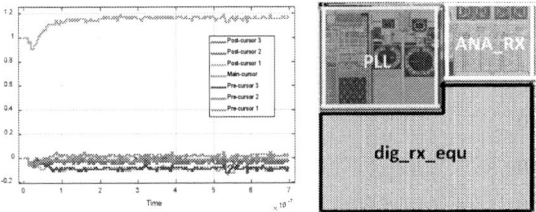

Fig. 4. Simulation result of FFE taps (left) and layout(right)

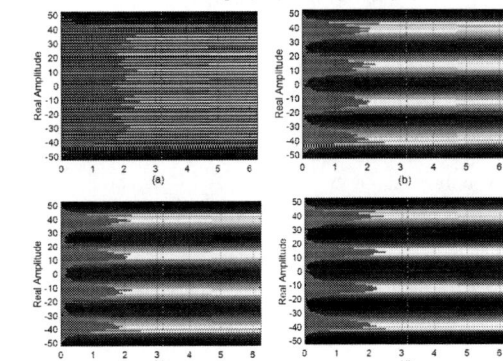

Fig. 5. Eye measurement result in different cases

TABLE I.   PERFORMANCE SUMMARY

	[1]	[3]	This work
Technology	16 nm FinFET	65 nm COMS	28 nm CMOS
Data rate	40-56 Gb/s	28 Gb/s	64 Gb/s
Channel loss	10 dB	30 dB	25 dB
Area	0.364 mm²	-	0.49 mm²
Power efficiency	4.1 pJ/bit	5.71 pJ/bit	5 pJ/bit

ACKNOWLEDGMENT

This work is supported by National Science Foundation of China Project, Grant No. 62090044.

REFERENCES

[1] J. Im et al., "A 40-to-56 Gb/s PAM-4 receiver with 10-tap direct decision-feedback equalization in 16 nm FinFET," in IEEE Int. Solid-State Circuits Conf. (ISSCC) Dig. Tech. Papers, Feb. 2017, pp. 114–115.

[2] P. J. Peng, J. F. Li, L. Y. Chen, and J. Lee, "A 56Gb/s PAM-4/NRZ transceiver in 40nm CMOS," in IEEE Int. Solid-State Circuits Conf. (ISSCC) Dig. Tech. Papers, Feb. 2017, pp. 110–111.

[3] A. K. M. D. Hossain, M. Mohammad, and M. Hossain, "Channel-adaptive ADC and TDC for 28 Gb/s PAM-4 digital receiver," IEEE J. Solid-State Circuits, vol. 53, no. 3, pp. 772–788, Mar. 2018.

978-1-6654-9270-6/22 $31.00 © 2022 IEEE

A Tunable Monopole Antenna for 5G Communication Applications

Liangfan Chen[1], Lu Zhao[1], Zihao Chen[1,2]

[1]School of Electronics and Information Engineering, Harbin Institute of Technology (Shenzhen), Shenzhen, China
[2]Peng Cheng Laboratory, Shenzhen, China

Abstract—A dual-band tunable monopole antenna is designed for 5G communication applications. The devised tuner consists of RF switch and RF capacitors of 0.3 pF, 0.5 pF, 1 pF, 2 pF and 5 pF, which enables the monopole antenna to be operated in different frequency bands. The proposed antenna is fabricated and measured. The measured -10 dB input impedance bandwidths of the proposed antenna are 1.32 GHz – 1.95 GHz and 1.98 GHz – 5.02 GHz, which can fully cover the 5G frequency spectrum in China.

Keywords—5G communication; Tunable antenna; RF switch; Tuner;

I. INTRODUCTION

The enormous growth of wireless communication services, such as the fifth generation (5G) communication and beyond 5G communication, has led to increasing demand for wide-band and low-profile antennas with stringent electrical characteristics. Tunable antennas are good candidates for the 5G and beyond 5G communication applications since they can realize various antenna functions by changing the operating frequency [1], radiation mode [2], polarization mode [3] or two or more of them [4]. Generally, the tunable antennas use diode, transistor, MEMS switch [5] and other technologies to change the structure of the antenna and improve the antenna performance. The frequency discrete tunable circular monopole antenna proposed by [6] has five operating frequency bands in 2.24-5.9 GHz in narrowband mode; [7] proposed a mobile terminal monopole tunable antenna based on digital tunable capacitor (DTC), which can work in the frequency bands of DCS / PCS, UTMS, lte1800 / 2600 and gsm850 / 900; [8] proposed a tunable MIMO antenna, which can work at 2.4 GHz – 2.95 GHz and 3 GHz – 4.5 GHz with return loss less than -6 dB. However, the impedance bandwidths of the antennas mentioned above are limited and cannot fully cover the 5G communication bandwidth.

A dual-band tunable monopole antenna is designed in this paper for 5G communication applications. Tuner with RF switch and capacitors is used to broaden the antenna bandwidth. In Section II, the structure of the proposed antenna with the tuners is demonstrated. In Section III, the simulated and measured results are shown and analyzed. Finally, conclusions are summarized in Section IV.

II. ANTENNA STRUCTURE

The structure of the proposed tunable monopole antenna is shown in Figure 1. The antenna consists of two metallic layers and one dielectric layer of FR4 substrate with thickness $h = 1$ mm, dielectric constant of 4.4 and loss tangent of 0.02. The antenna radiator on the top layer is composed of three quarter-wavelength monopoles to realize two resonant frequency bands.

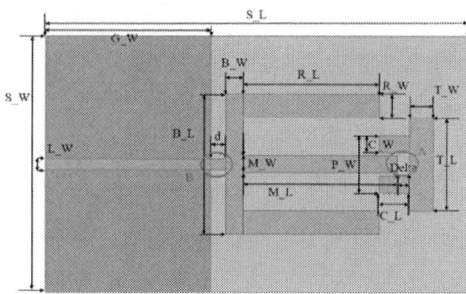

Fig. 1. The structure of proposed antenna.

TABLE I. ANTENNA STRUCTURE PARAMETERS

Name	Value(mm)	Name	Value(mm)
S_L	77	S_W	44.5
G_W	28	L_W	1.96
B_W	3	B_L	24
d	2.5	R_L	23
R_W	4	M_W	3
M_L	26	T_W	4
T_L	16	C_W	3
C_L	5	P_W	10
Delta	2	h	1

Two tuners are added to the point A and point B of the radiator to change the lower bound and the upper bound of the antenna input impedance bandwidth, respectively. The tuner used in this paper is based on high power RF switch QM11024 whose dimensions are 7mm × 7mm × 1mm. The dielectric substrate material used for the tuner is FR4, which is the same as the dielectric substrate material of the antenna. The PCB schematic diagram of the tuner is shown in Figure 2. The RF switch QM11024 is located in the center of the tuner board. The pads of 0402 encapsulated RF capacitors are arranged on one side of the RF switch, while the pads of control terminal, power supply and grounding terminal are on the other side of the RF switch. Points P1 and P2 in Figure 2 are two pads connecting the antenna radiator. The capacitors of 0.5pF, 1pF, 2pF and 5pF are connected to the tuner at point A and the capacitor of 0.3pF is connected to the tuner at point B, where the two tuners are operated independently.

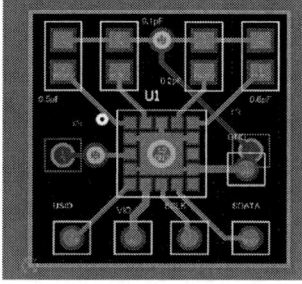

Fig. 2. Schematic diagram of antenna tuner PCB.

III. SIMULATED AND MEASURED RESULTS

The proposed antenna is fabricated to verify the feasibility of the antenna. Fig.3 (a) and Fig. 3 (b) show the front view and back view of the fabricated antenna, respectively.

 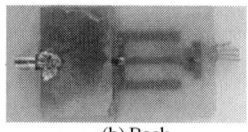

(a) Frontage   (b) Back

Fig. 3. Photographs of the fabricated tunable monopole antenna.

Fig.4 (a) shows the simulated and measured $|S_{11}|$ when the monopole antenna is unloaded with tuner. The measured -10dB input impedance bandwidths are from 1.68 GHz to 1.95 GHz and from 1.98 GHz to 4.31GHz. As shown in Fig. 4 (b), the upper bound of the operating bandwidth of the antenna with tuner at position B is increased to 5.02GHz which is higher than that of the antenna without tuner. Fig.5 shows the simulated and measured $|S_{11}|$ when tuner A works. We can find that as the capacitor of the tuner at position A changes from 0.5pF to 5 pF, the antenna resonant point of the lower frequency band decreases.

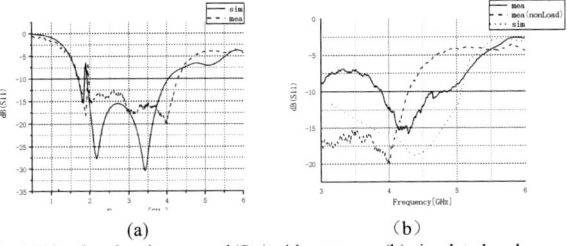

(a)   (b)

Fig. 4.   (a)Simulated and measured $|S_{11}|$ without tuner, (b) simulated and measured $|S_{11}|$ when the capacitor at point B is 0.3 pF.

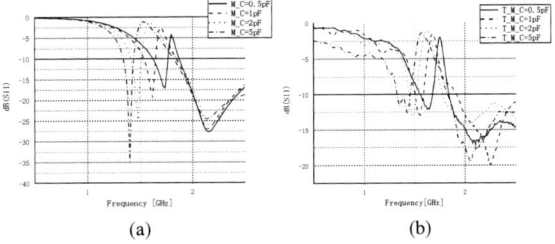

(a)   (b)

Fig. 5. Simulated and measured $|S_{11}|$ when tuner A works: (a) simulated $|S_{11}|$ when the capacitors at point A are changed from 0.5 pF to 5 pF, (b) measured $|S_{11}|$ when the capacitors at point A are changed from 0.5 pF to 5 pF.

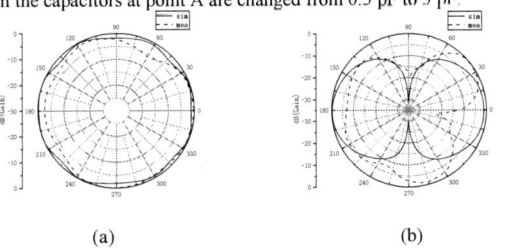

(a)   (b)

Fig. 6. Simulated and measured radiation patters for tunable monopole antenna: (a) radiation patterns in the E-plane at 1.72 GHz and (b) radiation patterns in the H-plane at 1.72 GHz.

TABLE II.   COMPARISON OF TUNABLE ANTENNAS

Index	Antenna	10 dB bandwidth	Antenna efficiency	Size
	Proposed	1.3GHz – 1.95 GHz 1.98 GHz – 5 GHz	0.4 - 0.96	44.5 mm × 77mm
[7]	Monopole + digital variable capacitance	1.5 GHz – 2.7 GHz 3.5 GHz – 3.8 GHz	0.6 – 0.79	44 mm × 10 mm 130 mm × 70 mm
[8]	MIMO + diode	2.4 GHz – 2.8 GHz 3 GHz – 4 GHz	0.6 – 0.8	75 mm × 150 mm
[9]	Monopole + tunable inductor	3.3 GHz – 5 GHz	0.5 – 0.75	68 mm × 148 mm
[10]	Monopole	3.4 GHz – 3.7 GHz 5.0 GHz – 5.8 GHz		150 mm × 75mm

As shown in the Fig. 6 (a) and Fig. 6 (b) show the radiation patterns of the tunable monopole antenna in the plane E and plane H at 1.72 GHz, respectively. The proposed antenna realizes a omnidirectional radiation in the E-plane which indicates that the tuners do not affect a lot the radiation patter of the monopole antenna. The discrepancies between the simulated and measured

results are mainly caused by the fabrication and soldering tolerance of the tuner.

Table II shows the comparison of working bandwidth of 5G tunable antenna between this paper and other literatures. By comparing with the 5G tunable antennas, the proposed antenna has obvious advantages in the operating bandwidth and antenna size.

IV. CONCLUSION

Based on the theory of tunable antenna and monopole antenna, a tunable monopole antenna with QM11024 RF switch is designed for 5G communication applications in this paper. By adding tuners at different positions of the antenna, the resonant frequency and the bandwidth of the antenna can be adjusted. In order to verify the feasibility of the designed antenna, the antenna is fabricated and tested. The center frequency obtained from the antenna test can be adjusted in the range of 1.4 GHz – 1.7 GHz, 3.38 GHz – 4.37 GHz, and the 10 dB bandwidth covers 1.3 GHz – 1.95 GHz and 1.98 GHz – 5. 02 GHz, which is attractive for the development of 5G frequency tunable antennas.

ACKNOWLEDGEMENT

This work was supported in part by the National Natural Science Foundation of China under Grant 61901139, in part by the Shenzhen Science and Technology Program ZDSYS20210623091808025, in part by the Shenzhen Science and Technology Program under Grant KQTD20210811090116029 and in part by the project "The Verification Platform of Multi-tier Coverage Communication Network for oceans (LZC0020)".

REFERENCES

[1] T. Ikeda, S. Saito and Y. Kimura, "A Frequency-Tunable Dual-Band Single-Layer Shorted Multi-Ring Microstrip Antenna Fed by an L-probe with Varactor Diodes," *2019 IEEE International Symposium on Antennas and Propagation and USNC-URSI Radio Science Meeting*, 2019, pp. 905-906.

[2] X. Ding, Z. Zhao, Y. Yang, Z. Nie and Q. H. Liu, "A Low-Profile and Stacked Patch Antenna for Pattern-Reconfigurable Applications," *IEEE Transactions on Antennas and Propagation*, vol. 67, no. 7, pp. 4830-4835, July 2019.

[3] A. Panahi, X. L. Bao, K. Yang, O. O'Conchubhair and M. J. Ammann, "A Simple Polarization Reconfigurable Printed Monopole Antenna," *IEEE Transactions on Antennas and Propagation*, vol. 63, no. 11, pp. 5129-5134, Nov. 2015.

[4] W. Lin and H. Wong, "Multipolarization-Reconfigurable Circular Patch Antenna With L-Shaped Probes," *IEEE Antennas and Wireless Propagation Letters*, vol. 16, pp. 1549-1552, 2017.

[5] S. Caporal del Barrio, E. Foroozanfard, A. Morris and G. F. Pedersen, "Tunable Handset Antenna: Enhancing Efficiency on TV White Spaces," *IEEE Transactions on Antennas and Propagation*, vol. 65, no. 4, pp. 2106-2111, April 2017.

[6] H. Boudaghi, M. Azarmanesh and M. Mehranpour, "A Frequency-Reconfigurable Monopole Antenna Using Switchable Slotted Ground Structure," *IEEE Antennas and Wireless Propagation Letters*, vol. 11, pp. 655-658, 2012.

[7] L. H. Trinh, F. Ferrero, L. Lizzi, R. Staraj and J. -M. Ribero, "Reconfigurable Antenna for Future Spectrum Reallocations in 5G Communications," *IEEE Antennas and Wireless Propagation Letters*, vol. 15, pp. 1297-1300, 2016.

[8] N. O. Parchin, H. J. Basherlou, Y. I. A. Al-Yasir, A. Ullah, R. A. Abd-Alhameed and J. M. Noras, "Frequency Reconfigurable Antenna Array with Compact End-Fire Radiators for 4G/5G Mobile Handsets," *2019 IEEE 2nd 5G World Forum (5GWF)*, 2019, pp. 204-207.

[9] H. J. Nam, S. Lim, Y. J. Yoon and H. Kim, "Tunable Triple-Band Antenna for Sub-6 GHz 5G Mobile Phone," *2020 IEEE International Symposium on Antennas and Propagation and North American Radio Science Meeting*, 2020, pp. 1455-1456.

[10] P. Yang, K. Yan, F. Yangl, L. Y. Zeng and S. Huang, "Reconfigurable Slot Antenna Design for 5G Smartphone with Metal Casing," *2018 IEEE International Symposium on Antennas and Propagation & USNC/URSI National Radio Science Meeting*, 2018, pp. 453-454.

PPBAM:A Preprocessing-based Power-Efficient Approximate Multiplier Design for CNN

Yi Hu[1], Tao Huang[1], Run Run [1], Li Yin [1] ,Guolin Li [2], Xiang Xie [1],*
[1] School of Integrated Circuits, Tsinghua University
[2] Department of Electronic Engineering, Tsinghua University
Email: xiexiang@tsinghua.edu.cn

Abstract—In the fields of CNN, there exists many multiply applications with one fixed operand. In view of such characteristics, this paper proposes a preprocessing-based power-efficient approximate multiplier (PPBAM) design for CNN. In the proposed design, the fixed operand is preprocessed to avoid additional dynamic power consumption due to repeated processing. To reduce the number of the partial products, the first '1' of both two operands are found and then the operands are truncated by a method named weak rounding. What's more, a sub multiplier array utilizing an approximate 4:2 compressor are proposed to calculate the truncation results with low power. The experimental results show that, with the same accuracy, on average, our design has a 30% improvement in power consumption compared with state-of-the-art approximate multiplier designs without additional latency and area.

Keywords—Approximate computing, power efficient, approximate 4:2 compressor, multiplier, CNN

I. INTRODUCTION

In recent years, with the rapid development of artificial intelligence technology, demand for computing resource is increasing in an overwhelming speed. Accordingly, power consumption has been one of the critical design constraints in designing digital system. Approximate computing (AC) is one of the approaches that can effectively reduce the power consumption, which can be exploited in error-resilient applications. Examples of these applications include audio and image processing, machine learning and data mining [1]. Among these applications, multiplication is most commonly used, which brings a large amount of calculation and requires tremendous space and power consumption in computing unit. This makes approximate multipliers good candidates for being employed in applications like CNN, which is robustness to noise, imprecision, and even error in data [2].

Based on various design angles, many approximate multiplier designs have been proposed [3]–[7]. However, few people design approximate multipliers according to the characteristics of data, but it is really important. In fact, there exists many multiply applications with one fixed operand in actual applications, and the data is not often completely independent. In this paper, we propose an approximate multiplier based on the characteristic of operands. Considering the characteristic of data, the fixed operands are preprocessed to reduce the power consumption due to repeated processing.

II. PROPOSED APPROXIMATE MULTIPLIER

In order to make our approximate multiplier design power-efficient in actual applications, the characteristics of data are studied. According to the research of Rob Banner in 2020 [8], we simply use 8-bit approximate multiplier to calculate a convolutional layer or full connection layer with little accuracy

This work was supported by National Key R&D Program of China(2019YFB2204800).

loss. According to our research, after quantization, 92% of high 3 bits of the feature map adjacent pixels in AlexNet [9] are the same on average. This is such a high rate that if we take advantage of this point, the dynamic power consumption of multiplier will be greatly reduced.

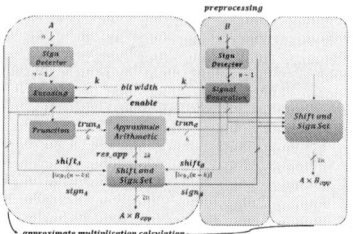

Fig. 1: Block diagram of the proposed approximate multiplier

The block diagram of the proposed approximate multiplier is depicted in Fig. 1. As illustrated in the figure, the proposed approximate multiplier is divided into two parts: preprocessing and approximate multiplication calculation. A and B are the two operands the same as previously assumed, and k represents the designer-defined value that specifies the bit width used in the sub approximate arithmetic module. For the fixed operand B, we directly store the preprocessed data $shift_B$,$trun_B$ and $enable$ after it is processed by the preprocessing unit. When operand A is input, we make use of A and the results of the preprocessing unit stored in advance to perform subsequent calculations. The $enable$ signal determines the subsequent calculation method.

As mentioned above, we need to truncate the input operand after the leading one detection, and then encode to obtain the corresponding shift amount $shift_A$. In the proposed design, the corresponding shift amount are directly obtained by using the input operands and the precision control signal k. After that, the shift amount becomes the control signal to truncate the input operand in order to obtain our approximate truncation result. As a comparison, the general method [3] requires to detect the first '1', and then convert it to a form of binary number through the look-up table. After that, the shift amount needs to be obtained through one-time addition and subtraction.

Because different bits have different weights, we need to reserve the more significant bits in high order. On account of the acknowledgement that all zeros before the first '1' are meaningless, for the n-bit input A and B, our design makes use of two leading one detector (LOD) circuit blocks to dynamically locate the most significant '1' in each of the two operands. Taking B as an example, as illustrated in Fig. 2, it is a general example of the approximate truncation process. Here, k is a designer-defined value which specifies the bit width used in the sub approximate arithmetic module. After the LOD circuit block, the location of the most significant '1' is used to select the following $k - 2$ consecutive number of bits based on the required accuracy. As revealed, our method only needs an extra AND gate.

Fig. 3 shows the schematic diagram of 4bit approximate arithmetic array using a 4-2 approximate compressor [6]. As described in the legend, we utilize different kind of circles to represent data and rectangular boxes to represent the calculation unit. Compared with the accurate calculation array, the proposed approximate calculation array reduces a half adder and a full adder, replaces a 4:2 compressor with an approximate 4:2 com-

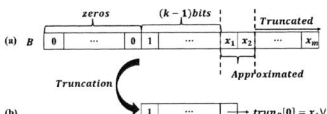

Fig. 2: A general example of the approximate truncation process. (a) Original number (b) Final approximate truncated result

pressor. In this way, the latency, area and energy consumption are greatly optimized.

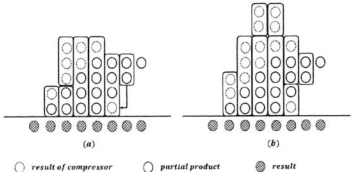

Fig. 3: Schematic diagram of 4bit multiplication calculation array. (a) Proposed approximate multiplication calculation array using approximate 4:2 compressor (b) Accurate multiplication calculation array

III. EXPERIMENTAL RESULTS

In this section, we compare the performance of PPBAM and various state-of-the-art approximate multipliers [1], [3], [4] with a bit width of 8 bit. Fig. 4 demonstrates the power and MRED(mean relative error distance) comparison of the proposed and other approximate multipliers. By observing the figure, it is simple to be aware of the fact that, the closer an approximate multiplier gets to the lower left corner, the better the multiplier is in respect of power consumption and MRED. Under this common view, we can draw a conclusion that PPBAM performs best in power and MRED in that the curve of PPBAM is at the the bottom left of all the other curves. When MRED is less than 10%, PPBAM has obvious advantages in power consumption. Meanwhile, PPBAM achieves higher accuracy because of the special utilization of approximate 4:2 compressor.

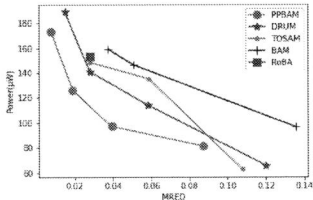

Fig. 4: Power and MRED comparison of the proposed and other approximate multipliers.

The power consumption of approximate multipliers applied in different convolutional networks are shown in TABLE I. Through TABLE I, compared with DRUM5, TOSAM(2,6) and RoBA, we can find that the power consumption of PPBAM is decreased by 23%, 52%, and 55%. Meanwhile, PPBAM has a improvement in accuracy compared with other designs.

IV. CONCLUSION

In this paper, we suggested a preprocessing-based power-efficient approximate multiplier design for convolutional neural

TABLE I. Power(μW) of approximate multipliers applied in convolution layer

Net	Power1	Power2	Power3	Power4
AlexNet [9]	37.6	48.4	76.1	87.8
vgg16 [10]	16.7	23.6	39.8	75.6
MobileNetV2 [11]	44.5	53.3	74.2	75.6
resnet18 [12]	27.1	37.7	72.8	75.5
average	31.5	40.8	65.7	69.9

networks. In the proposed approximate multiplier, fixed weights in convolutional layer are processed in advance because the fixed weights are used repeatedly for a convolution operation. What's more, the two operands of multiplication are handled with a specific method so that two consecutive truncated results will frequently remain unchanged, resulting in significant reduction in dynamic power consumption. That is because high bits of adjacent pixels of an input feature map are almost the same except the edge pixels. We analyzed PPBAM's error and power consumption characteristics empirically. Simulation results show that, on average, our design has a 30% improvement in power consumption compared with state-of-the-art approximate multiplier designs without additional latency and area.

REFERENCES

[1] S. Vahdat, M. Kamal, A. Afzali-Kusha and M. Pedram, "TOSAM: An Energy-Efficient Truncation- and Rounding-Based Scalable Approximate Multiplier," in IEEE Transactions on Very Large Scale Integration (VLSI) Systems, vol. 27, no. 5, pp. 1161-1173, May 2019, doi: 10.1109/TVLSI.2018.2890712.

[2] M. S. Kim, A. A. Del Barrio Garcia, H. Kim and N. Bagherzadeh, "The Effects of Approximate Multiplication on Convolutional Neural Networks," in IEEE Transactions on Emerging Topics in Computing, doi: 10.1109/TETC.2021.3050989.

[3] S. Hashemi, R. I. Bahar and S. Reda, "DRUM: A Dynamic Range Unbiased Multiplier for approximate applications," 2015 IEEE/ACM International Conference on Computer-Aided Design (ICCAD)

[4] R. Zendegani, M. Kamal, M. Bahadori, A. Afzali-Kusha and M. Pedram, "RoBA Multiplier: A Rounding-Based Approximate Multiplier for High-Speed yet Energy-Efficient Digital Signal Processing," in IEEE Transactions on Very Large Scale Integration (VLSI) Systems, vol. 25, no. 2, pp. 393-401, Feb. 2017, doi: 10.1109/TVLSI.2016.2587696.

[5] H. R. Mahdiani, A. Ahmadi, S. M. Fakhraie and C. Lucas, "Bio-Inspired Imprecise Computational Blocks for Efficient VLSI Implementation of Soft-Computing Applications," 2010 IEEE Transactions on Circuits and Systems I.

[6] A. G. M. Strollo, E. Napoli, D. De Caro, N. Petra and G. D. Meo, "Comparison and Extension of Approximate 4-2 Compressors for Low-energy Approximate Multipliers," in IEEE Transactions on Circuits and Systems I: Regular Papers, vol. 67, no. 9, pp. 3021-3034, Sept. 2020, doi: 10.1109/TCSI.2020.2988353.

[7] Reddy, K.M., Vasantha, M.H., Kumar, Y.N. and Dwivedi, D., 2019. Design and analysis of multiplier using approximate 4-2 compressor. AEU-International Journal of Electronics and Communications, 107, pp.89-97.

[8] Ron Banner, Itay Hubara, Elad Hoffer, Daniel Soudry:Scalable methods for 8-bit training of neural networks. NeurIPS 2018: 5151-5159

[9] Krizhevsky, Alex, Ilya Sutskever, and Geoffrey E. Hinton. "Imagenet classification with deep convolutional neural networks." Advances in neural information processing systems 25 (2012): 1097-1105.

[10] Simonyan K, Zisserman A. Very deep convolutional networks for large-scale image recognition[J]. arXiv preprint arXiv:1409.1556, 2014.

[11] Howard A, Zhmoginov A, Chen L C, et al. Inverted residuals and linear bottlenecks: Mobile networks for classification, detection and segmentation[J]. 2018.

[12] He, Kaiming, Xiangyu Zhang, Shaoqing Ren, and Jian Sun. "Deep residual learning for image recognition." In Proceedings of the IEEE conference on computer vision and pattern recognition, pp. 770-778. 2016.

2022 IEEE International Conference on
Integrated Circuits, Technologies and Applications

A 0.58-pJ/bit 56-Gb/s PAM-4 Optical Receiver Frontend with an Envelope Tracker for Co-Packaged Optics in 40-nm CMOS

Yue Yu[1], Da Ming[1], Min Tan[1,2,*]

[1] School of Optical and Electronic Information,
Huazhong Univ. of Sci. and Tech., Wuhan 430074, China
[2] Optics Valley Laboratory, Hubei 430074, China
Email*: mtan@hust.edu.cn

Abstract—A 56Gb/s PAM-4 optical receiver frontend with a envelope tracker for co-packaged optics (CPO) in 40-nm CMOS technology is presented. An inverter-based shunt-feedback transimpedance amplifier (TIA) is carefully optimized for low noise and high linearity, and a continuous time linear equalizer (CTLE) is adopted for high frequency boosting. A dB-linear variable gain amplifier (VGA) is adopted to realize decibel-linear gain control. An envelope tracker is used to extract the envelope of the received signal, which provides the control voltage for the automatic gain control loop. Post-layout simulations show that this work achieves a 56Gb/s PAM-4 optical receiver frontend with 0.58pJ/bit in 40-nm CMOS process.

Keywords—optical receiver, PAM-4, CMOS, envelope tracker

I. INTRODUCTION

Nowadays, the increasing demand of bandwidth-intense services like 5G and IoT is speeding up the exponential growth of internet traffic. It is noteworthy that a large proportion of the data communication is within the data center and optical links is expanding its footprint from long-haul to short distances. The bandwidth of conventional pluggable optical modules is limited by the lossy PCB channels to the payload ASIC. Co-packaged optics (CPO) could break bandwidth limitation and increase the energy efficiency by reducing the electrical channel length. With the integration of analog front-end and subsequent digital back-end, the CMOS-based receiver possess a huge competitive advantage in terms of costs and energy efficiency.

In this paper, we present a 56 Gb/s PAM-4 optical receiver frontend with a envelope tracker. Conventional designs either use eye-open monitors [1] or peak detectors [2]. The envelope tracker is used to adjust the gain of VGA, which greatly reduces the design complexity without sacrificing performance and is able to track the dynamics of the peak magnitudes.

II. CIRCUITS IMPLEMENTATION

As is shown in Fig.1, the proposed optical receiver frontend mainly consists of a two-stage front-end, post-amplifier chains and an envelope tracker. The input light is amplified and converted into proportional current by the photodetector firstly. Then the CMOS linear TIA and subsequent gain amplifier transform the current into voltage. The DCOC loop sinks the DC current throughout high optical input power levels.

A. S2D TIA and CTLE

Compared to common-gate stage, which is representative for its intelligent bandwidth performance, shunt-feedback TIA is more qualified for low-noise application. Besides, from the perspective of the optical link, RX with low-noise performance benefits to improve sensitivity, thus reducing the power budget as a whole. There is a transimpedance limit that demonstrates feedback resistor R_F decreases with the square of TIA bandwidth. However, larger R_F always means larger gain and

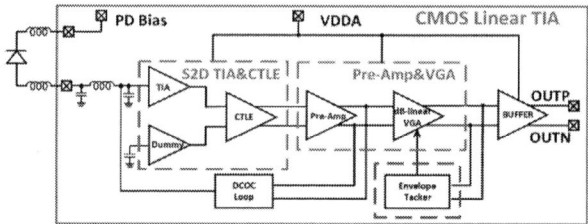

Fig. 1. Block diagram of proposed optical receiver frontend.

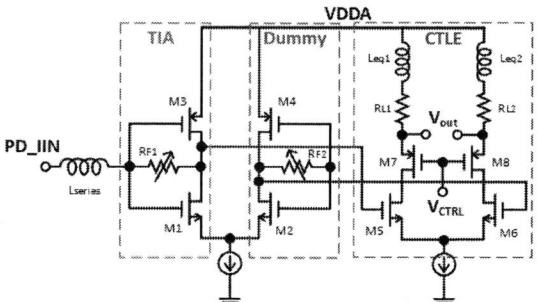

Fig. 2. Schematic of TIA, Dummy and CTLE.

lower noise.

In consideration of all the above properties, a low-bandwidth TIA followed by a CTLE is a common solution [3]. The operation principle is to exploit a narrowband TIA to meet the requirement of gain while establishing a lower noise floor, and then the CTLE provides a high frequency zero to boost the bandwidth.

To suppress the common-mode noise due to supply coupling and residual even-order harmonics, a pseudo-differential circuit topology with a replica structure is used. As shown in Fig 2, an inverter-based topology biased at moderate supply could provide high linear transimpedance per unit current for a low supply of 1.2V. The CTLE uses shunt inductive peaking techniques. Between the PAD and TIA, a series inductor is inserted to separate the two crucial parasitic capacitances.

B. Pre-Amplifer and dB-Linear VGA

Due to the output signal of CTLE is not fully differential, a pre-amplifier is inserted before VGA to amplify the differential signal further.

Conventional VGA adopts a variable transconductance or resistance to control gain that leads to non-dB-linear gain. In virtue of the inherent exponential characteristics of a transistor operating in subthreshold region, it is feasible to realize decibel-linear gain control [2]. As shown in Fig.3(a), M_{11} and M_{12}, which operate in the subthreshold region, are different from the differential pair M_9 and M_{10} operating in the saturation region. When V_{peak} varies from 750mV to 1050mV with a step of 100mV, the gain of dB-linear changes from 3.6dB to 1.3dB linearly as shown in Fig.3(b).

C. Envelope Tracker

A conventional peak detector comprises a power detector and integrator. However, this type of circuit suffers a precision limit as data rates increase, and its output is smaller than the input value. The envelope tracker includes a V-I converter, a current comparator, and a charge pump [5]. The V/I converter converts the difference between the current voltage and the previous, voltage into a proportional current. The current comparator determines the polarity of control signal, which is passed to the charge pump to determine the subsequent charge and discharge

978-1-6654-9270-6/22 $31.00 © 2022 IEEE
168

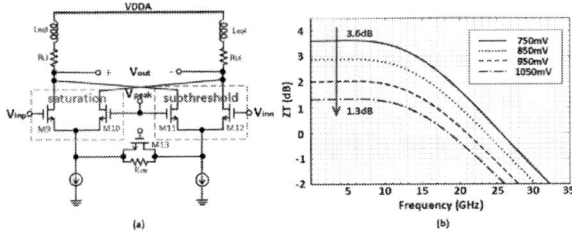

Fig. 3. (a) Schematic of dB-linear VGA and (b) Simulated dB-linear ac response of VGA.

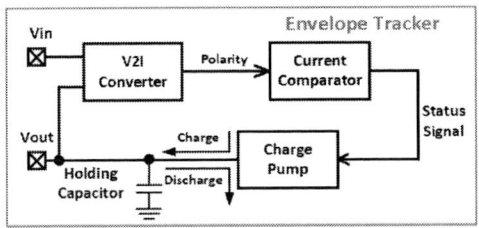

Fig. 4. Block diagram of envelope tracker.

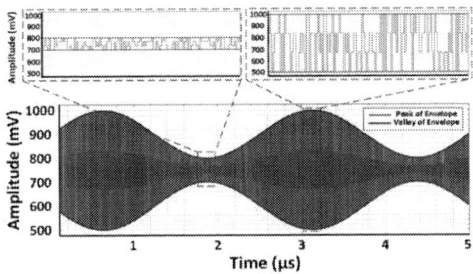

Fig. 5. Envelope detector simulation results in Verilog-A.

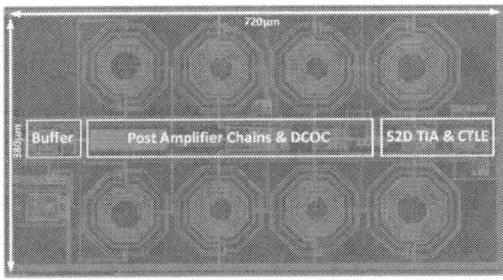

Fig. 6. Layout of proposed optical receiver frontend

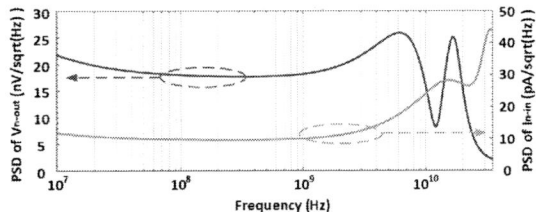

Fig. 7. Simulated power spectral density of optical receiver frontend noise

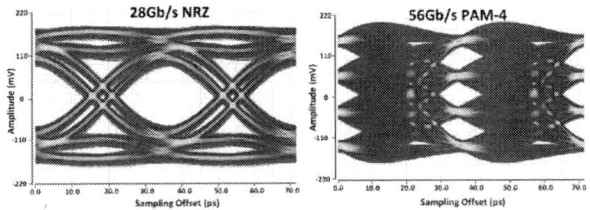

Fig. 8. Simulated 28Gb/s NRZ and 56Gb/s PAM-4 eye diagrams

TABLE.1 PERFORMANCE SUMMARY

Optical RX	[2]	[4]	[6]	This work*
Technology	40nm CMOS	0.25- μm SiGe	45nm SOI CMOS	40nm CMOS
Data rate (Gbps)	56 (PAM4)	56 (NRZ)	56 (PAM4)	56 (PAM4)
BW (GHz)	16	34	23	16.5
Noise (μArms)	2.56	2.08	1.8	1.68
Noise (pA/√Hz)	20.23	11.3	11.8	13.1
PD cap. (fF)	100	N/A	80	70
Gain (dBΩ)	62	66	74.4/45.5	63
Supply (V)	1.2	3.3	1	1.2
Power (mW)	120	205	36.6	32.3
Energy Eff. (pJ/bit)	2.41	3.66	0.65	0.58

* Post-layout simulations

IV. CONCLUSION

A 0.58-pJ/bit 56-Gbps PAM-4 optical receiver frontend for CPO is presented. Using the proposed envelope tracker, this design circumvents intrinsic disadvantage of conventional rectifier circuit, and tracks the envelope with little hardware overhead. This work provides a feasible solution for high-speed and low-noise applications with high energy efficiency.

ACKNOWLEDGMENT

This work is supported by the Innovation Project of Optics Valley Laboratory under Grant OVL2021BG005.

of the holding capacitor. The envelope tracker is realized in Verilog-A, and the simulated results is shown in Fig.5. As the envelope of the high-speed signal changes slightly, the peak detector can always follow the peak and valley values of the envelope.

III. POST-LAYOUT SIMULATION RESULTS

The proposed scheme is designed in TSMC's 40nm CMOS technology except that the envelope tracker is implemented in Verilog-A. Fig.6 shows the whole layout and the core area is 0.274mm². With the supply voltage of 1.2V, the total power is 32.3mW. The optical RX front-end has a transimpedance gain of 63-dBΩ with an input-referred noise current of 1.68-uArms. The PD and PAD capacitance is 70fF and 100fF, respectively. The bondwire inductance is 350pH. The post-layout simulated eye diagram at 28Gbps NRZ and 56Gbps PAM-4 from 2^{31}-1 PRBS input is shown in Fig.8(a) and (b), respectively. Table I compares the performance of the proposed design against prior published works.

References

[1] H. Won et al., "A 28-Gb/s Receiver With Self-contained Adaptive Equalization and Sampling Point Control Using Stochastic Sigma-Tracking Eye-Opening Monitor," *IEEE Transactions on Circuits and Systems I: Regular Papers,* vol. 64, no. 3, pp. 664-674, 2017.

[2] K.-L. Fu et al., "A 56Gbps PAM-4 optical receiver front-end," *IEEE Asian Solid-State Circuits Conference (A-SSCC),* 2017.

[3] D. Li *et al.,* "A Low-Noise Design Technique for High-Speed CMOS Optical Receivers," *IEEE Journal of Solid-State Circuits,* vol. 49, no. 6, pp. 1437-1447, 2014.

[4] G. Dziallas, A. Fatemi, A. Peczek, L. Zimmermann, A. Malignaggi, and G. Kahmen, "A 56-Gb/s Optical Receiver With 2.08-μA Noise Monolithically Integrated into a 250-nm SiGe BiCMOS Technology," *IEEE Transactions on Microwave Theory and Techniques,* pp. 1-1, 2021

[5] J.-H. Kim, J.-B. Shin, J.-Y. Sim, and H.-J. Park, "5-Gb/s Peak Detector Using a Current Comparator and a Three-State Charge Pump," *IEEE Transactions on Circuits and Systems II: Express Briefs,* vol. 58, no. 5, pp. 269-273, 2011

[6] Y. Xie *et al.,* "Low-Noise High-Linear 56Gbps PAM-4 Optical Receiver in 45nm SOI CMOS," *IEEE International Symposium on Circuits and Systems (ISCAS),* 2018

2022 IEEE International Conference on
Integrated Circuits, Technologies and Applications

A Novel Fold-Back Current Limiting Protection used in Sub-threshold LDO for Wireless Sensor Applications

Ziyue Chen, Yihui Shi, Ao Hu, Jiarui Xu, Guoyi Yu, Chao Wang

School of Optical and Electronic Information, Huazhong University of Science and Technology, Wuhan, China
Corresponding E-mail*: yuguoyi@hust.edu.cn

Abstract—This paper presents a novel fold-back current limiting protection circuit used in sub-threshold low dropout regulator (LDO) for wireless sensor applications. By sensing the output current and output voltage, then comparing the sampled results with VREF, the proposed protection circuit can control the gate voltage of the power transistor to limit the output current. The maximum output current and short current of the sub-threshold LDO can be limited to 330 mA and 90 mA, respectively, which can save up to 71.9% of power consumption in the case of an output short-circuit condition compared with traditional constant current limiting protection methods. In a 0.35-μm CMOS technology, the sub-threshold LDO can provide a 3.3 V output voltage stably under the 5 V input voltage when the load current changes from 1 μA to 200 mA, and the load regulation rate is only 0.019 mV/mA.

Keywords—*foldback, current limiting protection, sub-threshold LDO, load regulation rate*

I. INTRODUCTION

Recently, subthreshold LDOs have been widely used in the wireless sensor node applications due to its low power consumption achieved from subthreshold operation [1]. Since the endurance of subthreshold LDO power transistor is weaker than that of traditional LDOs, if the load current exceeds the limit, the power transistor will be more likely to break down and results in a large power consumption. Therefore, the current limit protection technology is essential to subthreshold LDO design.

Foldback current limit protection can effectively solve the problem of power consumption and heat dissipation of the sub-threshold LDO chips when the overcurrent shutdown. Once the output current of LDO I_{OUT} exceeds the maximum value I_{MAX}, the foldback current limit circuits will return current back to the low current state continuously [2]. The current limit methods proposed in [2]-[3] are both using a small size sensing transistor to sample the current of the power transistor. In [2], although a common drain structure is adopted to make the drain voltage of sensing transistor close to that of the power transistor, there is still a threshold voltage between the two. The accuracy of the sampling current is difficult to guarantee. Moreover, the accuracy of the inverting amplifier composed of two switching transistors is low, making it difficult to control the gate voltage of power transistor accurately [3]. The proposed of [4] can overcome those disadvantages, but an unbalance rail-to-rail current limit amplifier which used folded cascode structure are hardly conducive to implementation on sub-threshold low-power circuits.

This paper proposes an output current sensing circuit by adding a voltage comparator to the drain terminal of sensing transistor. The sampling accuracy of the I_{OUT} is ensured, as compared to the design in [2]. In the output voltage sensing circuit, a current mirror loop was utilized to convert the voltage change into the current change, thereby starting the foldback function, which further improves the reliability of the

overcurrent protection. And a well-designed current mirror chain as a subthreshold current source bias was also used to make the protection circuit more suitable for subthreshold LDO.

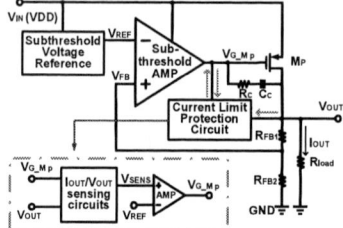

Fig. 1. The basic structure of subthreshold LDO with current limit protection circuit.

II. PROPOSED FOLD-BACK CURRENT LIMITING PROTECTION CIRCUIT

Fig.1 presents the architecture of proposed subthreshold LDO with a novel foldback current limiting protection circuit. The basic principle of the protection circuit is to sense the output current and output voltage, and convert the sampled results into a voltage V_{SENS}, then compare it with the reference voltage V_{REF} to control the gate voltage of the power transistor. Fig.2 shows the schematic of the proposed protection circuit consisting of an output current sensing circuit (Ms, M1-M8), an output voltage sensing circuit(M9-M14), a current limit amplifier CLA (M15-M21), and a sub-threshold current source bias circuit (Mb1-Mb11).

A. Output Current Sensing Circuit

The I_{OUT} is sampled by the sensing transistor M_S. M_S and M_P have the same gate, source and bulk voltage, so M_S can sense the I_{OUT}. To ensure that the drain voltage of M_S (V_{D_Ms}) is as equal as possible to the drain voltage of M_P (V_{D_Mp}), i.e., V_{OUT}, we introduce a voltage comparator consisting of M1-M5 to compare the V_{D_Ms} and the V_{OUT}. The comparator, M_s and M6 consist of a negative feedback loop. When V_{OUT} decrease, I_{M6} increase significantly but I_{MS} increase not as much as I_{M6}, so V_{D_Ms} decrease to balance I_{MS} and I_{M6}. As a result, the drain voltage of M_S (V_{D_Ms}) is close to the drain voltage of M_P (V_{D_Mp}). M6 is designed to have a larger W/L, so that $V_{SG\_M6} \approx |V_{th\_M6}|$. The comparator output V_{G_M6} can be set as V_{OUT} - $|V_{th\_M6}|$ through proper design. With the use of comparator, M_s can sense the I_{OUT} more accurately. A small sensing current I_S proportional to the I_{OUT} can be obtained by setting the $(W/L)_{Ms}$ to a very small value.

B. Output Voltage Sensing Circuit

The output voltage sensing circuit is composed of a current mirror loop and an NMOS M14 working in the deep linear region. The small series resistance R0 protect M14 from breakdown in case of high V_{OUT} attached directly to the gate of M14. When the LDO operates normally, I_S flows to the ground through M13 and M14, and generates a low voltage drop at point A through M14. M7 is turned on to obtain a low voltage of V_{SENS}. Then the CLA outputs a very low voltage. The protection module doesn't have any effect on the gate of the power PMOS transistor.

Once the I_{OUT} exceeds the I_{MAX}, the gate of the power PMOS transistor is controlled by the output of the CLA. When the overcurrent just happened, I_{D_M13} will basically not change since V_{OUT} doesn't change much. As the I_S increases, the I_{D_M14} also increases, causing the V_{SENS} to rise. When the R_{load} drops further. So that the output V_{G_Mp} of the

978-1-6654-9270-6/22 $31.00 © 2022 IEEE

CLA becomes higher, and the Mp is further turned off. At this time, the I_{OUT} returns to the lower current state, and the I_{OUT} maintains the I_{SHORT} state finally.

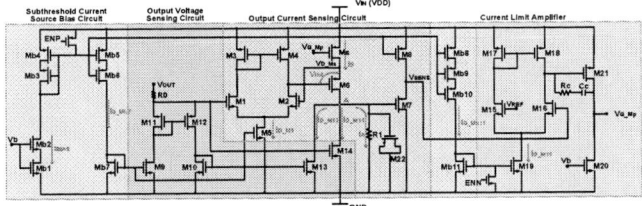

Fig. 2. The proposed current limiting protection circuit.

C. Sub-threshold Current Source Bias Circuit

V_{REF} and V_b are the bias voltages generated by the subthreshold voltage reference module of the LDO, where V_{REF}=1.64 V and V_b=580 mV. V_b provides gate voltage for M_{b1} and M_{b2} ($V_{th} \approx 0.74$V), so that they can all operate in the sub-threshold region, generating the subthreshold reference current I_{BIAS}. I_{BIAS} is first duplicated by a PMOS current mirror and then by a NMOS current mirror to obtain I_{D_M5}, which are the tail currents of the comparator. The relationship between the I_{BIAS} and I_{D_M5} is:

$$I_{D\_M5} = \frac{(W/2L)_{Mb6}}{(W/2L)_{Mb3}} \cdot \frac{(W/L)_{M5}}{(W/L)_{Mb7}} \cdot I_{BIAS} = \alpha \cdot \beta \cdot I_{BIAS} \quad (1)$$

Properly setting the values of the α and β can generate small accurate mirror currents. Similarly, I_{D_M19} can be analyzed.

III. SIMULATION RESULTS AND DISCUSSION

The proposed LDO circuit is implemented in a 0.35-μm CMOS technology. The layout of the proposed circuit is shown in Fig. 3. The total occupied areas are 0.84 mm².

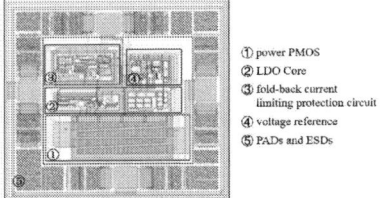

Fig. 3. The layout of the proposed LDO circuit

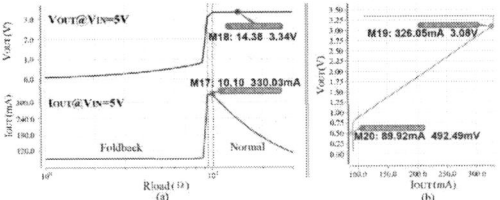

Fig. 4. The Simulation results of (a) Vout/Iout VS Rload and (b) Vout VS Iout.

Fig. 4 (a) shows that when the R_{load} changes from 30 Ω to 10 Ω, the LDO is in a normal operation condition, and the V_{OUT} is constant at 3.3 V. When the R_{load} continues to drop, the I_{OUT} remains at the maximum value 330 mA, and the V_{OUT} decreases linearly with the R_{load}. When the R_{load} decreases further from 9 Ω, the V_{OUT} and I_{OUT} accelerate decline at this time. Fig. 4 (b) shows that when overcurrent takes place, protection circuit limits I_{OUT} to 330 mA. As V_{OUT} drops to about 3 V, the foldback function starts up and further reduces I_{OUT}. Finally, I_{OUT} is limited to I_{SHORT}= 90 mA. The power consumption of the power transistor can be saved up to 71.9 % when the output is shorted to the ground or the R_{load} is very small.

Fig. 5 shows that the V_{OUT} changes only 3.80 mV while the I_{OUT} changes from 1μA to 200 mA, calculated the load regulation rate of the sub-threshold LDO is only 0.019 mV/mA. This result also shows that if there is no overcurrent, the protection module has no effect on the normal operation of the subthreshold LDO.

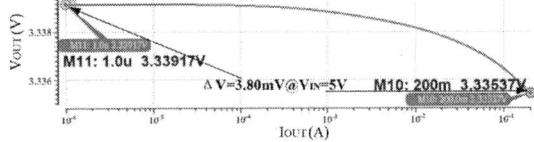

Fig. 5. The Simulation result of subthreshold LDO Load Regulation Rate.

Fig. 6 shows the transient response of the proposed LDO. The load current is changed from 50 mA to 100 mA and then to 50 mA again. The current rise and fall time are set to 0.5 μs. Simulation result shows that the recovery time is less than 20 μs.

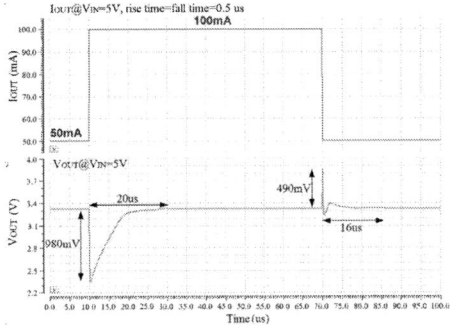

Fig. 6. The Simulation result of load transient response

The detailed performance comparison is listed in Table I. The proposed sub-threshold LDO has achieved the highest power saving in the overcurrent shutdown condition, and a lower load regulation rate in the normal operation state.

TABLE I.   PERFORMANCES SUMMARY AND COMPARISON

	This work	[3]	[4]
Process (nm)	350	350	350
V_{IN} (V)	3.5-5	0.9-1.1	2-3.3
V_{OUT} (V)	3.3	0.5	1.8
I_{MAX} (mA)	**330**	55	197
I_{SHORT} (mA)	90	40	77
I_Q (μA)	0.6	-	5.6
Power Consumption Saving	71.9%	27.3%	60.9%
Load Regulation (mV/mA)	0.019	-	-

IV. CONCLUSION AND FUTURE WORK

This paper presents a foldback current limiting protection circuit for subthreshold LDO. Simulation results show that when the I_{OUT} exceeds 330 mA the circuit will limit the I_{OUT} to 90 mA. The subthreshold LDO can achieve an approximately 71% power saving in the overcurrent shutdown condition, which is suitable for wireless sensor node applications.

REFERENCES

[1] B. S. Rikan et.al., "A low leakage retention LDO and leakage-based BGR with 120nA quiescent current," *Intl. SoC Design Conf.*, 2017, pp. 200-201.

[2] Lin Chuan et.al., "Design of current limiting circuit in low dropout linear voltage regulator," *Asia-Pacific Microwave Conf.*, 2005, pp. 4 pp.-.

[3] Y. C. Hung et.al., "A sub-1 V CMOS LDO regulator with multiple protections capabilities," *Intl. Symp. on Computer, Consumer and Control*, 2014, pp. 800-803.

[4] Jianping Guo et.al., "A fold-back current-limit circuit with load-insensitive quiescent current for CMOS low dropout regulator," *IEEE Intl. Symp. Circuits and Systs.*, 2009, pp. 2417-2420.

A Wideband and High Output Swing Analog Frontend Circuit for FMCW LiDAR

Yiyun Xie, Youze Xin, Bing Zhang*, Ruipeng Yang, Yaoxin Li, Li Geng*
School of Microelectronics, Xi'an Jiaotong University, Xi'an 710049, China

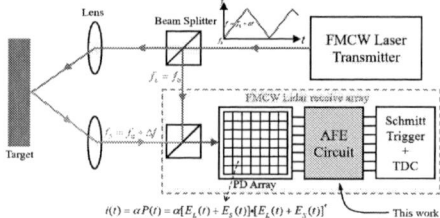

Fig. 1. Block diagram of FMCW LiDAR

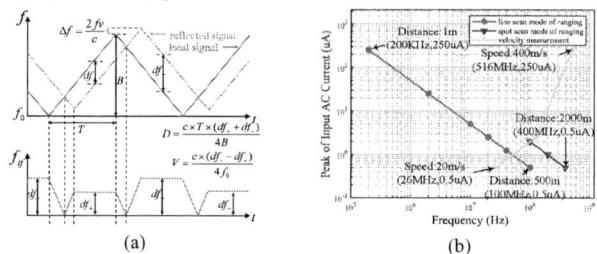

Fig. 2. (a) FMCW LiDAR operating principle. (b) Relationship between AC amplitude and frequency at different distances and speeds

Abstract—This paper proposes a low input impedance, high output swing column-level analog front-end (AFE) circuit for the frequency modulation continuous wave (FMCW) LiDAR. The AFE circuit adopts a shunt-feedback transimpedance amplifier (TIA) and an output swing compensation limiting amplifier (LA) to amplify the weak echo signal and realize the high output swing. The input direct current cancellation (IDCC) circuit is used to stabilize the output direct operating point, where the lag network is used to extend the frequency range of high gain, and the noise of the direct current is reduced by means of noise transfer. The proposed AFE circuit is implemented in a 55nm CMOS process, and the post-layout simulation results show that the circuit achieves a gain of more than 118.7 dBΩ and an output swing of higher than 840 mV in the frequency range of 26 MHz~400 MHz. The input-referred noise current is 12.45 pA/Hz$^{0.5}$ and the power consumption is 7 mW with a 1.2 V power supply.

Keywords—FMCW LiDAR, analog front-end, transimpedance amplifier, output swing, input direct current cancellation

I. INTRODUCTION

Light detection and ranging (LiDAR) has received more and more attention in many fields due to its advantages of large detection range and high spatial resolution, such as autonomous driving, intelligent robots [1][3][4]. Compared with the direct time of flight (DToF), frequency modulation continuous wave (FMCW) LiDAR has the advantages of long detection range, easy resistance to ambient light and interference from the same type of equipment. Due to the Doppler effect, FMCW LiDAR can measure distance and velocity in a single frame [2].

To extract the intermediate frequency (IF) information of the photocurrent, the AFE circuit needs to filter out the influence of the direct current, and has the characteristics of high bandwidth, high gain, and large output swing. The performance of current-mirror amplifier in [1] needs to improve at high bandwidth. The AFE circuit in [3] and [4] can't overcome the tradeoff between the bandwidth and the output swing in the low voltage domain, or handle the input direct current.

To satisfy the requirements of the FMCW LiDAR, this paper proposes a low input impedance, high gain, swing-compensated AFE circuit.

II. ARCHITECTURE AND PRINCIPLE OF FMCW LiDAR

Fig. 1 shows the FMCW LiDAR system architecture. The laser transmits a signal which frequency changes linearly within the frequency modulation period, and the reflected signal light is coherent with the local oscillator light. Due to the difference in transit time and the Doppler effect, there is a frequency difference between the two signals, as shown in Fig. 2(a). The photodetector detects the signal light and the local oscillator light to obtain an IF signal. After demodulation, the distance and speed information of the target can be obtained. The current generated in the detector can be expressed as:

$$i(t) = i_L + i_S + 2\sqrt{i_L i_S} \cos(\Delta\omega_{if} t + \Delta\varphi) \qquad (1)$$

Where i_L and i_S are the photocurrents generated when there is only local oscillator light or signal light, $\Delta\omega_{if}$ and $\Delta\varphi$ is the frequency difference and phase difference between the two beams. The third term in (1) is the AC component which generated by the coherence of the two beams. Different distances and speeds will change the amplitude and frequency of the AC component. With the change of distance and speed, the system selects different working modes to adapt it. The relationship between the amplitude and the frequency of the AC component is shown in Fig. 2(b), and the signals with minimum AC amplitude are at 26 MHz, 100 MHz and 400 MHz.

III. ARCHITECTURE OF PROPOSED AFE CIRCUIT

As shown in Fig. 3, the receiving array contains 300×300 photodiodes, and every 300 is a column. Each column contains an AFE circuit and a time-frequency domain conversion circuit. Each column only closes one switch during detection, and the corresponding photodiode starts to work and transmits the current signal to the AFE circuit. The AFE circuit cancels the direct current through IDCC and amplify the alternating current through the TIA and LA to reach the threshold of the Schmitt trigger.

A. Circuit Design of TIA and LA

Each column of circuit contains 300 photodiodes and their switches. The parasitic capacitance of the switches and the long metal wire in the layout greatly increase the input parasitic capacitance of the AFE circuit. The core amplifier of TIA adopts a three-stage symmetrical cascaded inverter-type to expand the bandwidth and gain, while ensuring the symmetry of the output. Large input parasitic capacitance can cause the TIA core amplifier to contribute a lot of noise. Under the limitation of power consumption, the noise can be reduced by increasing the g_m of the TIA input stage transistors.

The LA circuit adopts the same structure to expand the gain and bandwidth. The LA reduces the output impedance to drive a larger load capacitance by increasing the g_m of the transistors MN_{12} and MP_{12}. MN_{12} and MP_{12} output large current under large V_{GS}, but MN_{11} and MP_{11} cannot provide, so the output swing becomes small when the gain and bandwidth meet the requirements. To overcome the tradeoff between bandwidth and output swing, the swing compensation LA circuit is proposed. The resistor is connected in series with the drain terminals of MN_{12} and MP_{12}, respectively, which does not change the small signal characteristics. After reaches a certain output swing, the

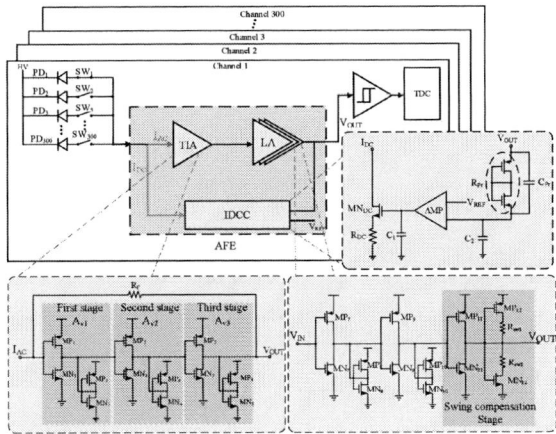

Fig. 3. Schematic of AFE circuit

voltage drop on the resistor pushes MN_{12} and MP_{12} into the linear region to reduce the output current and expand the overall output swing.

B. Circuit Design of IDCC

The IDCC loop controls the dc operating point of the TIA and LA by detecting the output dc voltage. IDCC needs a large low-frequency loop gain to reduce the output dc gain of the AFE closed-loop and suppress the influence of the input direct current. It also needs to have a small loop GBW to avoid degrading the high gain frequency range of the AFE circuit. To get above performance requires, huge resistor and capacitor are needed to form an extremely small pole, which cannot be realized in column-level circuits.

This paper proposes a lag network by using a pseudo-resistor, which introduces a pole-zero pair in the IDCC loop, reducing the high-frequency loop gain without reducing the high frequency phase. IDCC also uses the output resistance of the amplifier AMP and capacitance C_1 to generate a second pole, which decreases the GBW of the loop. The high gain and low GBW of the loop can be achieved without occupying large area and it reduce the difficulty of designing amplifiers in the IDCC.

This paper uses the noise transfer method to reduce the noise caused by the direct current, and a resistor R_{DC} is added to the source terminal of the direct current cancellation transistor MN_{DC}. The noise brought by direct current can be expressed as

$$\overline{i^2_{n,R_{DC}+MN_{DC}}} = \frac{8kT\gamma I}{V_{od,MN}} \times \left(\frac{V_{od,MN}}{V_R + V_{od,MN}}\right)^2 + \frac{8kTI}{V_R} \times \left(\frac{V_R}{V_R + V_{od,MN}}\right)^2 \quad (2)$$

where γ is the noise figure of the short channel MOS transistor, I is the input direct current, $V_{od,MN}$ is the overdrive voltage of MN_{DC}, and V_R is the voltage on the resistor R_{DC}. The noise figure of the resistor is less than γ. The sum of $V_{od,MN}$ and V_R is a fixed value and part of the noise from the MN_{DC} is transferred to the resistor. Simulation results show that the noise caused by input direct current is reduced by 20%.

IV. SIMULATION RESULTS

Fig. 4(a) shows the band-pass characteristic of the AFE output frequency response curve, the low frequency gain is 24 dBΩ, and the gain from 26 MHz to 400 MHz is above 118.7 dBΩ. Fig. 4(b) shows the output swings of each typical value under different detection distances and speeds, and the output swings are all greater than 840 mV.

Table I summarizes the performance of state-of-the-art AFE circuits in recent years. Compared with [1]-[3], the

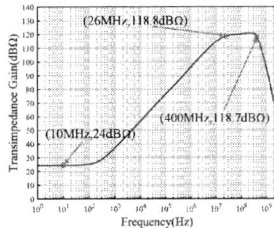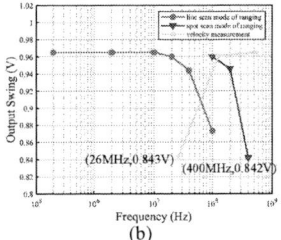

Fig. 4. (a) AFE output amplitude frequency characteristic curve. (b) Output swing of each typical value at different distances and speeds.

TABLE I.   COMPARISON WITH PRIOR WORKS

Parameter	TCAS-II [1]	SENSORS [2]	SENSORS [3]	This work*
Technology	CMOS 180nm	CMOS 180nm	CMOS 180nm	CMOS 55nm
Input parasitic capacitance(pf)	1.2	0.2	1.2	4.5
Input direct current (μA)	N/A	N/A	N/A	500~625
Transimpedance Gain (dBΩ)	106	140	100	121
Bandwidth (MHz)	50	1.1	450	400
Output swing(V) /Power supply(V)	1/3.3	1.2/3.3	N/A /3.3	0.84/1.2
Input referred noise current density (pA/sqrt(Hz))	1.52	N/A	2.59	12.45
Power consumption (mW)	8	211	6.6	7

*simulation results

proposed AFE circuit has larger gain and bandwidth with maximum input parasitic capacitance and can realize the cancellation of input direct current. Compared with [1]-[3], the designed AFE circuit can achieve large output swing at lower power supply. The large input parasitic capacitance increases the noise contributed by the TIA core amplifier, and the additional large input direct current also contributes large noise, which results in large noise in the AFE circuit.

V. CONCLUSION

This paper presents an AFE circuit with low input impedance, large output swing and direct current cancellation. The swing-compensated LA achieves a minimum output swing over 840 mV. The IDCC circuit cancels the direct current and reduces the noise. It also uses the lag network to avoid the reduction of the high gain frequency range.

ACKNOWLEDGMENT

The authors thank the supports from Xi'an Major Scientific and Technological Innovation Platform and Local Transformation Project of Scientific and Technological Achievements (20KYPT0001-11-1), the National Natural Science Foundation of China (61874085), the Postdoctoral Research Funding Project of Shaanxi Province (2018BSHEDZZ41).

REFERENCES

[1] R. Ma, M. Liu, H. Zheng and Z. Zhu, "A 77-dB Dynamic Range Low-Power Variable-Gain Transimpedance Amplifier for Linear LADAR," TCASII, vol. 65, no. 2, pp. 171-175, Feb. 2018.

[2] K. Hu, Y. Zhao, M. Ye, J. Gao, G. Zhao and G. Zhou, "Design of a CMOS ROIC for InGaAs Self-Mixing Detectors Used in FM/cw LADAR," Sensors, vol. 17, no. 17, pp. 5547-5557, Sept. 2017.

[3] H. Zheng, R. Ma, M. Liu and Z. Zhu, "A Linear-Array Receiver Analog Front-End Circuit for Rotating Scanner LiDAR Application," Sensors, vol. 19, no. 13, pp. 5053-5061, July. 2019.

[4] X. Wang et al., "A Low Walk Error Analog Front-End Circuit With Intensity Compensation for Direct ToF LiDAR," TCASI, vol. 67, no. 12, pp. 4309-4321, Dec. 2020.

A 0.3 V-4 V Input Voltage Range, 0.7 V Cold Start Boost Converter with 1 V Internal Voltage Supply Generator by Using 0.18 μm CMOS Process for Energy Harvesting Application

Zheng Lu, Shiquan Fan*, Weiqing Ma, Ying Xie, Li Geng

School of Microelectronics, Xi'an Jiaotong University

Abstract—In this paper, a wide input range boost converter is proposed. In consideration of the wide input voltage range, especially at very low input voltage, to guarantee the internal control circuit (ring oscillator and PFM controller) can operate correctly, an internal adaptive supply voltage generator is designed to produce 1 V supply voltage. The boost converter is fabricated with standard 0.18 μm 5P0 CMOS process. The active area of the boost converter is nearly 0.5 mm². Measured results show that the boost converter can cold start with 700 mV input voltage and operate with input voltage range of 0.3 V-4 V, which demonstrate the design concepts of boost converter well.

Keywords—Boost converter, wide input voltage range, cold start, PFM controller

I. INTRODUCTION

Harvesting energy from various ambient sources is a key technology for energy harvesting system, such as solar, vibration and RF energy [1]. However, the diversity and uncertainty of the environment energy lead to the lower and uncertain output voltage of energy harvesting system. Therefore, the energy harvesting system is not suitable to directly power the electronic devices. A boost converter with wide input voltage range is necessary for energy harvesting system.

When the input voltage changes in a wide range, in order to keep the output voltage constant, the duty cycle of traditional boost converter needs to be changed in a large range, which lead to problems such as complex controller design and poor system stability [2].

In recent years, several wide input voltage range circuits have been proposed. For isolated dc-dc converter, Y. -E. Wu proposed a novel bidirectional isolated dc–dc converter with high voltage gain and wide input voltage [3], which can be applied to bidirectional power conversion systems. A current-fed dual-inductor resonant full-bridge dc–dc isolated boost converter is proposed in [4], which is no need for additional start-up circuit. As for non-isolated boost converters, A. Gupta used a phase shedding control scheme for extended-duty-ratio (EDR) boost converter to achieve a high voltage gain [5]. Hysteresis-controlled structure is used to save power consumption in [6], which is most suitable for applications that demand wide input voltage and load ranges.

In this paper, a novel boost converter with internal voltage supply generator is proposed, which has 0.3 V -4 V wide input voltage and low power consumption. The internal voltage supply generator can provide stable 1 V supply voltage for internal control circuit in the whole input range.

II. IMPLEMENTATION OF PROPOSED BOOST CONVERTER

A. System architecture

Fig. 1 shows the entire system architecture of the hysteresis-controlled Boost converter with internal voltage supply generator. It mainly includes internal voltage supply generator,

This work was supported in part by the National Natural Science Foundation of China under Grant 62074124, in part by the Fundamental Research Funds for the Central Universities under Grant xzy012020011, and in part by the key lab of micro-nano electronics and system integration of Xi'an city.

hysteresis comparator, ring oscillator, boost power stage and pulse frequency modulation (PFM) control loop.

By comparing the feedback voltage, V_{FB} and the reference voltage V_{REF}, the hysteresis comparator will generate EN signal to control the ring oscillator, so as to cut off the power stage. An internal voltage supply generator is used to produce the 1V supply voltage for the ring oscillator and PFM controller in the whole 0.3V-4V wide input voltage range.

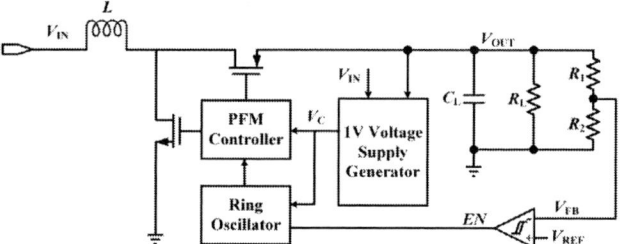

Fig. 1. System architecture of proposed boost converter.

B. Implementation of internal voltage supply generator

The detailed schematic of the 1V voltage supply generator circuit is shown in Fig. 2. It includes a two-stage LDO and adaptive voltage selection circuit (M_{P5}, M_{P6}). V_{IN} and V_{OUT} are the input voltage and output voltage of the boost converter, V_P and V_C is the input and output voltage of LDO. V_P can adaptively select the higher one between V_{IN} and V_{OUT}.

1) When $V_{IN} > V_{OUT}$, transistor M_{P5} is on, M_{P6} is off, V_P is close to V_{IN}.

2) When $V_{IN} < V_{OUT}$, transistor M_{P5} is off, M_{P6} is on, V_P is close to V_{OUT}.

3) When $V_{IN} = V_{OUT}$, V_P can acquire voltage through body diodes of transistor M_{P5} and M_{P6}. Although there will be a loss of threshold voltage at V_P by passing the voltage through the diode, in this case, V_{OUT} is usually large enough and V_P can still be guaranteed to have a voltage greater than 1V.

Through the cross-coupling connection transistors, the LDO can generate 1V supply voltage stably over the entire 0.3V-4V input voltage range.

The main structure of the LDO is composed of a common-source symmetrical amplifier, a power transistor M_P and feedback network. The on-chip resistor R_C and capacitor C_C are designed to guarantee loop stability.

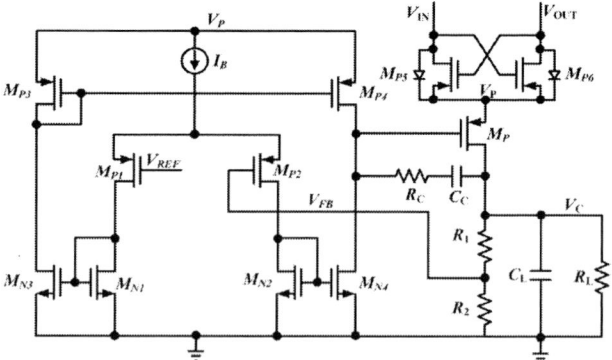

Fig. 2. Proposed 1V internal voltage supply generator.

C. Hysteresis comparator

Fig. 3 shows the transistor implementation of the hysteresis comparator, which is biased in subthreshold operating region. The current bias I_B is designed just 20 nA to reduce power consumption. The hysteresis has a nearly 30 mV trigger window voltage to guarantee the output voltage with as low as possible ripples. A buffer is employed to further shape the output voltage of the comparator, to ensure EN signal's validity.

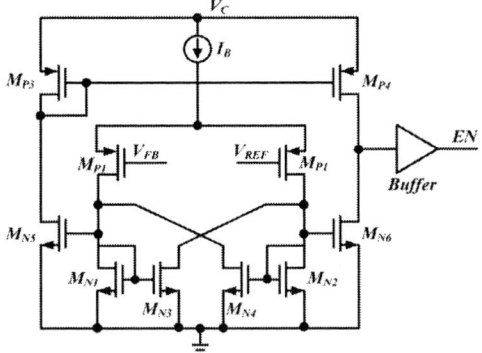

Fig. 3. Transistor implementation of the Hysteresis comparator.

D. Ring oscillator

The schematic of ring oscillator is shown in Fig. 4., including a two-input NAND, an inverter and a schmitt trigger. The frequency of oscillator can be modified through the off-chip capacitor C_D, and the frequency can be expressed

$$f_{OSC} = \frac{I_B}{2C_D \Delta V} \quad (1)$$

Where I_B (20 nA) is the bias current, and ΔV is the trigger window of the Schmitt trigger.

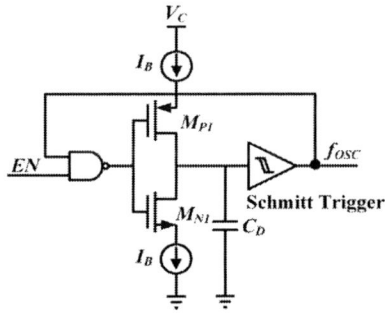

Fig. 4. Schematic implementation of ring oscillator.

III. EXPERIMENTAL RESULTS AND ANALYSIS

Fig. 5 shows the chip photo of the designed boost converter, and only standard 5 V CMOS devices are adopted, the chip active area is about 0.5 mm².

Fig. 6(a) shows the measured cold start process waveforms of the boost converter. The minimum of cold start input voltage is 0.7 V, the PFM signal has achieved its maximum duty cycle and the V_{OUT} is 2.7 V, internal voltage supply generator produces 1 V output to supply the control circuit. The load resistance R_L is 190 Ω and the efficiency of the boost converter is 94.3%.

Fig. 6(b) shows the measured results of stable waveforms of the boost at the minimum 0.3 V input voltage. The internal voltage supply generator is powered by the 1.2 V boost output and generate 1 V supply voltage as well.

Fig. 7(a) and (b) show the measured results of boost waveforms at 2 V input and 4 V input voltage. V_{OUT} can achieve 4 V, and the duty cycle of the PFM signal is very small (120 ns pulse width at 4 V input) at this time. The power stage is turned on intermittently to provide energy to the load.

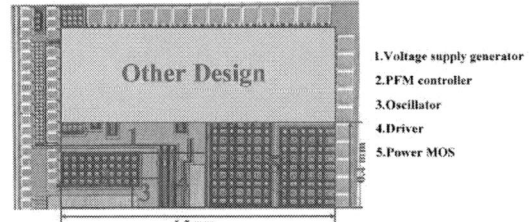

Fig. 5. Chip photo of the proposed boost converter.

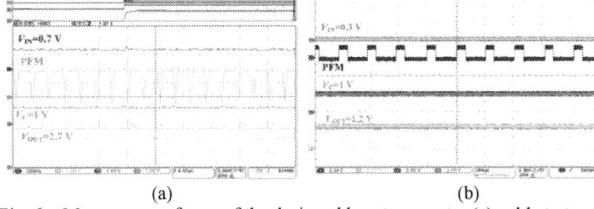

(a)   (b)

Fig. 6. Measure waveforms of the designed boost converter (a) cold startup at 0.7 V input voltage, and (b) minimum operating input voltage at 0.3 V.

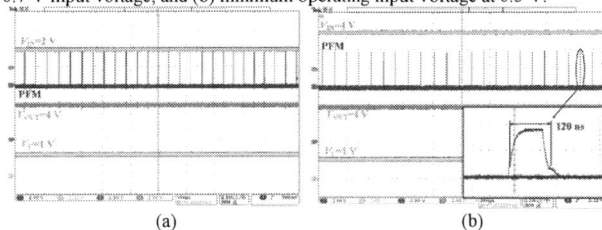

(a)   (b)

Fig. 7. Measure stable waveforms of the designed boost converter (a) 2 V input voltage, and (b) 4 V input voltage.

IV. CONCLUSION

In summary, a wide input voltage range, 0.7 V cold start boost converter with standard 0.18 μm 5P0 CMOS process is proposed and fabricated. A novel internal voltage supply generator with 1 V output is adopted to supply for the control circuit under the wide input voltage. Finally, measured results demonstrate that the boost converter can cold start at a minimum 0.7 V input voltage and operate under the whole 0.3 V-4 V input voltage range.

REFERENCES

[1] M. Chen, H. Yu, G. Wang and Y. Lian, "A Batteryless Single-Inductor Boost Converter With 190 mV Self-Startup Voltage for Thermal Energy Harvesting Over a Wide Temperature Range," in IEEE Transactions on Circuits and Systems II: Express Briefs, vol. 66, no. 6, pp. 889-893, June 2019, doi: 10.1109/TCSII.2018.2869328.

[2] Ortiz-Lopez, M. G., et al. "Modelling and analysis of switch-mode cascade converters with a single active switch." IET Power Electronics 1.4 (2008): 478-487.

[3] Y. -E. Wu and Y. -T. Ke, "A Novel Bidirectional Isolated DC-DC Converter with High Voltage Gain and Wide Input Voltage," in IEEE Transactions on Power Electronics, vol. 36, no. 7, pp. 7973-7985, July 2021, doi: 10.1109/TPEL.2020.3045986.

[4] V. K. Goyal and A. Shukla, "Isolated DC–DC Boost Converter for Wide Input Voltage Range and Wide Load Range Applications," in IEEE Transactions on Industrial Electronics, vol. 68, no. 10, pp. 9527-9539, Oct. 2021, doi: 10.1109/TIE.2020.3029479.

[5] A. Gupta, R. Ayyanar and S. Chakraborty, "Phase-Shedding Control Scheme for Wide Voltage Range Operation of Extended-Duty-Ratio Boost Converter," 2021 IEEE Applied Power Electronics Conference and Exposition (APEC), 2021, pp. 494-499, doi: 10.1109/APEC42165.-2021.9487308.

[6] S. Fan, R. Wei, L. Zhao, X. Yang, L. Geng and P. X. -. Feng, "An Ultralow Quiescent Current Power Management System with Maximum Power Point Tracking (MPPT) for Batteryless Wireless Sensor Applications," in IEEE Transactions on Power Electronics, vol. 33, no. 9, pp. 7326-7337, Sept. 2018, doi: 10.1109/TPEL.2017.2769708.

An AOT Buck Converter with Adaptive T$_{ON}$ Extender Achieving 2.5μs Settling Time in 4A Load Transient

Zhuang Zhang, Hanyu Shi, Danzhu Lu, Peng Cao, Zhiliang Hong*

State Key Laboratory of ASIC and System, Fudan University, Shanghai 200433, China

Abstract—An adaptive on-time (AOT) buck converter with constant switching frequency and fast transient response is presented. A frequency-locked loop (FLL) is used to achieve constant switching frequency. The on-time (T$_{ON}$) is adjusted by a T$_{ON}$ extender to achieve fast transient response. The proposed AOT buck converter is implemented in 0.18μm CMOS process. The simulation results show that the switching frequency is fixed at 1MHz under various load condition and the output voltage undershoot and settling time are only 50mV and 2.5μs, respectively during 4A load transient.

Keywords—DC-DC converter, adaptive on-time (AOT), fast transient response, fixed frequency

I. INTRODUCTION

Due to the rapid development of electronic devices, DC-DC converters with high power, high efficiency and fast transient response are in great demand. Unlike conventional pulse-width-modulated (PWM) control, constant on-time (COT) control can provide better transient response and maintain higher efficiency in light load current. However, conventional COT converter suffers from the limitations of transient response for the fixed T$_{ON}$ and the varying switching frequency. To overcome these drawbacks, various schemes have been investigated. The "fast adaptive on-time" control in [1] can extend T$_{ON}$ achieving fast transient response, but it requires extra off-chip capacitors and resistors. [2] proposes a T$_{ON}$ adjustment circuit to provide constant switching frequency, which in turn brings about extra overshooting and undershooting during load transient. [3] describes an adaptive quasi-adaptive T$_{ON}$ current mode control improving transient response, but this approach can't keep switching frequency constant. To realize COT converter with constant switching frequency and fast transient response, this paper proposes an adaptive on-time (AOT) buck converter with on-time generator (OTG) containing a frequency-locked (FLL) and a T$_{ON}$ extender.

II. THE PROPOSED AOT CONVERTER

A. Overall Architecture

Fig. 1 presents the architecture of the proposed AOT buck converter with OTG. Based on conventional on-time generator, the proposed FLL and T$_{ON}$ extender in Fig.1 are used in the OTG. The FLL detects the frequency difference between PWM pulses and the reference clock (CLK) to control T$_{ON}$ keeping switching frequency constant. The T$_{ON}$ extender senses the change of output voltage (V$_{OUT}$) during load step-up to extend T$_{ON}$ improving transient response.

B. FLL

Fig. 2 presents the structure of FLL including a phase frequency detector (PFD), a charge pump (CP) and a voltage-to-current (V to I) converter. When the frequency of PWM is higher than CLK, for example, the PFD detects the frequency

error and reduces the voltage V$_{CP}$ so that the current I$_{FLL}$ decreases and T$_{ON}$ increases until the frequency of PWM equals to CLK.

To ensure the stability of the AOT converter, the loop gain of FLL must be analyzed. The phase error to V$_{CP}$ gain, $G_{PFD\text{-}CP}(s)$, the V to I converter gain, $G_{VTI}(s)$ and charge current I$_{FLL}$ to phase error gain, $G_{CTP}(s)$ are expressed as (1), (2) and (3), respectively.

$$G_{PFD\_CP}(s) = \frac{\Delta V_{CP}}{\Delta \varphi} = \frac{I_{CP}(1+sC_2R_1)}{2\pi s^2 R_1 C_1 C_2 + s(C_1+C_2)} \quad (1)$$

$$G_{VTI}(s) = \frac{\Delta I_{RAMP}}{\Delta V_{CP}} = \frac{1}{R_{FLL}} \quad (2)$$

$$G_{CTP}(s) = \frac{\Delta \varphi}{\Delta I_{FLL}} = \frac{\int \Delta \omega_s dt}{\Delta I_{FLL}} \quad (3)$$

$$\Delta \omega_s = \frac{2\pi}{T_s + \Delta T_s} - \frac{2\pi}{T_s} \approx -\frac{2\pi}{T_s^2}\Delta T_s$$

$$\Delta T_s = \frac{\Delta T_{ON} V_{IN}}{V_{OUT}}$$

$$\Delta T_{ON} = \frac{V_{peak}C_{on}}{I_{FLL}+\Delta I_{FLL}} - \frac{V_{peak}C_{on}}{I_{FLL}} \approx -\frac{V_{peak}C_{on}\Delta I_{FLL}}{I_{FLL}^2}$$

Here, φ, ω_s, T_s and T_{ON} are the phase, angular frequency, period and on-time of the converter, respectively. According to (1), (2) and (3), transfer function $G_{FLL}(s)$ of the FLL can be expressed as (4).

$$G_{FLL}(s) = \frac{I_{CP}V_{OUT}}{T_s^2 R_{FLL}C_{on}V_{peak}V_{IN}} \frac{(1+sC_2R_1)}{s^3R_1C_1C_2+s^2(C_1+C_2)} \quad (4)$$

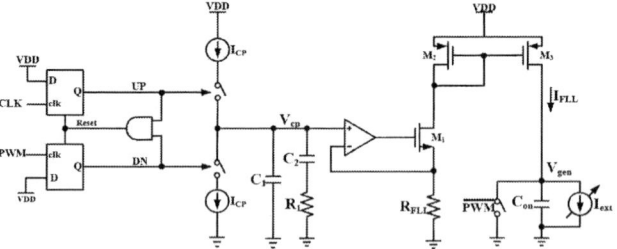

Fig. 1. Proposed AOT buck converter

The bode plots of the loop transfer function are shown in Fig. 3, which verifies that the unity gain frequency (UGF) and the phase margin (PM) of the FLL are 150kHz and 60°. Since the switching frequency is 1MHz in steady state which is 6.6 times higher than the UGF, the FLL can slowly adjust the switching frequency without instability issue.

Fig. 2. Structure of FLL

978-1-6654-9270-6/22 $31.00 © 2022 IEEE

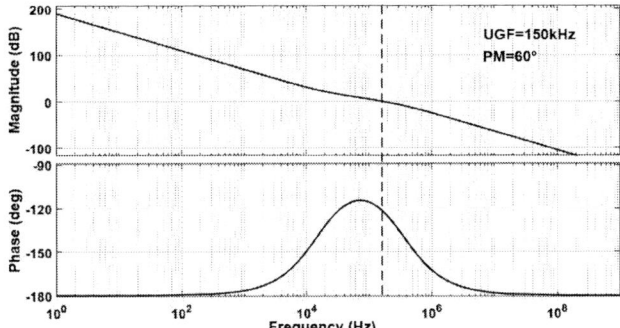

Fig. 3. Bode plots of FLL

C. T_{ON} Extender

Fig. 4 shows the structure of the proposed T_{ON} extender. The T_{ON} extender includes an RC filter and an operational transconductance amplifier (OTA). In load step-up transient, V_{OUT} drops down quickly, while V_{OUTF} keeps steady by the RC filter so that the output current I_{ext} of OTA increases. Then, the current charging capacitor C_{on} decreases resulting in more time for V_{gen} to reach V_{peak}. Therefore, T_{ON} is extended. Fig. 4 also shows the key operating waveforms of T_{ON} extender.

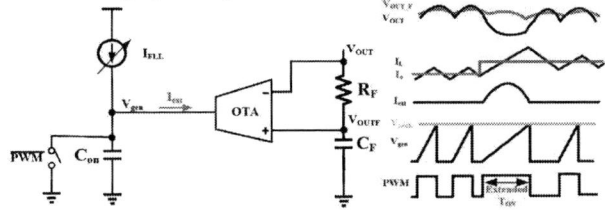

Fig. 4. Structure of T_{ON} extender and key waveforms of T_{ON} extender

III. SIMULATION RESULTS

The proposed AOT buck converter is implemented in 0.18μm CMOS process. The input voltage ranges from 5V to 12V and output voltage ranges from 0.33V to 1.2V. The maximum load current is 5A. For 12V-to-1.2V voltage conversion, the efficiency and switching frequency of the proposed converter are shown in Fig. 5. It is illustrated that the peak efficiency is 96.9% and switching frequency keeps constant in different load compared to conventional COT converter without FLL. Fig. 6 presents the transient performance of AOT converters with and without the proposed T_{ON} extender under 4A load transient. Fig. 6 shows that without the T_{ON} extender, the inductor current rises with a fixed on-time bringing about longer settling time and worse output voltage undershoot. The T_{ON} extender extends T_{ON} to make the inductor current reach the load current within one switching cycle. It's demonstrated that with the proposed T_{ON} extender, the settling time and V_{OUT} undershoot are improved by 79.2% and 37.5%, respectively.

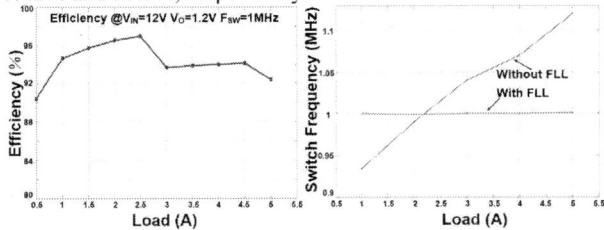

Fig. 5. Efficiency and switching frequency under varying load conditions

The performance of the proposed AOT buck converter is summarized in Table I and compared to state-of-art works.

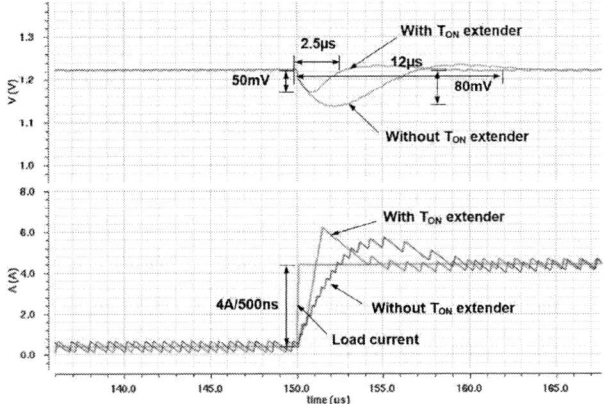

Fig. 6. Load step-up transient response with and without proposed T_{ON} extender

TABLE I. SUMMARY AND COMPARISON WITH STATE-OF-ART WORKS

Specifications	[4]	[5]	This work
Inductor(μH)/ Capacitor(μF)	2.2/10	1/4.7	2.2/47
Input voltage(V)	2.7-4.2	3.3	5-12
Output voltage(V)	1.2	1.05	0.33-1.2
Max. load current(A)	0.7	1.7	5
Fixed switching frequency	Yes	Yes	Yes
Transient step(A)/ Slew rate(A/μs)	0.3/N/A	1.4/0.7	4/8
Undershoot(mV)	48	95	50
Settling time(μs)	3	8	2.5
Peak efficiency(%)	95.7	89	96.9

IV. CONCLUSIONS

This paper proposes a AOT buck converter with constant switching frequency and fast transient response. Thanks to the FLL, switching frequency is fixed at 1MHz. T_{ON} is adjusted by the T_{ON} extender so that the V_{OUT} undershoot and the settling time are only 50mV and 2.5μs separately with 4A load step.

REFERENCES

[1] S. Bari, Q. Li and F. C. Lee, "A New Fast Adaptive On-Time Control for Transient Response Improvement in Constant On-Time Control," in IEEE Transactions on Power Electronics, vol. 33, no. 3, pp. 2680-2689, March 2018.

[2] K. Chen, J. Garrett, K. Peng, R. Hulfachor and M. Onabajo, "Buck Circuit Design With Pseudo-Constant Frequency and Constant On-Time for High Current Point-of-Load Regulation," in IEEE Transactions on Circuits and Systems I: Regular Papers, vol. 68, no. 10, pp. 4062-4075, Oct. 2021.

[3] C. -F. Nien et al., "A Novel Adaptive Quasi-Constant On-Time Current-Mode Buck Converter," in IEEE Transactions on Power Electronics, vol. 32, no. 10, pp. 8124-8133, Oct. 2017.

[4] K. Hu, S. Lin and C. Tsai, "A Fixed-Frequency Quasi-V² Hysteretic Buck Converter With PLL-Based Two-Stage Adaptive Window Control," in IEEE Transactions on Circuits and Systems I: Regular Papers, vol. 62, no. 10, pp. 2565-2573, Oct. 2015.

[5] W. Chen et al., "Pseudo-Constant Switching Frequency in On-Time Controlled Buck Converter with Predicting Correction Techniques," in IEEE Transactions on Power Electronics, vol. 31, no. 5, pp. 3650-3662, May.2016.

A 1000 fps Spiking Neural Network Tracking Algorithm for On-Chip Processing of Dynamic Vision Sensor Data

Chi Zhang[1,2], Lei Kang[1,2], Xu Yang[1], Guanghao Guo[1,2], Peng Feng[1,2], Shuangming Yu[1], Liyuan Liu[1,2*]

[1] State Key Laboratory of Superlattices and Microstructures, Institute of Semiconductors, Chinese Academy of Sciences
[2] Center of Materials Science and Optoelectronics Engineering, University of Chinese Academy of Sciences

Abstract—Dynamic vision sensor (DVS), an event-based camera, has attracted significant attention due to its unique characteristics. Unlike frame-based cameras, the data format of DVS makes it difficult for traditional algorithms to process it directly. On the other hand, as a new type of brain-like neural network, the spiking neural network is specially used to process spiking data, and it is well suited for this type of event-based camera. In addition, because of the rapid development of neuromorphic hardware in recent years, it is possible to deploy SNN applications on edge-side system-on-chip. Therefore, based on the characteristics of dynamic vision sensors, this paper designs a spike encoding module and an SNN for processing sensor information. We use selective search to accomplish object tracking by classifying targets and backgrounds. The SNN can achieve 98.66% classification accuracy on our synthetic test dataset, and the tracking algorithm can achieve over 1000 fps after quantizing and compiling the network to the hardware simulator.

Keywords—*dynamic vision sensor, spiking neural network, object tracking*

I. INTRODUCTION

Recently, event-based cameras have received much attention due to their bionic characteristics. Unlike previous sensors that image exposures with a fixed time, the pixels of an event-based camera generate a series of events by responding asynchronously to photocurrents, and these events usually also transmit signals outward in the form of spikes. A dynamic vision sensor (DVS) is one of the mature event-based cameras. The pixels of DVS asynchronously measure brightness changes, and the changing luminance can be captured quickly with a delay of tens of microseconds. Thus DVS can well reduce the motion blur phenomenon generated in conventional cameras and is suitable for tracking fast-moving objects [1].

Due to the unique spike format of DVS, there is not only information in the spatial dimension but also information in the time dimension. It is impossible to obtain enough information only through the spike at a single moment, so it is not easy to directly process visual information such as traditional convolutional neural networks (CNN). It is necessary to redesign a suitable algorithm to process data of DVS. Due to its same bionic brain-like characteristics, a spiking neural network (SNN) also processes spike data and is naturally suitable for processing data generated by DVS.

In this work, firstly we design a spike encoding module that can implement on both software and hardware. Then We propose a search method base on SNN to realize the tracking function without training on the temporarily missing DVS general target tracking data set. Finally, we verify the algorithm and get a 1000 fps tracking rate on the hardware simulator.

This work is supported by the National Key Research and Development Program of China under Grant 2019YFB2204303, the National Natural Science Foundation of China under Grant 62134004 and U21A20504, and the Youth Innovation Promotion Association Program Chinese Academy of Sciences under Grant 2021109.

II. DATA PREPROCESS

Currently, the most important problem in designing algorithms for DVS is the adaptation of data from sensors and processing modules. The output form of DVS is an asynchronous event data stream. However, in hardware implementation, it is impossible for the processor to keep up with the rapid event generation rate of DVS. And because of the unpredictability of the event sending rate, it is difficult to determine when to take out the results of the SNN, so we need a data preprocessing module to control the spike density.

Usually, DVS is read out by address-event representation (AER), which includes the address of the pixel in the array, the polarity of the spike, and the time stamp when the spike is generated. This work uses a specific DVS designed by our research group, and its read data format is also AER format. The spike encoding module we designed is used to restore the original AER format data into binary image frames like Fig.1. To facilitate binary encoding, we erase the polarity of the spike; whether it is an ON or OFF event, it is represented by 1, and no event is represented by 0. At the same time, a counter is used to count the number of accumulated spikes by setting a threshold N. Alternatively, we can also set a fixed time T to accumulate events. When the number of spikes reaches the threshold N or the accumulation time reaches T, we read out a frame of the binary image. In the accumulation process, even if a pixel generates multiple spikes, it is only marked as 1.

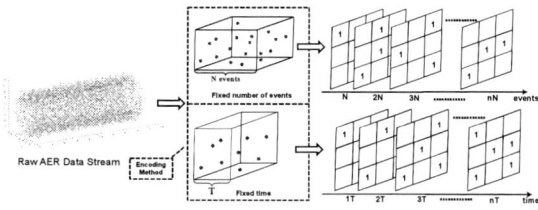

Fig. 1. AER data encoding.

With this module, the number of events contained in the binary image frame is relatively stable, and the SNN determines when to output a result by setting the number of input frames.

III. ALGORITHM DESIGN

A. Selective Search

Due to the super-high time resolution of DVS, we need to quickly realize the calculation on the hardware side. Different from commonly used neural network tracking methods such as the Siamese network[2], which requires complex network structure support and may need to update templates online, we design the algorithm for tracking gestures based on the selective search that can track selected target only by classification. First, at the beginning of tracking, we need to manually select the initial position of the target to be tracked $(x_0,\ y_0)$ (target center coordinates) and target size L to determine the search starting point. Then l candidate target boxes of size $L \times L$ will be cropped out of a box centered on the target point $(x_0,\ y_0)$ with a length and width of $2L$ in a series of binary image streams. The SNN needs to input n frames of encoded images to obtain an inference result. In the experiments of this paper, n is taken as 16. Secondly, each candidate box must be classified whether it is the target or the background, so after accumulating 16 frames of binary images, the images of each candidate box are input to obtain the inference result that it is the target class. After the calculation of the l candidate boxes is completed in turn, compare the number of spikes of each box on the target class. The more the number of spikes, the greater the probability that the candidate box is the target, and its center coordinate $(x_1,\ y_1)$ is selected as the target position, then the next frame is centered on it. In this way, the tracking task of the target is completed by the method of classification, and at the same time, the problem that the current DVS target tracking dataset is missing and cannot be trained for the target position is avoided.

The candidate box is randomly generated and fixed at the

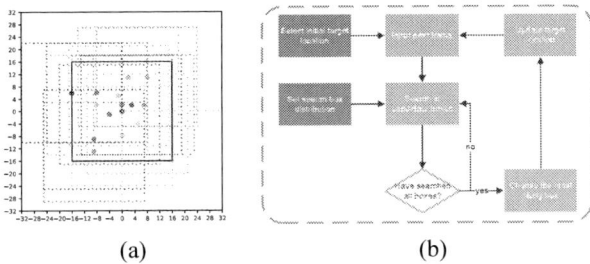

(a)   (b)

Fig. 2. (a) Candidate boxes distribution ($L=32, l=16, \sigma=8$), (b) workflow of tracking.

beginning of the search using a two-dimensional Gaussian distribution. Specifically, the probability density function of the offset $(\Delta x_i,\ \Delta y_i)$ relative to the target center position of the previous frame $(x_{t-1},\ y_{t-1})$ obeys the following equation:

$$f(\Delta x) = \frac{1}{\sigma\sqrt{2\pi}} e^{-\frac{\Delta x^2}{2\sigma^2}},\ as\ same\ as\ \Delta y \qquad (1)$$

In order to prevent the position of the candidate frame from overflowing the boundary, the overflow value needs to be truncated:

$$\Delta x_i = sign(\Delta x_i) \times L/2,\ if\ |\Delta x_i| > L/2$$
$$as\ same\ as\ \Delta y_i\ ,\ i \in \{1,2,...l\} \qquad (2)$$

where $sign(x)$ denotes the positive or negative signs of x. After filtering out the candidate box with the highest probability, add the corresponding offset to the original center point coordinates to get the coordinates of a new frame.

B. Network Structure

Since the target and background two classification network is used to detect the target, the required network structure is simple, and we defined the network with the following structure in Fig.3(a). We use Integrate-and-fire (IF) neurons to avoid the complex exponential calculations. This simple neuron model is also more conducive to simulation in hardware, and the neuron activation naturally replaces the ReLU function commonly used in CNNs. We only need to count the number of spikes during convolution and add the weight to the membrane potential.

Due to the lack of spike form dataset, we trained CNN by real-number image data first and then migrated to SNN using mapping rules in [3] to design the network. The network's training can be done easily using the proven training method of CNN networks, and the migration rules are adopted from the results in the paper.

IV. EXPERIMENT

We selected a gesture dataset Roshambo[4] for our experiment. To expand the dataset, we also manually captured some images using a grayscale sensor to synthesize the data in the form of DVS by the V2E tool[5] for training and test. After training the CNN, we transform the network to SNN by the method mentioned above. On this dataset, the SNN achieve a 98.66% classification accuracy.

After the network is trained on the software, we verified the network on the hardware platform to be deployed. This work uses a chip designed by our research group that can realize the SNN calculation of IF neurons. We build the behavior-level model of the hardware system on the software side. The overall process is shown in Fig.3(b). In our experiments, we use $L=32, l=16, \sigma=8$, and the network can achieve an object tracking of 1000 fps at 250M clock on the hardware simulator.

Input: binary image stream
Conv5-16, Spike count layer
Avg-pool/2
Conv5-16, Spike count layer
Avg-pool/2
Conv5-64, Spike count layer
FC-2

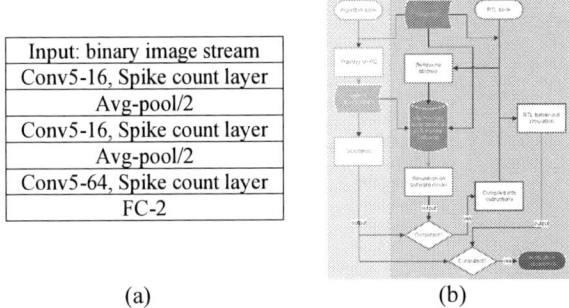

(a)   (b)

Fig. 3. (a) Network structure (Conv5-16 means 5x5 kernel size, and output channel is 16, FC-2 means fully connected layer with two outputs.), (b) Algorithm verification flow.

V. CONCLUSION

In this work, we complete the design of the algorithm on both software and hardware simulator. The proposed spike encoding module for DVS can stably control the information density algorithm and assist the SNN to output the results. The selective search method divides the candidate boxes into target and background. The algorithm achieves a tracking rate of 1000 fps on our hardware simulator.

REFERENCES

[1] G. Gallego *et al.*, "Event-Based Vision: A Survey," *IEEE Trans. Pattern Anal. Mach. Intell.*, vol. 44, no. 1, pp. 154–180, 2022.

[2] L. Bertinetto, J. Valmadre, J. F. Henriques, A. Vedaldi, and P. H. S. Torr, "Fully-Convolutional Siamese Networks for Object Tracking," in *Computer Vision – ECCV 2016 Workshops*, Cham, 2016, pp. 850–865.

[3] X. Yang, Z. Zhang, W. Zhu, S. Yu, L. Liu, and N. Wu, "Deterministic conversion rule for CNNs to efficient spiking convolutional neural networks," *Sci. China Inf. Sci.*, vol. 63, no. 2, p. 122402, Feb. 2020.

[4] I.-A. Lungu, F. Corradi, and T. Delbruck, "Live demonstration: Convolutional neural network driven by dynamic vision sensor playing RoShamBo," in *2017 IEEE International Symposium on Circuits and Systems (ISCAS)*, Baltimore, MD, USA, May 2017, pp. 1–1..

[5] Y. Hu, S.-C. Liu, and T. Delbruck, "v2e: From Video Frames to Realistic DVS Events," in *2021 IEEE/CVF Conference on Computer Vision and Pattern Recognition Workshops (CVPRW)*, Jun. 2021, pp. 1312–1321.

A Low Supply Sensitivity CMOS Temperature Sensor Using Dynamic-Distributing-Bias Circuit

Shichong Zhai[1,2], Wenchang Li[1,2], Jian Liu[1,2], Tianyi Zhang[1]

[1] Institute of Semiconductors, Chinese Academy of Sciences, Beijing, China
[2] University of Chinese Academy of Sciences, Beijing, China

Abstract—This work presents a low supply sensitivity CMOS temperature sensor using dynamic-distributing-bias circuits. A new hybrid PTAT/REF current generator is proposed to reduce the power consumption. Some techniques such as chopping, dynamic element matching (DEM) and ratiometric curvature correction are adopted to improve the temperature sensing accuracy. The sensor is designed and simulated in 0.153-µm CMOS process and occupies 0.07 mm² area. In the temperature range of −55 °C to 125 °C, the simulated temperature sensing accuracy is ±0.4°C after one-point calibration.

Keywords—CMOS Temperature Sensor, dynamic distributing bias, current mode

I. INTRODUCTION

Recently, temperature sensors integrated into passive radio frequency identification (RFID) tags are widely used in many applications such as perishable food monitoring and intelligent healthcare [1]. In these temperature-sensing RFID tags, the temperature sensor should have high accuracy such as ±1°C for food and environmental monitoring applications with single digit microwatt power consumption [2]. In addition, due to the poor stability of the DC voltage outputted by the electromagnetic wave energy harvesting circuit in RFID tag, the temperature sensor needs low supply sensitivity to meet the temperature sensing accuracy over the whole supply fluctuation range.

At present, the most common temperature sensing devices in temperature sensing RFID tags are bipolar junction transistors (BJT) because they can achieve high accuracy just using one point calibration with low power consumption. Recently, a current-mode temperature sensor using time-domain readout circuit realized ±0.85°C 3σ inaccuracy with 0.9 µW power consumption [3]. However, it has high voltage sensitivity due to the use of time-domain readout circuits. In order to save power consumption and area, a dynamic distributing current generator is used to produce proportional to absolute temperature (PTAT) and complementary to absolute temperature (CTAT) current dynamically [4]. However, it also has a relatively large voltage sensitivity due to the use of a simple supply-independent biasing circuit to generate PTAT or CTAT current.

In this paper, we propose a dynamic distributing bias current-mode temperature sensor. Unlike the above scheme in [4], we use a folded-cascode amplifier to build the hybrid PTAT/REF current generator which can reduce the power consumption and the voltage sensitivity. A special combination of PTAT and reference (REF) current is used to improve the used dynamic range of the analog to digital convertor (ADC). The sensor is designed and simulated in 0.153-µm CMOS process and occupies 0.07 mm² area. In the temperature range of −55°C to 125°C, the simulated temperature sensing accuracy is ±0.4°C after one point calibration.

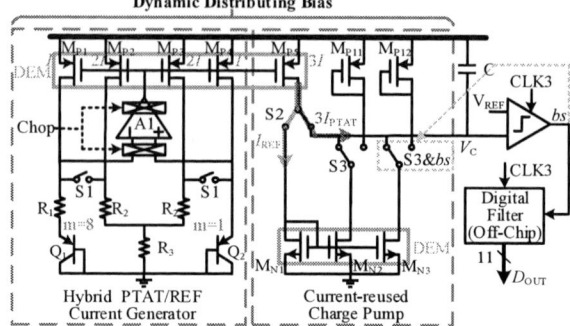

Fig. 1. Circuit schematic diagram of current-mode temperature sensor.

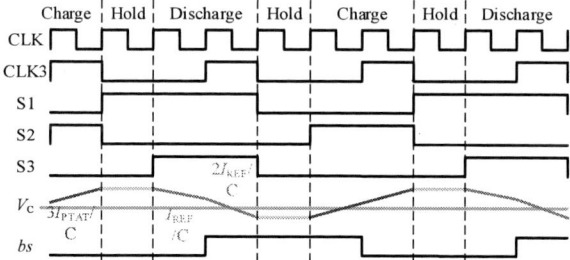

Fig. 2. Control signals of the temperature sensor.

II. DESIGN AND IMPLEMENTATION

As shown in Fig. 1, the proposed dynamic distributing bias temperature sensor is mainly composed of a hybrid PTAT/REF bias current generator, a current-reused charge pump, a capacitor and a comparator. The hybrid PTAT/REF bias current generator generates PTAT and REF currents dynamically under the control of S1. The current-reused charge pump is used to charge and discharge the capacitor by mirroring the PTAT or REF current. The control signals are shown in Fig. 2.

When the switch S1 is turned off, the collector voltages of transistor Q1 and Q2 are V_{BE1} and V_{BE2} respectively. Under the help of A1, the voltage on resistance R_1 is $\Delta V_{BE} = V_{BE1} - V_{BE2}$. The current flowing through R_1 is $I_{PTAT} = \Delta V_{BE} / R_1$ and the output current of M_{P5} is $3I_{PTAT}$. At this time, the switch S2 connects the M_{P5} and capacitor thus the capacitor is charged by this PTAT current. When the switch S1 is turned on, the current flowing through the resistor R_2 is $I_{CTAT} = V_{BE1} / (R_2 + R_3/2)$. The output current of M_{P5} is $I_{REF} = I_{PTAT} + I_{CTAT}$. This moment, the switch S2 connects the M_{P5} and M_{N1} thus the capacitor is discharged by this REF current through the current mirror M_{N2} and switch S3. Within one cycle of the comparator clock CLK3, the capacitor is charged and discharged one time respectively. The voltage on the capacitor C is compared with the reference voltage V_{REF} to generate the quantization signal bs. This signal controls the discharge current through the feedback circuit. When $bs=0$, the discharge current is I_{REF}. On the contrary, the discharge current is $2I_{REF}$ when $bs=1$. After N times of oversampling, the average charge in capacitor C is zero:

$$N \cdot \frac{3I_{PTAT}}{C} = N \cdot \frac{I_{REF}}{C} + M \cdot \frac{I_{REF}}{C} \qquad (1)$$

Where M is the number of high-level times in bs. By digitizing the pulse density of bs by using a counter, the digitized signal is:

$$X = \frac{M}{N} = \frac{3I_{PTAT} - I_{REF}}{I_{REF}} \qquad (2)$$

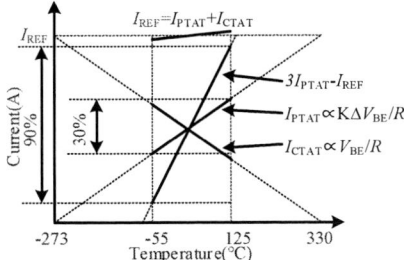

Fig. 5. Charge and discharge currents in this design

From this formula, we know that the temperature signal in the readout circuit is $3I_{PTAT}$ - I_{REF}. As shown in Fig. 3, the dynamic range utilization of the ADC is increased to 90%. The reference current is designed to have a slightly positive temperature coefficient to compensate the curvature in V_{BE} [5].

When the switch S1 is flipped, the output of the current generator is unstable. If this unstable current is delivered into the integrating capacitor, it will cause a large temperature error. Therefore, a holding period is added after the switch S1 is flipped. At this period, the control signals of S2 and S3 are low. That means the capacitor is not charged or discharged. As shown in Fig. 2, one cycle of CLK3 includes three CLK cycles. In these three cycles, the capacitor is charged, held and discharged once, respectively. In order to achieve better matching, the ratio of Q1 and Q2 is 1:8. M_{P1}-M_{P5} are cascode current mirrors which are not shown for simplicity. The DEM technology is used to reduce the mismatch of the MOSFETs. In addition, chopper stabilization technology is used in the amplifier A1 to reduce the influence of mismatch and 1/f noise. As the common mode voltage, the reference voltage V_{REF} does not need to be absolutely accurate. So it is generated by dividing the supply voltage of five diode-connected MOSFETs [6]. In order to ensure the temperature measurement accuracy under all corners, the capacitance of C is 14.4-pF. The oversampling rate of ADC is 2048 to achieve 11-bit resolution.

III. SIMULATION RESULTS

The proposed temperature sensor is designed and simulated in 0.153 μm CMOS technology. Fig. 4 shows the layout of the temperature sensor. The chip size is 0.1mm × 0.7mm. The simulation results show that the power consumption of the sensor is only 2.7 μW at 25°C when the supply voltage is 1.6 V.

Fig. 5(a) shows the simulation results of temperature accuracy in different corners (SS, TT, FF) after a single point calibration at 25°C. It shows that the systematic error of the temperature sensor is ±0.4°C in the range of -55 ~125°C.

Fig. 5(b) shows the simulation results of temperature accuracy in different supply voltage at 25°C. It can be seen from Fig. 5(b), when the supply voltage varies from 1.6 V to 2.0 V, the temperature variation is less than 0.1°C. It means that the simulated supply sensitivity is 0.25°C/V. Table I compares the performance with the prior art. Compared with other state-of-the-art temperature sensors, this sensor achieves relative high accuracy and low DC supply sensitivity with low area and power consumption.

IV. CONCLUSION

A dynamic distributing bias current-mode temperature sensor is proposed. It uses a folded-cascode amplifier to build the hybrid PTAT/REF current generator which can reduce the power consumption and the voltage sensitivity. By using a special combination of PTAT and REF currents, the dynamic

Fig. 3. Chip layout of temperature sensor

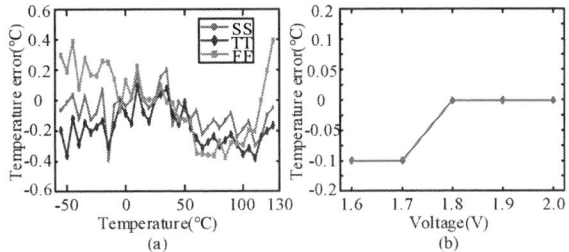

Fig. 4. (a) Temperature error in different corner (SS, TT, FF) after 25 °C single point calibration. (b) Temperature error in different supply voltage at 25°C

TABLE I. COMPARISON WITH STATE-OF-THE-ART TEMPERATURE SENSORS

Parameters	This Work	J. Yin [1]	K. Souri [2]	B. Wang [3]
Technology (μm)	0.153	0.18	0.16	0.18
Area (mm²)	0.07	1.1	0.08	0.1
Supply(V)	1.6 to 2.0	1.0	1.5 to 2.0	1.6 to 2.0
Nominal current(μA)*	1.7	0.9	3.4	0.54
Power (μW)*	2.7	0.9	5.1	0.9
DC supply sensitivity(°C/V)	0.25	N/A	0.5	0.7
Temp. Range (°C)	-55 to 125	-20 to 30	-55 to 125	-30 to 120
Inaccuracy (°C)	±0.4	±0.8	±0.15	±0.85
Resolution (m°C)	100	350	20	39

* Do not include clock and digital backend power;

range utilization of the ADC is increased to 90%. The sensor shows a simulated inaccuracy within ±0.4°C from −55°C to 125°C after one point calibration. It just consumes 2.7μW power consumption at 25°C. With the minimal supply sensitivity of 0.25°C/V, it proves to be suitable for wireless temperature-sensing applications.

REFERENCES

[1] J. Yin et al., "A system-on-chip EPC Gen-2 passive UHF RFID tag with embedded temperature sensor," *IEEE J. Solid-State Circuits*, vol. 45, no. 11, pp. 2404–2420, Nov. 2010.

[2] K. Souri, Y. Chae, and K. A. A. Makinwa, "A CMOS temperature sensor with a voltage-calibrated inaccuracy of ±0.15 °C (3σ) from -55 °C to 125 °C," *IEEE J. Solid-State Circuits*, vol. 48, no. 1, pp. 292–301, Jan. 2013.

[3] B. Wang, M. K. Law, C. Y. Tsui and A. Bermak, "A 10.6 pJ·K² resolution FoM temperature sensor using a stable multivibrator," *IEEE Trans. Circuits Syst. II Exp. Briefs*, vol. 65, no. 7, pp. 869-873, July 2018.

[4] Y. C. Hsu, C. L. Tai, M. C. Chuang, A. Roth and E. Soenen, "An 18.75 μW dynamic-distributing-bias temperature sensor with 0.87 °C (3σ) untrimmed inaccuracy and 0.00946 mm² area," *IEEE ISSCC*, pp. 102-103, 2017.

[5] M A P Pertijs, J H Huijsing, *Precision temperature sensors in CMOS technology*, Dordrecht: Springer Netherlands, 2006, pp.51–54.

[6] Z. Tang, Y. Fang, Z. Huang, X.-P. Yu, Z. Shi and N. N. Tan, "An untrimmed BJT-based temperature sensor with dynamic current-gain compensation in 55-nm CMOS process," *IEEE Trans. Circuits Syst. II Exp. Briefs*, vol. 66, no. 10, pp. 1613-1617, Oct. 2019.

A Reconfigurable SRAM Computing-in-Memory Macro Supporting Ping-Pong Operation and CIM pipeline for Multi-mode MAC operations

Kanglin Xiao[1,2], Xiaoxin Cui[1]*, Xin Qiao[1], Xin'an Wang[1,2], Yuan Wang[1]*

[1]Key Laboratory of Microelectronic Devices and Circuits (MoE), School of Integrated Circuits, Peking University
[2]Key Laboratory of Integrated Microsystem, School of ECE, Peking University Shenzhen Graduate School

Abstract—In this work, we present a reconfigurable SRAM computing-in-memory (CIM) macro supporting ping-pong operation and pipeline operation for multi-mode multiply-and-accumulate (MAC) operations. The macro can be reconfigured to execute AND or XNOR, offering great flexibilities to cover binary neural network (BNN), ternary neural network (TNN), and multi-bit operation through serially 1-bit AND operations. The main contributions include: (1) A reconfigurable scheme to map inputs and weight of 8T1C bit-cell, supporting three MAC operations; (2) An architecture integrated ping-pong operation and two-level CIM pipeline. Simulated in a standard 28-nm process, the proposed design shows good computing linearity and variations. The average energy efficiency of 1b-AND, BNN, and TNN MAC operation are 1533.7, 1522.9, and 1713.2 TOPS/W, respectively.

Keywords—computing-in-memory (CIM), multi-mode, ping-pong operation, pipeline design.

I. INTRODUCTION

Computing-in-memory (CIM) is an approach to overcome Von Neumann's bottleneck and improve energy efficiency. Prior SRAM CIM works [1]-[6] show significant energy efficiency and throughput advantages over conventional architectures. However, CIM has less flexibility which limits its application in deep neural networks (DNNs). Works[1]-[6] tried to improve the flexibility of CIM. [1] proposed a local computing cell and 6T cell-based CIM macro capable of 8-b multiply-and-accumulate (MAC) operation and near full-precision output for a variety of multibit neural networks (NN). [2] presents a programmable NN inference accelerator based on scalable CIM, overcoming the challenges of scalability. [3] proposed a ping-pong CIM to increase performance and reduce execution time, trying to map large NN models. 8T1C is a compact CIM bit-cell for robust charge-domain computing, which has fewer transistors and a moderate dynamic range. It was proposed in [5] for BNN implementation only, which lacks enough flexibility. To further increase the flexibility of CIM, in this paper, we proposed a 64Kb SRAM CIM macro with a reconfigurable mapping scheme of 8T1C bit-cell, supporting BNN, TNN, and multi-bit operations. We also proposed an architecture integrated ping-pong operation and two-level pipeline CIM, aiming to reduce executive time and increase the supportability of different scale NN models.

II. PROPOSED RECONFIGURABLE CIM MACRO

A. Macro architecture

Fig. 1 shows the proposed CIM architecture. It consists of a 256×256 8T1C bit-cell array, an input driver and configurator, a switch array for mode reconfiguration, a MUX array for pipeline design, a 1×128 4b SAR ADC array, and others. With different configurations, the macro supports 1b-AND MAC, BNN, TNN, and multi-bit MAC operations. All input lines (IP, IPB, IN, INB) and SRAM WLs for odd and even columns of each row are split, to achieve ping-pong operations. In CIM mode, the 256 input activations are fed to each row through row-wise buffers and multiplied with the weights in the 8T1C bit-cell in the charge domain. And then the accumulations are accomplished by capacitive coupling between 256 rows.

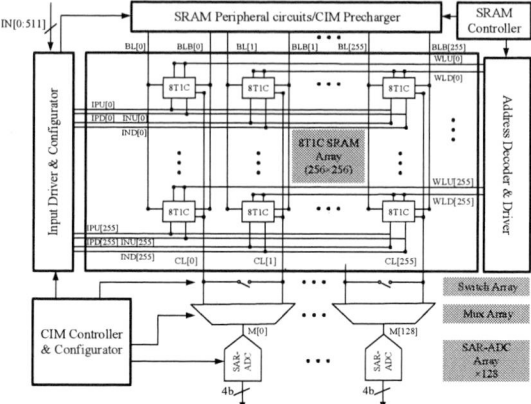

Fig.1. Proposed SRAM-CIM macro architecture

AND MAC Operation Table

Phase	Input	IP_U	IN_U	Weight	Q	QB	Result	Vc
Reset	×	GND	GND	1/0	1/0	0/1	×	GND
Eval.	0	GND	GND	0	GND	VDD	0	GND
	0	GND	GND	1	VDD	GND	0	GND
	1	VDDB	GND	0	GND	VDD	0	GND
	1	VDDB	GND	1	VDD	GND	1	VDDB

BNN MAC Operation Table

Phase	Input	IP_U	IN_U	Weight	Q	QB	Result	Vc
Reset	×	GND	GND	+1/-1	1/0	0/1	×	GND
Eval.	-1	GND	VDDB	-1	GND	VDD	+1	VDDB
	-1	GND	VDDB	+1	VDD	GND	-1	GND
	+1	VDDB	GND	-1	GND	VDD	-1	GND
	+1	VDDB	GND	+1	VDD	GND	+1	VDDB

Fig.2. Proposed mapping scheme of 8T1C bit-cell for AND and BNN MAC

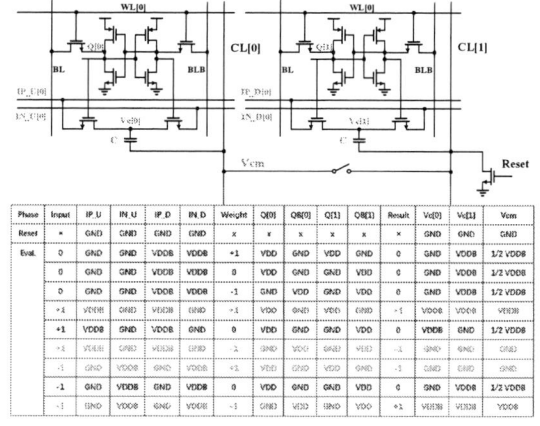

Phase	Input	IP_U	IN_U	IP_D	IN_D	Weight	Q[0]	QB[0]	Q[1]	QB[1]	Result	Vc[0]	Vc[1]	Vcm
Reset	×	GND	GND	GND	GND	×	x	x	x	x	×	GND	GND	GND
Eval.	0	GND	GND	VDDB	VDDB	+1	VDD	GND	GND	VDD	0	GND	VDDB	1/2 VDDB
	0	GND	GND	VDDB	VDDB	0	VDD	GND	GND	VDD	0	GND	VDDB	1/2 VDDB
	0	GND	GND	VDDB	VDDB	-1	GND	VDD	GND	VDD	0	GND	VDDB	1/2 VDDB
	+1	VDDB	GND	VDDB	GND	+1	VDD	GND	GND	VDD	-1	VDDB	GND	1/2 VDDB
	+1	VDDB	GND	VDDB	GND	0	VDD	GND	GND	VDD	0	VDDB	GND	1/2 VDDB
	+1	VDDB	GND	VDDB	GND	-1	GND	VDD	GND	VDD	-1	GND	GND	GND
	-1	GND	VDDB	GND	VDDB	+1	VDD	GND	GND	VDD	0	GND	VDDB	1/2 VDDB
	-1	GND	VDDB	GND	VDDB	-1	GND	VDD	GND	VDD	+1	VDDB	VDDB	VDDB

Fig.3. Proposed mapping scheme for TNN MAC

B. Reconfigurable mapping scheme based on 8T1C

Fig.2 and Fig.3 show the proposed reconfigurable mapping scheme of 8T1C bit-cell. With changes in CIM mapping of inputs and weights, the bit-cell supports three modes: 1b-AND, BNN, and TNN. 8T1C bit-cell computes through the pass gates and accumulation via the mechanism of capacitive coupling. The computing operation includes two phases: reset and evaluation. In the reset phase, both the top and bottom plates of cap C(~1.33fF) are reset to GND. After reset, the input is applied to IP_U/IN_U. Based on the selected mode, the top

---

This work was supported by the National Key Research and Development Program of China (Grant No. 2018YFE0203801). Corresponding authors: Xiaoxin Cui (cuixx@pku.edu.cn) and Yuan Wang (wangyuan@pku.edu.cn).

plate of C will be charged to VDDB (VDDB = 0.6V) or kept in GND according to the computing result.

C. Integrated Ping-pong and pipeline architecture

To reduce the executive time, we proposed an architecture integrated ping-pong operation and a two-level CIM pipeline. The ping-pong operation with simultaneous weight updating and CIM operation is for large NN models while the two-level pipeline between analog MAC and ADC readout is for small NN models. The proposed architecture is shown in Fig.4. It consists of two columns of bit-cells, a multiplexer, and a SAR-ADC. The WLs and input lines of two columns are split. In ping-pong mode, one column computes while another column updates the weights. When the longer CIM operation is complete, they switch working modes. The ping-pong principle is shown in Fig.5. In CIM pipeline mode, there are two levels: analog MAC and ADC readout. The two columns work in analog MAC mode alternately. Once a column completes analog MAC, the mux channel will be gated, and the column will switch to the ADC readout phase. The two-level pipeline principle is also shown in Fig.5. Both ping-pong and pipeline designs achieve a significant reduction in CIM executive time.

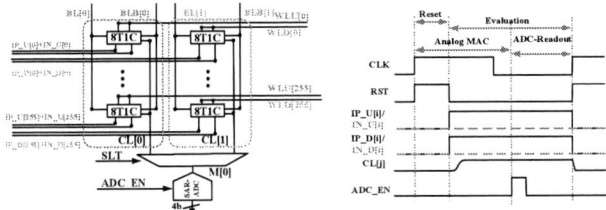

Fig.4. Proposed Ping-pong and two-level pipeline architecture

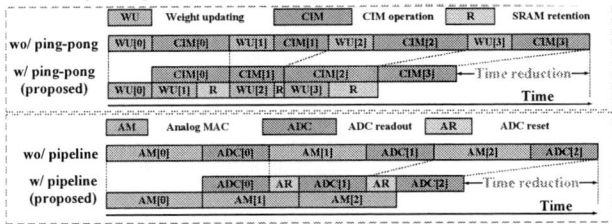

Fig.5. Ping-pong and wo-level pipeline operation principle

III. SIMULATION RESULTS

The proposed SRAM-CIM macro is designed and fulfilled in a standard 28-nm CMOS process. We conducted the following simulations. First, we simulate the transfer function without ADCs in Fig.6. In different combinations of temperature and process corner, the transfer function shows no obvious difference, which means temperature and process-related non-ideality has a small impact on its stability. Second, the overall transfer probability function is illustrated in Fig.7. The results are extracted from 50 points MC simulation. Fig.8 shows the energy breakdown in different CIM modes and the energy efficiency when sweeping the MAC value. The average energy efficiency (normalized to 1/1b) of 1b-AND MAC, BNN, and TNN(parallelism = 128) are 1533.7, 1522.9, and 1713.2 TOPS/W, respectively. Table.I. shows the comparison of recent SRAM-CIM works.

IV. CONCLUSION

In this work, we present an SRAM CIM macro with a reconfigurable mapping scheme based on 8T1C bit-cell, supporting BNN, TNN, and multi-bit MAC operations. An architecture integrated ping-pong operation and two-level CIM pipeline is proposed for different scale NN models to reduce executive time and increase performance. The proposed CIM

macro shows high energy efficiency, good linearity, and small variations in simulations.

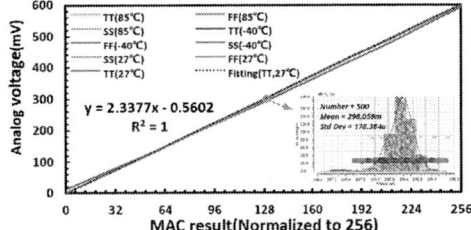

Fig.6. Simulated transfer function(wo/ ADC)

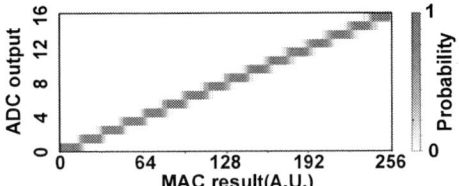

Fig.7. Transfer probability function (w/ ADC)

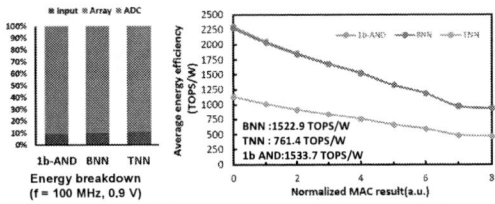

Fig.8. Energy breakdown and energy efficiency vs MAC result.

TABLE I. PERFORMANCE COMPARISON WITH OTHER CIMs

	[3]	[4]	[5]	[6]	Proposed
Process(nm)	65	65	65	40	28
Array size	2352 Kb	16 Kb	16 Kb	384 Kb	64 Kb
CIM mode	AND	BNN/TNN⁴	BNN	AND	AND/BNN/TNN
Input (b)	2/4/6/8	1/1.5	1	4/8	1/1.5
Weight (b)	1-8	1	1	4/8	1/1.5
Output (b)	32	3.46	5	12/20	4
Parallelism	128	256	256	16	512/256/128
Ping-pong	✓	✗	✗	✗	✓
CIM pipeline	✗	✗	✗	✓	✓
Energy Efficiency (TOPS/W)⁵	740.8⁶	604.5	671.5	964.5-1509³	1533.7(1b-AND)¹ 1522.9(BNN)¹ 1713.2(TNN)²
Simulated(S)/ measured (M)	M	M	M	M	S

¹Parallelism=256, ²Parallelism=128, ³Based on the result of 4/4b input/weight, ⁴Tenary input, binary weight, ⁵Normalized to 1/1b input/weight operation, not normalized to the technology, ⁶Macro level.

REFERENCES

[1] X. Si et al., "A Local Computing Cell and 6T SRAM-Based Computing-in-Memory Macro With 8-b MAC Operation for Edge AI Chips," *JSSC*, vol. 56, no. 9, pp. 2817–2831, Sep. 2021.

[2] H. Jia et al., "15.1 A Programmable Neural-Network Inference Accelerator Based on Scalable In-Memory Computing," *ISSCC*, Feb. 2021, vol. 64, pp. 236–238.

[3] J. Yue et al., "A 2.75-to-75.9TOPS/W Computing-in-Memory NN Processor Supporting Set-Associate Block-Wise Zero Skipping and Ping-Pong CIM with Simultaneous Computation and Weight Updating," *ISSCC*, Feb. 2021, vol. 64, pp. 238–240.

[4] S. Yin et al., "XNOR-SRAM: In-Memory Computing SRAM Macro for Binary/Ternary Deep Neural Networks," *JSSC*, vol. 55, no. 6, pp. 1733–1743, Jun. 2020.

[5] Z. Jiang et al., "C3SRAM: An In-Memory-Computing SRAM Macro Based on Robust Capacitive Coupling Computing Mechanism," *JSSC*, vol. 55, no. 7, pp. 1888–1897, Jul. 2020.

[6] J.-W. Su et al., "16.3 A 28nm 384kb 6T-SRAM Computation-in-Memory Macro with 8b Precision for AI Edge Chips," *ISSCC*, Feb. 2021, vol. 64, pp. 250–252.

Floorplanning and Power/Ground Network Design for A Programmable Vision Chip

Haozhe Xu[1,2,3], Siyuan Wei[2], Nan Qi[2,3], Peng Wu[1,3],
Jian Liu[2], Nanjian Wu[2,3], Liyuan Liu[2,3], Shuangming Yu[2,*]

[1] Aerospace Information Research Institute, Chinese Academy of Sciences, Beijing, China
[2] Institute of Semiconductors, Chinese Academy of Sciences, Beijing, China
[3] School of Electronic, Electrical and Communication Engineering, University of Chinese Academy of Sciences, Beijing, China
Email: yushuangming@semi.ac.cn

Abstract—A programmable vision chip integrates a high-speed image sensor and a vision processor on one single chip. It can be adapted to both traditional CV (computer vision) algorithms and deep learning algorithms efficiently. The chip is compute-intensive and memory-intensive, physical design of the chip encounters challenges. This article will introduce a floorplanning and power planning approach appropriate for this large-scale vision chip, limit the influence of IR drop, and finally compare the performance of this experimental result with state-of-the-art chips.

Keywords—floorplanning, power design, vision chip, physical design

I. INTRODUCTION

Floorplanning and power design are the foundation of physical design, it directly decides results and quality of placement and route. A programmable vision chip integrates a high-speed image sensor and a vision processor on one single chip. It can be adapted to both traditional CV (computer vision) algorithms and deep learning algorithms efficiently. Whether it is a traditional CV algorithm, or a neural network algorithm based on deep learning, it has a certain scale of computation, and image processing based on deep learning is characterized by compute-intensive and memory-intensive.

This paper will describe floorplanning and power planning for the vision chip with large storage capacity. Large-scale on-chip data memory and a large number of input and output interfaces are required to support it. 5MB of local memory is required to independently complete the corresponding algorithms without interacting with external memory. The vision chip also integrates 3 configurable analog IP blocks, as well as a coprocessor block for accelerated object detection. These determine that the core size will be large, which will cause IR drop. The chip is fabricated in a 65nm process. The experimental results show that the maximum clock frequency and die size are 200MHz and $10.2 \times 9.6 mm^2$ respectively.

II. FLOORPLAN

A. Architecture of the Chip

According to the functional division, the vision chip can be roughly divided into five parts: computing processor center, data transmission path, system control module, on-chip storage, and data interface, the architecture of the chip as shown in Fig .1. The computing processor center is mainly composed of a vector processor and a co-processor, which is responsible for algorithm calculation and controlled by the instruction programming system; the data transmission path is responsible for loading off-chip instructions, parameters, and images into on-chip memory; the system control module is the MCU, which

This work is supported in part by the National Key Research and Development Program of China under Grant 2019YFB2204303, the National Natural Science Foundation of China under Grant (U21A20504, U20A20205,62134004), and the Youth Innovation Promotion Association Program Chinese Academy of Sciences under Grant 2021109.

is the main control unit of the vision chip and undertakes data scheduling and status monitoring; on-chip memory contains data memory and some register-files, data memory holds the parameters and intermediate data required for the operation of the algorithm, and some temporary data will be temporarily stored in the register-files to achieve parallel processing and reuse of data; data interface includes High-speed data interface for transmitting images and QSPI interface for transmitting instructions[1].

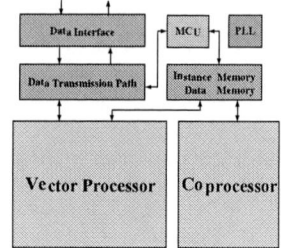

Fig.1. The architecture of the chip.

B. Floorplanning of IO and SRAM

The determination of die size is a top priority for floorplanning. Cell count, memory size, target utilization and the range of pads will have an effect on the dedication of the die size. The vision chip incorporates a million-level cell, a whole of 5MB of memory, and 50 signal pads. The ideal aspect ratio is 1, and the nearer the die form is to the square, the friendlier it is to the back-end implementation. After quite a few rounds of P&R iteration, the core size was finalized to be $9.6 \times 9.2 mm^2$. The distance from core to boundary is set to 300/200μm because a pad has a height of 135μm, and room is reserved for power rings.

The vision chip integrates three configurable IPs. High-speed data interface RX is positioned in the middle of the bottom edge of the chip, for receiving image data from the image sensor; High-speed data interface TX is used to output the data of the vision chip outward, together with the PLL clock IP placed in the middle of the top edge of the chip, The 50 I/O pads are distributed on the left and right sides of the chip boundary, input pad is evenly positioned on the left side of the chip, output and reset signal pads are evenly positioned on the right, and the pad of test clock is positioned subsequent to the PLL.

To achieve varieties of algorithms such as face detection image classification and object tracking, a complete of 5MB of on-chip data memory is required, inclusive of 4.5MB of vector processor (VP) data memory for global use, and local memory includes coprocessor memory, VP instruction memory, MCU instruction memory, resize memory and IO buffer memory. Multi-level distributed memory poses a challenge for floorplanning.

This paper proposes that the benefits of floorplanning are: Placing main data memory on the periphery is conducive to the data access of other modules; Leave the core region resources of the core to standard cell placement and establishment of the clock tree routing resources are more even and the utilization rate of the chip edge is effectively improved.

The floorplanning of chip is proven in Fig .2. VP-dm is positioned on the periphery of the core for global use, we put two processor cores in the center of the layout, VP-dm is placed round the processor in a circle manner, instruction register and image resizing block are positioned on the periphery. 4.5MB of VP data memory according to the VP instruction address is divided into high, medium, low three levels. Whether it is a vector processor, scalar processor, or co-processor, all data required for computation is integrated in 4.5MB SRAM, the final processing result is also stored back in this memory, for all processors to access data, so the SRAM is placed on the periphery of the core for reducing the length of path to access data. In the medium/low group, in order to make sure the shortest wiring, the equal wide variety of a team of SRAM, the top two orientation for R0, pin path down, the following two SRAM orientation for MX, the route of the pin is upwards, and the spacing between them is 100μm, making full use of the spacing of the core, place up to 7 units of SRAM in a row, for a whole size of 9421μm. Location of coprocessor memory is placed close to VP-dm；The instruction memory of MCU and VP is positioned in the lower-left corner of the core, assigning the duties of data scheduling and status monitoring.

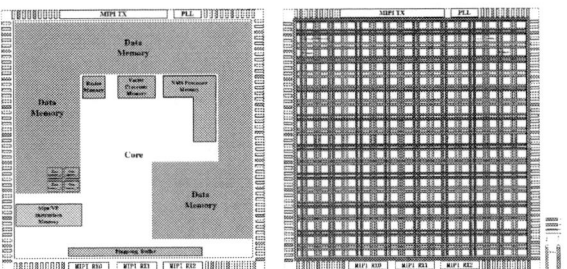

Fig.2. The floorplanning of the chip. Fig.3. The power planning of the chip

III. POWER/GROUND NETWORK DESIGN

Due to the significant IR-drop, supply voltage may be lower than the ideal reference. This impact may weaken the driving capability of logic gates, limit circuit performance and minimize the maximum clock frequency that can be performed.

This article will decrease the IR drop of large-scale programmable vision chips from the settings of power pads, power rings, and power stripes. The current flow direction is from PG IO pad to top PG net to low layer PG net and finally to standard cell, so in order to decrease the IR drop of the chip, it is imperative for both the PG IO pad and the top layer PG net design. Due to the limitation of chip area, side length and the minimum spacing of the pad, the number of PG IO cannot be too much, and spacing between pads is set to 100μm. A complete of 189 PG IO pads are placed, a total of 159 PVDD1 pads are placed, equal spacing is evenly positioned on the four sides of the chip, to ensure that each signal IO next to a set of PG IO pads and so that the chip core from any angle has a power input pad. The estimated chip power consumption is less than 1W, the core voltage of the chip is 0.9 ~ 1.1V, and the drive current of a single PG IO is 12mA (according to the standard unit manual), at least 93 PG IO pads are required.

PG nets are selected top metal for routing, top metal has a larger line width, smaller resistance. Under the condition of flowing through the same current, the voltage drop is smaller. Power ring is selected M9 and M8 two layers of metal wiring, considering the large area of the chip core, power rings need to be placed as much as possible, and consider the limitations of boundary, a whole of 4 power rings are placed ultimately. The setting of the power strips is hierarchical and wired with M6~M9 metal, firstly, current is transmitted from power ring to top metal stripe, power stripe is covered with M8 and M9 layers, width is set to 12μm maximum, spacing is set to 2μm, middle metal stripe uses M6, M7 routing, considering that routing of standard cell is M1 ~ M7, so the width cannot be set too large, width is set to 3μm, spacing is 11μm; Sroute setting is applied to connect the standard cell to the power stripes, which

completes power planning, from PG IO to standard cell. The power planning of chip is shown in Fig .3.

IV. EXPERIMENTAL RESULTS

Floorplan is completed, start placement, clock tree synthesis, route, and routing metal layers are M1 ~ M7, and clock concurrent optimization for the setup and hold rules to repair, eventually there are no violations. In EDA tools for IR drop simulation, the simulation consequences are below 82mV, and in Prime-Time simulation of timing and power consumption for MMMC, timing is simulated at high temperature, high voltage, low temperature, low voltage, in different corners, no timing violations, and power simulation result is 508mW in Fig.4.

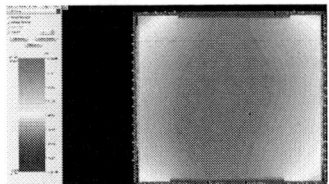

Fig.4. The simulation result of IR drop.

V. CONCLUSION

This brief proposed floorplanning and power/ground network design for large-scale programmable vision processors. On the processor, MCU is accountable for the master control, Data Memory assumes the duty of on-chip data storage, VP core and coprocessor correspond to the computing task. Processor is memory-intensive and applied at the edge, so there are strict requirements for die size and power consumption. This article targeted floorplanning that included global and local whole 5MB of multi-level distributed data memory, 50 signal IO, 189 PG IO, and 3 IPs. Power planning from PG IO to standard cell, hierarchical P/G design reduces the impact of IR drop on chip performance. The vision chip with the die size of $10.2 \times 9.6 mm^2$ in 65nm CMOS process can work at a clock frequency of 200 MHz.Table I. gives the performance of our chip and compares it with state-of-the-art designs.

TABLE I.   COMPARISON WITH STATE-OF-THE-ART DESIGNS

	Eyeriss-v2 '19[2]	ISSCC '21 [3]	ISSCC '21[4]	This work
Technology(nm)	65	28	22	65
Area (mm²)	2695k gates	3.8	7.56×8.2	10.2×9.6
On-chip SRAM	246 kB	872 kB	9 MB	5 MB
Clock frequency (MHz)	200	288	262.5	200
Voltage (V)	-	0.9 (Min 0.6)	0.8	0.9~1.1
Power （mW）	-	228	-	508 (Simulation)

REFERENCES

[1] Q. Luo et al. A Programmable and Flexible Vision Processor[J]. IEEE Transactions on Circuits and Systems II: Express Briefs, 2022, 69(9): 3884-3888.

[2] Chen Y H, Yang T J, Emer J, et al. Eyeriss v2: A flexible accelerator for emerging deep neural networks on mobile devices[J]. IEEE Journal on Emerging and Selected Topics in Circuits and Systems, 2019, 9(2): 292-308.

[3] Shah N, Olascoaga L I G, Zhao S, et al. 9.4 PIU: A 248GOPS/W Stream-Based Processor for Irregular Probabilistic Inference Networks Using Precision-Scalable Posit Arithmetic in 28nm[C]//2021 IEEE International Solid-State Circuits Conference (ISSCC). IEEE, 2021, 64:150-152.

[4] Eki R, Yamada S, Ozawa H, et al. 9.6 A 1/2.3 inch 12.3 Mpixel with On-Chip 4.97 TOPS/W CNN Processor Back-Illuminated Stacked CMOS Image Sensor[C]//2021 IEEE International Solid-State Circuits Conference (ISSCC). IEEE, 2021, 64: 154-156.

An Analytical Model for doping effect in charge-plasma-based TFET

Wenbo Li, Qian Xie*, Zheng Wang

University of Electronic Science and Technology of China

Abstract—In this paper, an analytical model for electrostatic doping effect in the charge-plasma-based TFET (CP-TFET) is proposed by solving Poisson's equation incorporating the mobile charge term. Closed forms of vertical potential and electrostatic doping concentration in CP-TFETs are developed. Meanwhile, the impacts of the electrode work functions and substrate thicknesses on them are analyzed. This predicted electrostatic doping concentration agrees well with the Sentaurus TCAD simulation results of the CP-TFET.

Keywords—analytical model, electrostatic doping, CP-TFET.

I. INTRODUCTION

Tunneling field-effect transistor (TFET) enables a steeper sub-threshold swing than the 60mV/dec limit of the conventional MOSFET [1]. However, physically doped TFETs with highly-doped abrupt junctions are challenging to be implemented. CP-TFETs have been proposed and investigated [2] based on the concept of charge plasma [3]. In CP-TFETs, physical doping is replaced by the electrostatic doping to form the drain electrostatic doping region and source electrostatic doping region in the undoped silicon. The electron and hole plasmas are induced by employing the electrodes with different work functions which tune the positions of Fermi level in the undoped silicon to the conduction band or valence band as a result of the band alignment near the interface.

An analytical model based on an insight into doping effect in CP-TFETs is proposed in this paper. We analyze the impacts of electrode work functions and substrate thicknesses for the equivalent electrostatic doping concentration. This model was simulated and verified by Sentaurus TCAD.

II. MODEL DEVELOPMENT

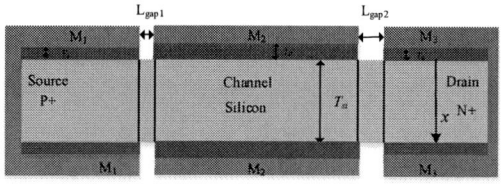

Fig. 1 Schematic of the studied CP-TFET.

Fig. 1 illustrates the cross-sectional of a symmetrical CP-TFET. The source, channel, and drain are physically undoped. The electrodes covering the source and drain with appropriate work functions excite charge in the substrate through thin insulating layers. The insulating material used in this CP-TFET is chosen as HfO_2 ($\varepsilon=22$) for the source and drain electrostatic doping regions to enhance the control of the charge in the substrate and reduce the charge leakage. The P^+ electrostatic doping source and N^+ electrostatic doping drain are framed on the intrinsic silicon substrate by using platinum (M_1) and aluminum (M_2) electrodes. The simulated electrostatic doping concentrations of the CP-TFET are shown in Fig. 2.

Corresponding Author: Qian Xie, email: xieqian@uestc.edu.cn

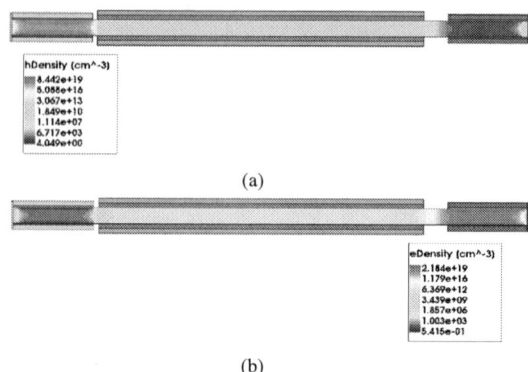

Fig. 2 The carrier concentration diagrams obtained from TCAD simulations. (a) the hole concentration, (b) the electron concentration.

As the silicon is physically undoped, the depletion charge can be ignored. The silicon Fermi level is mainly determined by the electrode work function. Setting the vertical center of silicon as the origin, we can write the vertical electrostatic doping concentration as a function of vertical potential:

$$N_p(x) = N_i e^{-q\varphi(x)/kT} \tag{1}$$

$$N_n(x) = N_i e^{q\varphi(x)/kT} \tag{2}$$

where N_i is the intrinsic carrier density. Poisson's equation with the electron as an example can be written as [4]:

$$\frac{d^2\varphi}{dx^2} = \frac{q}{\varepsilon_{si}} N_i e^{q\varphi/kT} \tag{3}$$

the ε_{si} is the silicon permittivity. Since the symmetrical structure, the boundary condition can be calculated as:

$$d\varphi / dx \,|_{x=0} = 0. \tag{4}$$

Proposing the vertical center potential $\varphi|_{x=0} = \varphi_m$ as the boundary condition, which will be solved later, we integrate the (1) and obtain:

$$\frac{d\varphi}{dx} = \sqrt{\frac{2kTN_i \left(e^{\frac{q\varphi}{kT}} - e^{\frac{q\varphi_m}{kT}} \right)}{\varepsilon_{si}}} \qquad \frac{T_{si}}{2} \leq x \leq 0 \tag{5}$$

$$\frac{d\varphi}{dx} = -\sqrt{\frac{2kTN_i \left(e^{\frac{q\varphi}{kT}} - e^{\frac{q\varphi_m}{kT}} \right)}{\varepsilon_{si}}} \qquad -\frac{T_{si}}{2} \leq x \leq 0. \tag{6}$$

Integrating (4) and (5), we obtain the vertical potential as a function of the position:

$$\frac{q(\varphi - \varphi_m)}{2kT} = -\ln\left[\cos\left(q\sqrt{\frac{N_i}{2\varepsilon_{si}kT}} e^{\frac{q\varphi_m}{2kT}} x \right) \right]. \tag{7}$$

Equation (6) is valid over the range $-T_{si} \leq x \leq T_{si}$. We propose boundary condition $\varphi|_{x=\pm T_{si}/2} = \varphi_b$, and consider the electric displacement continuity equation on the interface between the silicon and the insulating layer:

$$\frac{(-\phi_{ms} - \varphi_b)\varepsilon_i}{T_i} = \varepsilon_{si}\frac{d\varphi}{dx}|_{x=T_{si}/2} = -\varepsilon_{si}\frac{d\varphi}{dx}|_{x=-T_{si}/2} \tag{8}$$

here ϕ_{ms} is the difference in the work functions between metal and semiconductor, ε_i is the insulator permittivity, T_i is the thickness of the insulating layer. The right-hand side of (7) has been calculated from (3) and (4). We derive the following equation:

$$\frac{(-\varphi_{ms} - \varphi_b)\varepsilon_{ox}}{T_i} = \sqrt{2kTN_i(e^{\frac{q\varphi_b}{kT}} - e^{\frac{q\varphi_m}{kT}})\varepsilon_{si}} \tag{9}$$

$$\frac{q(\varphi_b - \varphi_m)}{2kT} = -\ln\left[\cos\left(q\sqrt{\frac{N_i}{2\varepsilon_{si}kT}}e^{\frac{q\varphi_m}{2kT}} \pm \frac{T_{si}}{2}\right)\right]$$

(10)

The φ_b and φ_m can be solved by (8) and (9). The vertical potential for both the electron and hole is derived as:

$$\varphi_p(x) = 2\frac{kT}{q}\ln\left[\cos\left(q\sqrt{\frac{N_i}{2\varepsilon_{si}kT}}e^{-\frac{q\varphi_m}{2kT}}x\right)\right] + \varphi_m \quad (11)$$

$$\varphi_n(x) = -2\frac{kT}{q}\ln\left[\cos\left(q\sqrt{\frac{N_i}{2\varepsilon_{si}kT}}e^{\frac{q\varphi_m}{2kT}}x\right)\right] + \varphi_m. \quad (12)$$

The electrostatic doping concentration can be written as:

$$N_{edp}(x) = N_i \exp\left\{2\ln\left[\cos\left(q\sqrt{\frac{N_i}{2\varepsilon_{si}kT}}e^{-\frac{q\varphi_m}{2kT}}x\right)\right] - \varphi_m\right\} \quad (13)$$

$$N_{edn}(x) = N_i \exp\left\{-2\ln\left[\cos\left(q\sqrt{\frac{N_i}{2\varepsilon_{si}kT}}e^{\frac{q\varphi_m}{2kT}}x\right)\right] + \varphi_m\right\}. \quad (14)$$

III. VERIFICATION AND DISCUSSIONS

The device structure is shown in Fig. 1. The gate length is set to 200 nm to avoid the short channel effect, and The insulating material used in the simulation is HfO$_2$. The main device variables considered in the simulated CP-TFET are demonstrated in Table I. The nonlocal BTBT model and Shockley-Read-Hall (SRH) recombination models are adopted. The modeled outcomes are compared with Sentaurus TCAD tool outcomes to verify and authenticate the results.

TABLE I. SIMULATION PARAMETERS

Variables	Value
Length of channel region	200nm
Length of source and drain region	50nm
Gate work-function	4.5eV
Thickness of insulating layer of source and drain region	1nm
Thickness of insulating layer of gate region	3nm
Length of underlap region L$_{gap1}$(L$_{gap2}$)	3nm(15nm)

In the P$^+$ source electrostatic doping region, as the work function of M$_1$ increases, the silicon Fermi level is closer to the conduction band. In the N$^+$ drain electrostatic doping region, as the work function of M$_2$ increases, the silicon Fermi level is closer to the valence band. Vertical potentials with different electrode work functions are shown in Fig. 3. A more significant difference between the electrode work functions and substrate provides a higher absolute value of electrical potentials. Note that, once the potential and electrostatic doping concentration exceeds a certain range, the model deviates from the simulation result due to adopting the Boltzmann approximation as shown in Fig. 4. However, it can be solved by using a non-approximate Fermi-Dirac distribution. Electrostatic doping concentrations with different electrode work functions are shown in Fig. 4. Electrostatic doping concentration increases as the absolute value of potential increases. A more significant difference between the electrode work function and substrate provides a higher electrostatic doping concentration slope. The different silicon thickness comparisons of the analytical model with the TCAD simulation results are shown in Fig. 5. The absolute value of center electric potential increases and the potential becomes steeper as silicon thickness decreases. Meanwhile, the silicon thickness has little effect on the value of the potential at the boundary.

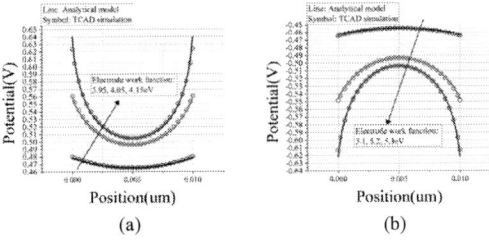

Fig. 3 The potential diagrams with different electrode work functions. (a) N$^+$ drain region, (b) P$^+$ source region.

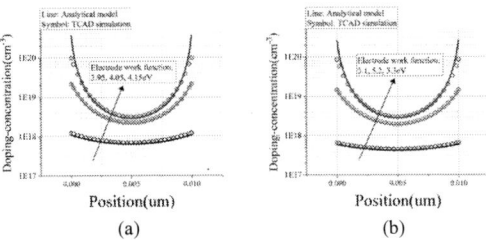

Fig. 4 The electrostatic doping concentration diagrams with different electrode work functions. (a) N$^+$ drain region, (b) P$^+$ source region.

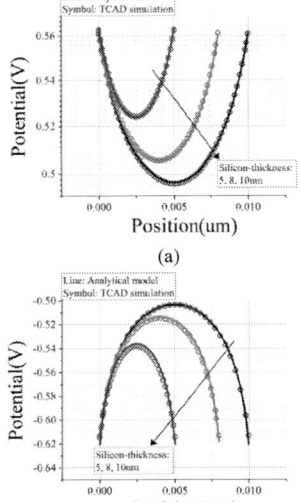

Fig. 5 The potential diagrams for different silicon thicknesses. (a) N$^+$ drain region, (b) P$^+$ source region.

IV. CONCLUSION

In summary, an analytical doping effect model for CP-TFET has been proposed and verified. It applies to different electrode work functions and substrate thicknesses without parameters fitting. The derivation and verification of the analytical model for electrostatic doping effect are helpful for the development of CP-TFET modeling.

REFERENCES

[1] J. Li, Q. Xie, A. Huang and Z. Wang, "Effects of Drain Doping Profile and Gate Structure on Ambipolar Current of TFET," 2020 IEEE 3rd International Conference on Electronics Technology (ICET), 2020, pp. 20-24, doi: 10.1109/ICET49382.2020.9119655.

[2] K. Eyvazi and M. A. Karami, "A new Junction-Less Tunnel Field-Effect Transistor with a SiO2/HfO2 stacked gate oxide for DC performance improvement," 2020 28th Iranian Conference on Electrical Engineering (ICEE), 2020, pp. 1-4, doi: 10.1109/ICEE50131.2020.9260621.

[3] R. J. E. Hueting, B. Rajasekharan, C. Salm and J. Schmitz, "The Charge Plasma P-N Diode," IEEE Electron Device Lett., vol. 29, no. 12, pp. 1367-1369, Dec. 2008. doi: 10.1109/LED.2008.2006864.

[4] Y. Taur and T. H. Ning, *Fundamentals of Modern VLSI Devices*, Cambridge, U.K.: Cambridge Univ. Press, 2009.

A Charge Pump Based 1.5A NMOS LDO with 1.0~6.5V Input Range and 110mV Dropout Voltage

Yifa Wang, Tong Wu, Jianping Guo*

School of Electronics and Information Technology

Sun Yat-sen University, Guangzhou, China

Email: guojp3@mail.sysu.edu.cn

Abstract—This paper presents a low dropout regulator (LDO) with ampere-level loading capability and wide input range. The NMOS power transistor with built-in charge pump was adopted to reduce the dropout voltage thus increase the power efficiency effectively. To realize a wide input voltage range, the fully-integrated charge pump can be configured adaptively for different input voltage. The proposed LDO has been designed and implemented in a 180nm CMOS technology. Experimental results show that the LDO has a wide input voltage range of 1.0~6.5 V, a wide output voltage range of 0.8~5.5 V, and a maximum output current of 1.5 A. In addition, the dropout voltage is only 110 mV under 1.5 A loading condition.

Keywords—Low dropout regulators (LDO), charge pump, wide input voltage range, low dropout voltage.

I. INTRODUCTION

Low dropout regulators (LDOs) are widely used for power management in modern electronic equipment. For example, in a 5G base station, LDOs are often used to be the post regulator after switching DC-DC regulators for their superior property of low noise, good power supply rejection, and fast transient response [1], [2]. In such a system, several LDOs are needed for powering digital processors, RF transmitters, and some noise-sensitive analog circuits. The power rails of these circuits are often very different. The supply voltage of a processor may be low to 0.8 V, while for the analog part such as power amplifier, it may be higher than 5 V for high output power or output swing. Generally speaking, different kinds of the LDO are chosen for different loading requirements. So that the performance can be the best, and the cost can be the lowest. However, this means several LDOs from different vendors are needed, and the system engineers need to understand many different LDOs to design different application circuits. As a result, the cost from the complicated supply chains and application circuits are often high. In that case, if there is a universal LDO with wide input/output voltage range and wide loading range, can fit all the requirements in a complicated system, it would be very helpful to reduce the system cost in many applications.

To achieve high power efficiency, low dropout voltage is required. Compared with PMOS, using NMOS as power transistor is more power efficient as NMOS features lower on resistance [3]–[5]. But an LDO with NMOS power transistor requires an additional high voltage rail to drive the gate of the power transistor. Without additional external voltage supply, a feasible solution is to use a charge pump to boost the supply voltage. In this paper, an NMOS LDO with built-in charge pump is proposed. The proposed LDO is universal for 1.0~6.5V input range, 0.8~5.5V output range, and 0~1.5A loading range.

Fig. 1 shows the proposed NMOS LDO with a charge pump system, where V_{IN} is the input voltage of the LDO, and it is the sole power supply for the whole LDO circuit. Under low input voltages V_{IN} is boosted to V_{DD_CP} by charge pump to supply the internal bandgap reference (BGR) and error amplifier (EA). At the same time, V_{DD_CP} is further boosted by charge pump system

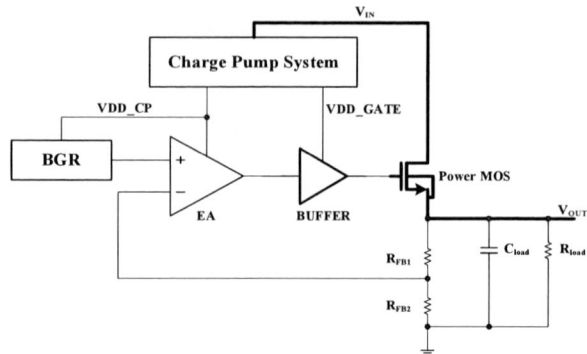

Fig. 1 Circuit diagram of the proposed NMOS LDO.

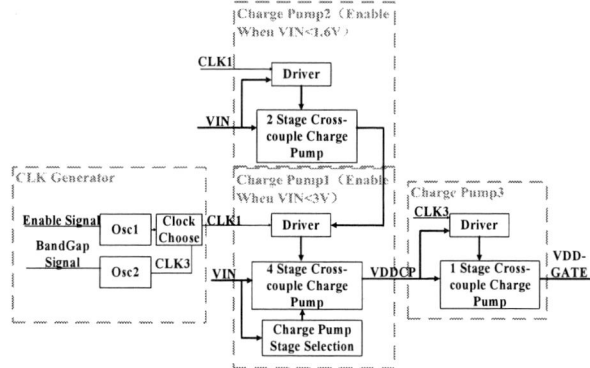

Fig. 2 Simplified block diagram of the proposed charge pump circuit.

to V_{DD_GATE} to provide voltage rail for the gate of the power transistor. The V_{DD_GATE} is high enough so that the LDO can work with very low dropout voltage.

II. PROPOSED CHARGE PUMP SYSTEM

Comparing with the switching DC-DC converters, charge pump can achieve fully-integrated operation. Cross-coupled charge pump is a widely used structure to boost the DC voltage [6]. Eliminating the voltage reduction of diode in Dickson charge pump (DCP), cross-coupled charge pump has a wider input range and a stronger boosting ability. It is worth noting that by changing the cascading stages, output voltage can be roughly adjusted.

The block diagram of the proposed charge pump circuit is shown in Fig. 2. The charge pump system can be divided into 3 sub-charge pumps (CP1, CP2, and CP3) and two oscillators (Osc1 and Osc2). They all use traditional cross-coupled charge pump structures, but have different series numbers. CP1 has four stages. It is enabled when the power supply voltage cannot supply the BGR and EA (lower than 3 V in this design), and the output voltage of CP1 is V_{DD_CP}. In particular, the number of charge pump stages enabled in CP1 is related to the input voltage, and V_{DD_CP} should be guaranteed to be between 3~5 V. CP2 consists of two stages. It starts up under low power supply voltage (lower than 1.6 V) to supply the clock driver of CP1. A higher clock-driver voltage can ensure the integrity of the clock signal and improve the loading capability and conversion efficiency of CP1. CP3 has only one stage. It is enabled when the voltage difference between V_{DD_CP} and the LDO output is insufficient (lower than $3V_{TH}$). CP3 continues to rise to V_{DD_GATE} at the basis of the V_{DD_CP} to provide a higher voltage rail for the gate of the power transistor.

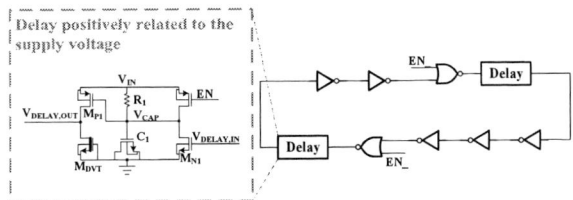

Fig. 3 The simplified schematic of Osc1.

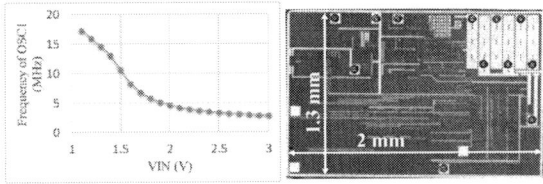

Fig. 4 Simulated f_{Osc1} vs. V_{IN} (left) and the chip micrograph (right)

The proposed charge pump also contains two oscillators. Osc1 is a ring oscillator and does not need reference voltage and bias current. So Osc1 can provide clock signal for charge pump when the BGR module is not working. Compared to Osc1, Osc2 is a relaxation oscillator to provide a precise clock signal. It needs reference voltage and bias current. It is worth noting that the clock frequency of Osc1 is inversely proportional to the supply voltage. This design solves the contradiction between the insufficient load capacity of the static frequency charge pump system under low voltage and the high dynamic power consumption or low efficiency under high voltage. In particular, under low-voltage conditions, the clock frequency provided by Osc2 is insufficient, even if the reference is already working normally, it still needs to provide a higher frequency clock from Osc1.

The simplified schematic of Osc1 is shown in Fig. 3, where a delay module positively dependent to the supply voltage is added. M_{P1} and M_{DVT} constitute an inverter with a threshold value of $V_{IN}-V_{THP}$. When $V_{DELAY,IN}$ is high level, M_{N1} turns on, and capacitor C_1 is discharged through M_{N1}. Then V_{CAP} is rapidly pulled down to ground, so that $V_{DELAY,OUT}$ is pulled high, and the circuit produces a delay. Similarly, when $V_{DELAY,IN}$ is low level, M_{N1} turns off, and C_1 is charged through the resistor R_1. When V_{CAP} exceeds $V_{IN}-V_{THP}$, $V_{DELAY,OUT}$ is pulled down to achieve the delay effect. The lower the V_{IN}, the shorter the delay time. Fig. 4 shows the simulated frequency of Osc1 under different V_{IN}. The simulation results verify that the frequency is decreased when V_{IN} increases.

III. MEASUREMENT RESULTS

The proposed LDO has been designed and fabricated in a 180nm CMOS technology. Fig. 4 shows the chip micrograph. The chip area is 2.6 mm². The input voltage ranges from 1.0~6.5 V and the output voltage ranges from 0.8~5.5 V. The measured dropout voltage is 110 mV under the maximum loading current of 1.5 A. The measured start-up waveforms are shown in Fig. 5. When V_{IN} is 1.0 V, the measured V_{DD_CP} is 3.5 V and V_{DD_GATE} is 3.0 V. When V_{IN} is 6.5 V, the measured V_{DD_CP} is 4.5 V and V_{DD_GATE} is 7.3 V. It is worth noting that if $V_{DD\_CP} - V_{OUT}$ is not sufficient, the sub-charge pump CP3 is enabled and V_{DD_GATE} is boosted. And when $V_{DD\_CP} - V_{OUT}$ is large enough, V_{DD_GATE} follows V_{DD_CP}. Hence, during the rising period of V_{OUT} when V_{IN} is 6.5V, there is a rapid rise of V_{DD_GATE}. The measured load transient responses under different VIN are shown in Fig. 6. The voltage spike is less than 36 mV when loading current changes

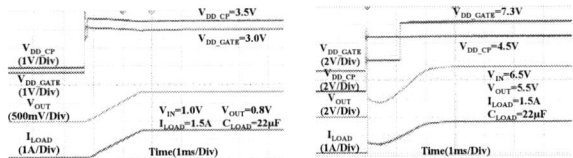

Fig. 5 Measured power-on process at 1.0V (left) and 6.5V (right) V_{IN}

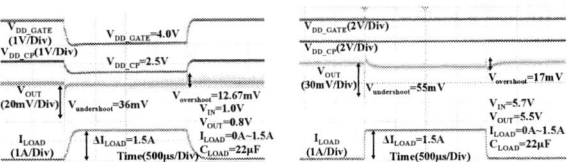

Fig. 6 Measured load transient response at 1.0V (left) and 5.7V (right) V_{IN}.

TABLE I. PERFORMANCE COMPARISON.

	[1]	[2]	[3]	[4]	[5]	This work
Type	Ana./pmos	Dig./pmos	Ana./nmos	Ana./nmos	Ana./nmos	Ana./nmos
Tech. (nm)	500	28	350	130	600	180
C_L (μF)	2.2	5	17	1	0~1000	22
V_{OUT} (V)	1.8	0.5~0.9	0.8~1.8	1.8	0.8~3.6	0.8~5.5
V_{IN} (V)	2.8~3.8	0.6~1	0.8~3.3	N/A	2.2~5.5	1~6.5
I_L (A)	3	1.5	6	1	3	1.5
V_{DO} @ I_{LMAX} (mV)	65	90	114	200	200	110
Load Reg. (mV/A)	<3	1	0.7	0.6	0.5	2.5
Line Reg. (mV/V)	<1	<1.5	0.12	0.23	0.3	0.93
under/over shoot (mV)	180 (0.01-3A)	70 (0.2-1.2A)	35.4 (0.4-5A)	11 (0-1A)	51 (0-3A)	36 (0-1.5A)

between 0 and 1.5 A. Table I summaries and compares the LDO performance. It can be found the proposed LDO achieves the widest input range, ultra-low dropout voltage, and small voltage spike under full load transient response.

IV. CONCLUSIONS

An NMOS LDO with built-in charge pump is proposed for ultra-low dropout. The charge pump can be configured adaptively dependent to V_{IN}, so that wide input/output and loading ranges can be achieved. In that case, the proposed LDO is a universal LDO for many systems.

ACKNOWLEDGMENT

This work was partly supported by the Natural Science Foundation of Guangdong Province, China (2022A1515011054).

REFERENCES

[1] X. Lai, J. Guo, and Z. Sun, "A 3-A CMOS low-dropout regulator with adaptive Miller compensation," *Analog Integr. Circ. Sig. Process.*, vol. 49, no. 1, pp. 5–10, Oct. 2006.

[2] X. Mao, Y. Lu, and R. P. Martins, "A scalable high-current high-accuracy dual-loop 4-phase switching LDO for microprocessors," *IEEE J. Solid-State Circuits*, vol. 57, no. 6, pp. 1841–1853, Mar. 2022.

[3] H. Cao, X. Yang, W. Li, Y. Ding, and W. Qu, "An impedance adapting compensation scheme for high current NMOS LDO design," *IEEE Trans. on Circuits and Syst. II: Express Briefs*, vol.68, no.7, pp. 2287–2291, Jul. 2021.

[4] K. Li, C. Yang, T. Guo, and Y. Zheng, "A multi-loop slew-rate-enhanced NMOS LDO handling 1-A-load-current step with fast transient for 5G applications," *IEEE J. Solid-State Circuits*, vol. 55, no. 11, pp. 3076–3086, Nov. 2020.

[5] H. Fan *et al.*, "An external capacitor-less low-dropout voltage regulator using a transconductance amplifier," *IEEE Trans. Circuits Syst. II Exp. Briefs*, vol. 66, no. 10, pp. 1748–1752, Oct. 2019.

[6] R. Pelliconi, D. Iezzi, A. Baroni, M. Pasotti and P. L. Rolandi, "Power efficient charge pump in deep submicron standard CMOS technology," in *IEEE J. Solid-State Circuits*, vol. 38, no. 6, pp. 1068–1071, Jun. 2003.

IPOCIM: Artificial Intelligent Processor with Adaptive Ping-pong Computing-in-Memory Architecture

Liang Chang, Chenglong Li, Xin Zhao, Shuisheng Lin, Jun Zhou

University of Electronics Science and Technology of China, Chengdu 611731, China

Abstract—Computing-in-memory (CIM) architecture is a promising solution toward energy-efficient artificial intelligent (AI) processor. Practically, the AI processor with CIM engine induces a series of issues including data updating and flexibility. For instance, in AI-oriented applications, the weight stored in the CIM must be reloaded due to the huge gap between limited capacity of CIM and growing weight parameter, which greatly reduces the computation efficiency of the AI processor. Moreover, the natural parallelism of CIM leads to the mismatch of various convolution kernel sizes in different networks and layers, which reduces hardware utilization efficiency. In this work, we explore a CIM engine with a ping-pong strategy as an alternative to traditional CIM macro and weight buffer, hiding the data update latency to enhance data reuse. In addition, we proposed a flexible CIM architecture adapting to different neural networks, namely IPOCIM, with a fine-grained data-flow mapping strategy. Based on the evaluation, IPOCIM achieves 1.4-7.1× performance improvement, and 2.2-6.1× energy efficiency, compared to baseline.

Keywords—*Computing-in Memory, Ping-Pong Computing, AI Processor, Circuit and System, Energy Efficiency*

I. INTRODUCTION

With the requirement of speed in data-intensive deep learning workloads, the hardware accelerators have been deployed into several edge computing scenarios, such as computer vision, speech recognition, automatic drive [1]. As edge computing applications generally require low power consumption, and real-time response on the device, in the demand of energy-efficient hardware accelerators [2]. However, traditional hardware accelerator suffers from limited memory bandwidth, energy consumption, and computing power, whereas the neural network applications require large memory and high parallel computing resources [3]. Consequently, it is challenging to implement an energy-efficient neural network processor for various deep-learning algorithms. In addition, the decline of Moore's law is pushing the hardware accelerator to contain more memory resource on-chip.

A potential solution is computing-in-memory (CIM) architecture to combine the memory and computation functions, where the memory array can support high parallel computation operations [4]. By supporting the computation, the peripheral circuits or/and bit-cell structure should be modified. Recently, SRAM has been used to develop CIM processors due to mature technology process [5]. Several CIM circuits has been developed via analog- and digital computation schemes. However, the scalability of CIM computing array (CIM macro) is a critical issue to design the AI processor, leading to update parameters from buffers and off-chip memory. In addition, the data mapping strategy and CIM macro utilization influence the performance of AI processors, which is rarely mentioned in the previous studies.

In this work, we analyze the problem of current CIM architecture and provide a ping-pong computing solution to CIM macro structure. Based on our observation, we develop a ping-pong computation engine based CIM architecture, namely

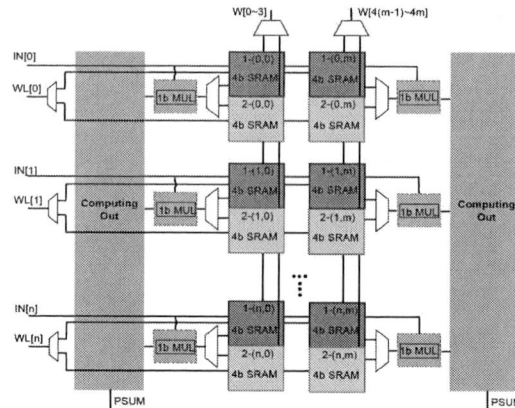

Fig. 1. Macro Circuit of ping-pong computing manner.

IPOCIM, which is equipped with ping-pong CIM macro to alternative computing engine to hide the data update latency. Also, the data mapping strategy is discussed to maintain the high speed and energy efficiency of the proposed IPOCIM.

II. IPOCIM ARCHITECTURE AND DESIGN

In order to guarantee continuous logic operations without interruption, a ping-pong computing structure is designed as shown in the Fig. 1, where green and blue colors mark a group of memory with ping-pong function in CIM macro. For the writing and reading of this group of memory, it is controlled by a pair of MUXs with inverted selection signals, and reading and writing are carried out alternately to hide the weight update time to ensure no interruption in the computing. Additionally, the ping-pong memory shares a group of peripheral circuits, which decreases the area overhead.

Fig. 2 (a) illustrates the proposed IPOCIM architecture, which contains a global shared input buffer, 4 CIM cores, controllers, and configuration modules. Specifically, the input buffer adopts a ping-pong structure to hide the update time of input feature map, while the configuration module configures current data flow. Each CIM core includes a psum buffer, a special processing unit (SPU), and a CIM group. The SPU is responsible for prefetching and processing the input and output data in convolution. The CIM group accelerates the core computational work, which includes several CIM macros with a ping-pong structure shown in Memory centric computing architecture Fig. 3(b). In the CIM core, the calculation results of CIM group are truncated and encapsulated to a vector in the psum pack module. Then, the vector is sent to complete the loop accumulation between the psum buffer and the SPU adder. Finally, through corresponding activation and pooling operations, the output feature map is obtained. In addition, we observe that the power consumption of off-chip data transfer is 30~700× larger than that of on chip logic operations. IPOCIM employs random linear coding (RLC) module to compress the off-chip interaction data for reducing the transmission power.

III. EXPERIMENTAL RESULTS

For the proposed IPOCIM, we select CIM size as 32 × 256 × 4 in column, row and macro number, respectively. Based on this, we implement every component except CIM macro via Verilog HDL modeling and synthesize them through Synopsys Design Compiler with T28nm library to get the power consumption and area, respectively. As for the off-chip memory and on-chip buffer, we model them with CACTI simulator. An

978-1-6654-9270-6/22 $31.00 © 2022 IEEE

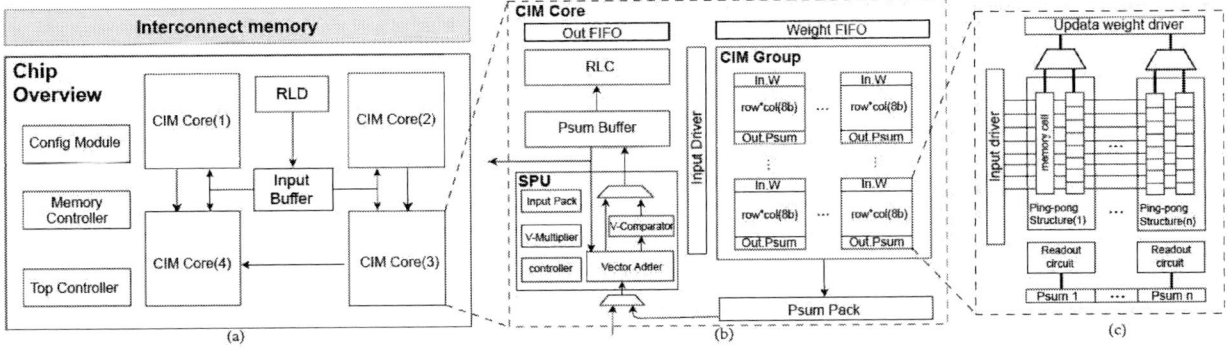

Fig. 2. IPOCIM architecture. (a)Overview of whole architecture. (b)Main components of CIM core. (c)The structure of the ping-pong CIM array.

in-house CIM-based AI processor simulator is built as the evaluation platform. At the 8-bit precision, the IPOCIM can provide a peak performance of 3.28 TOPS with ultra-low power. In order to give a more intuitive performance comparison, we select the universal CPU (Intel Core i5), GPU (Nvidia RTX 2080Ti) and TPU hardware platforms as baselines for evaluation. Besides, a state-of-the-art CIM-based chip and CIM-pingpong are included for comparison [6]. We run the different DNNs through the PyTorch to obtain the latency with the help

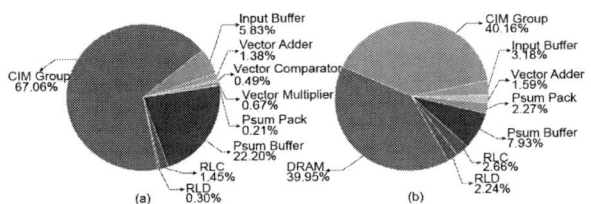

Fig. 3. (a) Area breakdown. (b) Power breakdown.

of inside APIs.

The area and power breakdown are shown in Fig. 3. The whole chip only occupies 5.43 mm2 in area and 307.4 mW in power at the technology of 28 nm. For each CIM macro, it only occupies 0.23 mm2 (3.42%) in area and 8.25 mW (2.55%) in power due to the advantage of analog domain computing of CIM and advanced technology, which outperforms conventional MAC-based PEs dramatically. From the area breakdown, the main area stems from CIM Groups and Buffer, for that they are the major computing and store logic. Besides, other logics only occupy less than 5% area due to compact architecture design. As for the power composition, the overall trend is consistent with the area breakdown.

Fig. 4 demonstrates the comparisons of energy efficiency. Overall, the proposed architecture outperforms baselines and achieves average 934.3×, 239.6×, 19.6× and 2.9× higher energy efficiency over CPU, GPU, TPU and CIM-based architecture.

IV. CONCLUSION

Typically, CIM architecture is recognized as high energy efficiency with high parallel computation, which can be used to develop AI processor. Several operations of the neural network can be accelerated, including matrix multiply-accumulation, and pooling operations. This work enhanced the CIM macro with ping-pong engine to efficient update NN parameters. Based on the ping-pong engine, we analyzed the corresponding mapping strategy to further improve the performance of the

proposed IPOCIM in terms of speed and energy efficiency. Besides, there are several factors influence the scalability of IPOCIM such as survival time of data resource. Based on the evaluation, our proposed IPOCIM can achieve about 3 times acceleration compared with previous CIM-based work. In our future work, we may deeply analyze the factors of design space

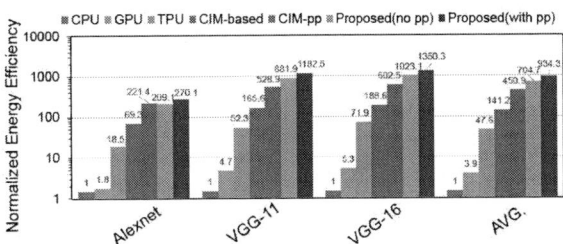

Fig. 4. IPOCIM Energy efficiency comparison, with CPU, GPU, TPU

exploration and flexible data update technique.

V. ACKNOWLEDGEMENT

This work was supported by the National Natural Science Foundation of China under Grant No. 62104025, and State Key Laboratory of Computer Architecture (ICT, CAS) under Grant No. CARCHB202117. Jun Zhou is the corresponding author.

REFERENCES

[1] Y. Gong,et al. Raodat: An energy-efficient reconfigurable ai-based object detection and tracking processor with online learning. In 2021 IEEE Asian Solid-State Circuits Conference (A-SSCC), pages 1–3, 2021.

[2] L. Chang, et al. Towards intelligent edge computing: Application, architecture, circuit, and device perspective. In 2021 IEEE International Conference on Integrated Circuits, Technologies and Applications (ICTA), pages 125–126, 2021.

[3] Y. Chih et al. 16.4 an 89tops/w and 16.3tops/mm2 all-digital sram-based full-precision compute-in memory macro in 22nm for machine learning edge applications. In 2021 IEEE International Solid- State Circuits Conference (ISSCC), volume 64, pages 252–254, 2021.

[4] J. Su et al. 16.3 a 28nm 384kb 6tsram computation-in-memory macro with 8b precision for ai edge chips. In 2021 IEEE International Solid- State Circuits Conference (ISSCC), volume 64, pages 250–252, 2021.

[5] S Yin, et al. XNOR-SRAM: In Memory Computing SRAM Macro for Binary/Ternary Deep Neural Networks. IEEE Journal of Solid-State Circuits, 55(6):1733–1743, 2020.

[6] Y. Chiu, et al. A 4-kb 1-to-8-bit configurable 6t sram-based computation-in memory unit-macro for cnn-based ai edge processors. IEEE Journal of Solid-State Circuits, 55(10): 2790–2801, 2020.

CVD Monolayer tungsten-based PMOS Transistor with high performance at V_{ds} = -1 V

Xin Wang[1,2], Yanqing Wu[2]

[1] Wuhan National High Magnetic Field Center and School of Optical and Electronic Information, Huazhong University of Science and Technology, Wuhan, China
[2] Institute of Microelectronics and Key Laboratory of Microelectronic Devices and Circuits (MOE), Peking University, Beijing, China

Abstract—Two-dimensional (2D) semiconducting materials channels enable ultimate scaling of transistors and will help Moore's Law Scaling for decades. In this paper, we reported p-type WSe_2 transistors using monolayer (~0.85 nm) channels by molten-salt-assisted chemical vapor deposition. The transfer-free back-gate devices fabricated based on 100 nm SiO_2/Si substrate exhibit highest on current at V_{ds} = -1 V among transistors of monolayer p-WSe_2, and a high on/off ratio up to 10^8.

Keywords—TDMs, CVD, p-type, transistor

I. INTRODUCTION

Two-dimensional (2D) semiconducting materials have received increasing attention for electronic applications due to their excellent transport properties and ultrathin body nature. High performance n-type MoS_2 field-effect transistors have been reported in many experiments. However, almost all high-performance transistors are bilayer or even more layers, or mechanical exfoliate flakes. What's more challenging is that there are few p-type transistors with comparable performance to n-type transistors, especially for single-layer CVD-grown two-dimensional materials. Monolayer tungsten selenide (WSe_2) has been studied extensively in this context. WSe_2 are tungsten-based transition metal chalcogenides (TMDCs).

In this paper, we synthesized monolayer tungsten-based transition metal chalcogenides WSe_2, which domain sizes above 200 µm. We fabricated monolayer (sub 1 nm) channel p-type WSe_2 transistors. The monolayer WSe_2 p-type transistor based on 100 nm SiO_2/Si substrate exhibit on/off ratio beyond 10^8, on current up to 140 µA/µm at Vds = -1 V of 140-nm-long channel for the first time among CVD monolayer WSe_2.

II. FABRICATION PROCESS

A. CVD growth and physical characterization

Monolayer WSe_2 were grown on 100 nm SiO_2/Si substrate by low pressure chemical vapor deposition (LPCVD) in 1 inch diameter quartz tube furnace as depicted in Fig. 1. The TMDCs material synthesis used molten salt-assisted LPCVD method reported in our works before. WSe_2 was grown at 890 ℃ maintaining 15 min with KCl and the growth resulted in not only large size monolayer crystals about 100 µm but also monolayer film above 1 mm.

As depicted in Fig. 2 (a), the size of monolayer WSe_2 single crystals can reach 221 µm. The morphology of CVD monolayer (~0.85 nm) WSe_2 has been measured by atomic force microscope (AFM) as shown in Fig. 2 (b). As depicted in Fig. 3, WSe_2 were characterized by using Raman and photoluminescence (PL) spectroscopy. The Raman spectra for monolayer WSe_2 show a peak at 250 cm⁻¹, which are in agreement with previous reports. The PL spectra show peaks at an energy of 1.6 eV for WSe_2.

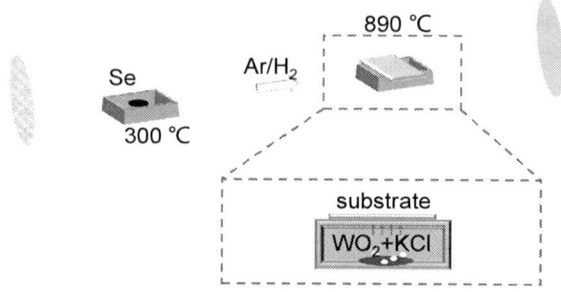

Fig. 1. Schematic of the LPCVD setup for the synthesis of WSe_2.

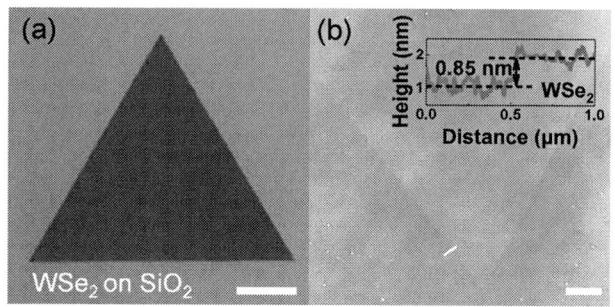

Fig. 2. (a) Optical microscopic image of WSe_2 grown on 100 nm SiO_2. (b) AFM image of WSe_2. The scale bar are 50 and 2 µm, respectively.

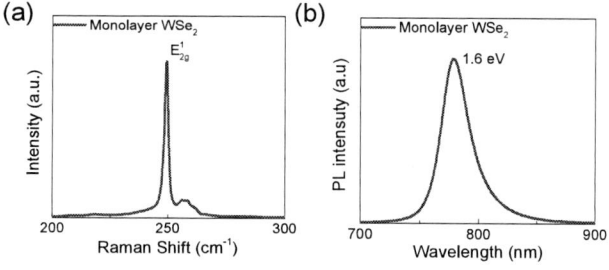

Fig. 3. (a) Raman and (b) PL spectrums of transfer-free monolayer WSe_2.

B. Device fabrication and physical characterization

The bottom-gate transfer-free WSe_2 transistors were fabricated on a low-resistivity Si substrate with 100 nm SiO_2. First, inductively coupled plasma (ICP) was used to etch the isolation region to patterning the active area. A 300 °C post-annealing process step was performed to remove PMMA residual to good contact. The S/D ohmic contact metal is 20 nm nickel Platinum (Pt) and 60 nm gold (Au).

III. RESULTS AND DISCUSSION

The field-effect mobilities were extracted by using the expression $\mu=[dI_{ds}/dV_{bg}]\times[L_{ch}/WC_iV_{ds}]$ in the linear region, where L_{ch} is the channel length and W is the channel width, and the hole mobility is 57 cm²/Vs.

Fig. 4 shows the transfer characteristic of WSe_2 transistors with channel-length varying from 1000 nm to 140 nm at V_{ds} = -1 V. The on/off ratio is from 3×10^7 to 10^8 with channel-length scaling and the on-state current can be up to 140 µA/µm. To further verify the electrical performance, we fabricated 33 devices based on these monolayer WSe₂ and Transfer Length Measurement (TLM) on-state current at V_{ds} = -1 V is shown in Fig. 5.

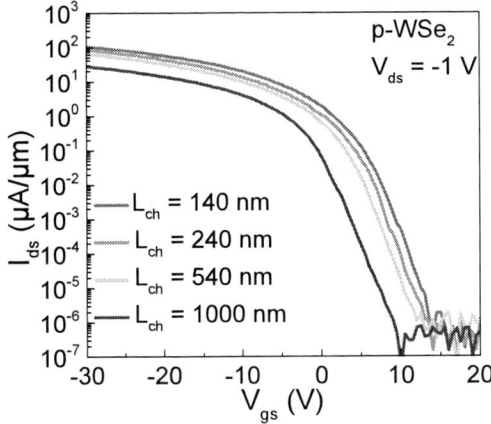

Fig. 4. Transfer characteristics of monolayer WSe₂ TFTs with different channel-length from 140 nm to 1000 nm at Vds = -1 V.

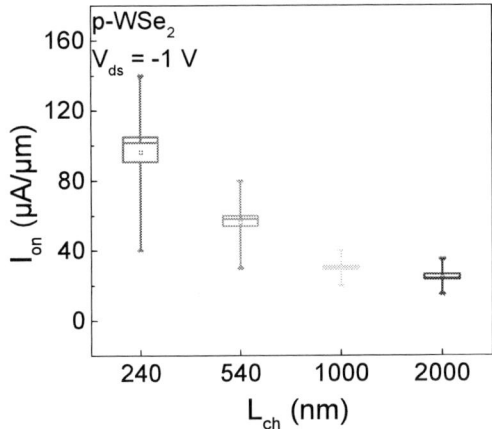

Fig. 5. Device-to-device variation of Ion with different channel length for 33 monolayer p-WSe₂ TFTs. Ion extracted in the case of Vds = -1 V.

As shown in Fig. 6, the on current increases from 15 to 140 μA/μm as the channel length decreases from 3 μm to 0.14 μm at V_{ds} = -1 V, 140 μA/μm of 140-nm-long is the largest on current among CVD monolayer WSe p-type transistors at V_{ds} = -1 V.

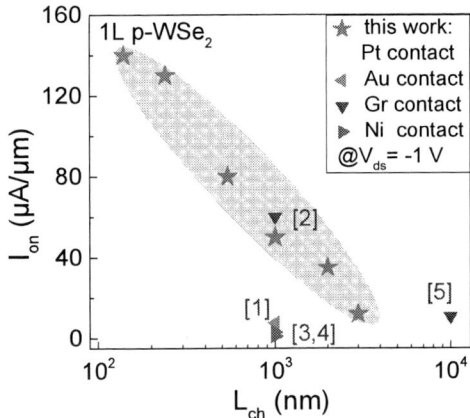

Fig. 6. Benchmarks of Lch dependent Ion ratio for monolayer p-WSe₂-based transistors with different contact metal at Vds = -1 V.

IV. CONCLUSION

In summary, this letter reports the synthesis of large single-crystal monolayer WSe₂ on 100 nm SiO_2/Si substrate with high-performance transistor characteristics. The field-effect mobility is 57 cm²/Vs. Transistors fabricated on the p-WSe₂ show a much higher on current compared with previous work. This work demonstrates the potential of CVD p-WSe₂ for future high-performance short-channel electronic device.

ACKNOWLEDGMENT

This work was supported by the National Key Research and Development Program of China (2021YFA1202903), the NSFC (Grant Nos. 62090034, 62104012 and 61927901).

REFERENCES

[1] H. Zhou, et al, "Large Area Growth and Electrical Properties of p-Type WSe₂ Atomic Layers," *Nano Lett.* Washington, vol. 15, pp. 709-713, November 2014.

[2] C-H. Yeh, et al, "Graphene−Transition Metal Dichalcogenide Heterojunctions for Scalable and Low-Power Complementary Integrated Circuits," *ACS Nano.* Washington, vol. 14, pp. 985-992, January 2020.

[3] X. Zhang, et al, "Defect-Controlled Nucleation and Orientation of WSe₂ on hBN: A Route to Single-Crystal Epitaxial Monolayers," *ACS Nano.* Washington, vol. 13, pp. 3341-3352, February 2019.

[4] A. Kozhakhmetov, et al, "Scalable Substitutional Re-Doping and its Impact on the Optical and Electronic Properties of Tungsten Diselenide," *Adv. Mater.* Weinheim, vol. 32, pp. 2005159, November 2020.

[5] J. Chen, et al, "Chemical Vapor Deposition of Large-Sized Hexagonal WSe₂ Crystals on Dielectric Substrates," *Adv. Mater.* Weinheim, vol. 27, pp. 6722-6727, September 2015.

Efficient AVS3 Intra Prediction Hardware Design for Real-time Applications

Yucheng Jiang[1,2], Haifeng Guo[2], Junhao Zheng[2], Jingsheng Wang[2], Songping Mai[1]

[1] Tsinghua Shenzhen International Graduate School
[2] Peng Cheng Laboratory

Abstract—a hardware-efficient hybrid greedy CU (coding unit) partition algorithm for AVS3 intra prediction, which has advantages over the traditional regression algorithm on both scheduling complexity and resource consumption, is presented. Compared with the NVidia hardware acceleration of HEVC, the proposed algorithm achieves 21% performance improvement on AI (all-intra) configuration for UHD 4K video encoding.

Keywords—real-time application, hardware design, video coding, intra prediction

I. INTRODUCTION

High-definition video compression is the basis of further video analysis in real-time applications, such as robotic vision. AVS3 proposed by the AVS working group in China is the next generation of AVS2 with about 24% bit-rate reduction [1]. The sophisticated coding tree and the diverse prediction modes increase the encoding complexity which means more difficulties for hardware design and more time-consumption for software encoding. Considerable related works have been done to reduce the complexities of the MD (mode decision) process including CU partition decision and CU prediction modes decision algorithm [2-5]. In this paper, we designed a novel intra-CU partition algorithm and its hardware architecture. The proposed algorithm simplifies the complexity of the partition scheduling process dramatically which is essential for its hardware implementation, especially for real-time applications.

II. ORIGINAL PARTITION ALGORITHM

For each LCU (largest CU), the MD process needs to traverse all of the coding tree nodes to find the minimum RD (rate-distortion) cost combination. During the traverse procedure, the depth-first processing of one coding tree node can proceed only when all of its sub nodes' decisions are completed.

A. Basic Intra Prediction Pipeline

For each CU size, we need to execute the RDO (rate-distortion optimization) procedure five times to find the prediction mode with the minimum RD cost among the five candidates provided by the RMD (rough mode decision) module. Considering the circuit throughput and the area, we adopt a pipeline structure for the intra-prediction module as shown in Fig 1. The pipeline latency T can be calculated using the following equation [6], where t_i represents the latency of the corresponding stage.

$$T = \sum_{i=0}^{7} t_i + 4 \cdot \max\{t_0, t_1, ..., t_7\} \quad (1)$$

B. Hardware Implementation Analysis

This work was supported by the Science, Technology and Innovation Commission of Shenzhen Municipality (WDZC20200820160650001) and Guangdong Province Science and Technology Program (2019B010143003).

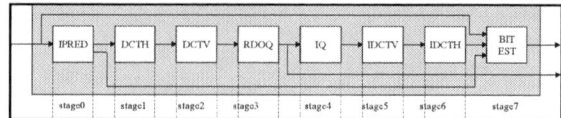

Fig. 1. Intra prediction pipeline

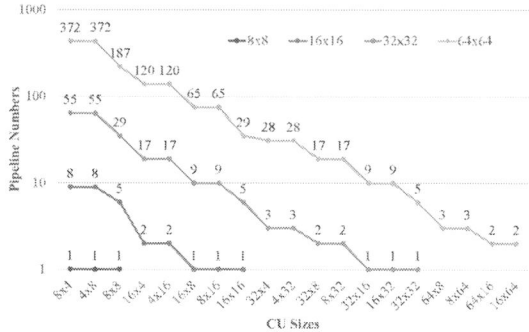

Fig. 2. Numbers of basic pipelines needed in regression MD design

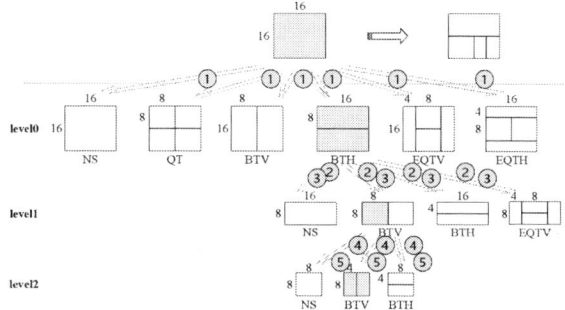

Fig. 3. Demonstration of the greedy CU partition flow

To meet the requirements of real-time compression, all of the coding tree nodes start simultaneously in a parallel design. According to our derived results, for the size 64x64 LCU (the yellow line in Fig 2), more than 1000 basic pipelines are needed in total which complicates the hardware schedule.

III. HYBRID GREEDY PARTITION ALGORITHM

A. Hybrid Greedy Algorithm Analysis

Unlike the regression algorithm, the greedy algorithm is a top-down procedure and can just find the locally optimal combination at the current level. Fig 3 shows the three-level depth greedy algorithm flow for the size 16x16 LCU. There is no need to further split the unselected nodes in greedy design. Firstly, at level 0, all six split modes are evaluated at the same time; after the best prediction mode is decided for each split mode, the one with the minimum RD cost will be selected as the best partition which is BTH in Fig 3; the other split modes will be discarded. Other levels' best split modes are decided in the same way. After finishing the decision of the last level, we can get the suboptimal partition result for the current LCU. Because of the greedy algorithm's unidirectional character, it can be pipelined in hardware design for throughput improvement. To avoid the dramatic performance degradation caused by the greedy process at level 0, we merged the first two levels in greedy design and traverse all of the possible split modes at these two levels, just like the depth-first algorithm. As shown in Fig 4, the total number of basic pipeline is far less than the regression design to deal with the size 64x64 LCU.

TABLE I. BD-RATE(Y) PERFORMANCE OF DIFFERENT ENCODING ALGORITHMS COMPARED WITH AVS3 HPM4.0 UNDER AI CONFIGURATION

Test Sequence		*X265*			*Basic version BD-Rate(Y)*	*hybrid greedy version BD-Rate(Y)*	*HEVC NVidia acceleration BD-Rate(Y)*
		Placebo BD-Rate(Y)	*Medium BD-Rate(Y)*	*Ultrafast BD-Rate(Y)*			
*UHD 4K 3840*2160*	*Tango2*	16.14%	22.79%	31.07%	19.73%	22.17%	44.60%
	ParkRunning3	4.95%	13.81%	21.63%	1.86%	1.31%	24.61%
	Campfire	18.11%	28.44%	47.92%	19.61%	18.69%	43.01%
	DaylightRoad2	12.80%	18.19%	37.78%	16.68%	25.40%	40.63%
	Average	13.00%	20.81%	34.60%	14.47%	16.89%	38.21%

The larger positive BD-Rate(Y), the worse luma encoding performance.

TABLE II. SOFTWARE ENCODING TIME-SAVING RATIO OF THE PROPOSED ALGORITHM COMPARED WITH THE BASIC VERSION

Test Sequence		T_{enc} *QP=27*	T_{enc} *QP=32*	T_{enc} *QP=38*	T_{enc} *QP=45*
4K	*Tango2*	76.31%	74.64%	72.72%	70.88%
	ParkRunning3	76.20%	75.40%	74.25%	72.65%
	Campfire	78.38%	75.05%	72.84%	71.29%
	DaylightRoad2	78.20%	75.26%	73.60%	72.48%
	Average	77.27%	75.09%	73.35%	71.83%

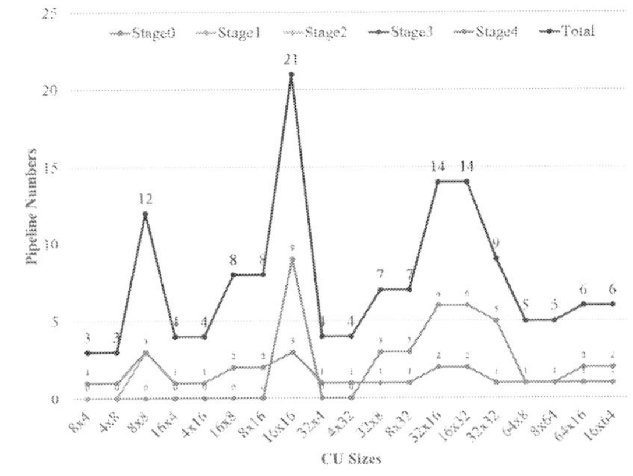

Fig. 4. Numbers of basic pipelines needed in hybrid greedy MD design

B. Innovations for Line buffer in Hardware Architecture

For each CU, the generation of its prediction pixels relies on the reference pixels stored in the line buffers. Conceivably, the line buffers at CU and LCU levels cause data dependencies that should be carefully designed for hardware coding efficiency. As for reference pixels on the left side, we made a compromise that using the reconstruction pixels at current stage as reference pixels. The LCU processing by the final stage is the fifth one behind the LCU in the first stage in our five-level algorithm. So if the width of the video content is five times longer than the LCU width which is satisfied for 4K content, we can choose the final reconstructed pixels as the upper reference pixels for performance consideration.

IV. EXPERIMENTAL RESULTS

We conducted several comparison tests on AI configuration using AVS3 HPM4.0 as an anchor. The algorithms under test include the basic version, the proposed hybrid greedy version, the different classes of X265 software, and the HEVC NVidia hardware acceleration version. The algorithms' performances are measured with BD-rate. Comparison results of the UHD 4K videos are shown in Table I. It is reasonable for the performance degradation of the hardware designs compared with the software designs. However, we still maintain more than 21% coding performance improvement over the HEVC NVidia acceleration version.

We synthesized 19 kinds of the basic intra prediction pipelines using Catapult 10.5c with Nangate OpenCell 45nm Library. By weighting the pipeline numbers of the traditional regression MD design (yellow line in Fig 2) and the hybrid greedy MD design (dark blue line in Fig 4) with the corresponding area results respectively, we can conclude that 75.7% of the circuit area has been reduced.

Table II shows the experimental results of the time-saving. It can easily be found that more than 74% of the encoding time are reduced for UHD 4K videos.

V. CONCLUSION

In this paper, we replaced the traditional regression algorithm in the AVS3 intra-MD process with the hybrid greedy algorithm which was much more efficient in hardware design. Optimizations of the pipeline architecture and line buffer have been discussed in detail. As a result, the MD algorithm's scheduling difficulty has been dramatically reduced, and its advantages in latency reduction and resource-saving are significant. More than 75% of circuit area and 74% of encoding time consumption were reduced with our hybrid greedy algorithm.AI configuration experimental results show that the proposed design achieves 21% performance improvement compared with the HEVC NVidia hardware acceleration version for UHD 4K videos. We will further optimize the real-time AVS3 hardware implementation for real-time applications based on this intra-MD structure.

REFERENCES

[1] J. Zhang, C. Jia, M. Lei, S. Wang, S. Ma, et al., "Recent Development of AVS Video Coding Standard: AVS3," 2019 Picture Coding Symposium (PCS), 2019, pp. 1-5.

[2] L. Shen, Z. Zhang and P. An, "Fast CU size decision and mode decision algorithm for HEVC intra coding," in IEEE Transactions on Consumer Electronics, vol. 59, no. 1, pp. 207-213.

[3] J. Wang, B. Ji, H. Wang, and L. Cheng, "Prediction mode grouping and coding bits grouping based on texture complexity for Fast HEVC intra-coding." in Journal of Real-Time Image Processing, vol. 18, 2021, pp. 839-856.

[4] M. Li, Ch. Zhu, Y. Li, X. Huang, H. Jia, et al., "A low-complexity hardware-oriented mode decision scheme based on rate-distortion estimation," 2014 IEEE International Conference on Multimedia and Expo Workshops (ICMEW), 2014, pp. 1-6.

[5] Y. Xu, and X. Huang,"Hardware-Oriented Fast CU Size and Prediction Mode Decision Algorithm for HEVC Intra Prediction," 2019 IEEE 5th International Conference for Convergence in Technology (I2CT), 2019, pp. 1-5.

[6] C. Zhu, H. Jia, S. Zhang, X. Huang, X. Xie, et al., "On a Highly Efficient RDO-Based Mode Decision Pipeline Design for AVS," in IEEE Transactions on Multimedia, vol. 15, no. 8, pp. 1815-1829, Dec. 2013.

Thermal Fatigue Analysis of Microbumps in a 3D TSV Integration Device

Yuqing Lu[1], Jun Wang[1,2]*

[1]Department of Materials Science, Fudan University, Shanghai 201114, China
[2]Yiwu Research Institute of Fudan University, Yiwu, Zhejiang 322000, China

Abstract—Copper microbumps are generally used in three-dimensional (3D) through silicon vias (TSV) integration devices. Because the structure of 3D TSV integration is complex, the thermal fatigue of microbumps may take place due to higher stresses during thermal cycles. In this study, a typical 3D TSV integration was analyzed by finite element method to evaluate the thermal fatigue life of microbumps in different locations based on Coffin-Manson model. To keep the accuracy of analysis, the elastoplastic and Anand constitutive relationships were adopted for the copper microbumps and the micro solder balls, respectively. The results revealed that the critical microbump with lower fatigue life is under the memory module, and the microbumps under the peripheral chips have much better fatigue performance.

Keywords—3D integration, finite element analysis (FEA), microbumps, fatigue life.

I. INTRODUCTION

Recently, the rapid development of information technology makes a higher demonds for the integration and performance of devices. The three-dimensional (3D) through silicon vias (TSV) integration devices are developed according to the rule of "More Than Moore" on the basis of the advanced packaging technologies. Compared with the conventional package, 3D integration contains the chips with various functions or stacked-die modules, forming a heterogeneous integration and increasing the integration density greatly[1].

Microbumps and TSV interconnections are widely used in 3D packaging, which contribute to advantages of small size, light weight, low power and high performance, etc. [2]. However, high thermal stresses due to the mismatch of material coefficient of thermal expansion (CTE) under thermal loadings and complex structures of 3D integration can cause thermal fatigue of components, especially the weaker parts, such as the microbumps [3]. It is also important to identify the location of most dangerous microbumps.

In this study, a typical 3D TSV integration was analyzed by ABAQUS, a famous finite element software, to evaluate the thermal fatigue life of microbumps in different locations based on Coffin-Manson equation. To improve the analysis accuracy, we adopted the elastoplastic model for the copper microbumps, Anand model for the micro Sn3.5Ag solder balls and the homogenized material properties for the redistribution layer. The finite element analysis (FEA) was carried out to demonstrate the variation of the stresses and plastic strain in the integration device. Thereby, the fatigue lives of microbumps in different locations were predicted through the Coffin-Manson model. According to the analysis results, the critical microbumps and design rules were discussed.

II. FINITE ELEMENT MODELLING

A. Modeling

A typical 3D integration with TSV interconnections is shown as Fig. 1, which includes four chips and a stacked memory module connected by microbumps. All components are integrated on a silicon interposer which is mounted on the substrate with Sn3.5Ag solder balls that embedded in the underfill layer. The whole integration is covered by EMC.

Some simplifications are assumed in the modelling. Firstly, the whole model was reduced to a quarter (shown in Fig. 2) owing to the geometry symmetry. Secondly, the tiny structures, such as the barrier layers of TSV etc., were ignored because the microbumps were concerned. Finally, the redistribution layer (RDL) was homogenized by the basic mixed rule of copper and polyimide (PI). The 3D quarter model was established in ABAQUS, which contains 729253 hexahedral elements in total.

The model was subjected to the load of thermal cycling. The temperature ranges from -55°C to 125°C with ramp rate 0.2°C/s and dwell time 500s at the highest and lowest temperatures. The stress-free temperature is assumed to 25°C.

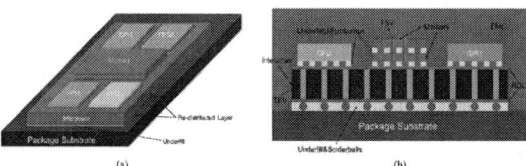

Fig. 1 Internal structures of the integration with (a) isometric view, and (b) the cross section of the device

Fig. 2 The 3D quarter model

B. Material Properties

Since the device consists of different materials, three kinds of material models were used in the analysis. The elastoplastic model is for copper microbumps and the viscoplasticity of Anand model is for Sn3.5Ag solder balls. The other materials are assumed to be elastic. The RDL was supposed as an uniform mixture of 50% copper and 50%PI and its material parameters can be estimated by mixed rules [4]. The modulus and CTE of materials are listed in TABLE I.

TABLE I. MATERIAL PROPERTIES

Material	Elastic Modulus (GPa)	CTE (ppm/K)	Poisson Ratio
Cu	130	17	0.343
RDL	66.75	28.55	0.32
EMC	23@293K	11	0.34
	22.2@333K		
	21@373K		
	20@393K		
	18.4@403K		
Si	131	2.8	0.3
Sn3.5Ag	-0.075T+52.58	$21.86+20.39\times10^{-3}$T	0.36
Substrate	26	15	0.22
Underfill	26	39	0.35

The bilinear isotropic hardening elastoplastic model is used for the copper microbumps, in which the yield stress is 240MPa

---

* Corresponding author: jun_wang@fudan.edu.cn

and the tangent modulus is 12GPa [5].

The Sn3.5Ag solder balls were governed by Anand model, illustrated in equation (1)-(3), whose parameters are listed in TABLE II [6].

$$\dot{\varepsilon}^p = A\exp\left(-\frac{Q}{RT}\right)\left[\sinh\left(\xi\frac{\bar{\sigma}}{s}\right)\right]^{\frac{1}{m}} \tag{1}$$

$$\dot{s} = \left\{h_0\left|1-\frac{s}{s^*}\right|^a \text{sign}\left(1-\frac{s}{s^*}\right)\right\}\dot{\varepsilon}^p \tag{2}$$

$$s^* = \hat{s}\left[\frac{\dot{\varepsilon}^p}{A}\exp\left(\frac{Q}{RT}\right)\right]^n \tag{3}$$

where s_0 and h_0 are related to temperature and strain rate, and the expressions and parameters can be found in [6].

TABLE II. PARAMETERS OF ANAND MODEL

Parameter	A	Q/R	$\hat{s}$	ξ	m	n	a
Value	$177016s^{-1}$	10278K	52.4MPa	7	0.207	0.0177	16

III. RESULTS AND DISCUSSION

A. Global Deformation and Stresses in the Microbumps

The displacement fields of the model at -55°C and 125°C are shown in Fig. 3. The displacement value at 125°C is larger than that at -55°C.

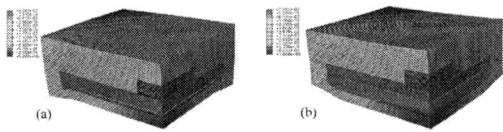

Fig. 3 Global displacement fields at (a) -55°C and (b) 125°C

The stress distributions of microbumps between memories (upper microbumps) and microbumps between the lower memory and RDL (lower microbumps) are respectively shown in Fig. 4 and Fig. 5, attached with the enlarged figure of the critical microbump.

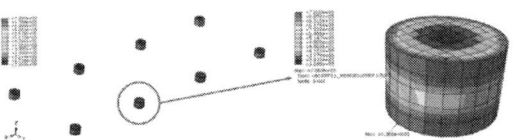

Fig. 4 The stress distribution of the upper microbumps at 125°C

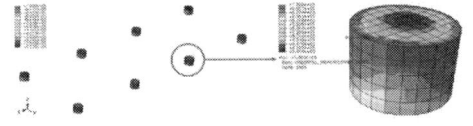

Fig. 5 The stress distribution of lower microbumps at 125°C

B. Fatigue Life Prediction of Copper Microbumps

The Von Mises stress and PEEQ (the equivalent plastic strain) of critical microbumps in different locations are demonstrated in Fig.6 and Fig.7. The increase step of PEEQ curves is gradually steady.

Coffin-Manson fatigue model has been widely used to predict the fatigue life of ductile metal. The fatigue life N_f is mainly dependent on the material and increase of equivalent plastic strain per cycle, which is expressed as

$$N_f = C\left(\Delta\varepsilon_p\right)^m \tag{4}$$

where the fatigue ductility coefficient C=0.163 and the exponent m=-2.43 for electro-deposited copper [7].

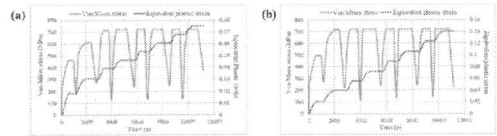

Fig. 6 The stress and equivalent plastic strain in (a) the upper critical microbump and (b) the lower critial microbump

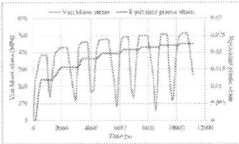

Fig. 7 The stress and equivalent plastic strain in the lower critical microbump under the chip of CPU

Therefore, the fatigue lives of microbumps can be predicted based on equivalent plastic strain variation given by FEA and Coffin-Manson model. The results are listed in TABLE IV, except the very long life of the critical microbump under the chip of CPU. According to the results, the upper microbumps have a much better fatigue performance than the lower microbumps.

TABLE III. FATIGUE LIFE PREDICTION

Component	Increase of equivalent plastic strain per cycle	Fatigue life (cycles)
The upper critical microbump	0.01480	4554
The lower critical mcrobump	0.03373	615

IV. CONCLUSIONS

The fatigue lives of copper microbumps in a typical 3D TSV integration device under thermal cyclic loadings were analyzed based on FEA and Coffin-Manson fatigue model in this study. The computational results revealed that the most critical microbump with low fatigue life is located under the memory module, which is near the center of the integration. It is different from that of a two-dimensional package where the risk bumps are always at the far corner of the device. This may be related to the stress redistribution caused by the complex structure of the 3D packaging.

ACKNOWLEDGMENT

Yuqing Lu gratefully acknowledges the Wangdao Project of Fudan University (No. 21072).

REFERENCES

[1] J. H. Lau, "Recent Advances and Trends in Advanced Packaging," *IEEE Transactions on Components, Packaging and Manufacturing Technology*, vol. 12, no. 2, pp. 228-252, Feb. 2022.

[2] J. H. Lau, "Overview and outlook of through-silicon via (TSV) and 3D integrations", *Microelectronics International*, vol. 28, no. 2, pp. 8-22, 2011.

[3] J. U. Knickerbocker et al., "2.5D and 3D technology challenges and test vehicle demonstrations," *2012 IEEE 62nd Electronic Components and Technology Conference*, 2012, pp. 1068-1076.

[4] D. K. Hale, "The physical properties of composite materials, " *Journal of Material Science*, vol. 11, pp. 2105-2141, 1976.

[5] Y. Yao et al., "Fatigue Life Predictions of SnPb Solder Ball in A Ceramic PoP Device," *2021 22nd International Conference on Electronic Packaging Technology (ICEPT)*, 2021, pp. 1-5.

[6] X. Chen et al., "Prediction of Stress-Strain Relationship With an Improved Anand Constitutive Model For Lead-Free Solder Sn-3.5Ag," *IEEE Transactions on Components and Packaging Technologies*, vol. 28, no. 1, pp. 111-116, Mar. 2005.

[7] T. H. Low and J. H. L. Pang, "Modeling plated copper interconnections in a bumpless flip chip package," *Proceedings of the 5th Electronics Packaging Technology Conference (EPTC 2003)*, 2003, pp. 791-796.

A 224-Gb/s Inverter-Based TIA with Interleaved Active-Feedback and Distributed Peaking in 28-nm CMOS for Silicon Photonic Receivers

Sikai Chen [1,2], Jintao Xue [4,5], Leliang Li [2,3], Guike Li [2,3], Zhao Zhang [2,3], Jian Liu [2,3], Liyuan Liu [2,3], Binhao Wang [4,5], Yingtao Li [1,*] Nan Qi [2,3,**]

[1] Lanzhou University, Lanzhou, China
[2] State Key Laboratory of Superlattices and Microstructures, Institute of Semiconductors, Chinese Academy of Sciences, Beijing, China
[3] Center of Materials Science and Optoelectronics Engineering, University of Chinese Academy of Sciences, Beijing, China
[4] School of Future Technology, University of Chinese Academy of Sciences, Beijing, China
[5] State Key Laboratory of Transient Optics and Photonics, Xi'an Institute of Optics and Precision Mechanics, Chinese Academy of Sciences, Xi'an, China

*Email: *ytli@lzu.edu.cn and **qinan@semi.ac.cn

Abstract—This paper presents the design and simulations of a 224-Gb/s inverter-based transimpedance amplifier (TIA) designed in 28-nm CMOS process. The co-design of the photodiode and CMOS TIA efficiently optimize the optical receiver. The TIA achieves the bandwidth enhancement with the distributed peaking and interleaved active-feedback (IAFB) technology. The simulations result show that the proposed TIA has 39-dBΩ transimpedance gain and 70-GHz bandwidth (BW). Overall, it achieves a clear 224-Gb/s PAM4 eye diagram. The total power consumption is 20.7mW at 0.9-V supply voltage. And the input referred noise current is 8.2uA$_{rms}$.

Keywords— *Transimpedance amplifier (TIA), optical receiver, CO-packaged optics (CPO), distributed peaking, interleaved active feedback (IAFB).*

I. INTRODUCTION

With the development of cloud services and mobile computing, the ever-growing demand for high-bandwidth transceivers in switches has driven the use of Co-packaged optics (CPO) in hyper-scale datacenter interconnects. For 800G-1.6T CPOs, the 224Gb/s per-channel data-rate is required.

The first-stage TIA essentially determines the overall bandwidth, noise and linearity of an optical receiver. In advanced CMOS processes, the core transistor supply voltage drops to 0.9V and below. Limited by the voltage headroom, design challenges arise for CML circuits at low power supply. On the other side, the shrink improves the power efficiency of inverter-based circuits, since the shrink brings larger transistor f_t, which could be meet by both PMOS and NMOS at the same time for both optimized V_{GS}-V_{TH}. Besides, the f_t of the PMOS is close to that of NMOS, which makes the inverter-based circuits have higher gain-bandwidth [1], [4].

The traditional inverter-based TIA cannot provide sufficient BW, because of insufficient single-stage open loop gain. Previously, inductance peaking [2], active feedback [3], and other techniques have been used to expand the bandwidth to up to 112Gb/s. In this paper, by co-designing with the silicon photonic PD and employing the fast-transient LDO similar to [5], we present a high-speed 224Gb/s linear TIA optimized for

This work is supported in part by the National Key Research and Development Program of China (Grant No. 2021YFB2206504) and in part by the National Natural Science Foundation of China (Grant No. 62235017).

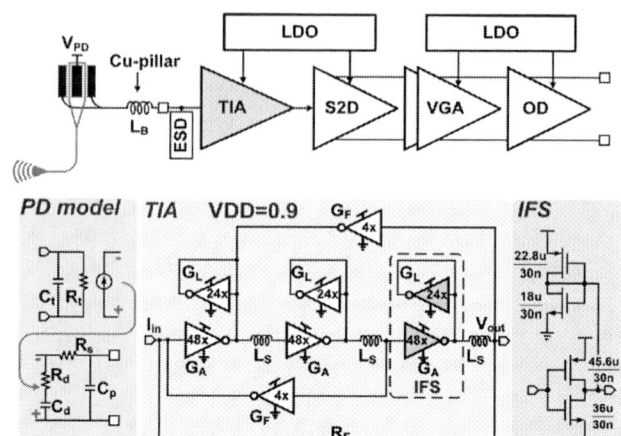

Fig. 1. The optical receiver including co-designed photodetector (PD), proposed TIA and other circuits.

28nm and below advanced CMOS process. A novel interleaved active-feedback (IAFB) topology is proposed, incorporating with inverter-based standard-cells and distributed series inductors.

II. CO-DESIGN WITH SI-PHOTONIC PD

The schematic of the Si-Ge photodiode is shown in Fig. 1. The Evanescent-coupled waveguide structure optimizes photocurrent responsivity and bandwidth simultaneously by decoupling light absorption and carrier transit time. The epitaxial n-type Ge as an absorption region is sandwiched between p-type doped Si. The metal contact to n-type doped Ge, indicated by a solid bar in the middle, is connected to a voltage supply while other two parallel metal contacts transport the photocurrent to CMOS circuitry as signals.

The design consideration of a PD in a high-speed optical receiver front-end is often multifaceted. An accurate and validated equivalent circuit model of a PD is significant for subsequent CMOS circuit design. Also, the PD model is an efficient tool for design iteration. As shown in Fig.1, the PD equivalent circuit model consists of carrier transit time effect and electrical parasitics, where the carrier transit dynamics is modelled by a RC input circuit to describe the movement of photocarriers in Ge and Si, and the electrical parasitics include the absorption region (R_d, C_d, R_s), substrate and interconnects (C_p, L_B, R_p). The parameters in the equivalent circuit model can be extracted by curve fitting PD small signal responses.

III. REALIZATION OF TIA

Fig.1 shows an optical receiver including co-designed PD, proposed TIA, S2D, variable gain amplifier (VGA), and output driver (OD). The S2D uses the Shorted inverter load topology, and VGA&OD consist of the cascaded second-order inverter-based cherry-hooper amplifier with similar distributed peaking technology for high BW. The TIA is the key component in determining the optical receiver's performance. Therefore, this paper focuses on the TIA stage

The TIA is composed of a feedback resistor R_F, the interleaved feedforward stage (IFS), and interleaved active feedback network G_F. Traditional inverter-based TIAs are limited by f_t and cannot provide sufficient open-loop gain, which affects the dominant pole at the input of the TIA. To solve this problem, we add distributed series inductance peaking technology. The series inductors L_S are used to distribute IFS capacitance, and generate sufficiently large gain boosting out of the band. Furthermore, to ensure adequate in-band flatness and minimize group delay variation in TIA response, we introduced IAFB technology composed of the G_F amplifier and IFS. Moreover, with the feedback gain $\beta = G_F/G_L$ increasing, dominant poles move and damping factor reduce which generates peaking in TIA frequency response.

In order to control the stability of the loops and simplify the analysis, we will analyze the two loops separately. The IAFB loop poses a low threat to stability. Considering in-band flatness and stability, β is equal to 0.16 which is good. For the loop generated by R_F, we need to balance the size of open loop gain and open loop bandwidth to ensure the desired phase margin requirement.

IV. SIMULATION RESULTS

The proposed TIA fabricated in 28-nm CMOS consumes 20.7mW at 0.9V supply voltage. The PD capacitance is 35fF, and resistance R_S is 10Ω. Bumping inductance is 90pH caused by copper pillar, and the C_{PAR} of ESD and PAD is 90fF.

Fig. 2 compares the frequency response of the inductance peaking technique and the IAFBs technique. In comparison to

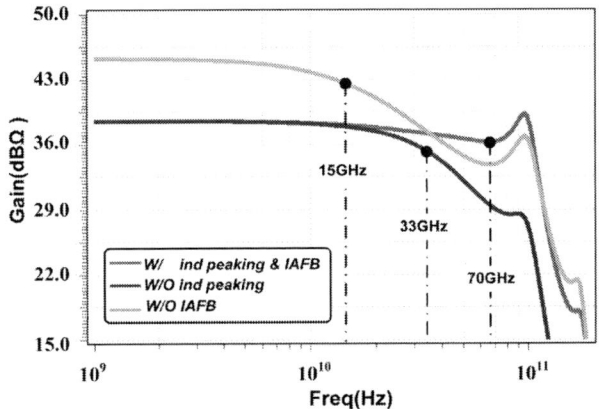

Fig. 2. IAFB and distributed inductive peaking extend the bandwidth of the TIA containing the photodetector model.

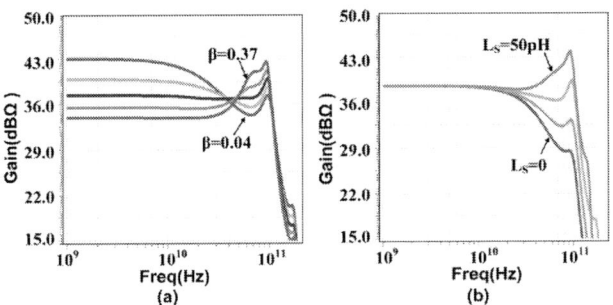

Fig. 3. Simulated frequency response for various (a) $\beta = G_F/G_L$ and (b) L_S.

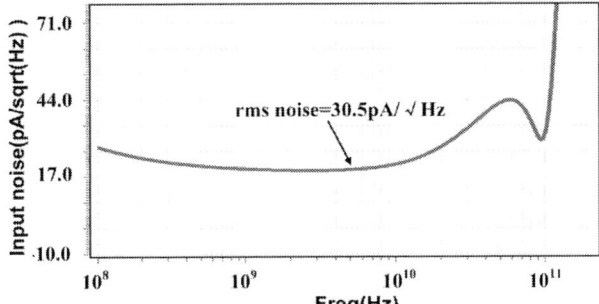

Fig. 4. Simulated input referred noise of TIA.

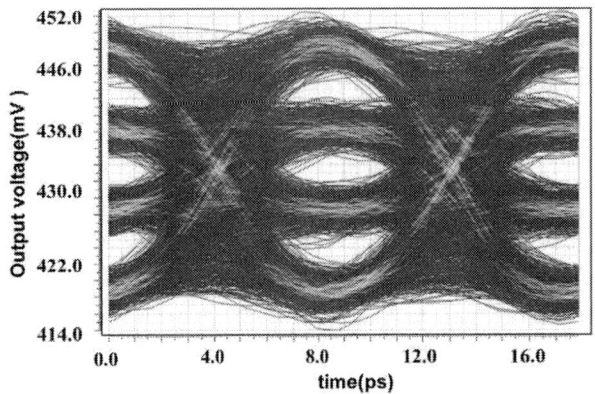

Fig. 5. Simulated PD+TIA 224Gb/s PAM4 eye diagram with transient noise.

TABLE I. PERFORMANCE COMPARISON WITH RECENT PUBLISHED TIAS

	[2]	[3]	[4]	This work
Technology	28-nm CMOS	65-nm CMOS	22-nm FinFET	28-nm CMOS
Data rate (Gb/s)	112	25	128	224
Modulation	PAM4	NRZ	PAM4	PAM4
BW(GHz)	60	13.6	45.5	70
Noise (uArms)	4.7	3.28	2.7	8.2
Supply(V)	0.9	1.1	0.8	0.9

using only one technology, using both the technologies increases the bandwidth enhancement ratio of 4.6 and 2.12. Fig. 3 shows the simulation results of transimpedance gain for varying β and L_S. With the increase of β, 7dB gain boosting has been got at 70GHz. Similarly, enhancing L_S earns 16dB boosting at 90GHz. The simulated single-ended input referred noise of the TIA is shown Fig. 4. The simulate noise is 8.2uA$_{rms}$ and 31pA/$\sqrt{Hz}$. The 224Gb/s PAM4 output eye diagram is shown in Fig. 5. When the input signal current is 300uA$_{pp}$, the total eye amplitude is 30mV$_{pp}$. Then, the eye will be further amplifier by VGA&OD and output. Table I compares the performance of the recent published TIAs.

V. CONCLUSIONS

A 224Gb/s PAM4 linear transimpedance amplifier is designed in a 28-nm CMOS process, which illustrates that the inductance peaking technique and the IAFB technique can effectively extend bandwidth. The simulation results show that MS-SF-TIA with various techniques has 39dBΩ transimpedance gain and 70GHz bandwidth at the cost of 20.7mW.

REFERENCES

[1] D. Patel, et al, "A 112 Gb/s -8.2 dBm Sensitivity 4-PAM Linear TIA in 16nm CMOS with Co-Packaged Photodiodes," 2022 IEEE Custom Integrated Circuits Conference (CICC), 2022, pp. 1-2.

[2] H. Li, et al, "A 112 Gb/s PAM4 Linear TIA with 0.96 pJ/bit Energy Efficiency in 28 nm CMOS," ESSCIRC 2018 - IEEE 44th European Solid State Circuits Conference (ESSCIRC), 2018, pp. 238-241.

[3] M. Moayedi Pour Fard, et al, "1.23-pJ/bit 25-Gb/s Inductor-Less Optical Receiver with Low-Voltage Silicon Photodetector," in IEEE JSSC, vol. 53, no. 6, pp. 1793-1805, June 2018.

[4] S. Daneshgar, et al, "A 128 Gb/s, 11.2 mW Single-Ended PAM4 Linear TIA With 2.7 µArms Input Noise in 22 nm FinFET CMOS," in IEEE JSSC, vol. 57, no. 5, pp. 1397-1408, May 2022.

[5] H. Liu et al., "A Fast-Transient Capacitor-Less Low-Dropout Regulator for Wideband Optical Transceivers," 2021 IEEE ICTA, 2021, pp. 257-258.

A 0.3-μW,2.1-μVrms Neural Recording Chopper Amplifier with Low Noise DC-Servo-Loop

Yuchen Bao, Weijian Chen, Zhixian Li,Yongsen Chen, Yanhan Zeng
School of Electronics and Communication Engineering, Guangzhou University

Abstract—This paper presents a low noise and low power circuit for neural recording. A Capacitively-Coupled Chopper Instrumentation Amplifier (CCIA) with embedded DC feedback is proposed to reduce the noise of system. Implemented a continuous-time low-pass filter (LPF) at the output of the system and utilized bulk-feedback techniques to increase its output swing. Furthermore, the DC-block and Chopper-Capacitor-Chopper Integrator Based DC Servo Loop (C3IB-DSL) are combined to reduce the interferences. According to experiment, the circuit consumes only 0.3 μW at 1.2 V. In addition, the input-referred noise reached 2.1 μVrms and the noise efficiency factor (NEF) 3.6 at the same time. The proposed CCIA was simulated in a 180n CMOS process.

Keywords—CCIA, embedded DC feedback, low power, low noise, neural recording

I. INTRODUCTION

Neuronal activity recording is one of the most important tools for predicting cardiovascular disease and figure out people's physical and mental problems comprehensively. However, there is still many challenges, such as electrode offset (EOS), noise and so on. Besides, long-term monitoring and wearable action require lower power and small chip area.

In addition to the noise of the amplifier, the DSL used to eliminate EOS will also increase the input reference noise (IRN) by splitting the charge with capacitor C_{hp}. When the DSL is enabled, the IRN is increased from 0.7 to 6.7 $Vrms$ [1]. In [2], in order to reduce noise from DSL, the conventional integrator in the DSL is replaced by a G_m-C LPF. But the source degradation technique used to increase the differential mode input range of the G_m-C LPF also increases its current consumption. Moreover, the LPF must be incorporated by neural recording circuit with chopper technology, which increases the area and power consumption.To minimize the area, [3] integrates the functions of amplifications, filtering, and buffering into one block, but the Class-AB output stage limits its output swing.

To solve these shortages, We propose a CCIA with embedded DC feedback. To further reduce the noise of DSL at low power consumption, we combine C3IB-DSL with embedded DC feedback technology. A Source-Follower-C based LPF is implemented in the output stage, using bulk-feedback to provide large output swings. This paper is organized as follows. Section II gives a brief description of the whole architecture, then analyzes the circuit in detail and describes its principle. Simulation results are discussed in section III. Brief conclusion is provided in section IV.

II. SYSTEM ARCHITECTURE AND CIRCUIT IMPLEMENTATION

Fig.1 shows the overall architecture of the proposed neural recording circuit. It is composed of a two-stage miller compen-

This work was upported in part by the Special Projects in Key Fields of Guangdong Education Department under Grant No. 2022ZDZX1019, and in part by the National Science Foundation of China under Grant No. 62141414 and 61704037.Yanhan Zeng is the corresponding author (yanhanzeng@gzhu.edu.cn).

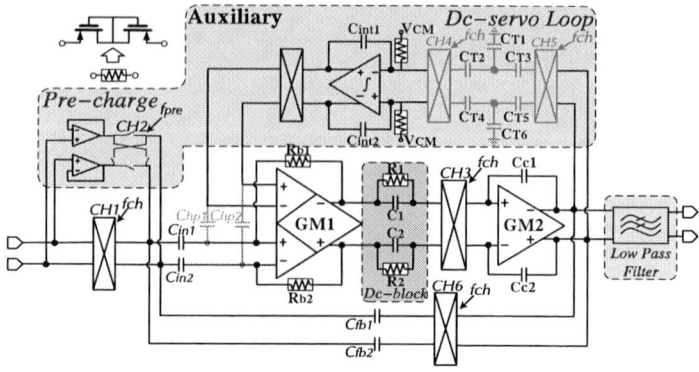

Fig. 1: The proposed neural recording circuit

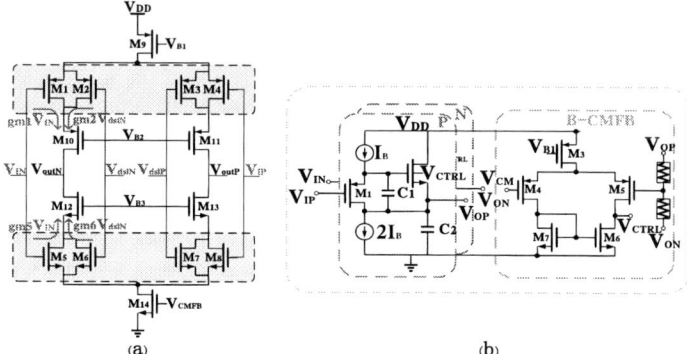

Fig. 2: (a) The proposed IA, (b) Schematic of the Low pass filter with the CMFB circuits

sated operational amplifier, a negative feedback loop (NFL), a ripple reduction block, a LPF and auxiliary circuits. The $C_{in1,2}$ and $C_{fb1,2}$ form the NFL which defines the pass band gain of the CCIA as $\frac{C_{in1,2}}{C_{fb1,2}}$. Pre-Charge to boost input impedance. To eliminate ripple at the output of the CCIA caused by the offset of the G_{m1} and chopper CH3, a DC-block is applied. The DSL is used to reject the EOS. LPF is a key part of the system that filters out out-of-band noise and interference.

A. CCIA with Embedded DC Feedback to reduce noise from DSL

The circuit of proposed IA can be seen in Fig.2(a). A cascade amplifier with complementary input is used to provide a boosted transconductance, which obtains larger open-loop gain and lower IRN compared with other structures. In Fig.1, we can see that the proposed structures unlike previous works which used C_{hp}. As shown in Fig.2(a), the voltage from the DSL is converted into current by $M_{2,6}$ and then summed with the current from $M_{1,4}$ and $M_{5,8}$. The capacitor C_{hp} in the traditional DSL increases the IRN via charge dividing, it can be reduced by reducing the C_{hp}. But the minimum C_{hp} will be limited by the parasitic capacitance. The IRN of the CCIA, $\overline{v_{n,in}^2}$ can be expressed as:

$$
\begin{aligned}
\overline{v_{n,in}^2} &= \left(\frac{C_{in1,2}+C_{fb1,2}}{C_{in1,2}}\right)^2 \overline{v_{n,in,Gm1}^2} + \overline{I_{n,Rfb}^2} \times \overline{R_{in,eq}^2} \\
&\simeq \left(\frac{C_{tot}}{C_{in1,2}}\right)^2 \left[\frac{8kTn}{g_{m1,4}+g_{m5,8}} + \frac{8kTn\,(g_{m2,3}+g_{m6,7})}{(g_{m1,4}+g_{m5,8})^2}\right. \\
&\quad \left. + 2\overline{v_{n,out,DSL}^2}\left(\frac{g_{m2,3}+g_{m6,7}}{g_{m1,4}+g_{m5,8}}\right)^2\right] + 8kT\frac{1}{R_{fb1,2}}R_{in,eq}^2,
\end{aligned}
\tag{1}
$$

where $\overline{v_{n,in,Gm1}^2}$ is the IRN of G_{m1}, $\overline{I_{n,Rfb}^2}$ is the noise current of $R_{fb1,2}$, $R_{in,eq}$ is the equivalent SC input resistance, n is the

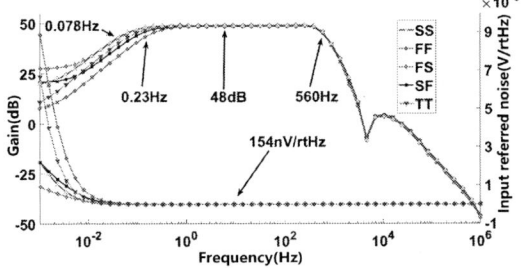

Fig. 3: Corner simulations of frequency response and input-referred noise

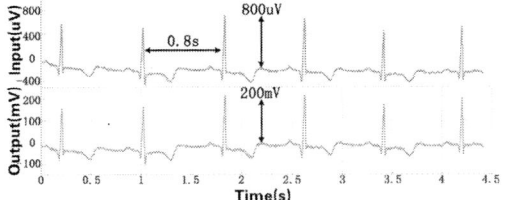

Fig. 4: The input and output signal of neural recording circuit

sub-threshold slope factor. It can be seen in Equation (1), the noise of the DSL is reduced by the factor $\frac{g_{m2,3}+g_{m6,7}}{g_{m1,4}+g_{m5,8}}$.

B. Source-Follower-C Based LPF to provide large swing

As shown in Fig.1, a LPF is implemented at the output of the system. The detailed circuit of this filter is shown in Fig.2(b). In order to improve the output swing, NMOS and PMOS input versions are cascaded to reduce the difference between the input and output voltages. The transfer function of the filter can be expressed as:

$$H\left(s\right) = \frac{g_{m1}g_{m2}}{s^2 c_1 c_2 + s c_2 g_{m1} + g_{m1}g_{m2}}. \quad (2)$$

Nevertheless, a simple cascade cannot completely eliminate voltage differences. As shown in Fig.2(b), a bulk-feedback path is added to adjust the threshold voltage of the NMOS transistor, which eliminates the voltage difference between input and output. As a result, both input and output swings are maximized.

C. Chopper-Capacitor-Chopper Integrator Based DC Servo Loop to eliminate offset

The electrode DC offset is modulated to high frequency by the Chopper modulation, which can not be eliminated by the feedback structure formed of pseudo resistance and capacitance. Therefore, the DSL is proposed. However, switched-capacitor integrators are used in conventional DSL with a high noise level. This paper implements a C3IB-DSL to solve this problem. As shown in Fig.1, choppers CH4,5 and the T-type capacitor networks are operating in continuous time instead of periodically sampling and holding voltage. As compared with switched-capacitor integrators, the C3IB-DSL in this work does not exhibit aliasing of KT/C noise and op-amp wideband noise in the baseband, thus it greatly reduces the in-band noise caused by DSL.

D. Pre-Charge to boost input impedance

The pre-charge mainly consists of a unity gain amplifier, which transmits the signal from the electrode to IA through the non-overlapping channels. The operation is divided into pre-charge phase and chopping phase. In the pre-charge phase, the chopper CH2, controlled by f_{pre}, receives the signal. At the same time, CH1 is in an inoperative state. The pre-charge

module replaces the electrodes to provide charge to C_{in} and increases R_{in} of IA. The chopping phase is the opposite of pre-charge phase.

III. SIMULATION RESULTS AND DISCUSSION

To verify the features, the proposed neural recording circuit is implemented and simulated in a 180nm CMOS process. The overall power consumption is 0.307 μW. Fig.3 shows the frequency response and IRN of the circuit. The mid-band gain is 48 dB with a -3dB bandwidth of 560 Hz, and high pass corner frequencies range from 78 to 230 mHz depending on the Process Corner. The IRN from 1 to 200 Hz of the circuit with and without DSL are 2.1 $\mu Vrms$ and 1.9 $\mu Vrms$, respectively. When the DSL is enabled, the noise density is 154 nV/sqrt(Hz). Fig.4 shows the key-node voltages of CCIA with the ECG data of the MIT-BIH Arrhythmia Database. Benefited from the gain of the circuit, the ECG signal is amplified from around 800 μV to 200 mV.

Table I compares the performance of the circuit with several works proposed in recent years. It can be seen that the neural recording circuit proposed in this brief achieves the lowest power consumption and NEF among these works. In addition, a low noise is achieved, respectively.

TABLE I. PERFORMANCE COMPARISON

Reference	[4] 2017	[3] 2018	[5] 2021	This work
Technology	40nm	130nm	180nm	180nm
Supply voltage(V)	1.2	2	1.2	1.2
Power(μW)	2	1.8	2.76	0.307
Gain(dB)	-	34.6	-	48
IRN(μVrms)	2 (1-200Hz)	3.2 (0.9-350Hz)	3.2 (0.5-400Hz)	2.1 (1-200Hz)
NEF	7	6.25	9.4	3.63
CMRR(dB)	-	95	>110	>77

IV. CONCLUSION

This paper proposes an 0.3 μW, 2.1 μVrms circuit for neural sensing. In the CCIA, particular attention has been paid to reducing the noise from DSL by embedding the output of the DSL inside the main amplifier and combined with C3IB-DSL. Simulation results prove that our design can significantly reduce the noise caused by DSL. Owing to the low power consumption, the proposed chopper amplifier is suitable for long-term neuronal recording applications.

REFERENCES

[1] Q. Fan, F. Sebastiano, J. H. Huijsing and K. A. A. Makinwa, "A 1.8 W 60 nV / Hz Capacitively-Coupled Chopper Instrumentation Amplifier in 65 nm CMOS for Wireless Sensor Nodes," in IEEE Journal of Solid-State Circuits, vol. 46, no. 7, pp. 1534-1543, July 2011.

[2] D. Luo, M. Zhang and Z. Wang, "A Low-Noise Chopper Amplifier Designed for Multi-Channel Neural Signal Acquisition," in IEEE Journal of Solid-State Circuits, vol. 54, no. 8, pp. 2255-2265, Aug. 2019.

[3] Y. Hsu, Z. Liu and M. M. Hella, "A 1.8 μ W 65 dB THD ECG Acquisition Front-End IC Using a Bandpass Instrumentation Amplifier With Class-AB Output Configuration," in IEEE Transactions on Circuits and Systems II: Express Briefs, vol. 65, no. 12, pp. 1859-1863, Dec. 2018.

[4] H. Chandrakumar and D. Marković, "A High Dynamic-Range Neural Recording Chopper Amplifier for Simultaneous Neural Recording and Stimulation," in IEEE Journal of Solid-State Circuits, vol. 52, no. 3, pp. 645-656, March 2017.

[5] S. Zhang, X. Zhou, C. Gao and Q. Li, "An AC-Coupled Instrumentation Amplifier Achieving 110-dB CMRR at 50 Hz With Chopped Pseudoresistors and Successive-Approximation-Based Capacitor Trimming," in IEEE Journal of Solid-State Circuits, vol. 56, no. 1, pp. 277-286, Jan. 2021.

A 56Gb/s De-serializer with PAM-4 CDR for Chiplet Optical-I/O

Yunqi Yang[1,2], Ming Zhong[2], Qianli Ma[2,3], Ziyi Lin[2,4], Leliang Li[2,3], Guike Li[2,3], Liyuan Liu[2,3], Jian Liu[2,3], Nanjian Wu[2,3], Haikun Jia[4], Xinghui Liu[1], Nan Qi[2,3,*]

[1] Liaoning University, Shenyang, China
[2] State Key Laboratory of Superlattices and Microstructures, Institution of Semiconductors, Chinese Academy of Sciences, Beijing, China
[3] Center of Materials Science and Optoelectronics Engineering, University of Chinese Academy of Sciences, Beijing, China
[4] School of Integrated Circuits, Tsinghua University, Beijing, China
[*] Email: qinan@semi.ac.cn

Abstract—**This paper presents a 56Gb/s de-serializer with PAM-4 CDR for chiplet optical-I/O in 28nm CMOS. There are two channels in this chip. Each channel consists of a high-performance analog front end (AFE) and a half-rate clock and data recovery (CDR) circuit based on a digital phase interpolator and digital loop filter. To provide 28-GHz clock signals to both channels, a clock distribution circuit is integrated. Experimental results show that the proposed de-serializer recovers a 56Gb/s PAM-4 input signal with channel loss, achieving an output swing of 1.01-Vppd and 760ps RMS jitter.**

Keywords—*Optical-I/O, de-serializer, equalization, clock and data recovery (CDR), phase interpolation*

I. INTRODUCTION

The new generation datacenter is developing towards large-scale and resource pooling. As the payload asks for longer reach and higher bandwidth (BW) interconnects, the optical I/O (OIO) draws more attention. By interfacing optical transceiver chiplets to the payload IC, the co-packaged optics enables direct fiber attachment to the chip-edge with extremely high-density BW and significantly power reduction.

Shown in Fig. 1, the payload is co-packaged with OIO and other functional chiplets. In each OIO receiver path, the input optical signal is converted into current by a photodetector (PD), and then gets amplified by the TIA to form sufficient swing voltage. An integrated de-serializer retimes the data and suppresses the input inter-symbol interference (ISI), interfacing to the payload IC's parallel data bus [1]. For FPGA and GPU payload ICs, the total bandwidth reaches 1.2Tb/s, which asks for multiple OIOs with the per-channel speed reaching 50Gb/s.

II. DE-SERIALIZER

Fig. 2(a) depicts the block diagram of the proposed de-serializer used in the OIO. The data-path is composed of the analog front-end (AFE), half-rate sampling bang-bang phase detector (BBPD), 1-to-4 demultiplexer (DeMUX) and an output driver. A continuous-time linear equalizer (CTLE) compensates for the input channel loss up to 6dB, while the following stage main amplifiers (MA) provides variable gain up to 10dB to maintain 300mV$_{PP}$ voltage swing at the PD input. Fig. 2(c) shows the 56Gb/s PAM-4 input with 4.5dB channel loss at 14GHz, which leads to significant inter-symbol interference (ISI). Fig. 2(d) shows the AFE output with optimum equalization and around 14dB cascaded gain. Sufficient voltage swing with much improved signal integrity is achieved with the help of the AFE.

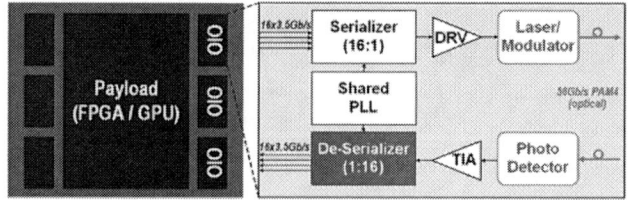

Fig. 1. The proposed PAM-4 de-serializer used in an optical-I/O chiplet.

A time-interleaved BBPD down-samples the 28 GBd 4-level data to 4-way 14 Gb/s MSB and LSB outputs. The 4-way data stream is further demultiplexed to 16-way 3.5Gb/s before outputting through the 50Ω driver.

III. DIGITAL CDR

Considering to the dense integration and low power, a shared PLL is typically employed to provide the 28GHz clock for all channels. In each channel, the clock is further synchronized to the local input data. In the clock distribution path, an open-drain current-mode logic (CML) buffer drives the long trace, leaving its load resistors at the far-end of the T-line. High common-mode noise rejection and low transient noise is achieved. The divider converts differential clock into quadrature phases, which then gets phase interpolation (PI) to achieve the clock and data recovery. The BBPD uses 4-phase 14GHz clock for sampling, which is further divided to 7GHz for DeMUX and 875MHz for the digital loop filter (DLF).

In this work, a digital PAM-4 CDR is employed to avoid the area consuming analog loop filter, which is essential for multi-channel integration and the flexible configuration. The CDR adopts a 2× oversampling BBPD. Both the data and edge information are collected. A 1:16 DeMUX is employed to further demultiplex the UP and DN signals to match the speed with DLF. The function of the DLF in the CDR loop is to perform low-pass filtering to limit the loop bandwidth within 10MHz. A series of processes are used to generate control words to control the phase interpolator, thereby adjusting the clock phase for data recovery. The CDR offers effective regulation of the clock phase and fast adjustment of the loop for data and clock recovery.

IV. EXPERIMENTAL RESULTS

The proposed de-serializer is fabricated in a 28-nm CMOS, and each channel occupies 0.571×0.19 mm^2 area as shown in Fig. 3. To evaluate the performances, a differential 56Gb/s PAM-4 PRBS-31 data and 28GHz differential clock pulse are generated from the pulse pattern generator (Keysight M8045A) both with 300mV$_{ppd}$ swing. The de-serializer chip can be configured via a standard I2C interface from the computer. The output recovered clock and retimed data are separately connected to the sampling oscilloscope (Keysight N1092C) to observe the eye-diagram. As shown in Fig. 4, the 3.5Gb/s recovered data features a clear eye diagram, with 279.5 ps eye width and 945 fs RMS jitter. The recovered 7GHz clock is measured by the signal source analyzer (R&S FSUP). As shown in Fig. 5, the recovered 7GHz clock exhibits 760 fs RMS jitter at the time domain, while featuring -102.34dBc/Hz, -118.9dBc/Hz and −124.15dBc/Hz phase noise at a 100kHz, 1MHz and 10MHz frequency offset, respectively.

V. CONCLUSION

This paper presented a 56Gb/s de-serializer with a PAM-4 digital CDR based on a phase-interpolator in 28nm CMOS. By collaborating wideband AFE with EQ, low-power clock distribution and digital PI-based CDR, high quality data deserialization is achieved at small chip area cost. Experiment results show 760 fs low jitter data and -124.15dBc/Hz low phase noise clock.

This work is supported in part by the National Key Research and Development Program of China (Grant No. 2021YFB2206504) and in part by the National Natural Science Foundation of China (Grant No. 62235017).

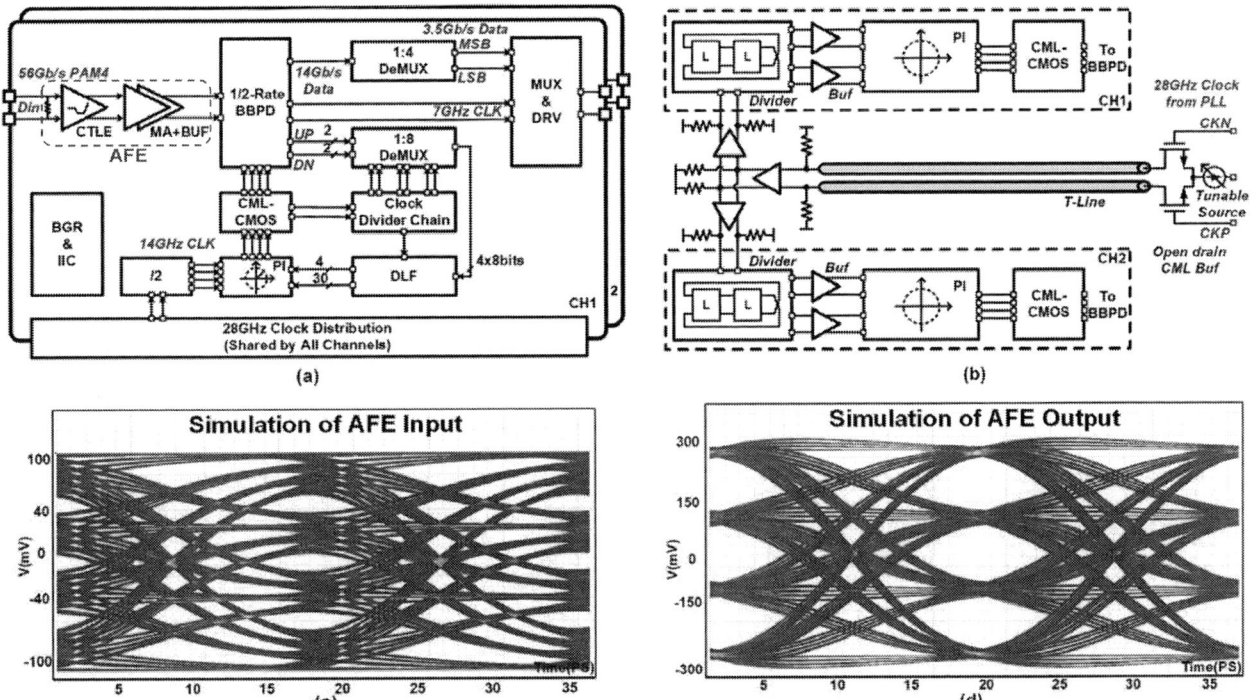

Fig. 2. Details of the de-serializer: (a) System block diagram, (b) schematic of clock distribution, (c) input data eye-diagram, (d) AFE output data eye-diagram.

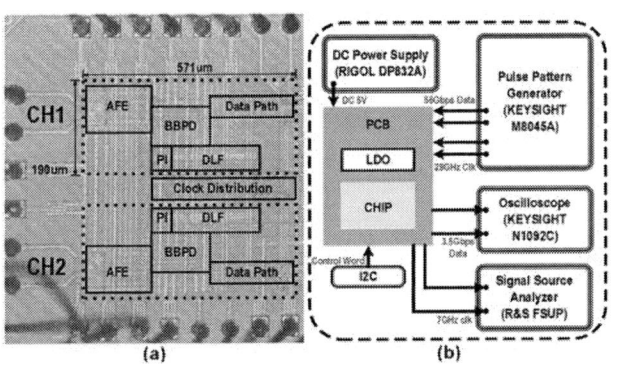

Fig. 3. Chip implementation (a) Chip micrograph, (b) Measurement setup.

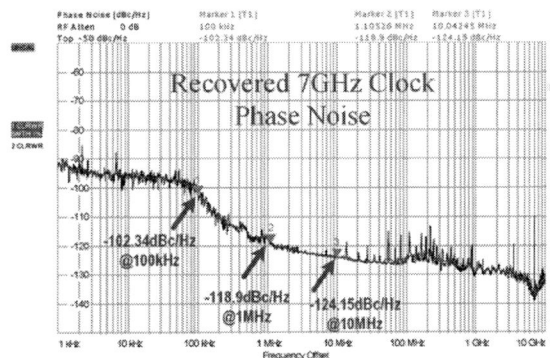

Fig. 5. Measured phase noise of the recovered clock.

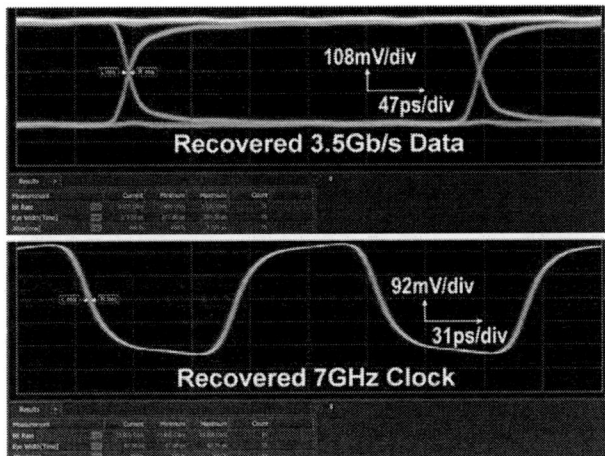

Fig. 4. Measured eye diagrams of the recovered data and clock.

TABLE I. PERFORMANCE SUMMARY AND COMPARISON

	This work	[2]	[3]	[4]
Data Rate (Gb/s)	**56**	22-26.5	56	28
Supply (V)	**0.9**	1.2	\	\
Phase Interpolator	**yes**	no	no	yes
Active area (mm²)	**0.108**	0.75	1.60	0.52
Recovered clock jitter$_{RMS}$ (ps)	**0.76 @7GHz**	1.28 @3.8GHz	0.52 @28GHz	0.949 @14GHz
Technology	**28nm CMOS**	65nm CMOS	40nm CMOS	28nm CMOS

REFERENCES

[1] G. Hou and B. Razavi, "A 56-Gb/s 8-mW PAM4 CDR/DMUX With High Jitter Tolerance," IEEE JSSC, vol. 57, no. 9, March 2022.

[2] S. -H. Chu et al., "A 22 to 26.5 Gb/s Optical Receiver with All-Digital Clock and Data Recovery in a 65 nm CMOS Process," in IEEE JSSC, vol. 50, no. 11, Nov. 2015.

[3] J. Lee et al., "Design of 56 Gb/s NRZ and PAM4 SerDes Transceivers in CMOS Technologies," in IEEE JSSC, vol. 50, no. 9, Sep. 2015.

[4] J. Liang, et al., "On-Chip Jitter Measurement Using Jitter Injection in a 28 Gb/s PI-Based CDR," in IEEE JSSC, vol. 53, no. 3, March 2018.

Author Index

A
An Fengwei   66
An Sirui   114

B
Bao Jiayu   135
Bao Juncheng   137
Bao Rongxin   36
Bao Xiue   137
Bao Yuchen   200
Bi Xiaojun   154
Bian Zhongjian   90

C
Cai Hao   48,70,72,90,160
Cai Jialin   80
Cao Jifang   76
Cao Lei   3
Cao Peng   176
Chang Chengjun   150
Chang Liang   190
Chen Bing   76
Chen Binglu   24
Chen Chi   100
Chen Chun-Zhang   104
Chen Chuqi   108
Chen Fei   24
Chen Jiayi   131
Chen Jiezhi   24
Chen Lei   66
Chen Liangfan   164
Chen Lin   50
Chen Peng   146
Chen Renwei   34,78
Chen Sihao   62
Chen Sikai   198
Chen Song   56
Chen Weijian   200
Chen Weiliang   129
Chen Xiang   158
Chen Xingyu   133
Chen Yan   86

Chen Yongsen   200
Chen Yongzhen   162
Chen Zhuo   102
Chen Zihao   164
Chen Zixin   32
Chen Ziyang   133
Chen Ziyue   170
Cheng Lin   156
Chi Baoyong   84
Crupi Giovanni   137
Cui Xiaoxin   182
Cui Xuecheng   76

D
Deng Huipeng   144
Deng Longge   30
Deng Wei   84
Ding Chaoyang   125
Ding Jiale   13
Ding Zixuan   133
Donato Nicola   137
Dong Junchen   54
Du Xiang   114
Du Yuxuan   102
Duan Baoxing   38
Duan Mengqi   64,127

F
Fan Shiquan   174
Fan Taiyang   104
Fang Shuyue   110,121
Fang Xiaotong   24
Fang Yuqing   50
Farrukh Fasih Ud Din   44,125
Feng Fei   86
Feng Guangyin   152
Feng Haigang   110,121
Feng Peng   178
Feng Tian   13
Feng Xinjie   162
Fu Jiawei   72
Fu Yuyang   58

G

Gai Weixin	52
Gan Weizhuo	3
Gao Xiang	86
Ge Haitao	98
Geng Li	64,68,127,148,172,174
Gu Youzhi	162
Gugliandolo Giovanni	137
Guo Guanghao	178
Guo Haifeng	194
Guo Jianping	188
Guo Peng	24
Guo Zhuoqi	64,68,127,148

H

Han Aoze	133
Han Ke	15
Han Su	74
Han Tingting	46
Hao Yue	30,114
He Dian	38
He Haitao	54
He Junxian	62
He Weifeng	150
He Yuhui	112
Hong Zhiliang	176
Hou Bin	30
Hu Ao	170
Hu Ertao	135
Hu Jiahao	38
Hu Jinrui	110,121
Hu Kuan	100
Hu Linxuan	104
Hu Yi	166
Huang Cece	52
Huang Lei	92
Huang Tao	166
Huang Xiangrong	84
Huang Yuhao	88
Huang Yukang	7
Huang Zhongxian	38

J

Ji Xuansheng	74
Ji Yafei	52
Jia Haikun	84,202
Jia Zhengzhe	104
Jiang Dong	7
Jiang Hanjun	44
Jiang Yucheng	194
Jin Gaofeng	86
Jing Xu	131

K

Kang Dingxuan	131
Kang Lei	178
Kang Qiushi	40
Kang RuiYuan	100
Kang Yi	56
Kao C.R.	20
Ke Jiacong	152

L

Li Chenglong	190
Li Dan	5
Li Daokun	15
Li Dejian	102
Li Dengquan	13
Li Dongxu	123
Li Fule	60
Li Ge	40
Li Guangyao	17
Li Guike	198,202
Li Guolin	166
Li Guolong	156
Li Hao	5
Li Haoyang	58
Li Jia	5
Li Jindong	158
Li Jinkai	137
Li Leliang	198
Li Leliang	202
Li Ping	62
Li Qiang	36
Li Qing	38
Li Qingxuan	50
Li Shumeng	119

Li Wenbo	186	Lu Yuqing	196
Li Wenchang	180	Lu Zheng	174
Li Yaoxin	172	Luo Changhao	54
Li Yi	58	Luo Weilin	100
Li Yingtao	198	Luo Xun	86
Li Zhaoshi	129	Luo Zhiqiang	60
Li Zheng	32	Lv Dexuan	64
Li Zhi	150	**M**	
Li Zhicheng	116	Ma Qianli	202
Li Zhijian	46	Ma Shuaizhe	5
Li Zhixian	200	Ma Weiqing	174
Li Zhuoneng	64,68,127	Ma Xiaohua	30,114
Liang Chenglong	127,148	Mai Songping	194
Liang Wenhui	112	Mao Shuman	92,94
Liao Huabing	84	Meng Jialin	50
Liao Xufeng	11,42	Mi Minhan	114
Lin Jiahui	36	Miao Min	54
Lin Luhua	106	Ming Da	168
Lin Shuisheng	190	**N**	
Lin Ziyi	202	Niu Fanfan	40
Liu Bo	48,160	Niu Liting	44,125
Liu Dong	76	**O**	
Liu Heng	123	Okada Kenichi	32
Liu Jian	180,184,198,202	Ouyang Keqing	17
Liu Jiaxin	36	**P**	
Liu Leibo	129	Pan Dongfang	156
Liu Lianxi	11,42	Pan Quan	104
Liu Lingyun	148	Pan Zhiming	22
Liu Liyuan	178,184,198,202	Pang Jian	17,32
Liu Shubin	13	**Q**	
Liu Xihao	64,68,127	Qi Nan	184,198,202
Liu Xincai	42	Qiao Xin	182
Liu Xinghui	202	**R**	
Liu Xuhui	104	Ran Nianquan	5
Liu Yao	96	Rhee Woogeun	74,106
Liu Yi	135	Run Run	166
Liu Yueduo	36	**S**	
Liu Yujie	94	Shan Weiwei	98,102
Lu Danzhu	176	Shen Chongfei	102
Lu Hao	30	Shen Qiao	48
Lu Yicheng	98	Shen Yi	13
Lu Yifan	58	Shen Zixuan	112

Shi Bao	84	Wang Min	142
Shi Cong	62	Wang Mingyu	96,144
Shi Hanyu	176	Wang Peng	60
Shi Yihui	170	Wang Pengfei	114
Shirane Atsushi	32	Wang Qiaoyu	32
Song Yifan	56	Wang Shucai	1
Song Zhulu	116	Wang Tengxiao	62
Song Ziqi	38	Wang Tianyu	50
Su Fukun	119	Wang Weilun	22
Sun Jianjun	17	Wang Xin	192
Sun Jiayun	116	Wang Xin'an	182
Sun QingQing	50	Wang Xinpeng	133
Sun Yanan	150	Wang Yanchao	104
T		Wang Yanjie	152
Tan Min	168	Wang Yifa	188
Tang Xian	119,123	Wang Yiyang	125
Tang Xiaoqiang	80	Wang Yu	135
Tao Li	131	Wang Yuan	182
Tian Mi	46	Wang Zhaohao	142
Tian Min	62	Wang Zhaojin	116
Tian Qin	26	Wang Zhehan	131
Tian Xiaoyun	90	Wang Zheng	22,186
Tong Yi	133,135	Wang Zhihua	60,74,84,106
W		Wang Ziwei	74
Wan Hengzhi	28	Wei Shaojun	129
Wan Tianqing	58	Wei Siyuan	184
Wan Xiang	135	Wu Haige	160
Wang Binhao	198	Wu Hailong	158
Wang Bowen	106	Wu Hanming	104
Wang Chao	112,170	Wu Jiangfeng	162
Wang Chenxi	40,142	Wu Jixuan	24
Wang Dengfeng	150	Wu Kaikai	1
Wang Haibing	62	Wu Nanjian	184,202
Wang Han	110,121	Wu Peng	184
Wang Hongyi	1	Wu Pengcheng	72
Wang Huanpeng	92,94	Wu Pengfei	22
Wang Jian	144	Wu Qingzhi	92,94
Wang Jingsheng	194	Wu Tong	188
Wang Jun	196	Wu Yanqing	192
Wang Kai	116	Wu Yu-ang	70
Wang Keping	9	Wu Zhenhua	3
Wang Lei	133	**X**	

Xiao Jian	48	Yang Ruipeng	172
Xiao Kanglin	182	Yang Shiheng	36
Xiao Lanxiang	66	Yang Xin	52
Xiao Shanlin	88	Yang Xu	178
Xiao Zhilong	94	Yang Yongkui	7
Xie Jinyu	154	Yang Yunqi	202
Xie Kenie	24	Yang Yuye	5
Xie Qian	186	Yao Enyi	7
Xie Xiang	166	Yao Yifan	108
Xie Ying	174	Ye Huafeng	144
Xie Yiyun	172	Yin Huaxiang	3
Xie Yuhang	42	Yin Kang	100
Xin Youze	172	Yin Li	166
Xing Xinpeng	110,121	Yu Chao	146
Xiong Yuhao	68	Yu Guoyi	112,170
Xiong Zhiyong	156	Yu Haiyang	20
Xu Dongfan	104	Yu Jianyi	62
Xu Hao	46	Yu Shuangming	178,184
Xu Haoqing	3	Yu Xiao	76
Xu Haozhe	184	Yu Xuanchi	86
Xu Jiarui	112,170	Yu Ying	11
Xu Rongqing	135	Yu Yucheng	146
Xu Ruimin	94	Yu Yue	168
Xu Tianqi	119	Yu Zhiyi	88,96,144
Xu Xiao	131	Yuan Haoyun	137
Xu Yuehang	92,94	**Z**	
Xu Zeyang	88	Zeng Yanhan	200
Xu Zihang	160	Zhai Shichong	180
Xue Jialong	46	Zhan Chenchang	34,78,108
Xue Jintao	198	Zhan Dongshen	104
Xue Zhongming	64,68,127,148	Zhan Xuepeng	24
Y		Zhang Bing	172
Yan Bin	17	Zhang Chi	178
Yan BO	94	Zhang Chun	44,60,125
Yan Na	46	Zhang David Wei	50
Yan Yu	135	Zhang Hao	48,133,135,160
Yang Binwei	78	Zhang Miaocheng	133
Yang Fei	100	Zhang Mingkang	36
Yang Hao	17	Zhang Qingzhe	9
Yang Jun	98	Zhang Renyuan	48,160
Yang Lei	110,121	Zhang Shuqiang	17
Yang Ling	30,58	Zhang Tianyi	180

Zhang Tieliang	52	Zhao Yuansheng	112
Zhang Weiyi	44,125	Zhao Yudi	54
Zhang Xuanhao	160	Zheng Junhao	194
Zhang Yifei	34	Zhong Ming	202
Zhang Yongchao	64,38,127	Zhong Zhengqing	62
Zhang Zehao	36	Zhou Jun	190
Zhang Zhao	198	Zhou Kaize	102
Zhang Zhengyang	104	Zhou Likun	30
Zhang Zhuang	176	Zhou Mingyang	70
Zhao Dixian	26,28	Zhou Xiong	36
Zhao Jiahao	74	Zhu Bowen	82
Zhao Kai	54	Zhu Ruimin	88
Zhao Lu	164	Zhu Shiquan	94
Zhao Shangzhou	64,68,127	Zhu Weiqiang	46
Zhao Shunpeng	56	Zhu Zhangming	13
Zhao Wanqing	5	Zhu Zihao	36
Zhao Weisheng	142	Zhuang Haoyu	36
Zhao Xin	190	Zhuang Junjie	96
Zhao Ying	100	Zou Tenghao	46
Zhao Yuanfu	52	Zuo Sheng	96